Lecture Notes in Mathematics 2063

T0241017

For further volumes:
http://www.springer.com/series/304

Irina Mitrea • Marius Mitrea

Multi-Layer Potentials and Boundary Problems

for Higher-Order Elliptic Systems in Lipschitz Domains

Springer

Irina Mitrea
Temple University
Department of Mathematics
Philadelphia, PA, USA

Marius Mitrea
University of Missouri
Department of Mathematics
Columbia, MO, USA

ISBN 978-3-642-32665-3 ISBN 978-3-642-32666-0 (eBook)
DOI 10.1007/978-3-642-32666-0
Springer Heidelberg New York Dordrecht London

Lecture Notes in Mathematics ISSN print edition: 0075-8434
ISSN electronic edition: 1617-9692

Library of Congress Control Number: 2012951623

Mathematics Subject Classification (2010): 35C15, 35G45, 35J48, 35J58, 35J08, 35B30, 31B10, 31B30

Printed on acid-free paper

Springer is part of Springer Science+Business Media (www.springer.com)

Dedicată cu mult drag familiei noastre

Preface

Layer potentials constitute one of the most powerful tools in the treatment of boundary value problems associated with (elliptic and parabolic) partial differential equations (PDEs). They have been traditionally employed in the context of second-order PDE, and one of the main goals of this monograph is to systematically develop a multilayer theory applicable to the higher-order setting. This extension of the classical theory is carried out in the context of arbitrary Lipschitz domains and includes mapping properties of such multilayers associated with complex, matrix-valued, constant coefficient, homogeneous elliptic systems, in function spaces suitably adapted to the higher-regularity case (of Besov and Triebel–Lizorkin type), Carleson measure estimates, non-tangential maximal function estimates, jump-relations, etc., which turn out to be just as versatile and effective as their second-order counterparts. In particular, this theory applies to such basic differential operators like the Laplacian, the bi-Laplacian, the polyharmonic operator, and the Lamé system of elasticity, though the gist of the present work is constructing, for the first time, a comprehensive theory (of Calderón–Zygmund type) for singular integral operators of multilayer type associated with generic higher-order PDEs and to discuss some of the implications of this multilayer theory to the well-posedness of boundary value problems for higher-order PDEs. As such, one of the main purposes of this monograph is to address an obvious gap/discrepancy/imbalance in the present literature between the second- and higher-order case.

The intended audience consists of any mathematically trained scientist with an interest in boundary value problems and partial differential equations. While this is an original research monograph, significant effort has been put in to make the material as reasonably accessible as possible. In particular, this monograph should also be useful to junior scientists working in the area of PDE.

Philadelphia, PA
Columbia, MO
June 2012

Irina Mitrea
Marius Mitrea

Contents

Chapter 1
Introduction

One of the main goals of the present monograph is to develop the framework of a theory for the multiple layer (or multi-layer, for short) potential operators arising in the treatment of boundary value problems associated with a higher-order, matrix-valued (complex) constant coefficient, elliptic differential operator

$$Lu = \sum_{|\alpha|=|\beta|=m} \partial^\alpha A_{\alpha\beta}\, \partial^\beta u \tag{1.1}$$

(where $m \in \mathbb{N}$) in a Lipschitz domain $\Omega \subset \mathbb{R}^n$. As such, our work falls within the scope of the program outlined by A.P. Calderón in his 1978 ICM plenary address in which he advocates the use of layer potentials *for much more general elliptic systems [than the Laplacian]*—see p. 90 in [19].

Recall that the classical *harmonic* double layer potential operator in a Lipschitz domain $\Omega \subset \mathbb{R}^n$, along with its principal-value version on $\partial\Omega$, are, respectively, defined as

$$(\mathscr{D}_\Delta f)(X) := \frac{1}{\omega_{n-1}} \int_{\partial\Omega} \frac{\langle \nu(Y), Y - X\rangle}{|X - Y|^n} f(Y)\, d\sigma(Y), \quad X \in \mathbb{R}^n \setminus \partial\Omega, \tag{1.2}$$

$$(K_\Delta f)(X) := \lim_{\varepsilon\to 0+} \frac{1}{\omega_{n-1}} \int\limits_{\substack{Y\in\partial\Omega \\ |X-Y|>\varepsilon}} \frac{\langle \nu(Y), Y - X\rangle}{|X - Y|^n} f(Y)\, d\sigma(Y), \quad X \in \partial\Omega, \tag{1.3}$$

where $\nu = (\nu_j)_{1\leq j\leq n}$ is the outward unit normal defined a.e. with respect to the surface measure σ on $\partial\Omega$, and ω_{n-1} denotes the surface area of the unit sphere in $\mathbb{R}^n$. The modern study of these operators originates with [18, 20, 27], and the fundamental role they play in the treatment of boundary value problems

I. Mitrea and M. Mitrea, *Multi-Layer Potentials and Boundary Problems*, Lecture Notes in Mathematics 2063, DOI 10.1007/978-3-642-32666-0_1,

for the Laplacian in C^1 domains has been brought to prominence in the celebrated 1979 Acta Mathematica paper by E. Fabes et al. [40]. There, these authors have established many basic properties such as the existence of the principal-value singular integrals, non-tangential maximal function estimates, boundedness, jump relations, etc. At the core of their approach for proving well-posedness results for the Laplacian is the fact that, for bounded C^1 domains, the operator

$$K_\Delta : L^p(\partial\Omega) \longrightarrow L^p(\partial\Omega) \tag{1.4}$$

is compact, for each $p \in (1,\infty)$. Although many other striking advances (an excellent account of which may be found in [68]) have been made since its publication, the paper [40] continues to serve as a blue-print in the study of many other types of elliptic differential operators of second order (including systems).

Here we take the next natural step and explore the extent to which a parallel theory can be developed for *multi-layer potentials* associated with constant coefficient, elliptic, higher-order differential operators, defined as in (1.1), in Lipschitz and C^1 domains. This multi-layer potential theory should have the same trademark features as the one corresponding to second-order operators—indeed, the latter should eventually become a particular case of the former.

In this process, one is confronted with a number of significant issues from the very outset, as several traditional concepts need to be changed or adapted in order to accommodate operators of higher order. In this regard, one of the first tasks is to identify the correct analogues of such classical spaces as $L^p(\partial\Omega)$ (Lebesgue), $L^p_1(\partial\Omega)$ (Sobolev), $H^p_1(\partial\Omega)$ (Hardy–Sobolev), $B^{p,q}_s(\partial\Omega)$ (Besov), $C^\alpha(\partial\Omega)$ (the class of Hölder spaces), $F^{p,q}_s(\partial\Omega)$ (Triebel–Lizorkin), $\mathrm{BMO}(\partial\Omega)$ (John–Nirenberg), $\mathrm{VMO}(\partial\Omega)$ (Sarason), etc., which are relevant in the treatment of the higher-order Dirichlet problem formulated for a differential operator L as in (1.1) in a Lipschitz domain Ω. The question of determining the correct function spaces on which singular integral operators (of potential type) associated with L are well-defined and bounded, was asked by A.P. Calderón on page 95 of [19]. In Problem 7 on page xvii in [106], N.M. Rivière also raises the issue of "*prescribing classes of boundary data which give existence and uniqueness* [for boundary problems for Δ^2 on C^1 domains]."

Inspired by the classical work of [3, 131], as well as the more recent considerations in [103, 127], we shall employ *Whitney arrays* adapted to various types of scales on which the smoothness of scalar-valued functions on Lipschitz surfaces is traditionally measured. To define these, we shall use the same basic recipe, irrespective of the nature of the original template space. Namely, given a Lipschitz domain $\Omega \subset \mathbb{R}^n$ with outward unit normal $\nu = (\nu_j)_{1\leq j\leq n}$ and surface measure σ, along with a number $m \in \mathbb{N}$ and a space of scalar functions $X(\partial\Omega) \hookrightarrow L^1_{loc}(\partial\Omega)$, we let $\dot{X}_{m-1}(\partial\Omega)$ be the space of all families $\dot{f} = \{f_\alpha\}_{|\alpha|\leq m-1}$, indexed by multi-indices $\alpha \in \mathbb{N}^n_0$ of length $\leq m-1$, satisfying the following properties:

Table 1.1 Function spaces of Whitney arrays

$X(\partial\Omega)$	$\dot{X}_{m-1}(\partial\Omega)$	Introduced by
$L^p(\partial\Omega)$	$\dot{L}^p_{m-1,0}(\partial\Omega)$	Cohen–Gosselin [25] if $m = n = 2$
$L^p_1(\partial\Omega)$	$\dot{L}^p_{m-1,1}(\partial\Omega)$	Verchota [127]
$B^{p,q}_s(\partial\Omega)$	$\dot{B}^{p,q}_{m-1,s}(\partial\Omega)$	Maz'ya et al. [76] when $p = q > 1$
$C^\alpha(\partial\Omega)$	$\dot{C}_{m-1,\alpha}(\partial\Omega)$	Agmon [3] when $n = 2$
$H^p_1(\partial\Omega)$	$\dot{H}^p_{m-1,1}(\partial\Omega)$	Pipher–Verchota [103] if $n = 3$, $m = 2$
$\mathrm{BMO}(\partial\Omega)$	$\dot{\mathrm{BMO}}_{m-1}(\partial\Omega)$	Pipher–Verchota [103] if $n = 3$, $m = 2$

$$f_\alpha \in X(\partial\Omega) \quad \forall\, \alpha \in \mathbb{N}_0^n \text{ with } |\alpha| \leq m-1, \quad \text{and} \tag{1.5}$$

$$\begin{cases} \displaystyle\int_{\partial\Omega} (\nu_j f_{\alpha+e_k} - \nu_k f_{\alpha+e_j})\, \varphi \, d\sigma = \int_{\partial\Omega} f_\alpha \, \partial_{\tau_{kj}} \varphi \, d\sigma, \\ \forall\, \alpha \in \mathbb{N}_0^n : |\alpha| \leq m-2, \ \ 1 \leq j,k \leq n, \quad \forall\, \varphi \in C_c^\infty(\mathbb{R}^n). \end{cases} \tag{1.6}$$

Here $\mathbb{N}_0 := \mathbb{N} \cup \{0\}$, $e_j := (\delta_{j\ell})_{1\leq \ell \leq n}$, $1 \leq j \leq n$, and $\partial_{\tau_{jk}}$ is the tangential derivative operator on $\partial\Omega$ given by $\nu_j \partial_k - \nu_k \partial_j$, for each $j,k \in \{1,\dots,n\}$. Whenever $X(\partial\Omega)$ is equipped with a quasi-norm, we shall endow its associated Whitney array space with the quasi-norm

$$\|\dot{f}\|_{\dot{X}_{m-1}(\partial\Omega)} := \sum_{|\alpha|\leq m-1} \|f_\alpha\|_{X(\partial\Omega)}, \quad \forall\, \dot{f} = \{f_\alpha\}_{|\alpha|\leq m-1} \in \dot{X}_{m-1}(\partial\Omega). \tag{1.7}$$

Table 1.1 illustrates what this procedure, associating to a given space of scalar-valued functions $X(\partial\Omega)$ a Whitney array space $\dot{X}_{m-1}(\partial\Omega)$, yields in a number of cases which have been previously considered in the literature. Succinctly put, one of the goals of this monograph is to systematically develop a theory for such function spaces. This includes various characterizations, density results, as well as interpolation formulas, such as

$$\left(\dot{L}^p_{m-1,0}(\partial\Omega)\,,\, \dot{L}^p_{m-1,1}(\partial\Omega)\right)_{s,q} = \dot{B}^{p,q}_{m-1,s}(\partial\Omega) \tag{1.8}$$

for $1 < p < \infty$, $0 < q < \infty$, $0 < s < 1$ (established in Theorem 3.39), and

$$\Big(\big[\dot{L}^{p_0}_{m-1,0}(\partial\Omega)\,,\, \dot{L}^{p_1}_{m-1,1}(\partial\Omega)\big]_\theta\,,\, \big[\dot{L}^{\widetilde{p_0}}_{m-1,0}(\partial\Omega)\,,\, \dot{L}^{\widetilde{p_1}}_{m-1,1}(\partial\Omega)\big]_{\widetilde{\theta}}\Big)_{\eta,q}$$

$$= \dot{B}^{p,q}_{m-1,s}(\partial\Omega) \tag{1.9}$$

whenever

$$
\begin{aligned}
&\theta,\widetilde{\theta}\in(0,1)\ \text{ with }\ \theta\neq\widetilde{\theta},\quad p_0,p_1,\widetilde{p}_0,\widetilde{p}_1\in(1,\infty),\\
&p\in(1,\infty)\ \text{ with }\ \frac{1}{p}=\frac{1-\theta}{p_0}+\frac{\theta}{p_1}=\frac{1-\widetilde{\theta}}{\widetilde{p}_0}+\frac{\widetilde{\theta}}{\widetilde{p}_1}, \qquad (1.10)\\
&\eta\in(0,1),\ \ q\in(0,\infty],\ \text{ and }\ s=(1-\eta)\theta+\eta\widetilde{\theta},
\end{aligned}
$$

(established in Corollary 3.46).

Having identified and studied the natural function spaces of higher-order smoothness on Lipschitz surfaces, the next order of business is to find the correct concept of multi-layer associated with an elliptic, constant coefficient differential operator L as in (1.1). If E denotes a (matrix-valued) fundamental solution for the operator L, we define the action of the double multi-layer potential operator $\dot{\mathscr{D}}$ on a Whitney array $\dot{f}=\{f_\delta\}_{|\delta|\leq m-1}$ on $\partial\Omega$ by setting, for each $X\in\mathbb{R}^n\setminus\partial\Omega$,

$$
\dot{\mathscr{D}}\dot{f}(X):=-\sum_{\substack{|\alpha|=m\\|\beta|=m}}\sum_{k=1}^{m}\sum_{\substack{|\delta|=m-k\\|\gamma|=k-1\\\gamma+\delta+e_j=\alpha}}\frac{\alpha!(m-k)!(k-1)!}{m!\gamma!\delta!}\times \qquad (1.11)
$$

$$
\times\int_{\partial\Omega}\big\langle\nu_j(Y)A_{\alpha\beta}(\partial^{\beta+\gamma}E)(X-Y),f_\delta(Y)\big\rangle\,d\sigma(Y).
$$

In the same context, we define the principal-value multiple layer $\dot{K}$ by the formula

$$
\dot{K}\dot{f}:=\Big\{\big(\dot{K}\dot{f}\big)_\gamma\Big\}_{|\gamma|\leq m-1} \qquad (1.12)
$$

where, for each multi-index $\gamma\in\mathbb{N}_0^n$ of length $\leq m-1$, and σ-a.e. point $X\in\partial\Omega$, we have set

$$
\big(\dot{K}\dot{f}\big)_\gamma(X):=\sum_{\substack{|\alpha|=m\\|\beta|=m}}\sum_{\ell=1}^{|\gamma|}\sum_{\substack{\delta+\eta+e_k=\alpha\\|\delta|=\ell-1,|\eta|=m-\ell}}\sum_{\substack{\theta+\omega+e_j=\gamma\\|\theta|=\ell-1,|\omega|=|\gamma|-\ell}}\frac{\alpha!|\delta|!|\eta|!\gamma!|\theta|!|\omega|!}{m!\,\delta!\,\eta!\,|\gamma|!\,\theta!\,\omega!}\times
$$

$$
\times\lim_{\varepsilon\to0^+}\int_{\substack{Y\in\partial\Omega\\|X-Y|>\varepsilon}}\Big\langle A_{\alpha\beta}\partial_{\tau_{kj}(Y)}\Big[(\partial^{\delta+\omega+\beta}E)(X-Y)\Big],\,f_{\theta+\eta}(Y)\Big\rangle\,d\sigma(Y)
$$

$$-\sum_{\substack{|\alpha|=m\\|\beta|=m}}\sum_{\ell=|\gamma|+1}^{m}\sum_{\substack{\delta+\eta+e_k=\alpha\\|\delta|=\ell-1,|\eta|=m-\ell}}\frac{\alpha!|\delta|!|\eta|!}{m!\,\delta!\,\eta!}\times \tag{1.13}$$

$$\times\lim_{\varepsilon\to 0+}\int_{\substack{Y\in\partial\Omega\\|X-Y|>\varepsilon}}\Big\langle\nu_k(Y)A_{\alpha\beta}(\partial^{\delta+\beta}E)(X-Y),f_{\gamma+\eta}(Y)\Big\rangle\,d\sigma(Y).$$

In spite of their seemingly intricate nature, definitions (1.11)–(1.13) are most natural. For example, they reduce precisely to (1.2)–(1.3) when $L=\Delta$, and they also contain as particular cases the multi-layer potentials introduced by S. Agmon in [3]. In fact, all double layer potential operators historically considered in connection with basic PDE's in mathematical physics, such as the Lamé system of elastostatics and the Stokes system of hydrodynamics, fall under scope of formulas (1.11)–(1.13). Most importantly, these operators satisfy properties similar to those proved by E. Fabes, M. Jodeit Jr. and N.M. Rivière in [40] for the classical harmonic layer potentials (1.2)–(1.3) and this opens the door for extending the applicability of the classical boundary layer method for higher-order PDE's (a point we shall return to later).

Formulas (1.11)–(1.13) are new and, as is already apparent from these definitions, the case of higher-order operators is algebraically much more sophisticated than the case of second-order operators, which has been studied at length in the literature. Remarkably, the Calderón–Zygmund-type theory for such multi-layer potentials developed in this monograph is virtually analogous (in power and scope) to its more traditional counterpart pertaining to singular integral operators associated with second-order PDE's. For the purpose of this introduction, let us highlight some of these analogies by quoting several basic theorems established here, which extend well-known results from the second-order case.

Theorem 1.1 (Non-tangential maximal function estimates). *Let Ω be a bounded Lipschitz domain in $\mathbb{R}^n$ and let L be a homogeneous, matrix-valued constant (complex) coefficient, elliptic differential operator of order $2m$ as in* (1.1)*. Consider the double multi-layer potential operator $\dot{\mathscr{D}}$ associated with L and Ω as in* (1.11)*. Then for each $p\in(1,\infty)$ there exists $C=C(\Omega,p,L)>0$, a finite constant, such that if $s\in\{0,1\}$,*

$$\big\|\mathscr{N}(\nabla^{m-1+s}\,\dot{\mathscr{D}}\dot{f})\big\|_{L^p(\partial\Omega)}\leq C\|\dot{f}\|_{\dot{L}^p_{m-1,s}(\partial\Omega)}, \tag{1.14}$$

for every Whitney array $\dot{f}=\{f_\alpha\}_{|\alpha|\leq m-1}\in\dot{L}^p_{m-1,s}(\partial\Omega)$.

Above, $\mathscr{N}$ is the non-tangential maximal operator (cf. (2.13)), and ∇^k is the collection all partial derivatives of order k. Moving on, given an arbitrary function

u in Ω, let us define its higher-order boundary trace $u\Big|_{\partial\Omega}^{m-1} := \Big(\partial^\alpha u\Big|_{\partial\Omega}\Big)_{|\alpha|\leq m-1}$, where the restriction to the boundary is taken in the non-tangential limit sense.

Theorem 1.2 (Principal-value integral operators and jump relations). *Retain the same setting as in Theorem 1.1. For each Whitney array $\dot{f} \in \dot{L}^p_{m-1,0}(\partial\Omega)$, $1 < p < \infty$, the expression $\dot{K}\dot{f}(X)$ in* (1.12)–(1.13) *is meaningful for σ-almost every point $X \in \partial\Omega$. Also, for each $p \in (1,\infty)$ and $s \in \{0,1\}$, the singular integral operator*

$$\dot{K} : \dot{L}^p_{m-1,s}(\partial\Omega) \longrightarrow \dot{L}^p_{m-1,s}(\partial\Omega), \tag{1.15}$$

is well-defined, linear and bounded, and (with I denoting the identity operator) the following jump-relation holds:

$$\dot{\mathscr{D}}\dot{f}\Big|_{\partial\Omega}^{m-1} = (\tfrac{1}{2}I + \dot{K})\dot{f}, \qquad \forall\, \dot{f} \in \dot{L}^p_{m-1,0}(\partial\Omega). \tag{1.16}$$

The formulation of the theorem below makes use of the notion of Carleson measure (defined in (2.43)).

Theorem 1.3 (BMO-to-Carleson mapping property). *Consider an elliptic homogeneous differential operator L of order $2m$ with matrix-valued (complex) constant coefficients. Let Ω be a bounded Lipschitz domain in $\mathbb{R}^n$ and let $\dot{\mathscr{D}}$ be the double multi-layer potential operator associated with L and Ω as in* (1.11)*. Then for each $\dot{f} \in \dot{\mathrm{BMO}}_{m-1}(\partial\Omega)$,*

$$|\nabla^m \dot{\mathscr{D}}\dot{f}(X)|^2 \,\mathrm{dist}\,(X,\partial\Omega)\,dX \quad \textit{is a Carleson measure in } \Omega \tag{1.17}$$

and, for some finite geometric constant $C > 0$, the following naturally accompanying estimate holds:

$$\left(\sup_{\substack{r>0\\ Y\in\partial\Omega}} \frac{1}{r^n}\int_{\Omega\cap B(Y,r)} |\nabla^m \dot{\mathscr{D}}\dot{f}(X)|^2 \,\mathrm{dist}\,(X,\partial\Omega)\,dX\right)^{1/2} \leq C\|\dot{f}\|_{\dot{\mathrm{BMO}}_{m-1}(\partial\Omega)}. \tag{1.18}$$

To state our next result, recall that given a bounded Lipschitz domain Ω in $\mathbb{R}^n$, the restriction to the boundary map, $\mathrm{Tr}\, u := u\big|_{\partial\Omega}$, originally defined for functions $u \in C^0(\overline{\Omega})$, extends to a bounded linear operator

$$\mathrm{Tr} : B^{p,q}_{s+1/p}(\Omega) \longrightarrow B^{p,q}_s(\partial\Omega), \tag{1.19}$$

if $\frac{n-1}{n} < p \leq \infty$, $0 < q < \infty$, and $\max\big\{0, (n-1)(\frac{1}{p}-1)\big\} < s < 1$. In addition, the operator in (1.19) is onto and

$$\Big\{u \in B^{p,q}_s(\Omega) : \mathrm{Tr}\, u = 0\Big\} \text{ is the closure of } C^\infty_c(\Omega) \text{ in } B^{p,q}_{s+1/p}(\Omega). \tag{1.20}$$

Finally, similar results are valid for Triebel–Lizorkin spaces.

When $1 \le p,q \le \infty$, these claims have been proved in [58, 62]. The above, more general, version of these results has been obtained in [73]. For the problems we have in mind, we are naturally led to considering traces from spaces exhibiting a larger amount of smoothness than the above results could handle. Thus, the very nature of such trace results changes in the higher smoothness setting. For now, given $m \in \mathbb{N}$, we define the higher-order trace operator by setting

$$\mathrm{tr}_{m-1}(u) := \Big\{\mathrm{Tr}\,[\partial^\alpha u]\Big\}_{|\alpha|\le m-1}, \tag{1.21}$$

whenever meaningful.

Theorem 1.4 (Mapping properties of multi-layers on Besov and Triebel–Lizorkin spaces). *Let Ω be a bounded Lipschitz domain in $\mathbb{R}^n$, L an elliptic differential operator of order $2m$ as in* (1.1)*, and consider the double multi-layer potential operator defined in* (1.11)*. Then, if the indexes p,q,s satisfy $\frac{n-1}{n} < p \le \infty$, $0 < q \le \infty$, $\max\{0, (n-1)(\frac{1}{p}-1)\} < s < 1$, the operator*

$$\dot{\mathscr{D}} : \dot{B}^{p,q}_{m-1,s}(\partial\Omega) \longrightarrow B^{p,q}_{m-1+s+1/p}(\Omega) \tag{1.22}$$

is well-defined, linear and bounded. Furthermore,

$$\mathrm{tr}_{m-1} \circ \dot{\mathscr{D}} = \tfrac{1}{2}I + \dot{K} \ \ in \ \ \dot{B}^{p,q}_{m-1,s}(\partial\Omega). \tag{1.23}$$

Finally, analogous results are valid for Triebel–Lizorkin spaces in Ω.

Let us momentarily digress in order to explain a remarkable application of the above result to the trace theory for higher-order smoothness spaces in Lipschitz domains. Part of our motivation stems from the observation that the higher-order trace operator tr_{m-1} plays a key role in the context of Theorem 1.4. Specifically, we shall prove the following result.

Theorem 1.5 (Multi-trace theory). *Let $\Omega \subset \mathbb{R}^n$ be a bounded Lipschitz domain and assume that $0 < p,q \le \infty$, $\max\{0, (n-1)(\frac{1}{p}-1)\} < s < 1$. Also, fix $m \in \mathbb{N}$. Then the higher-order trace operator* (1.21) *induces a well-defined, linear and bounded mapping*

$$\mathrm{tr}_{m-1} : B^{p,q}_{m-1+s+1/p}(\Omega) \longrightarrow \dot{B}^{p,q}_{m-1,s}(\partial\Omega). \tag{1.24}$$

Moreover, this operator is onto, in fact it has a bounded linear right-inverse

$$\mathscr{E}_{m-1} : \dot{B}^{p,q}_{m-1,s}(\partial\Omega) \longrightarrow B^{p,q}_{m-1+s+1/p}(\Omega), \tag{1.25}$$

and its null-space, i.e.,

$$\Big\{u \in B^{p,q}_{m-1+s+1/p}(\Omega) : \mathrm{tr}_{m-1}(u) = 0\Big\}, \tag{1.26}$$

may be described in the case when $\max\{p,q\}<\infty$ *as*

$$\textit{the closure of } C_c^{\infty}(\Omega) \textit{ in } B^{p,q}_{m-1+s+1/p}(\Omega). \tag{1.27}$$

Finally, similar results are valid on the Triebel–Lizorkin scale.

Theorem 1.5 is proved in § 3.2. Here we would only like to comment on the role played by our multi-layer potentials in the construction of a linear, bounded, right-inverse for the operator (1.24). Concretely, let $\dot{\mathscr{D}}^{\pm}$ be the versions of the double layer potential associated with the polyharmonic operator $L:=\Delta^m$ in the Lipschitz domains $\Omega_+ := \Omega$ and $\Omega_- := \mathbb{R}^n\setminus\overline{\Omega}$, respectively (assuming that the exterior version of the double multi-layer is suitably truncated at infinity). Also, set $\mathscr{R}_{\Omega_\pm}$ for the operators restricting distributions in $\mathbb{R}^n$ to $\Omega_\pm$, and let $E_{\Omega_\pm}$ stand for Rychkov's universal extension operator for the Besov and Triebel–Lizorkin scales (extending distributions in $\Omega_\pm$ to $\mathbb{R}^n$ with preservation of smoothness; cf. [108]). If we now set

$$\mathscr{E}_{m-1} := \dot{\mathscr{D}}^{+} - \mathscr{R}_{\Omega_+}\circ E_{\Omega_-}\circ\dot{\mathscr{D}}^{-} \tag{1.28}$$

it can be verified, based on (1.22)–(1.23), written for both domains Ω_+ and Ω_-, that the operator $\mathscr{E}_{m-1}$ maps $\dot{B}^{p,q}_{m-1,s}(\partial\Omega)$ boundedly into the space $B^{p,q}_{m-1+s+1/p}(\Omega)$ and $\mathrm{tr}_{m-1}\circ\mathscr{E}_{m-1} = (\frac{1}{2}I+\dot{K})-(-\frac{1}{2}I+\dot{K}) = I$ on $\dot{B}^{p,q}_{m-1,s}(\partial\Omega)$, as desired. Compared with other, more traditional methods, used in the study of the trace operator, this is a conceptually new approach, which highlights the important role played by the multi-layer theory developed in this monograph.

Returning to the main topic of discussion, it should be noted that, given a differential operator L of order $2m$, there are infinitely many ways of expressing it as in (1.1), corresponding to various choices of the coefficient tensor $A:=\big(A_{\alpha\beta}\big)_{\alpha,\beta}$. In turn, each such choice of the coefficient tensor A leads to (typically) different operators $\dot{\mathscr{D}}=\dot{\mathscr{D}}_A$ and $\dot{K}=\dot{K}_A$ (associated with A as in (1.11) and (1.12)–(1.13)). Hence, schematically,

$$\begin{array}{c}\text{a given differential}\\ \text{elliptic operator } L\end{array} \Longrightarrow \begin{array}{c}\text{infinitely many writings of}\\ \text{the operator } L \text{ as in (1.1)}\end{array} \Longrightarrow \begin{array}{c}\text{infinitely many}\\ \text{multi-layers.}\end{array} \tag{1.29}$$

Remarkably, these distinctions play no role in the broad picture of the Calderón–Zygmund-type theory developed here, in the sense that all the theorems stated so far are valid for *any* multiple layer potential, corresponding to any choice of a tensor coefficient in the writing of the differential operator L.

In contrast to the case of the double multi-layer potential operators discussed so far, the so-called `single multi-layer operator` associated with a homogeneous, elliptic, matrix-valued (complex) constant coefficient, higher-order system L in a bounded Lipschitz domain $\Omega\subset\mathbb{R}^n$, i.e.,

$$(\dot{\mathscr{S}}\Lambda)(X) := \left\langle \Lambda(\cdot), \left\{ \partial^\alpha_\cdot [E(X-\cdot)] \Big|_{\partial\Omega} \right\}_{|\alpha|\leq m-1} \right\rangle, \quad X \in \Omega, \tag{1.30}$$

is uniquely determined by the differential operator L itself, and not by the choice of the coefficient tensor used to represent L as in (1.1). For this genre of integral operator, we shall prove several basic mapping properties such as

$$\dot{\mathscr{S}} : \left(\dot{B}^{p',q'}_{m-1,1-s}(\partial\Omega)\right)^* \longrightarrow B^{p,q}_{m-1+s+1/p}(\Omega), \tag{1.31}$$

$$\dot{S} := \mathrm{tr}_{m-1} \circ \dot{\mathscr{S}} : \left(\dot{B}^{p',q'}_{m-1,1-s}(\partial\Omega)\right)^* \longrightarrow \dot{B}^{p,q}_{m-1,s}(\partial\Omega), \tag{1.32}$$

boundedly, whenever $1 < p, q < \infty$, $0 < s < 1$, $\frac{1}{p} + \frac{1}{p'} = \frac{1}{q} + \frac{1}{q'} = 1$. The reader is referred to the discussion in § 5 for a more detailed analysis and further results.

The second major aim of this monograph is to explore the extent to which multi-layer singular integral operators, of potential type, are useful in the treatment of boundary value problems for a homogeneous elliptic operator L of higher order (say $2m$, where $m \in \mathbb{N}$), with constant, matrix-valued, coefficients in a Lipschitz domain $\Omega \subset \mathbb{R}^n$. For the sake of the present discussion, let us formulate several basic problems of this nature. For example, consider

$$\begin{cases} Lu = 0 \ \text{ in } \Omega, \\ \mathcal{N}(\nabla^{m-1+j}u) \in L^p(\partial\Omega), & j \in \{0,1\}, \\ u\Big|^{m-1}_{\partial\Omega} = \dot{f} \in \dot{L}^p_{m-1,j}(\partial\Omega), \end{cases} \tag{1.33}$$

which reduces to the Dirichlet problem when $j = 0$ and to the so-called Regularity problem when $j = 1$. The strategy for proving that these problems are well-posed is to search for a solution represented in the form

$$u(X) = \dot{\mathscr{D}}\Big[\Big(\tfrac{1}{2}I + \dot{K}\Big)^{-1} \dot{f}\,\Big](X), \qquad X \in \Omega. \tag{1.34}$$

As such, the crux of the matter becomes inverting the operator $\frac{1}{2}I + \dot{K}$ on the Whitney–Lebesgue space $\dot{L}^p_{m-1,0}(\partial\Omega)$, in the case when $j = 0$, and the Whitney–Sobolev space $\dot{L}^p_{m-1,1}(\partial\Omega)$, in the case when $j = 1$. Granted this, our earlier estimates for multi-layer potentials then ensure that the function u is indeed a solution to (1.33) which satisfies

$$\|\mathcal{N}(\nabla^{m-1+j}u)\|_{L^p(\partial\Omega)} \leq C(\Omega, L, p)\|\dot{f}\|_{\dot{L}^p_{m-1,j}(\partial\Omega)}. \tag{1.35}$$

Variants of (1.33), involving boundary data from Whitney–Besov spaces, include the inhomogeneous Dirichlet problem in which the solution is sought in a higher-order Besov space in Ω, i.e.,

$$\begin{cases} Lu = w \in B^{p,q}_{-m-1+s+1/p}(\Omega), \\ u \in B^{p,q}_{m-1+s+1/p}(\Omega), \\ \mathrm{tr}_{m-1} u = \dot{f} \in \dot{B}^{p,q}_{m-1,s}(\partial\Omega), \end{cases} \tag{1.36}$$

where

$$\tfrac{n-1}{n} < p < \infty, \quad 0 < q < \infty, \quad (n-1)\max\left\{\tfrac{1}{p} - 1\,,\,0\right\} < s < 1, \tag{1.37}$$

and the inhomogeneous Dirichlet problem in which the solution is sought in a higher-order Triebel–Lizorkin space in Ω, i.e.,

$$\begin{cases} Lu = w \in F^{p,q}_{-m-1+s+1/p}(\Omega), \\ u \in F^{p,q}_{m-1+s+1/p}(\Omega), \\ \mathrm{tr}_{m-1} u = \dot{f} \in \dot{B}^{p,p}_{m-1,s}(\partial\Omega), \end{cases} \tag{1.38}$$

with p, q, s as before. Here, the crucial aspects are the boundedness and invertibility of the multi-layers on Besov and Triebel–Lizorkin spaces.

Other basic types of boundary problems involve boundary conditions of Neumann type. One example from this category which we shall be concerned with reads as follows:

$$\begin{cases} Lu = 0 \quad \text{in } \Omega, \\ u \in B^{p,p}_{m-1+s+1/p}(\Omega), \\ \partial^A_\nu u = \Lambda \in \left(\dot{B}^{p',p'}_{m-1,1-s}(\partial\Omega)\right)^*, \end{cases} \tag{1.39}$$

where $1 < p, p' < \infty$, $1/p + 1/p' = 1$, $0 < s < 1$ and ∂_{ν_A} denotes the so-called conormal derivative for the operator L (written as in (1.1)). A precise definition is given in § 5.4 and here we want to mention that the distinguishing feature of $\partial^A_\nu u$ is the property that for any reasonable null-solution u of L in Ω one has

$$\mathscr{B}(u, v) = \int_{\partial\Omega} \langle \partial^A_\nu u, \mathrm{tr}_{m-1} v\rangle \, d\sigma, \qquad \forall\, v \in C^\infty_c(\mathbb{R}^n), \tag{1.40}$$

where $\mathscr{B}$ is a bilinear form naturally associated to the differential operator L. For the purpose of this introduction we shall illustrate this concept by briefly discussing the case $m = 2$ (i.e., when L a fourth-order operator). First, consider the biharmonic operator in a bounded Lipschitz domain $\Omega \subset \mathbb{R}^n$ and consider the family of bilinear forms $\mathscr{B}_\theta$ associated to Δ^2 (and indexed by $\theta \in \mathbb{R}$) given by

$$\mathscr{B}_\theta(u,v) = \frac{1}{1+2\theta+n\theta^2} \times \tag{1.41}$$
$$\times \sum_{i,j=1}^{n} \int_\Omega \big[(\partial_i\partial_j + \theta\,\delta_{ij}\,\Delta)u\big](X)\big[(\partial_i\partial_j + \theta\,\delta_{ij}\,\Delta)v\big](X)\,dX,$$

where u, v are any two reasonably behaved (real-valued) functions in Ω. Set

$$N_\theta(u) := \frac{\partial(\Delta u)}{\partial\nu} + \frac{1}{2(1+2\theta+n\theta^2)} \sum_{i,j=1}^{n} \partial_{\tau_{ij}}\Big(\frac{\partial}{\partial\nu}\partial_{\tau_{ij}} u\Big),$$
$$M_\theta(u) := \frac{2\theta+n\theta^2}{1+2\theta+n\theta^2}\Delta u + \frac{1}{1+2\theta+n\theta^2}\frac{\partial^2 u}{\partial\nu^2}. \tag{1.42}$$

Then, if $v \in C_c^\infty(\mathbb{R}^n)$ and u is a reasonably behaved biharmonic function in Ω, identity (1.40), on the one hand, and direct integrations by parts, on the other hand, give

$$\int_{\partial\Omega} \langle \partial_\nu^A u, \mathrm{tr}_{m-1} v\rangle\, d\sigma = \mathscr{B}_\theta(u,v) = \int_{\partial\Omega} \Big[M_\theta(u)\frac{\partial v}{\partial\nu} - N_\theta(u)v\Big]\,d\sigma. \tag{1.43}$$

Therefore,

$$L = \Delta^2 \text{ for the choice of bilinear form } \mathscr{B}_\theta \Rightarrow \partial_\nu^A u := \Big\{\Big(\partial_\nu^A u\Big)_j\Big\}_{0\le j\le n} \tag{1.44}$$
$$\text{where } \Big(\partial_\nu^A u\Big)_0 := -N_\theta(u) \text{ and } \Big(\partial_\nu^A u\Big)_j := \nu_j M_\theta(u) \text{ for } 1 \le j \le n.$$

More generally, consider the *anisotropic plate bending operator*, i.e.

$$Lu = \sum_{i,j,k,\ell=1}^{2} \frac{\partial^2}{\partial x_k \partial x_\ell}\Big(a_{ijk\ell}\frac{\partial^2 u}{\partial x_i \partial x_j}\Big), \tag{1.45}$$

in a Lipschitz domain $\Omega \subset \mathbb{R}^2$, where

$$A = \Big(a_{ijk\ell}\Big)_{i,j,k,\ell} \text{ has constant, real, entries satisfying } a_{ijk\ell} = a_{k\ell ij}. \tag{1.46}$$

Above in (1.45) the function u models the displacement vector of the plate and, with $\tau := (\tau_1, \tau_2) = (-\nu_2, \nu_1)$ denoting the unit tangent vector to $\partial\Omega$,

$$N_A(u) := \sum_{i,j=1}^{2} \nu_j \frac{\partial}{\partial x_i}\left[\sum_{k,\ell=1}^{2} a_{ijk\ell}\frac{\partial^2 u}{\partial x_k \partial x_\ell}\right] + \partial_\tau\left[\sum_{i,j,k,\ell=1}^{2} a_{ijk\ell}\frac{\partial^2 u}{\partial x_k \partial x_\ell}\nu_i\tau_j\right], \tag{1.47}$$

$$M_A(u) := \sum_{i,j,k,\ell=1}^{2} a_{ijk\ell}\frac{\partial^2 u}{\partial x_k \partial x_\ell}\nu_i\nu_j,$$

stand, respectively, for the *Kirchhoff transverse force* and the *normal bending moment*, respectively. Also,

$$\sum_{i,j,k,\ell=1}^{2} a_{ijk\ell}\frac{\partial^2 u}{\partial x_k \partial x_\ell}\nu_i\tau_j \tag{1.48}$$

stands for the *twisting moment*, while

$$\sum_{k,\ell=1}^{2} a_{ijk\ell}\frac{\partial^2 u}{\partial x_k \partial x_\ell} \tag{1.49}$$

denotes the *symmetric bending moment*. In particular,

$$\sum_{i,j=1}^{2} \nu_j \frac{\partial}{\partial x_i}\left[\sum_{k,\ell=1}^{2} a_{ijk\ell}\frac{\partial^2 u}{\partial x_k \partial x_\ell}\right] \tag{1.50}$$

is the *transverse shear force*. The operators N_A, M_A play a similar role in the current case as the operators N_θ, M_θ did in the case of the biharmonic operator. Indeed, for any reasonably behaved null-solution u of the anisotropic plate bending operator L in Ω, a direct calculation gives

$$\begin{aligned}\mathscr{B}_A(u,v) &:= \sum_{i,j,k,\ell=1}^{2} \int_\Omega \langle a_{ijk\ell}\partial_i\partial_j u, \partial_k\partial_\ell v\rangle\, dX \\ &= \int_{\partial\Omega}\Big[M_A(u)\frac{\partial v}{\partial \nu} - N_A(u)v\Big]\, d\sigma,\end{aligned} \tag{1.51}$$

for every $v \in C_c^\infty(\mathbb{R}^2)$, which is analogous to (1.43). Furthermore, when specialized to

$$a_{ijk\ell} := \frac{1}{1+2\theta+2\theta^2}\Big[\delta_{ik}\delta_{j\ell} + (2\theta+2\theta^2)\delta_{ij}\delta_{k\ell}\Big], \tag{1.52}$$

the operators N_A, M_A in (1.47) become, respectively, N_θ and M_θ in (1.42) corresponding to the case $n = 2$. To conclude, the definition of the conormal derivative operator for the anisotropic plate bending operator (1.45) corresponding to the bilinear form $\mathscr{B}_A$ in (1.51) is

$$\partial_\nu^A u := \{-N_A(u)\,,\, \nu_1 M_A(u)\,,\, \nu_2 M_A(u)\}\,. \tag{1.53}$$

Moving on, another version of the Dirichlet problem for the higher-order differential operator L, is the limiting case $p = \infty$ and $j = 0$ of (1.33), i.e.,

$$\begin{cases} Lu = 0 \ \text{ in } \Omega, \\ \displaystyle\sup_{0<r<\kappa,\, X\in\partial\Omega} \frac{1}{r^n}\int_{\Omega\cap B(X,r)} |\nabla^m u(Y)|^2 \operatorname{dist}(Y,\partial\Omega)\, dY < +\infty, \\ u\Big|_{\partial\Omega}^{m-1} = \dot f \in \mathrm{B\dot{M}O}_{m-1}(\partial\Omega), \end{cases} \tag{1.54}$$

(for some sufficiently small $\kappa = \kappa(\Omega) > 0$) with the naturally accompanying estimate

$$\left(\sup_{\substack{0<r<\kappa\\ X\in\partial\Omega}} \frac{1}{r^n}\int_{\Omega\cap B(X,r)} |\nabla^m u(Y)|^2 \operatorname{dist}(Y,\partial\Omega)\, dY\right)^{1/2} \le C\,\|\dot f\|_{\mathrm{B\dot{M}O}_{m-1}(\partial\Omega)}. \tag{1.55}$$

Once again, a reasonable approach is to look for a solution represented as in (1.34). In this approach, the estimates advertised in Theorem 1.3 play a basic role.

Finally, we wish to record the formulation of the Dirichlet problem for a higher-order differential operator L with boundary data from the Whitney–Hardy–Sobolev space $\dot H^{p,at}_{m-1,1}(\partial\Omega)$. Specifically, this reads

$$\begin{cases} Lu = 0 \ \text{ in } \Omega, \\ \mathcal{N}(\nabla^m u) \in L^p(\partial\Omega), \\ u\Big|_{\partial\Omega}^{m-1} = \dot f \in \dot H^{p,at}_{m-1,1}(\partial\Omega), \end{cases} \tag{1.56}$$

where $p \in (\frac{n-1}{n}, 1]$.

In order to put matters into a broader perspective, let us now review some important relevant earlier work, pertaining to the various problems formulated so far. As alluded to earlier, the case of (1.33) corresponding to $L = \Delta$ and Ω a bounded C^1 domain in $\mathbb{R}^n$ has been dealt with by E. Fabes, M. Jodeit Jr. and N. Rivière in [40]. The fact that (1.33) has at least one solution when $L = \Delta^2$ and Ω is a bounded C^1 domain in $\mathbb{R}^2$, has first been established by J. Cohen and J. Gosselin in [25]. A few years later, B. Dahlberg, C. Kenig and G. Verchota [30] were able to solve (1.33) for $L = \Delta^2$ when $p = 2$, when $\Omega \subset \mathbb{R}^n$ is an arbitrary Lipschitz domain. L^p-versions, $1 < p < \infty$, of these results in C^1 domains in $\mathbb{R}^n$ have been subsequently obtained by G. Verchota in [126]. In turn, the results in [126] have been extended to the polyharmonic operator in [127] in Lipschitz domains in $\mathbb{R}^n$ for $2 - \varepsilon < p < 2 + \varepsilon$, for some $\varepsilon > 0$ which depends on Ω, p and L. Let us also mention here that, using a different approach, R. Selvaggi and I. Sisto [111] have dealt with the case of L^p-Dirichlet problem for the biharmonic operator, in the domain above the graph of a function with small Lipschitz constant. Employing Rellich type inequalities, boundary Gårding inequalities and dilation invariant estimates for solutions to (1.33), J. Pipher and G. Verchota established in [104] the well-posedness of (1.33) when L is as in (1.1) with real, scalar coefficients, $2 - \varepsilon < p < 2 + \varepsilon$, and Ω is a bounded Lipschitz domain in $\mathbb{R}^n$. See also [128], where G. Verchota has extended these results to the case of higher-order systems. The parabolic version of the main results in [104] have been worked out by R. Brown and W. Hu in [14]. In [115–117], Z. Shen developed a new approach for establishing well-posedness results for (1.33) in the case $j = 0$, $L = \Delta^m$ or a second-order system with real coefficients, and Ω a bounded Lipschitz domain in $\mathbb{R}^n$, $n \geq 4$, in the range $2 - \varepsilon < p < \frac{2(n-1)}{n-3} + \varepsilon$ for some $\varepsilon = \varepsilon(\Omega, L, p) > 0$. A related well-posedness result for (1.33) with $j = 1$ in the case when $L = \Delta^2$ may be found in [69]. The problem (1.33) for $j = 0$ has also been studied using Mellin transform techniques. For example, in [37], L. Diomeda and B. Lisena have considered the case $L = \Delta^2$ and Ω is a square in $\mathbb{R}^2$. In [38], the same authors have treated the case when $L = \Delta^2$ and Ω is the first quadrant in $\mathbb{R}^2$. See also [105] where R. Pisani and M. Tucci have dealt with (1.33) for $L = \Delta^m$, $m \in \mathbb{N}$, $m \geq 2$, and Ω the first quadrant in $\mathbb{R}^2$. The reader is also referred to [70] for a wealth of related results.

Let us also mention that the version of (1.33) formally corresponding to $p = \infty$ has been studied by J. Pipher and G. Verchota in [102, 103], when $L = \Delta^m$, $m \geq 2$, and Ω a bounded Lipschitz domain in $\mathbb{R}^n$ if $n = 2, 3$, or a C^1 domain in $\mathbb{R}^n$ if $n \geq 2$. In this setting, these authors establish the following weak maximum modulus principle. If $\Delta^m u = 0$ in Ω, $\nabla^{m-1} u\big|_{\partial\Omega} \in L^\infty(\partial\Omega)$ and $\mathcal{N}(\nabla^{m-1} u) \in L^2(\partial\Omega)$, then

$$\|\nabla^{m-1} u\|_{L^\infty(\Omega)} \leq C(\partial\Omega)\|\nabla^{m-1} u\|_{L^\infty(\partial\Omega)}. \tag{1.57}$$

In particular, the above estimate implies the solvability of (1.33) when $j = 0$, $L = \Delta^m$, $m \in \mathbb{N}$, $m \geq 2$, $2 - \varepsilon < p < \infty$, and Ω bounded Lipschitz domain in $\mathbb{R}^3$. Other related L^∞-estimates are as follows. When specialized to the case of the bi-Laplacian, Theorem 3.3 on p. 329 of J. Nečas' book [98] yields that if $\Omega \subset \mathbb{R}^3$ satisfies a uniform exterior ball condition then, with $\mathbf{G}$ denoting the Green operator for the bi-Laplacian with homogeneous Dirichlet boundary conditions (i.e., the solution operator $w \mapsto u$ for the problem $\Delta^2 u = w$ in Ω and $u = \partial_\nu u = 0$ on $\partial\Omega$), one has

$$\nabla \mathbf{G} : W^{-1,2}(\Omega) \longrightarrow L^\infty(\Omega) \text{ boundedly.} \tag{1.58}$$

See also J. Seo's paper [113] for related estimates. More recently, S. Mayboroda and V. Maz'ya have proved in [75] that if $\Omega \subset \mathbb{R}^3$ is an arbitrary bounded open set then

$$\nabla \mathbf{G} : L^{\frac{3}{2},1}_{-1}(\Omega) \longrightarrow L^\infty(\Omega), \tag{1.59}$$

where $L^{p,q}_{-1}(\Omega)$ consists of first order derivatives of functions from the Lorentz space $L^{p,q}(\Omega)$. Moreover, as pointed out in [75], it is not generally the case that the function $\nabla \mathbf{G}u$ is continuous on $\overline{\Omega}$ for each $u \in L^{3/2,1}_{-1}(\Omega)$ if the open set Ω lacks any smoothness.

Concerning the Dirichlet boundary value problem stated in (1.56), J. Pipher and G. Verchota have shown in [102, 103] that (1.56) is well-posed when $L = \Delta^m$, $m \in \mathbb{N}$, $m \geq 2$, $p = 1$ and Ω is a bounded Lipschitz domain in $\mathbb{R}^3$.

As regards the problem (1.54), a well-posedness result when $L = \Delta$ and Ω is a bounded C^1 domain in $\mathbb{R}^n$, has first been established by E. Fabes and C. Kenig in [42]. On the other hand, the version of the above result corresponding to $L = \Delta^2$ and Ω a bounded C^1 domain in $\mathbb{R}^2$, has been proved by J. Cohen in [24]. The BMO–Carleson measure estimate (1.55) for the solution of (1.54) has been obtained by J. Pipher and G. Verchota in [103] in the case when $L = \Delta^2$ and Ω is a bounded Lipschitz domain in $\mathbb{R}^3$.

Let us now comment on the inhomogeneous Dirichlet problems with boundary data from Besov spaces (cf. (1.36) and (1.38)). From the classical work of S. Agmon, N. Douglis and L. Nirenberg [5] such problems are well-understood in the case of smooth domains. See also Theorem 14 on p.145 in [44] (as well as the discussion on p. 134 in [107]) for an extension to the entire scale of Besov and Triebel–Lizorkin spaces in C^∞ domains. Following the successful treatment by D. Jerison and C. Kenig [58] of the inhomogeneous Dirichlet problem for $L = \Delta$ with Sobolev-Besov data on Lipschitz domains, V. Adolfsson and J. Pipher in [2] have extended the scope of this theory as to include the case of the biharmonic operator in the three-dimensional setting. The work in [2] has been further extended by I. Mitrea, M. Mitrea, and M. Wright in [88], where a larger, sharp, range of indices for the spaces involved was considered for Lipschitz domains in $\mathbb{R}^3$. A very general result of this type has been proved in [76] by V. Maz'ya, M. Mitrea and T. Shaposhnikova for Lipschitz domains whose unit normal has small oscillations, in the higher dimensional setting, for systems with (complex) bounded measurable coefficients sufficiently close to VMO, in the case when $1 < p = q < \infty$ and the solution belongs to a weighted Sobolev space. The particular case when $p = q = \infty$ deserves special mention, since Hölder spaces occur on the Besov scale precisely for this end-point. This corresponds to the boundary value problem

$$\begin{cases} Lu = 0 \ \text{ in } \Omega, \\ u \in C^{m-1+s}(\overline{\Omega}), \\ u\Big|_{\partial\Omega}^{m-1} = \dot{f} \in \dot{C}_{m-1,s}(\partial\Omega), \end{cases} \tag{1.60}$$

where $s \in (0,1)$. Theorem 14.2 on p. 238 in Agmon's paper [3] proves that the problem (1.60) is well-posed in the case when the domain Ω is assumed to be of class C^{1+r} for some $r \in (\frac{1}{2},1)$. The reader may also consult [54] for a very general uniqueness statement for (1.38) corresponding to $p = q = 1/s = 2$ and $L = \Delta^m$, formulated in arbitrary bounded open subsets in $\mathbb{R}^n$.

Our last set of comments pertains to the Neumann problem for the bi-Laplacian, i.e.,

$$\begin{cases} \Delta^2 u = 0 \quad \text{in } \Omega, \\ \mathcal{N}(\nabla\nabla u) \in L^p(\Omega), \\ M_\theta(u) = f \in L^p(\partial\Omega), \\ N_\theta(u) = \Lambda \in \left(L_1^{p'}(\partial\Omega)\right)^*, \end{cases} \tag{1.61}$$

where $\theta \in \mathbb{R}$, $1 < p, p' < \infty$, $1/p + 1/p' = 1$, and M_θ, N_θ are as in (1.42). This has been solved by G. Verchota in [129] in the case when $\Omega \subset \mathbb{R}^n$, $n \geq 2$, bounded Lipschitz domain, for $2 - \varepsilon < p < 2 + \varepsilon$. Shortly thereafter, Z. Shen has proved in [118] L^p versions of Verchota's results when Ω is a bounded Lipschitz domain in $\mathbb{R}^n$, $n \geq 4$, and $\frac{2(n-1)}{n+1} - \varepsilon < p < 2$. Another recent development is the work in [89], where I. Mitrea and G. Verchota have employed Mellin transform techniques to study spectral properties of the 2×2 matrix of singular integral operators with entries of the type (1.42) when Ω is an infinite sector in $\mathbb{R}^2$ of arbitrary aperture. This is relevant in the context of the Neumann problem (1.61) for the biharmonic operator on curvilinear polygons.

Here we continue this line of work and establish several new well-posedness results. The main tools are the multi-trace theory in arbitrary Lipschitz domains and the Calderón–Zygmund theory for multi-layer operators of potential type associated with higher-order operators, developed here. For the purpose of this introduction we wish to quote two such results pertaining to the bi-Laplacian. The first result deals with the inhomogeneous Dirichlet problem in Lipschitz domains in $\mathbb{R}^n$ with boundary data from Whitney–Besov spaces and may be regarded as a higher dimensional version (as well as a refinement) of the work of V. Adolfsson and J. Pipher in [2] for Lipschitz domains in $\mathbb{R}^3$.

Theorem 1.6 (The inhomogeneous Dirichlet problem). *Assume that $\Omega \subset \mathbb{R}^n$, $n \geq 2$, is a bounded Lipschitz domain with connected boundary. Then there exists $\varepsilon > 0$ such that the inhomogeneous Dirichlet problem*

$$\begin{cases} u \in B^{p,q}_{s+\frac{1}{p}+1}(\Omega), \\ \Delta^2 u = w \in B^{p,q}_{s+\frac{1}{p}-3}(\Omega), \\ \left(\operatorname{Tr} u\,,\, \operatorname{Tr}(\nabla u)\right) = \dot{f} \in \dot{B}^{p,q}_{1,s}(\partial\Omega), \end{cases} \tag{1.62}$$

is well-posed whenever $0 < q \le \infty$ *while* $p \in (1, \infty)$ *and* $s \in (0, 1)$ *satisfy*

$$\tfrac{n-3-\varepsilon}{2} < \tfrac{n-1}{p} - s < \tfrac{n-1+\varepsilon}{2} \quad \textit{when } n \ge 4, \tag{1.63}$$

$$0 < \tfrac{1}{p} - \left(\tfrac{1-\varepsilon}{2}\right)s < \tfrac{1+\varepsilon}{2} \quad \textit{when } n \in \{2, 3\}. \tag{1.64}$$

In the class of Lipschitz domains, this well-posedness result is sharp if $n \in \{4, 5\}$. *Moreover, if* $w = 0$ *then the unique solution* u *of* (1.62) *admits an integral representation of the form* (1.34) *for a suitable choice of double multi-layer for the bi-Laplacian.*

Furthermore, given any exponents $p \in (1, \infty)$, $q \in (0, \infty]$, *and* $s \in (0, 1)$, *there exists* $\varepsilon > 0$, *depending only on* p, q, s, *and the Lipschitz character of* Ω, *with the property that if the outward unit normal* ν *to* Ω *satisfies*

$$\limsup_{r \to 0+} \left\{ \sup_{X \in \partial\Omega} ⨍_{B(X,r)\cap\partial\Omega} ⨍_{B(X,r)\cap\partial\Omega} \left|\nu(Y) - \nu(Z)\right| d\sigma(Y) d\sigma(Z) \right\} < \varepsilon \tag{1.65}$$

then the problem (1.62) *is well-posed. As a consequence, the problem* (1.62) *is well-posed for any* $p \in (1, \infty)$, $q \in (0, \infty]$, *and* $s \in (0, 1)$ *if*

$$\nu \in \mathrm{vmo}(\partial\Omega), \tag{1.66}$$

hence, in particular, if Ω *is a* C^1 *domain.*

Finally, similar results are valid for the inhomogeneous Dirichlet problem on Triebel–Lizorkin spaces, i.e., for

$$\begin{cases} u \in F^{p,q}_{s+\frac{1}{p}+1}(\Omega), \\ \Delta^2 u = w \in F^{p,q}_{s+\frac{1}{p}-3}(\Omega), \\ (\operatorname{Tr} u, \operatorname{Tr}(\nabla u)) = \dot{f} \in \dot{B}^{p,p}_{1,s}(\partial\Omega). \end{cases} \tag{1.67}$$

It should be also noted that Theorem 1.6 is of the same flavor as the well-posedness results obtained by D. Jerison and C. Kenig in [58] for the inhomogeneous Dirichlet problem for the Laplacian in Lipschitz domains in $\mathbb{R}^n$, formulated on Sobolev spaces and diagonal Besov spaces $B^{p,p}_\alpha$ with $1 \le p \le \infty$. Although the methods employed by D. Jerison and C. Kenig in [58] yield sharp results in this case, they rely in an essential fashion on the maximum principle and, as such, do not readily adapt to other types of problems, such as higher-order PDE's, or boundary conditions of Neumann type. In fact, the latter issue was singled out as open problem # 3.2.21 in Kenig's book [68]. In the case of the Laplacian this has been subsequently solved by E. Fabes, O. Mendez and M. Mitrea in [41] via an approach which relies on a systematic use of singular integral operators. The reader is also referred to the work of M. Mitrea and M. Taylor in [91] for an extension to variable coefficient differential operators of second order, and to the work of S. Mayboroda and M. Mitrea in [73] for a refinement (having to do with the consideration of Besov spaces with $p < 1$) which, in particular, has led to a solution of a conjecture formulated by D.-C. Chang, S. Krantz and E. Stein in [22, 23], regarding the regularity of the harmonic Green potentials on Hardy spaces in Lipschitz domains.

The second well-posedness result we wish to quote here pertains to the inhomogeneous Neumann problem for the bi-Laplacian in Lipschitz domains in $\mathbb{R}^n$ with boundary data from duals of Whitney–Besov spaces.

Theorem 1.7 (The inhomogeneous Neumann problem). *Assume that $\Omega \subset \mathbb{R}^n$, $n \geq 2$, is a bounded Lipschitz domain with connected boundary and outward unit normal $\nu = (\nu_j)_{1\leq j\leq n}$. For each real parameter $\theta > -\frac{1}{n}$ consider the inhomogeneous Neumann problem for the biharmonic operator formulated as follows:*

Given $w \in \left(B^{p',q'}_{2-s+1/p'}(\Omega)\right)^*$ and $\Lambda \in \left(\dot{B}^{p',q'}_{1,1-s}(\partial\Omega)\right)^*$ satisfying the compatibility condition

$$\big\langle \Lambda, (P, \nabla P)\big\rangle = \big\langle w, P\big\rangle \ \text{ for each polynomial } P \text{ of degree} \leq 1, \tag{1.68}$$

(with $\langle\cdot,\cdot\rangle$ natural duality brackets) find a function u satisfying

$$u \in B^{p,q}_{1+s+1/p}(\Omega), \quad \Delta^2 u = w \ \text{ as distributions in } \Omega, \tag{1.69}$$

and the boundary condition (written using the summation convention over repeated indices)

$$\left(-\partial_\nu \Delta u - \frac{\partial_{\tau_{ij}}\left[\nu_\ell \nu_i \partial_j \partial_\ell u\right]}{1+2\theta+n\theta^2}, \left(\frac{2\theta+n\theta^2}{1+2\theta+n\theta^2}\, \nu_r \Delta u + \frac{\nu_r \nu_j \nu_\ell \partial_j \partial_\ell u}{1+2\theta+n\theta^2}\right)_{1\leq r\leq n}\right) = \Lambda. \tag{1.70}$$

Then there exists $\varepsilon > 0$ with the property that the inhomogeneous boundary value problem (1.68)–(1.70) is well-posed, with uniqueness understood modulo polynomials of degree ≤ 1, whenever the exponents $p, p', q, q' \in (1,\infty)$, $s \in (0,1)$ satisfy $1/p + 1/p' = 1/q + 1/q' = 1$ as well as

$$\tfrac{-n+1-\varepsilon}{2} < -\tfrac{n-1}{p} + s < \tfrac{-n+3+\varepsilon}{2} \ \text{ when } n \geq 4, \tag{1.71}$$

$$-\tfrac{1+\varepsilon}{2} < -\tfrac{1}{p} + \left(\tfrac{1-\varepsilon}{2}\right)s < 0 \ \text{ when } n \in \{2,3\}. \tag{1.72}$$

Moreover, if $w = 0$ then the problem (1.68)–(1.70) admits a solution of the form

$$u = \dot{\mathscr{S}}\widehat{\Lambda}, \quad \text{for a suitable } \widehat{\Lambda} \in \left(\dot{B}^{p',q'}_{1,1-s}(\partial\Omega)\right)^*, \tag{1.73}$$

where $\dot{\mathscr{S}}$ is the single multi-layer associated with the operator $L = \Delta^2$ as in (1.30).

Furthermore, given any exponents $p, q \in (1, \infty)$ *and* $s \in (0, 1)$, *there exists* $\varepsilon > 0$, *depending only on* p, q, s, *and the Lipschitz character of* Ω, *with the property that if the outward unit normal* ν *to* Ω *satisfies* (1.65) *then the problem* (1.68)–(1.70) *is well-posed. As a consequence the boundary value problem* (1.68)–(1.70) *is well-posed for any exponents* $p, q \in (1, \infty)$ *and* $s \in (0, 1)$ *if* $\nu \in \mathrm{vmo}(\partial\Omega)$ *(hence, in particular, if* Ω *is a* C^1 *domain).*

Finally, similar results hold for the inhomogeneous Neumann problem formulated in Triebel–Lizorkin spaces.

The reader is referred to the discussion in § 6 for other theorems of similar nature, including well-posedness results boundary value problems in which the size of the solution is measured in terms of the non-tangential maximal function.

Acknowledgements The first named author was supported in part by the US NSF Grant DMS 1201736.

Chapter 2
Smoothness Scales and Calderón–Zygmund Theory in the Scalar-Valued Case

While one of the main goals of this monograph is the systematic development of a Calderón–Zygmund theory for multi-layer type operators associated with higher-order operators with matrix-valued coefficients, the starting point is the consideration of the scalar-valued case. As such, the aim of this introductory chapter is to present an account of those aspects of the scalar theory which are most relevant for the current work.

2.1 Lipschitz Domains and Non-tangential Maximal Function

We start by recalling that, given a metric space $(\mathscr{X}, d)$, a function $f : \mathscr{X} \to \mathbb{R}$ is called `Lipschitz` provided there exists some finite constant $M \in [0, \infty)$ such that $|f(x) - f(y)| \leq M\, d(x, y)$ for all $x, y \in \mathscr{X}$. By an `unbounded Lipschitz domain` Ω in $\mathbb{R}^n$ we shall simply understand the upper-graph region for a Lipschitz function $\varphi : \mathbb{R}^{n-1} \to \mathbb{R}$, i.e.,

$$\Omega = \{X = (x', x_n) \in \mathbb{R}^{n-1} \times \mathbb{R} : \varphi(x') < x_n\}. \tag{2.1}$$

Below, and elsewhere, $\mathbb{N}_0 := \mathbb{N} \cup \{0\}$.

Definition 2.1. A nonempty, open, bounded subset Ω of $\mathbb{R}^n$ is called a `bounded Lipschitz domain` (respectively, `bounded domain of class` $C^{\ell+\alpha}$ for some $\ell \in \mathbb{N}_0$, $\alpha \in [0, 1)$) if for any $X_0 \in \partial\Omega$ there exist $r, h > 0$ and a coordinate system $(x_1, \ldots, x_n) = (x', x_n)$ in $\mathbb{R}^n$ which is isometric to the canonical one and has origin at X_0, along with a function $\varphi : \mathbb{R}^{n-1} \to \mathbb{R}$ which is Lipschitz (respectively, of class $C^{\ell+\alpha}$) and for which the following property holds. If $\mathscr{C}(r, h)$ denotes the open cylinder

I. Mitrea and M. Mitrea, *Multi-Layer Potentials and Boundary Problems*, Lecture Notes in Mathematics 2063, DOI 10.1007/978-3-642-32666-0_2,

$$\{X = (x', x_n) \in \mathbb{R}^{n-1} \times \mathbb{R} : |x'| < r \text{ and } -h < x_n < h\} \subseteq \mathbb{R}^n, \tag{2.2}$$

then

$$\Omega \cap \mathscr{C}(r, h) = \big\{X = (x', x_n) \in \mathbb{R}^{n-1} \times \mathbb{R} : |x'| < r \text{ and } \varphi(x') < x_n < h\big\}. \tag{2.3}$$

An `atlas` for $\partial\Omega$ is a finite collection of cylinders $\{\mathscr{C}_k(r_k, h_k)\}_{1 \le k \le N}$ (with associated Lipschitz maps $\{\varphi_k\}_{1 \le k \le N}$) covering $\partial\Omega$. Having fixed such an atlas, the `Lipschitz character` of Ω is defined as the quartet consisting of numbers N, $\max\{\|\nabla\varphi_k\|_{L^\infty(\mathbb{R}^{n-1})} : 1 \le k \le N\}$, $\min\{r_k : 1 \le k \le N\}$, and $\min\{h_k : 1 \le k \le N\}$.

In the context of the above definition, it turns out that condition (2.3) further entails

$$\partial\Omega = \partial\overline{\Omega}, \tag{2.4}$$

as well as

$$\partial\Omega \cap \mathscr{C}(r, h) = \big\{X = (x', x_n) \in \mathbb{R}^{n-1} \times \mathbb{R} : |x'| < r \text{ and } x_n = \varphi(x')\big\} \tag{2.5}$$

and

$$\big(\mathbb{R}^n \setminus \overline{\Omega}\big) \cap \mathscr{C}(r, h) = \big\{X = (x', x_n) \in \mathbb{R}^{n-1} \times \mathbb{R} : |x'| < r \text{ and } -h < x_n < \varphi(x')\big\}. \tag{2.6}$$

Throughout, by a `Lipschitz domain` we shall understand an open set which is either a bounded Lipschitz domain or an unbounded Lipschitz domain, in the sense previously defined. As is well-known, for a Lipschitz domain Ω, the surface measure σ is well-defined on $\partial\Omega$ and may be described as

$$\sigma = \mathscr{H}^{n-1} \lfloor \partial\Omega, \tag{2.7}$$

where $\mathscr{H}^{n-1}$ stands for the $(n-1)$-dimensional Hausdorff measure in $\mathbb{R}^n$. For each index $p \in (0, \infty]$ we shall denote by $L^p(\partial\Omega)$ the Lebesgue space of σ-measurable, p-th power integrable functions on $\partial\Omega$ (with respect to σ). In addition, we shall denote by $L^p_{comp}(\partial\Omega)$ the subspace of $L^p(\partial\Omega)$ consisting of functions which vanish outside of a compact subset of $\partial\Omega$, and by $L^p_{loc}(\partial\Omega)$ the space of locally p-th power integrable functions on $\partial\Omega$.

As a consequence of the classical Rademacher theorem, the outward pointing normal vector $\nu = (\nu_1, \cdots, \nu_n)$ to a given Lipschitz domain $\Omega \subset \mathbb{R}^n$ exists at σ-almost every point on $\partial\Omega$. In fact, in the case when $\Omega \subset \mathbb{R}^n$ is the domain lying above the graph of a real-valued Lipschitz function φ defined in $\mathbb{R}^{n-1}$, we have

$$\nu(X) = \frac{(\nabla\varphi(x'), -1)}{\sqrt{1+|\nabla\varphi(x')|^2}} \quad \text{and} \quad d\sigma(X) = \sqrt{1+|\nabla\varphi(x')|^2}dx', \qquad \text{for} \quad X = (x', \varphi(x')) \in \partial\Omega. \tag{2.8}$$

Given a Lipschitz domain $\Omega \subset \mathbb{R}^n$, by a `surface ball` $S_r(X)$ (or $\Delta(X,r)$) we shall understand a set of the form $B_r(X) \cap \partial\Omega$, with $X \in \partial\Omega$ and $0 < r < \operatorname{diam}(\partial\Omega)$. When the center is already specified, or is of no particular importance, we simplify the notation by writing S_r. When equipped with the surface measure and the Euclidean distance, $\partial\Omega$ becomes a space of homogeneous type (in the sense of Coifman and Weiss [28]). Hence, the associated Hardy–Littlewood maximal operator

$$\mathscr{M}f(X) := \sup_{r>0} \frac{1}{\sigma\big(S_r(X)\big)} \int_{S_r(X)} |f(Y)|\, d\sigma(Y), \qquad X \in \partial\Omega, \tag{2.9}$$

is bounded on $L^p(\partial\Omega)$ for each $p \in (1,\infty]$. Furthermore, there exists some constant $C = C(\partial\Omega) \in (0,\infty)$ such that

$$\sigma\Big(\{X \in \partial\Omega : \mathscr{M}f(X) > \lambda\}\Big) \le C\lambda^{-1}\|f\|_{L^1(\partial\Omega)}, \tag{2.10}$$

for every $f \in L^1(\partial\Omega)$ and $\lambda > 0$.

Given a Lipschitz domain Ω, we shall set

$$\Omega_+ := \Omega \quad \text{and} \quad \Omega_- := \mathbb{R}^n \setminus \overline{\Omega}. \tag{2.11}$$

Then, for a fixed parameter $\kappa > 0$ define the `non-tangential approach regions` with vertex at $X \in \partial\Omega$ (corresponding to $\Omega_\pm$) as

$$R_\kappa^\pm(X) := \big\{Y \in \Omega_\pm : |X-Y| < (1+\kappa)\operatorname{dist}(Y,\partial\Omega)\big\}, \tag{2.12}$$

and, further, the `non-tangential maximal operator` of a given function u in $\Omega_\pm$ by

$$\mathscr{N}_\kappa^\pm(u)(X) := \sup\{|u(Y)| : Y \in R_\kappa^\pm(X)\}. \tag{2.13}$$

When unambiguous, we agree to drop the superscripts $\pm$. As we shall see momentarily, the dependence of $R_\kappa^\pm$ and $\mathscr{N}_\kappa^\pm$ on κ plays only an auxiliary role (and will be eventually dropped as well). We elaborate on this issue in the case when $\Omega \subset \mathbb{R}^n$ is the domain lying above the graph of a real-valued Lipschitz function φ defined in $\mathbb{R}^{n-1}$, i.e.,

$$\Omega := \big\{X = (x', x_n) \in \mathbb{R}^n : x_n > \varphi(x')\big\}, \quad \text{where} \qquad \varphi : \mathbb{R}^{n-1} \longrightarrow \mathbb{R} \ \text{has} \ M := \|\nabla\varphi\|_{L^\infty(\mathbb{R}^{n-1})} < \infty, \tag{2.14}$$

as the case of a bounded Lipschitz domain requires only minor alterations. To this end, it is convenient to also introduce the following variant of (2.12)–(2.13):

$$\widetilde{\mathcal{N}}_\theta(u)(X) := \sup\big\{|u(Y)| : Y \in X + \Gamma_\theta\big\}, \qquad \theta \in (0, 1/M), \tag{2.15}$$

where

$$\Gamma_\theta := \{Z = (z', z_n) \in \mathbb{R}^n : |z'| < \theta\, z_n\} \tag{2.16}$$

is a circular, upright cone, with vertex at the origin in $\mathbb{R}^n$, of aperture $\arctan\theta$. Elementary geometrical considerations show that for every $\theta \in (0, 1/M)$ there exists $\kappa = \kappa(\theta, M) > 0$ such that

$$X + \Gamma_\theta \subset R_\kappa(X) \text{ for every } X \in \partial\Omega. \tag{2.17}$$

Conversely, if κ is sufficiently large (e.g., $\kappa > -1 + \sqrt{1 + M^2}$ will do) then there exists $\theta = \theta(\kappa) \in (0, 1/M)$ such that (2.17) holds. After these preparations, we are ready to prove the following result.

Proposition 2.2. *Let Ω be an unbounded Lipschitz domain as in* (2.14). *Then for every $\kappa, \kappa' > 0$, $\theta, \theta' \in (0, 1/M)$, and $p \in (0, \infty)$, one has*

$$\|\mathcal{N}_\kappa u\|_{L^p(\partial\Omega)} \approx \|\mathcal{N}_{\kappa'} u\|_{L^p(\partial\Omega)} \approx \|\widetilde{\mathcal{N}}_\theta u\|_{L^p(\partial\Omega)} \approx \|\widetilde{\mathcal{N}}_{\theta'} u\|_{L^p(\partial\Omega)}, \tag{2.18}$$

uniformly in the real-valued function u defined in Ω.

Proof. Pick an arbitrary $\theta \in (0, 1/M)$ and fix some $\kappa > 0$ such that (2.17) holds. Our immediate goal is to prove that there exists $C = C(M, n, \theta, \kappa) > 0$ such that

$$\|\mathcal{N}_\kappa u\|_{L^p(\partial\Omega)} \leq C\, \|\widetilde{\mathcal{N}}_\theta u\|_{L^p(\partial\Omega)}, \tag{2.19}$$

for any real-valued function u defined in Ω. To this end, we shall adapt a point-of-density argument of Fefferman and Stein [43]. Specifically, fix $\lambda > 0$ and consider the open subsets of $\partial\Omega$ given by

$$\mathcal{O}_\kappa := \Big\{X \in \partial\Omega : \mathcal{N}_\kappa u(X) > \lambda\Big\}, \quad \widetilde{\mathcal{O}}_\theta := \Big\{X \in \partial\Omega : \widetilde{\mathcal{N}}_\theta u(X) > \lambda\Big\}. \tag{2.20}$$

In particular,

$$A := \partial\Omega \setminus \widetilde{\mathcal{O}}_\theta \tag{2.21}$$

is closed. For each $\gamma \in (0, 1)$, let A^*_γ be the collection of points of (global) γ-density for the set A, i.e.,

$$A^*_\gamma := \{X \in \partial\Omega : \sigma(A \cap \Delta(X, r)) \geq \gamma\sigma(\Delta(X, r)), \ \forall\, r > 0\}, \tag{2.22}$$

where $\Delta(X, r) := B(X, r) \cap \partial\Omega$.

We now claim that there exists $\gamma \in (0,1)$ such that

$$\mathcal{O}_\kappa \subseteq \partial\Omega \setminus A^*_\gamma. \tag{2.23}$$

To justify this inclusion, fix an arbitrary point $X \in \mathcal{O}_\kappa$. Then $X \in \partial\Omega$ and there exists $Y = (y', y_n) \in R_\kappa(X)$ such that $|u(Y)| > \lambda$. Fix next $0 < \theta_o < \theta$ and let the point $W = (w', \varphi(w')) \in \partial\Omega$ be such that $Y \in W + \Gamma_{\theta_o}$. We claim that there exists $t = t(M, \theta) > 0$ such that

$$Z \in \Delta(W, t|Y - W|) \Longrightarrow Y \in Z + \Gamma_\theta. \tag{2.24}$$

Indeed, let $Z = (z', \varphi(z')) \in \partial\Omega$ be such that $|Z - W| < t|Y - W|$, with t to be specified below. Since $Y \in W + \Gamma_{\theta_o}$, it follows that $|y' - w'| < \theta_o(y_n - \varphi(w'))$. Hence,

$$\begin{aligned} |y' - z'| &\le |y' - w'| + |w' - z'| < \theta_o(y_n - \varphi(w')) + |w' - z'| \\ &= \theta_o(y_n - \varphi(z')) + \theta_o(\varphi(z') - \varphi(w')) + |w' - z'| \\ &\le \theta_o(y_n - \varphi(z')) + (1 + \theta_o M)\,|w' - z'|. \end{aligned} \tag{2.25}$$

On the other hand,

$$\begin{aligned} |z' - w'|^2 + (z_n - \varphi(w'))^2 &< t^2|Y - W|^2 = t^2\Big(|y' - w'|^2 + (y_n - \varphi(w'))^2\Big) \\ &\le t^2(1 + \theta_o^2)(y_n - \varphi(w'))^2 \\ &\le 2t^2(1 + \theta_o^2)(y_n - \varphi(z'))^2 + 2t^2M^2(1 + \theta_o^2)|z' - w'|^2, \end{aligned} \tag{2.26}$$

which further yields

$$|z' - w'| \le \sqrt{\frac{2t^2(1 + \theta_o^2)}{1 - 2t^2M^2(1 + \theta_o^2)}}\,\Big(y_n - \varphi(z')\Big) \tag{2.27}$$

granted that $0 < t < M^{-1}[2(1 + \theta_o^2)]^{-1/2}$. Assuming that this is the case, a combination of (2.25) and (2.27) gives

$$|y' - z'| < \left(\theta_o + (1 + \theta_o M)\sqrt{\frac{2t^2(1 + \theta_o^2)}{1 - 2t^2M^2(1 + \theta_o^2)}}\right)\Big(y_n - \varphi(z')\Big). \tag{2.28}$$

This implies

$$|y' - z'| < \theta(y_n - \varphi(z')), \tag{2.29}$$

i.e., $Y \in Z + \Gamma_\theta$, provided t is small enough. This finishes the proof of (2.24).

Our next claim is that, with X, Y, W, t, as before,

$$\Delta(W, t|Y-W|) \subseteq \Delta(X, (2+t+\kappa)|Y-W|). \tag{2.30}$$

To see this, we note that if $Z \in \partial\Omega$ and $|Z-W| < t|Y-W|$ then

$$\begin{aligned} |X-Z| &\le |X-Y| + |Y-W| + |W-Z| \\ &\le (1+\kappa)\,\mathrm{dist}\,(Y, \partial\Omega) + (1+t)|Y-W| \\ &= (2+t+\kappa)|Y-W|. \end{aligned} \tag{2.31}$$

In concert, (2.24), (2.30) and the fact that $|u(Y)| > \lambda$ yield

$$\Delta(W, t|Y-W|) \subseteq \widetilde{\mathscr{O}}_\theta \cap \Delta\big(X, (2+t+\kappa)|Y-W|\big), \tag{2.32}$$

so that

$$\begin{aligned} \frac{\sigma\Big(\widetilde{\mathscr{O}}_\theta \cap \Delta\big(X, (2+t+\kappa)|Y-W|\big)\Big)}{\sigma\Big(\Delta\big(X, (2+t+\kappa)|Y-W|\big)\Big)} &\ge \frac{\sigma\Big(\Delta(W, t|Y-W|)\Big)}{\sigma\Big(\Delta\big(X, (2+t+\kappa)|Y-W|\big)\Big)} \\ &\ge c\left(\frac{t}{2+t+\kappa}\right)^{n-1}, \end{aligned} \tag{2.33}$$

where $c \in (0,1)$ is a constant which depends only on Ω and n. In particular, if we set $r := (2+t+\kappa)|Y-W|$, then

$$\frac{\sigma\Big(A \cap \Delta(X, r)\Big)}{\sigma\Big(\Delta(X, r)\Big)} \le 1 - c\left(\frac{t}{2+t+\kappa}\right)^{n-1}. \tag{2.34}$$

Thus, if we select γ such that

$$1 - c\Big(\frac{t}{2+t+\kappa}\Big)^{n-1} < \gamma < 1, \tag{2.35}$$

then (2.34) entails $X \notin A^*_\gamma$. This proves the claim (2.23).

Moving on, let $\mathscr{M}$ be the Hardy–Littlewood maximal operator as in (2.9). Then, based on (2.23) and (2.10), we may write

$$\begin{aligned} \sigma(\mathscr{O}_\kappa) &\le \sigma(\partial\Omega \setminus A^*_\gamma) = \sigma\Big(\{X \in \partial\Omega : \mathscr{M}(\chi_{\partial\Omega\setminus A})(X) > 1-\gamma\}\Big) \\ &\le \frac{C}{1-\gamma}\sigma(\partial\Omega \setminus A) = C(\gamma, M)\sigma(\widetilde{\mathscr{O}}_\theta). \end{aligned} \tag{2.36}$$

Hence,

$$\sigma(\mathscr{O}_\kappa) \le C(\kappa, \theta, M)\sigma(\widetilde{\mathscr{O}}_\theta), \tag{2.37}$$

and (2.19) readily follows from this. Thanks to (2.17), the opposite inequality in (2.19) holds as well so, all in all,

$$\forall\, \theta \in (0, 1/M)\ \exists\, \kappa = \kappa(\theta, M) > 0 \text{ so that } \|\mathscr{N}_\kappa u\|_{L^p(\partial\Omega)} \approx \|\widetilde{\mathscr{N}}_\theta u\|_{L^p(\partial\Omega)}. \tag{2.38}$$

Since whenever $0 < \kappa_1 < \kappa_2 < \infty$, we have $R_{\kappa_1}(X) \subset R_{\kappa_2}(X)$, for each $X \in \partial\Omega$, a slight variant of the above reasoning shows that

$$\|\mathscr{N}_{\kappa_1} u\|_{L^p(\partial\Omega)} \approx \|\mathscr{N}_{\kappa_2} u\|_{L^p(\partial\Omega)}. \tag{2.39}$$

Now (2.18) follows easily from (2.38) and (2.39). □

Remark. Proposition 2.2 continues to hold for a *bounded Lipschitz domain* Ω, with only minor alterations. In the latter setting, the definition of the non-tangential approach regions remains unchanged and, indeed,

$$\|\mathscr{N}_\kappa u\|_{L^p(\partial\Omega)} \approx \|\mathscr{N}_{\kappa'} u\|_{L^p(\partial\Omega)} \tag{2.40}$$

for every $\kappa, \kappa' > 0$ and $0 < p < \infty$. As for the concept of non-tangential cone, we continue to assume that their (common) aperture, encoded by the parameter θ, is not too large (we shall refer to such value of θ as admissible). Besides this requirement, we will, nonetheless, incorporate a couple of genuinely new features. First, the cones need to be uniformly truncated, at a sufficiently small height as to ensure their containment in Ω. Second, the axis of the cone is taken along the vertical direction relative to the coordinate system used to describe $\partial\Omega$ locally as a piece of a Lipschitz graph. We are going to denote such a family of non-tangential cones $\{\Gamma_\theta(X)\}_{X \in \partial\Omega}$, with θ the parameter describing the uniform aperture of the cones. If $\widetilde{\mathscr{N}}_\theta$ is now defined much as before, relative to this family of cones then, for every $p \in (0, \infty)$ and any admissible θ, θ', we also have

$$\|\widetilde{\mathscr{N}}_\theta u\|_{L^p(\partial\Omega)} \le C\, \|\widetilde{\mathscr{N}}_{\theta'} u\|_{L^p(\partial\Omega)} + C \sup_{X \in \mathscr{O}} |u(X)|, \tag{2.41}$$

for some finite constant $C > 0$ and some relatively compact subset $\mathscr{O}$ of Ω, independent of the function u. Similar estimates also hold for $\widetilde{\mathscr{N}}_\theta u$ and $\mathscr{N}_\kappa u$. ■

In the sequel, it is going to be useful to derive global integrability properties for a function u in Ω, based on the integrability properties of $\mathscr{N} u$ on $\partial\Omega$. While Proposition 2.3 below, which addresses this issue, has been first proved in [36], we choose to present an alternative argument, discovered during some discussions with S. Hofmann.

Proposition 2.3. *Let Ω be a Lipschitz domain in $\mathbb{R}^n$ and assume that the measurable function u, taking finite values at almost every point in Ω, satisfies $\mathcal{N}u \in L^p(\partial\Omega)$ for some $p \in (0,\infty)$. Then*

$$\|u\|_{L^{\frac{np}{n-1}}(\Omega)} \le C(\Omega,p)\|\mathcal{N}u\|_{L^p(\partial\Omega)}. \tag{2.42}$$

Let us first dispense with a number of prerequisites. Given a Lipschitz domain $\Omega \subset \mathbb{R}^n$, a Borelian measure μ on Ω is called a `Carleson measure` provided

$$\|\mu\|_{Car} := \sup\Big\{R^{1-n}\mu(B(X,R)\cap\Omega) : X \in \partial\Omega,\ 0 < R < \operatorname{diam}(\partial\Omega)\Big\} \tag{2.43}$$

is finite. In the sequel, we shall refer to $\|\mu\|_{Car}$ as the `Carleson constant` of μ. Two properties of Carleson measures are going to be of importance for us at this stage. First,

$$f \in L^n(\Omega) \Longrightarrow \begin{cases} \mu := |f|\,dX \text{ is a Carleson measure on } \Omega \\ \text{and the estimate } \|\mu\|_{Car} \le C\|f\|_{L^n(\Omega)} \text{ holds.}\end{cases} \tag{2.44}$$

This can be readily verified from definitions. Second, if μ is a Carleson measure in Ω, then there exists $C = C(\Omega) > 0$ such that

$$\Big|\int_\Omega F\,d\mu\Big| \le C\|\mu\|_{Car}\|\mathcal{N}F\|_{L^1(\partial\Omega)}. \tag{2.45}$$

When $\Omega = \mathbb{R}^n_+$ this is well-known (see, e.g., [21]). Here we present an argument well-suited to the setting of Lipschitz domains. To begin with, it suffices to show that, for each $\lambda > 0$,

$$\mu(\{X \in \Omega : |F(X)| > \lambda\}) \le C\|\mu\|_{Car}\,\sigma(\{Q \in \partial\Omega : \mathcal{N}_\kappa F(Q) > \lambda\}) \tag{2.46}$$

To prove this, for an arbitrary open subset $\mathcal{O}$ of $\partial\Omega$ define the associated "tent" region

$$\begin{aligned} T_\kappa(\mathcal{O}) &:= \Omega \setminus \Big[\bigcup_{P\in\partial\Omega\setminus\mathcal{O}} R_\kappa(P)\Big] \\ &= \{X \in \Omega : \operatorname{dist}(X,\mathcal{O}) \le (1+\kappa)^{-1}\operatorname{dist}(X,\partial\Omega\setminus\mathcal{O})\}, \end{aligned} \tag{2.47}$$

and observe that

$$\{X \in \Omega : |F(X)| > \lambda\} \subset T_\kappa(\{Q \in \partial\Omega : \mathcal{N}_\kappa F(Q) > \lambda\}). \tag{2.48}$$

Another property of the tent regions which is going to be of importance for us here is as follows. Decompose a given open subset $\mathcal{O}$ of $\partial\Omega$ into a finite-overlap family of Whitney surface balls $\{S_k\}_k$. Next, for each surface ball $\Delta := B_R(Q) \cap \partial\Omega$, $Q \in \partial\Omega$, $0 < R < \operatorname{diam}\Omega$ and each fixed $t > 0$, consider the Carleson region

$$\mathscr{C}_t(\Delta) := B_{tR}(Q) \cap \Omega. \tag{2.49}$$

Then choosing t large enough ensures

$$T_\kappa(\mathcal{O}) \subset \bigcup_k \mathscr{C}_t(S_k). \tag{2.50}$$

Using (2.48) and (2.50) with $\mathcal{O} := \{X \in \partial\Omega : \mathcal{N}_\kappa F(X) > \lambda\}$ we may now write

$$\begin{aligned}
\mu(\{X \in \Omega : |F(X)| > \lambda\}) &\le \mu(T_\kappa(\{X \in \partial\Omega : \mathcal{N}_\kappa F(X) > \lambda\})) \\
&\le \mu(\cup_k \mathscr{C}_t(S_k)) \le \sum_k \mu(\mathscr{C}_t(S_k)) \\
&\le C\,\|\mu\|_{Car} \sum_k \sigma(S_k) \\
&\le C\,\|\mu\|_{Car}\, \sigma(\{X \in \partial\Omega : \mathcal{N}_\kappa F(X) > \lambda\}). \qquad (2.51)
\end{aligned}$$

This finishes the proof of (2.45).

After this preamble, we are ready to present the

Proof of Proposition 2.3. By homogeneity, the general case is easily reduced to the case when $p = 1$, which we shall assume from now on. Let us prove (2.42) under the additional assumption that $u \in L^{n/(n-1)}(\Omega)$. This implies (cf. (2.44)) that

$$\mu := |u|^{\frac{1}{n-1}}\,dX \ \text{ is a Carleson measure on } \Omega \text{ and } \|\mu\|_{Car} \le C\,\|u\|_{L^{\frac{n}{n-1}}(\Omega)}. \tag{2.52}$$

Based on (2.45) and (2.52) we may therefore write

$$\begin{aligned}
\int_\Omega |u|^{\frac{n}{n-1}}\,dX &= \int_\Omega |u|\,d\mu \le C\,\|\mu\|_{Car}\|\mathcal{N} u\|_{L^1(\partial\Omega)} \\
&\le C\,\|\mathcal{N} u\|_{L^1(\partial\Omega)} \Big(\int_\Omega |u|^{\frac{n}{n-1}}\,dX\Big)^{\frac{1}{n}}. \qquad (2.53)
\end{aligned}$$

Thus, if $\int_\Omega |u|^{\frac{n}{n-1}}\,dX < \infty$, (2.42) follows from (2.53).

There remains to explain how to a posteriori eliminate the assumption that the function $u \in L^{\frac{n}{n-1}}(\Omega)$. Let $\{\mathcal{O}_j\}_{j\in\mathbb{N}}$ be a nested family of open, relatively compact subsets of Ω, such that $\cup_j \mathcal{O}_j = \Omega$. Also, for each $j \in \mathbb{N}$, set

$$E_j := \{X \in \Omega : |u(X)| \le j\}. \tag{2.54}$$

Then $u_j := \chi_{\mathcal{O}_j}\chi_{E_j}u \in L^\infty(\Omega)$ has compact support and, by the argument in the first part of the proof, there exists $C > 0$ independent of j such that

$$\|u_j\|_{L^{\frac{n}{n-1}}(\Omega)} \leq C\|\mathcal{N}u_j\|_{L^1(\partial\Omega)} \leq C\|\mathcal{N}u\|_{L^1(\partial\Omega)}. \tag{2.55}$$

Then the desired conclusion follows from Lebesgue's Monotone Convergence Theorem, by letting $j \to \infty$ in (2.55). □

We wish to briefly elaborate on a related concept. Given a Lipschitz domain $\Omega \subset \mathbb{R}^n$, a Borelian measure μ on Ω is called a `vanishing Carleson measure` provided it is a Carleson measure and

$$\lim_{R\to 0+}\Big(\sup\Big\{r^{1-n}\mu(B(X,r)\cap\Omega) : X \in \partial\Omega,\ 0 < r < R\Big\}\Big) = 0. \tag{2.56}$$

Assume that Ω is a bounded Lipschitz domain in $\mathbb{R}^n$ and set

$$\rho(X) := \operatorname{dist}(X, \partial\Omega), \qquad X \in \mathbb{R}^n. \tag{2.57}$$

Then, for any real-valued, measurable function f in Ω, it is elementary to check that

$$|f| \leq C\rho^\theta \text{ for some } \theta > -1/n \Longrightarrow \begin{cases} |f|\,dX \text{ is a vanishing} \\ \text{Carleson measure on } \Omega. \end{cases} \tag{2.58}$$

Going further, let us define the `non-tangential boundary trace` of a function u defined in $\Omega_\pm$ as

$$u\Big|_{\partial\Omega}(X) := \lim_{\substack{Y\to X\\ Y\in R_\kappa^\pm(X)}} u(Y), \qquad X \in \partial\Omega, \tag{2.59}$$

whenever meaningful.

We shall frequently use the Divergence Theorem in the context of Lipschitz domains. A version well-suited for our purposes reads as follows.

Proposition 2.4. *Let Ω be a Lipschitz domain in $\mathbb{R}^n$ and assume that $U = (u_1, \ldots, u_n)$ is a vector field with components in $L^1_{loc}(\Omega)$ and such that* $\operatorname{div} U = \sum_{j=1}^n \partial_j u_j \in L^1(\Omega)$ *(with the divergence operator considered in the sense of distributions). If, in addition, $\mathcal{N}U \in L^1(\partial\Omega)$ and $U\big|_{\partial\Omega}$ exists at σ-a.e. point, then*

$$\int_{\partial\Omega} \nu\cdot\big(U\big|_{\partial\Omega}\big)\,d\sigma = \int_\Omega \operatorname{div} U\,dX, \tag{2.60}$$

where ν and σ denote, respectively, the outward unit normal and surface measure on $\partial\Omega$.

Furthermore, if Ω is a bounded Lipschitz domain then the same type of result holds in $\mathbb{R}^n \setminus \overline{\Omega}$ under the additional decay assumption

$$|U(X)| = \mathcal{O}(|X|^{-(n-1+\varepsilon)}), \qquad \textit{for some } \varepsilon > 0, \quad \text{as } |X| \to \infty. \tag{2.61}$$

This is a particular case of a result valid for a much more general class of domains established in [84].

2.2 Sobolev Spaces on Lipschitz Boundaries

Assume that $\Omega \subset \mathbb{R}^n$ is a Lipschitz domain and consider the first-order tangential derivative operators $\partial_{\tau_{jk}}$ acting on a compactly supported function ψ of class C^1 in a neighborhood of $\partial\Omega$ by

$$\partial_{\tau_{jk}}\psi := \nu_j(\partial_k\psi)\Big|_{\partial\Omega} - \nu_k(\partial_j\psi)\Big|_{\partial\Omega}, \qquad j,k = 1,\dots,n. \tag{2.62}$$

Lemma 2.5. *Suppose that $\Omega \subset \mathbb{R}^n$ is a Lipschitz domain and $\varphi, \psi \in C_c^1(\mathbb{R}^n)$. Then for every $j,k \in \{1,\dots,n\}$,*

$$\int_{\partial\Omega} \varphi\,(\partial_{\tau_{jk}}\psi)\,d\sigma = \int_{\partial\Omega} (\partial_{\tau_{kj}}\varphi)\,\psi\,d\sigma. \tag{2.63}$$

Proof. Assuming first that $\varphi, \psi \in C_c^\infty(\mathbb{R}^n)$, we use successive integration by parts in order to write

$$\begin{aligned}
\int_{\partial\Omega} \varphi\,(\partial_{\tau_{jk}}\psi)\,d\sigma &= -\int_{\partial\Omega} \varphi\,\Big(\nu_k\partial_j\psi - \nu_j\partial_k\psi\Big)\,d\sigma \\
&= \int_{\Omega} \Big(\partial_j(\varphi\,\partial_k\psi) - \partial_k(\varphi\,\partial_j\psi)\Big)\,dX \\
&= \int_{\Omega} \Big(\partial_k(\psi\,\partial_j\varphi) - \partial_j(\psi\,\partial_k\varphi)\Big)\,dX \\
&= \int_{\partial\Omega} (\partial_{\tau_{kj}}\varphi)\,\psi\,d\sigma.
\end{aligned} \tag{2.64}$$

Finally, the more general case when $\varphi, \psi \in C_c^1(\mathbb{R}^n)$ is easily reduced to the previous situation using a standard mollifying and limiting argument. □

We shall now make use of (2.63) in order to prove that, if $\psi \in C_c^1(\mathbb{R}^n)$, then $\partial_{\tau_{jk}}\psi$ actually depends only on $\psi|_{\partial\Omega}$. More specifically, we have:

Lemma 2.6. *Assume that $\Omega \subset \mathbb{R}^n$ is a Lipschitz domain and that $\psi \in C_c^1(\mathbb{R}^n)$ satisfies $\psi \equiv 0$ on $\partial\Omega$. Then $\partial_{\tau_{jk}}\psi \equiv 0$ on $\partial\Omega$ for every $j,k \in \{1,\dots,n\}$.*

Proof. If ψ is as above and $\varphi \in C_c^1(\mathbb{R}^n)$ is arbitrary, (2.63) gives that

$$\int_{\partial\Omega} \varphi\,(\partial_{\tau_{jk}}\psi)\,d\sigma = 0. \tag{2.65}$$

Then the density result below finishes the proof of the lemma. □

Lemma 2.7. *If $\Omega \subset \mathbb{R}^n$ is a Lipschitz domain then the inclusion*

$$\{\varphi|_{\partial\Omega} : \varphi \in C_c^\infty(\mathbb{R}^n)\} \hookrightarrow L^p(\partial\Omega) \quad \textit{has dense range} \tag{2.66}$$

for every $p \in [1, \infty)$.

Proof. Denote by $\mathrm{Lip}_c(\partial\Omega)$ the space of Lipschitz functions with compact support on $\partial\Omega$. In this regard, two properties are going to be important for us. First, $\mathrm{Lip}_c(\partial\Omega) \hookrightarrow L^p(\partial\Omega)$ densely (see [56] for a proof of this elementary result). Second, any function $f \in \mathrm{Lip}_c(\partial\Omega)$ may be extended to a compactly supported Lipschitz function F in $\mathbb{R}^n$ with control of the Lipschitz constant. See, e.g., [6] for a proof in the more general context of quasi-metric spaces (as well as other pertinent references on this topic). Next, mollify F in order to produce a sequence $\{F_\varepsilon\}_{\varepsilon>0}$ of C^∞ smooth functions, supported in a fixed compact subset of $\mathbb{R}^n$, which converges uniformly to F. Clearly, this can be used to conclude that the embedding (2.66) has dense range. □

Inspired by (2.63), for every $f \in L^1_{loc}(\partial\Omega)$ we next define the functional $\partial_{\tau_{kj}} f$ by setting

$$\partial_{\tau_{kj}} f : C_c^1(\mathbb{R}^n) \ni \psi \mapsto \int_{\partial\Omega} f\,(\partial_{\tau_{jk}}\psi)\,d\sigma. \tag{2.67}$$

Indeed, when $f \in L^1_{loc}(\partial\Omega)$ has $\partial_{\tau_{kj}} f \in L^1_{loc}(\partial\Omega)$, the following integration by parts formula holds:

$$\int_{\partial\Omega} f\,(\partial_{\tau_{jk}}\psi)\,d\sigma = \int_{\partial\Omega} (\partial_{\tau_{kj}} f)\,\psi\,d\sigma, \qquad \forall\,\psi \in C_c^1(\mathbb{R}^n). \tag{2.68}$$

Let us also point out that the "distributional" version of $\partial_{\tau_{kj}}$ from (2.67) agrees with the "point-wise" definition considered in (2.62).

Moving on, for each $p \in [1, \infty]$ we then define the L^p-based Sobolev type space of order one on $\partial\Omega$ as

$$L_1^p(\partial\Omega) := \Big\{ f \in L^p(\partial\Omega) : \partial_{\tau_{jk}} f \in L^p(\partial\Omega),\ \ j,k = 1,\dots,n \Big\}, \tag{2.69}$$

which becomes a Banach space when equipped with the natural norm

$$\|f\|_{L_1^p(\partial\Omega)} := \|f\|_{L^p(\partial\Omega)} + \sum_{j,k=1}^{n} \|\partial_{\tau_{jk}} f\|_{L^p(\partial\Omega)}. \tag{2.70}$$

Finally, given $p \in [1, \infty]$, we shall denote by $L^p_{1,loc}(\partial\Omega)$ the space of all measurable functions $f : \partial\Omega \to \mathbb{R}$ with the property that $\xi f \in L^p_1(\partial\Omega)$ for every $\xi \in C^1_c(\mathbb{R}^n)$.

Proposition 2.8. *Assume that* $\Omega \subset \mathbb{R}^n$ *is a Lipschitz domain and that* $1 < p, p' < \infty$ *satisfy* $\frac{1}{p} + \frac{1}{p'} = 1$. *Then*

$$L^p_1(\partial\Omega) = \Big\{ f \in L^p(\partial\Omega) : \textit{ there exists a constant } c > 0 \textit{ such that if } \varphi \in C^1_c(\mathbb{R}^n)$$
$$\textit{then } \left| \int_{\partial\Omega} f\,(\partial_{\tau_{jk}}\varphi)\,d\sigma \right| \leq c\|\varphi\|_{L^{p'}(\partial\Omega)} \textit{ for } j,k = 1,\dots,n \Big\}. \tag{2.71}$$

Proof. Fix a function f belonging to the right-hand side of (2.71) and note that the functional $\Lambda_{jk} : \{\varphi|_{\partial\Omega} : \varphi \in C^\infty_c(\mathbb{R}^n)\} \to \mathbb{R}$, given by $\Lambda_{jk}(\varphi|_{\partial\Omega}) := \int_{\partial\Omega} f \partial_{\tau_{jk}}\varphi\, d\sigma$, is well-defined (by Lemma 2.6), linear and bounded (by assumptions). Based on this, (2.66) and Riesz's Representation Theorem we may then conclude that $f \in L^p_1(\partial\Omega)$. This proves the right-to-left inclusion in (2.71). Since the opposite inclusion is clear from (2.67)–(2.69), the proof of the lemma is finished. □

We continue our discussion of the Sobolev space $L^p_1(\partial\Omega)$ by presenting the following pull-back result.

Proposition 2.9. *Let* $\Omega \subset \mathbb{R}^n$ *be the domain lying above the graph of a Lipschitz function* $\phi : \mathbb{R}^{n-1} \to \mathbb{R}$. *Then, for each* $p \in (1, \infty)$,

$$f \in L^p_1(\partial\Omega) \iff f(\cdot, \phi(\cdot)) \in L^p_1(\mathbb{R}^{n-1}), \tag{2.72}$$

with equivalence of norms.

Proof. As a preamble, we shall establish two auxiliary results. Let p' be the Hölder conjugate exponent of p. We first claim that given $\psi \in C^\infty_c(\mathbb{R}^{n-1})$ then

$$\exists\ G_\varepsilon \in C^\infty_c(\mathbb{R}^n) \quad \text{so that} \quad G_\varepsilon(\cdot, \phi(\cdot)) \to \psi \text{ in } L^{p'}_1(\mathbb{R}^{n-1}) \quad \text{as} \quad \varepsilon \to 0^+. \tag{2.73}$$

To justify this claim, fix $\psi \in C^\infty_c(\mathbb{R}^{n-1})$ and consider $\Psi \in \mathrm{Lip}_c(\mathbb{R}^n)$ given by

$$\Psi(x', x_n) := \psi(x')\theta(x_n - \phi(x')), \tag{2.74}$$

where $\theta \in C^\infty_c(\mathbb{R})$ is such that $\theta \equiv 1$ near zero. If P is a standard mollifier in $\mathbb{R}^n$, we set $P_\varepsilon(x) := \frac{1}{\varepsilon^n}P(x/\varepsilon)$ for $\varepsilon > 0$, and $G_\varepsilon := \Psi * P_\varepsilon \in C^\infty_c(\mathbb{R}^n)$. Then, using (2.74), we obtain

$$G_\varepsilon(x', \phi(x')) - \psi(x') = \int_{\mathbb{R}^n} \Big[\Psi((x', \phi(x')) - Y) - \Psi(x', \phi(x'))\Big] P_\varepsilon(Y)\,dY. \tag{2.75}$$

A straightforward application of the mean value theorem in (2.75) readily gives that

$$\|G_\varepsilon(\cdot,\phi(\cdot)) - \psi\|_{L^\infty(\mathbb{R}^{n-1})} \le C\varepsilon\|\nabla\Psi\|_{L^\infty(\mathbb{R}^n)} \to 0 \quad \text{as} \quad \varepsilon \to 0^+, \tag{2.76}$$

which further implies

$$G_\varepsilon(\cdot,\phi(\cdot)) \to \psi \quad \text{in} \quad L^{p'}(\mathbb{R}^{n-1}), \quad \text{as} \quad \varepsilon \to 0^+. \tag{2.77}$$

Next, let $1 \le j \le n-1$ and $Y = (y', y_n) \in B_\varepsilon(0)$. Since $\theta \equiv 1$ near zero, it follows that

$$\partial_j[G_\varepsilon(x',\phi(x')) - \psi(x')] = \int_{\mathbb{R}^n} (\partial_j\psi)(x'-y') - (\partial_j\psi)(x')]P_\varepsilon(Y)\,dY, \tag{2.78}$$

if $\varepsilon > 0$ is small enough. Thus, much as before $\|\partial_j[G_\varepsilon(\cdot,\phi(\cdot)) - \psi]\|_{L^\infty(\mathbb{R}^{n-1})} \to 0$ as $\varepsilon \to 0^+$ and, ultimately,

$$\partial_j[G(\cdot,\phi(\cdot))] \to \partial_j\psi \quad \text{in} \quad L^{p'}(\mathbb{R}^{n-1}), \quad \text{as} \quad \varepsilon \to 0^+. \tag{2.79}$$

Now (2.73) follows from (2.77) and (2.79).

The second claim is that, if the functions $f \in L^p_1(\partial\Omega)$ and $G \in C^\infty_c(\mathbb{R}^n)$ then for any $1 \le j \le n-1$ we have

$$\Big|\int_{\mathbb{R}^{n-1}} f(x',\phi(x'))\partial_j[G(x',\phi(x'))]\,dx'\Big|$$
$$\le C\|\partial_{\tau_{jn}} f\|_{L^p(\partial\Omega)}\|G(\cdot,\phi(\cdot))\|_{L^{p'}_1(\mathbb{R}^{n-1})}. \tag{2.80}$$

To prove (2.80), set $g := G\Big|_{\partial\Omega} \in L^{p'}_1(\partial\Omega)$ and recall that, for each $j \in \{1,\dots, n-1\}$,

$$\partial_{\tau_{jn}} g = \nu_j(\partial_n G)\Big|_{\partial\Omega} - \nu_n(\partial_j G)\Big|_{\partial\Omega} \quad \text{on} \quad \partial\Omega. \tag{2.81}$$

Hence, using (2.8) and integration by parts, we may write

$$\begin{aligned}
\int_{\partial\Omega} \partial_{\tau_{jn}} f(X)g(X)\,d\sigma(X) &= -\int_{\partial\Omega} f(X)(\nu_j\partial_n G - \nu_n\partial_j G)(X)\,d\sigma(X)\\
&= -\int_{\partial\Omega} f(X)(\partial_{\tau_{jn}} g)(X)\,d\sigma(X)\\
&= -\int_{\mathbb{R}^{n-1}} f(x',\phi(x'))\Big[\partial_j\phi(x')(\partial_n G)\big(x',\phi(x')\big)\\
&\qquad + (\partial_j G)\big(x',\phi(x')\big)\Big]dx'\\
&= -\int_{\mathbb{R}^{n-1}} f\big(x',\phi(x')\big)\partial_j\big[G(x',\phi(x'))\big]\,dx'.
\end{aligned} \tag{2.82}$$

Consequently, (2.80) follows from (2.82) and the Hölder inequality.

With (2.73) and (2.80) in hand, we are ready to prove the left-to-right implication in (2.72). Concretely, fix $f \in L_1^p(\partial\Omega)$, $\psi \in C_c^\infty(\mathbb{R}^{n-1})$, and consider $G_\varepsilon \in C_c^\infty(\mathbb{R}^n)$, $\varepsilon > 0$, as in (2.73). Employing (2.73) and (2.80) we may then conclude that for each $1 \leq j \leq n-1$,

$$\begin{aligned}\Big|\int_{\mathbb{R}^{n-1}} f(x',\phi(x'))\partial_j\psi(x')\,dx'\Big| &= \lim_{\varepsilon\to 0+}\Big|\int_{\mathbb{R}^{n-1}} f(x',\phi(x'))\partial_j[G_\varepsilon(x',\phi(x'))]\,dx'\Big| \\ &\leq C\|\partial_{\tau_{jn}}f\|_{L^p(\partial\Omega)}\lim_{\varepsilon\to 0+}\|G_\varepsilon(\cdot,\phi(\cdot))\|_{L_1^{p'}(\mathbb{R}^{n-1})} \\ &= C\|\partial_{\tau_{jn}}f\|_{L^p(\partial\Omega)}\|\psi\|_{L_1^{p'}(\mathbb{R}^{n-1})}. \end{aligned} \tag{2.83}$$

This shows that $f(\cdot,\phi(\cdot)) \in L_1^p(\mathbb{R}^{n-1})$, as desired.

Turning our attention to the right-to-left implication in (2.72), let $f : \partial\Omega \to \mathbb{R}$ be such that $f(\cdot,\phi(\cdot)) \in L_1^p(\mathbb{R}^{n-1})$. Then, there exists $C > 0$ such that, for each $\psi \in C_c^\infty(\mathbb{R}^{n-1})$, we have

$$\Big|\int_{\mathbb{R}^{n-1}} f(x',\phi(x'))\partial_j\psi(x')\,dx'\Big| \leq C\|f(\cdot,\phi(\cdot))\|_{L_1^p(\mathbb{R}^{n-1})}\|\psi\|_{L^{p'}(\mathbb{R}^{n-1})}. \tag{2.84}$$

Since functions in $\mathrm{Lip}_c(\mathbb{R}^{n-1})$ can be arbitrarily well approximated, in the norm of $L_1^{p'}(\mathbb{R}^{n-1})$, by functions in $C_c^\infty(\mathbb{R}^{n-1})$, it follows that (2.84) holds, in fact, for each $\psi \in \mathrm{Lip}_c(\mathbb{R}^{n-1})$. Next, pick an arbitrary function $G \in C_c^1(\mathbb{R}^n)$ and set $g := G\big|_{\partial\Omega}$. In particular, $g \in \mathrm{Lip}_c(\partial\Omega)$ and $g(\cdot,\phi(\cdot)) \in \mathrm{Lip}_c(\mathbb{R}^{n-1})$. Using the last three identities in (2.82) then invoking (2.84) yields, for each $1 \leq j \leq n-1$,

$$\Big|\int_{\partial\Omega} f(X)\partial_{\tau_{jn}}g(X)\,d\sigma(X)\Big| \leq C\|f(\cdot,\phi(\cdot))\|_{L_1^p(\mathbb{R}^{n-1})}\|g\|_{L^{p'}(\partial\Omega)}. \tag{2.85}$$

This proves that $\partial_{\tau_{jk}}f \in L^p(\partial\Omega)$ if $1 \leq j \leq n-1$ and $k = n$.

We aim to prove the same conclusion in the case when the indices $j,k \in \mathbb{N}$ satisfy $1 \leq j,k \leq n-1$. To this end, retaining the same setting as above we write that $\partial_{\tau_{jk}}g = \nu_j(\partial_k G)\big|_{\partial\Omega} - \nu_k(\partial_j G)\big|_{\partial\Omega}$ for $1 \leq j,k \leq n-1$. Based on (2.8) we have

$$\begin{aligned}\int_{\partial\Omega} f(X)\partial_{\tau_{jk}}g(X)\,d\sigma(X) &= \int_{\mathbb{R}^{n-1}} f(x',\phi(x'))\Big[\partial_j\phi(x')\partial_k[G(x',\phi(x'))]\Big]\,dx' \\ &\quad - \int_{\mathbb{R}^{n-1}} f(x',\phi(x'))\Big[\partial_k\phi(x')\partial_j[G(x',\phi(x'))]\Big]\,dx'. \end{aligned} \tag{2.86}$$

Consider next $H_\varepsilon \in C_c^\infty(\mathbb{R}^{n-1})$, $\varepsilon > 0$, such that $H_\varepsilon \to G(\cdot, \phi(\cdot))$ as $\varepsilon \to 0^+$ in $L_1^{p'}(\mathbb{R}^{n-1})$ and $\operatorname{supp} H_\varepsilon$ is contained in a fixed compact set in $\mathbb{R}^{n-1}$. Then (2.86) gives

$$\int_{\partial\Omega} f(X)\partial_{\tau_{jk}} g(X)\, d\sigma(X) = \lim_{\varepsilon\to 0+} \int_{\mathbb{R}^{n-1}} f(x', \phi(x'))\partial_j[\phi(x')\partial_k H_\varepsilon(x')]\, dx' \tag{2.87}$$

$$- \lim_{\varepsilon\to 0+} \int_{\mathbb{R}^{n-1}} f(x', \phi(x'))\partial_k[\phi(x')\partial_j H_\varepsilon(x')]\, dx'.$$

Note that for each $\varepsilon > 0$ and each $j \in \{1, \ldots, n-1\}$ we have that the function $\phi(\cdot)\partial_j H_\varepsilon(\cdot) \in \operatorname{Lip}_c(\mathbb{R}^{n-1})$. Consequently, (2.84) and (2.87) further imply that there exists $C = C(\partial\Omega) > 0$ such that

$$\begin{aligned} \left|\int_{\partial\Omega} f(X)\partial_{\tau_{jk}} g(X)\, d\sigma(X)\right| &\le C\, \|f(\cdot, \phi(\cdot))\|_{L_1^p(\mathbb{R}^{n-1})} \lim_{\varepsilon\to 0+} \|H_\varepsilon\|_{L^{p'}(\mathbb{R}^{n-1})} \\ &= C\, \|f(\cdot, \phi(\cdot))\|_{L_1^p(\mathbb{R}^{n-1})} \|G(\cdot, \phi(\cdot))\|_{L^{p'}(\mathbb{R}^{n-1})} \\ &\le C\, \|f(\cdot, \phi(\cdot))\|_{L_1^p(\mathbb{R}^{n-1})} \|g\|_{L^{p'}(\partial\Omega)}. \end{aligned} \tag{2.88}$$

Now (2.88) gives that $\partial_{\tau_{jk}} f \in L^p(\partial\Omega)$, $1 \le j, k \le n-1$, and this completes the proof of the fact that $f \in L_1^p(\partial\Omega)$. □

Corollary 2.10. *Let Ω be a bounded Lipschitz domain in $\mathbb{R}^n$. Then,*

$$\operatorname{Lip}(\partial\Omega) \hookrightarrow L_1^p(\partial\Omega) \quad \text{and} \quad C^\infty(\mathbb{R}^n)\Big|_{\partial\Omega} \hookrightarrow L_1^p(\partial\Omega) \quad \text{densely} \tag{2.89}$$

whenever $1 < p < \infty$.

Proof. Since $\operatorname{Lip}_c(\mathbb{R}^{n-1}) \hookrightarrow L_1^p(\mathbb{R}^{n-1})$ densely, the first part of (2.89) follows from Proposition 2.9. With this in hand, for the second part of (2.89) it suffices to show that any $f \in \operatorname{Lip}(\partial\Omega)$ can be approximated in $L_1^p(\partial\Omega)$ by functions in $C^\infty(\mathbb{R}^n)\Big|_{\partial\Omega}$. To see this, fix f and, as in the past, construct $F \in \operatorname{Lip}_c(\mathbb{R}^n)$ such that $F\Big|_{\partial\Omega} = f$. Then a standard mollification argument yields the desired conclusion. □

Going further, we set

$$\nabla_{tan} f := \Big(\sum_{k=1}^{n} \nu_k \partial_{\tau_{kj}} f\Big)_{1\le j\le n}, \qquad \forall\, f \in L_1^p(\partial\Omega), \tag{2.90}$$

Proposition 2.11. *Consider a Lipschitz domain Ω in $\mathbb{R}^n$ with surface measure σ and outward unit normal $\nu = (\nu_1, \ldots, \nu_n)$. Then for each function $f \in L_1^p(\partial\Omega)$,*

$$\partial_{\tau_{jk}} f = \nu_j (\nabla_{tan} f)_k - \nu_k (\nabla_{tan} f)_j, \qquad j,k = 1,\ldots,n, \tag{2.91}$$

σ-a.e. on $\partial\Omega$.

Proof. By Corollary 2.10, it suffices to establish (2.91) in the case in which $f := F\big|_{\partial\Omega}$ for some $F \in C_c^1(\mathbb{R}^n)$. Assuming that this is the case, (2.62) gives

$$\partial_{\tau_{jk}} f = \nu_j (\partial_k F)\Big|_{\partial\Omega} - \nu_k (\partial_j F)\Big|_{\partial\Omega}. \tag{2.92}$$

Consequently, for each $j, k \in \{1, \ldots, n\}$, we have

$$\begin{aligned} \nu_j (\nabla_{tan} f)_k - \nu_k (\nabla_{tan} f)_j &= \nu_j \sum_{\ell=1}^n \nu_\ell \partial_{\tau_{\ell k}} f - \nu_\ell \sum_{r=1}^n \nu_r \partial_{\tau_{rj}} f \\ &= \nu_j \Big(\sum_{\ell=1}^n \nu_\ell^2\Big)(\partial_k F)\Big|_{\partial\Omega} - \nu_j \nu_k \Big(\sum_{\ell=1}^n \nu_\ell (\partial_\ell F)\Big|_{\partial\Omega}\Big) \\ &\quad - \nu_k \Big(\sum_{r=1}^n \nu_r^2\Big)(\partial_j F)\Big|_{\partial\Omega} + \nu_j \nu_k \Big(\sum_{r=1}^n \nu_r (\partial_r F)\Big|_{\partial\Omega}\Big), \end{aligned} \tag{2.93}$$

where the first equality follows from (2.90) and the second equality is a direct consequence of (2.62). Using that $|\nu| = 1$, then (2.93) and (2.62) further imply

$$\nu_j (\nabla_{tan} f)_k - \nu_k (\nabla_{tan} f)_j = \nu_j (\partial_k F)\Big|_{\partial\Omega} - \nu_k (\partial_j F)\Big|_{\partial\Omega} = \partial_{\tau_{jk}} f. \tag{2.94}$$

Therefore the identity (2.91) holds as stated. □

Among the consequences of (2.90), we note here that for each $p \in (1, \infty)$,

$$|\nabla_{tan} f| \approx \sum_{j,k=1}^n |\partial_{\tau_{jk}} f|, \qquad \text{pointwise } \sigma\text{-a.e. on } \partial\Omega, \tag{2.95}$$

$$\|\nabla_{tan} f\|_{L^p(\partial\Omega)} \approx \sum_{j,k=1}^n \|\partial_{\tau_{jk}} f\|_{L^p(\partial\Omega)} \approx \sum_{j=1}^{n-1} \|\partial_{\tau_{jn}} f\|_{L^p(\partial\Omega)}, \tag{2.96}$$

$$\|f\|_{L_1^p(\partial\Omega)} \approx \|f\|_{L^p(\partial\Omega)} + \|\nabla_{tan} f\|_{L^p(\partial\Omega)}, \tag{2.97}$$

uniformly in $f \in L_1^p(\partial\Omega)$.

Corollary 2.12. *Let Ω be a bounded Lipschitz domain in $\mathbb{R}^n$, with surface measure σ, and assume that $1 < p, p' < \infty$ satisfy $1/p + 1/p' = 1$ and that $j, k \in \{1, \dots, n\}$. Then*

$$\int_{\partial\Omega} (\partial_{\tau_{jk}} f)\, g\, d\sigma = \int_{\partial\Omega} f\, (\partial_{\tau_{kj}} g)\, d\sigma \tag{2.98}$$

for every $f \in L^p_1(\partial\Omega)$ and $g \in L^{p'}_1(\partial\Omega)$.

Proof. By Corollary 2.10, it can be assumed that $g \in C^\infty_c(\mathbb{R}^n)\Big|_{\partial\Omega}$, in which case (2.98) follows from (2.68). □

For each $1 < p < \infty$, $L^p_1(\partial\Omega)$ is a Banach space, densely embedded into $L^p(\partial\Omega)$ (cf. (2.89)). Furthermore, since the mapping

$$J : L^p_1(\partial\Omega) \longrightarrow \Big[L^p(\partial\Omega)\Big]^{1+\frac{(n-1)n}{2}}, \qquad Jf := \Big(f, (\partial_{\tau_{jk}} f)_{1\le j,k\le n}\Big), \tag{2.99}$$

is bounded both from above and below, its image is closed. Now, $L^p_1(\partial\Omega)$ is isomorphic to the latter space and, hence, is reflexive. Thus, if for each $1 < p < \infty$, we set

$$L^p_{-1}(\partial\Omega) := \Big(L^{p'}_1(\partial\Omega)\Big)^*, \qquad 1/p + 1/p' = 1, \tag{2.100}$$

it follows that

$$\Big(L^p_{-1}(\partial\Omega)\Big)^* = L^{p'}_1(\partial\Omega), \qquad 1/p + 1/p' = 1. \tag{2.101}$$

Corollary 2.13. *Let Ω be a bounded Lipschitz domain in $\mathbb{R}^n$, $1 < p < \infty$ and fix two indices $j, k \in \{1, \dots, n\}$. Then the operator*

$$\partial_{\tau_{jk}} : L^p_1(\partial\Omega) \longrightarrow L^p(\partial\Omega) \tag{2.102}$$

extends in a (unique) compatible fashion to a bounded, linear mapping

$$\partial_{\tau_{jk}} : L^p(\partial\Omega) \longrightarrow L^p_{-1}(\partial\Omega). \tag{2.103}$$

Proof. For every $f \in L^p(\partial\Omega)$, set

$$\langle \partial_{\tau_{jk}} f, g\rangle := \int_{\partial\Omega} f\, \partial_{\tau_{kj}} g\, d\sigma, \qquad \forall\, g \in L^{p'}_1(\partial\Omega), \tag{2.104}$$

where $1/p + 1/p' = 1$. Then the desired conclusion follows from Corollary 2.12. □

Corollary 2.14. *Let Ω be a bounded Lipschitz domain in $\mathbb{R}^n$ and fix $p \in (1,\infty)$. Then for every $f \in L^p_{-1}(\partial\Omega)$ there exist $g_0, g_{jk} \in L^p(\partial\Omega)$, $1 \le j,k \le n$ (not necessarily unique) with the property that*

$$f = g_0 + \sum_{j,k=1}^{n} \partial_{\tau_{jk}} g_{jk} \quad \text{in } L^p_{-1}(\partial\Omega). \tag{2.105}$$

Furthermore,

$$\|f\|_{L^p_{-1}(\partial\Omega)} \approx \inf\Big[\|g_0\|_{L^p(\partial\Omega)} + \sum_{j,k=1}^{n} \|g_{jk}\|_{L^p(\partial\Omega)}\Big], \tag{2.106}$$

where the infimum is taken over all representations of f as in (2.105).

Proof. Let $p' \in (1,\infty)$ be such that $1/p + 1/p' = 1$. If $f \in L^p_{-1}(\partial\Omega)$ is regarded as a functional $f : L^{p'}_1(\partial\Omega) \to \mathbb{R}$, then $f \circ J^{-1} : \operatorname{Im} J \to \mathbb{R}$ is well-defined, linear and bounded (where J is as in (2.99) with p' in place of p). At this stage, the Hahn–Banach Theorem in conjunction with Riesz's Representation Theorem ensure the existence of $g_0, g_{jk} \in L^p(\partial\Omega)$ such that (2.105)–(2.106) hold. □

Proposition 2.15. *Let $\Omega \subset \mathbb{R}^n$ be a bounded Lipschitz domain and assume that the function $u \in C^1(\Omega)$ is such that $\mathcal{N}(\nabla u) \in L^p(\partial\Omega)$ for some $p \in (1,\infty)$. Then u has a non-tangential limit at almost every boundary point on $\partial\Omega$,*

$$u\Big|_{\partial\Omega} \in L^p_1(\partial\Omega) \quad \text{and} \quad \sum_{j,k=1}^{n} \|\partial_{\tau_{jk}}(u\lfloor_{\partial\Omega})\|_{L^p(\partial\Omega)} \le C\|\mathcal{N}(\nabla u)\|_{L^p(\partial\Omega)}. \tag{2.107}$$

Furthermore,

$$\Big\| u\Big|_{\partial\Omega}\Big\|_{L^p_1(\partial\Omega)} \le C\|\mathcal{N}(\nabla u)\|_{L^p(\partial\Omega)} + C\|\mathcal{N}u\|_{L^p(\partial\Omega)}. \tag{2.108}$$

If, in addition, each $\partial_j u$, $1 \le j \le n$, also has a non-tangential limit at almost every boundary point on $\partial\Omega$, then

$$\partial_{\tau_{jk}}(u\lfloor_{\partial\Omega}) = \nu_j(\partial_k u)\Big|_{\partial\Omega} - \nu_k(\partial_j u)\Big|_{\partial\Omega}, \qquad \forall\, j,k \in \{1,\dots,n\}. \tag{2.109}$$

Proof. Assume that $\mathcal{N}_\kappa(\nabla u) \in L^p(\partial\Omega)$ for some $\kappa > 0$. We first aim to show that (2.59) exists for a.e. $X \in \partial\Omega$. By Proposition 2.2 and the remark following it, there is no loss of generality in assuming that κ is sufficiently large. In particular, we can assume that there exists $\theta > 0$ with the property that $\Gamma_\theta(X) \subset R_\kappa(X)$ for every $X \in \partial\Omega$. Since

$$\begin{aligned}|u(P) - u(Q)| &\le \widetilde{\mathcal{N}}_\theta(\nabla u)(X)\,|P-Q| \\ &\le \mathcal{N}_\kappa(\nabla u)(X)\,|P-Q|, \qquad \forall\, P,Q \in \Gamma_\theta(X),\end{aligned} \tag{2.110}$$

it follows that u is Lipschitz in each cone $\Gamma_\theta(X)$ provided $\mathcal{N}_\kappa(\nabla u)(X) < \infty$. Now, $\mathcal{N}_\kappa(\nabla u) \in L^p(\partial\Omega)$ ensures that the latter condition is satisfied for σ-a.e. $X \in \partial\Omega$ (where, as usual, σ denotes the surface measure on $\partial\Omega$). Hence,

$$u(X) := \lim_{\substack{Y\to X\\ Y\in\Gamma_\theta(X)}} u(Y) \ \text{ exists for } \sigma\text{-a.e. } X \in \partial\Omega. \tag{2.111}$$

In order to prove a similar result with $R_\kappa(X)$ in place of $\Gamma_\theta(X)$, we shall avail ourselves of a remark made on p. 93 in [59] which entails the following. Given $\varepsilon > 0$, $X \in \partial\Omega$ and $P \in R_\kappa(X) \cap B(X,\varepsilon)$, there exists $Q \in \Gamma_\theta(X) \cap B(X,\varepsilon)$ and a rectifiable path γ joining P and Q such that $\gamma \subset R_{\kappa'}(X)$ for some $\kappa' > 0$ and length $(\gamma) \le C\varepsilon$. We may then write

$$\begin{aligned}|u(P)-u(X)| &\le |u(P)-u(Q)| + |u(Q)-u(X)|\\ &\le C\varepsilon\,\mathcal{N}_{\kappa'}(\nabla u)(X) + |u(Q)-u(X)|.\end{aligned} \tag{2.112}$$

Now, $\mathcal{N}_{\kappa'}(\nabla u) \in L^p(\partial\Omega)$ forces $\mathcal{N}_{\kappa'}(\nabla u)(X) < \infty$ for σ-a.e. $X \in \partial\Omega$. Also, (2.111) implies that $|u(Q)-u(X)|$ can be made as small as desired by taking ε to be small. Consequently, $|u(P)-u(X)|$ can be made arbitrarily small provided ε is small enough. This proves that u has a non-tangential limit at a.e. boundary point. The fact that $u\lfloor_{\partial\Omega} \in L^p(\partial\Omega)$ and

$$\|u\lfloor_{\partial\Omega}\|_{L^p(\partial\Omega)} \le C\,\|\mathcal{N}(\nabla u)\|_{L^p(\partial\Omega)} + C\,\|\mathcal{N}u\|_{L^p(\partial\Omega)} \tag{2.113}$$

are simple consequences of (2.110) and (2.111). For further use, let us also point out here that (2.110) implies that $\mathcal{N}u \in L^p(\partial\Omega)$.

Consider next a sequence of domains Ω_j, $j \in \mathbb{N}$, enjoying the following properties:

(i) Each Ω_j is a C^∞ domain, $\Omega_j \subset \Omega_{j+1} \subset \Omega$ for every j and $\Omega = \cup_{j\in\mathbb{N}}\Omega_j$;
(ii) There exist $\kappa > 0$ and bi-Lipschitz homeomorphism $\Lambda_j : \partial\Omega \to \partial\Omega_j$, $j \in \mathbb{N}$, such that $\Lambda_j(X) \in R_\kappa(X)$ for all $X \in \partial\Omega$ and $\Lambda_j(X) \to X$ as $j \to \infty$;
(iii) If $\nu^j = (\nu^j_1, \dots, \nu^j_n)$ is the outward unit normal vector and σ_j is the surface measure on $\partial\Omega_j$, then $\nu^j(\Lambda_j(\cdot)) \to \nu$ as $j \to \infty$ pointwise σ-a.e. on $\partial\Omega$;
(iv) There exists a sequence $(J_j)_j$ of non-negative, measurable functions on $\partial\Omega$, bounded away from zero and infinity uniformly in j, satisfying $J_j(X) \to 1$ as $j \to \infty$ for σ-a.e. $X \in \partial\Omega$, and for which the following change of variable formula

$$\int_{\partial\Omega_j} f(X)\,d\sigma_j(X) = \int_{\partial\Omega} f(\Lambda_j(X))J_j(X)d\sigma(X), \tag{2.114}$$

holds for any $f \in L^1(\partial\Omega_j)$.

For the construction of a sequence of domains $(\Omega_j)_{j\in\mathbb{N}}$ satisfying properties (i)–(iv), we refer the reader to [86, 125]. In the sequel, we shall write $\Omega_j \nearrow \Omega$ in order to indicate that the family $\{\Omega_j\}_{j\in\mathbb{N}}$ satisfies the properties (i)–(iv) above. One can also construct a family of smooth, bounded domains approximating a given Ω from the *exterior* (i.e., with the inclusions in (i) reversed). We abbreviate by $\Omega_j \searrow \Omega$ the fact that $(\Omega_j)_{j\in\mathbb{N}}$ is such a family.

Going further, if $\Omega_j \nearrow \Omega$, for each $1 \le k, \ell \le n$ and $j \in \mathbb{N}$ we set

$$\partial_{\tau^j_{kl}} := \nu^j_k \partial_\ell - \nu^j_\ell \partial_k. \tag{2.115}$$

Also, we fix $1 \le k, \ell \le n$ along with $g \in C^\infty_c(\mathbb{R}^n)$. On the one hand, (2.114), the properties of u and Lebesgue's' Dominated Convergence Theorem allow us to write

$$\int_{\partial\Omega} u(\partial_{\tau_{k\ell}} g)\,d\sigma = \lim_{j\to\infty} \int_{\partial\Omega_j} u(\partial_{\tau^j_{k\ell}} g)\,d\sigma_j. \tag{2.116}$$

On the other hand, for each j we have $u|_{\partial\Omega_j} \in \mathrm{Lip}\,(\partial\Omega_j)$, so integrating by parts on $\partial\Omega_j$, then changing variables back to $\partial\Omega$, yields

$$\begin{aligned}
\Big|\int_{\partial\Omega_j} u(\partial_{\tau^j_{k\ell}} g)\,d\sigma_j\Big| &= \Big|\int_{\partial\Omega_j} g(\partial_{\tau^j_{k\ell}} u)\,d\sigma_j\Big| \qquad\qquad (2.117)\\
&= \Big|\int_{\partial\Omega} \nu^j_\ell(\Lambda_j(X))(\partial_k u)(\Lambda_j(X)) g(\Lambda_j(X)) J_j(X)\,d\sigma(X)\\
&\quad - \int_{\partial\Omega} \nu^j_k(\Lambda_j(X))(\partial_\ell u)(\Lambda_j(X)) g(\Lambda_j(X)) J_j(X)\,d\sigma(X)\Big|,\\
&\le C\,\|\mathcal{N}_\kappa(\nabla u)\|_{L^p(\partial\Omega)} \|g\|_{L^{p'}(\partial\Omega)},
\end{aligned}$$

where $1/p + 1/p' = 1$. This implies that $\partial_{\tau_{k\ell}} u \in L^p(\partial\Omega)$ for all $k, \ell \in \{1, \dots, n\}$ and the estimate in (2.107) holds. Thus, $u|_{\partial\Omega} \in L^p_1(\partial\Omega)$ and (2.108) holds.

Finally, for two arbitrary indices $j, k \in \{1, \dots, n\}$ and $\psi \in C^1_c(\mathbb{R}^n)$ we may invoke the Divergence Theorem twice in order to write

$$\begin{aligned}
\int_{\partial\Omega} u(\partial_{\tau_{kj}} \psi)\,d\sigma &= \int_{\partial\Omega} u(\nu_k \partial_j \psi - \nu_j \partial_k \psi)\,d\sigma\\
&= \int_{\Omega} \Big(\partial_k u \partial_j \psi - \partial_j u \partial_k \psi\Big)\,dX\\
&= \int_{\partial\Omega} \Big(\nu_j (\partial_k u)|_{\partial\Omega} - \nu_k (\partial_j u)|_{\partial\Omega}\Big)\psi\,d\sigma.
\end{aligned} \tag{2.118}$$

That the hypotheses of Proposition 2.4 are verified, is ensured by the properties of u and Proposition 2.3. Since ψ is arbitrary, (2.118) proves (2.109). □

Corollary 2.16. *Let $\Omega \subset \mathbb{R}^n$ be a bounded Lipschitz domain with surface measure σ. If the functions $u, v \in C^1(\Omega)$ are such that $\mathcal{N}(\nabla u) \in L^p(\partial\Omega)$ and $\mathcal{N}(\nabla v) \in L^{p'}(\partial\Omega)$ with $p, p' \in (1,\infty)$ conjugate exponents, then*

$$\int_{\partial\Omega} u\,(\partial_{\tau_{jk}} v)\,d\sigma = \int_{\partial\Omega} (\partial_{\tau_{kj}} u)\,v\,d\sigma. \tag{2.119}$$

Proof. This is a direct consequence of Proposition 2.15 and Corollary 2.13. □

2.3 Brief Review of Smoothness Spaces in $\mathbb{R}^n$

Recall that, for each $p \in (1,\infty)$ and $s \in \mathbb{R}$, the `Bessel potential space` $L^p_s(\mathbb{R}^n)$ is defined by

$$\begin{aligned} L^p_s(\mathbb{R}^n) &:= \Big\{(I-\Delta)^{-s/2} g \,:\, g \in L^p(\mathbb{R}^n)\Big\} \\ &= \Big\{\mathcal{F}^{-1}(1+|\xi|^2)^{-s/2}\mathcal{F} g \,:\, g \in L^p(\mathbb{R}^n)\Big\} \end{aligned} \tag{2.120}$$

and is equipped with the norm

$$\|f\|_{L^p_s(\mathbb{R}^n)} := \|\mathcal{F}^{-1}(1+|\xi|^2)^{s/2}\mathcal{F} f\|_{L^p(\mathbb{R}^n)}, \tag{2.121}$$

where $\mathcal{F}$ denotes the Fourier transform in $\mathbb{R}^n$. As is well-known, when the smoothness index is a natural number, say $s = k \in \mathbb{N}$, this can be identified with the classical Sobolev space

$$W^{k,p}(\mathbb{R}^n) := \Big\{f \in L^p(\mathbb{R}^n) : \|f\|_{W^{k,p}(\mathbb{R}^n)} := \sum_{|\gamma|\le k} \|\partial^\gamma f\|_{L^p(\mathbb{R}^n)} < \infty\Big\}, \tag{2.122}$$

i.e.,

$$L^p_k(\mathbb{R}^n) = W^{k,p}(\mathbb{R}^n), \qquad k \in \mathbb{N}_0,\ 1 < p < \infty. \tag{2.123}$$

For further reference, we define here the Hölder space $C^s(\mathbb{R}^n)$, $s > 0$, $s \notin \mathbb{N}$, consisting of functions f for which

$$\|f\|_{C^s(\mathbb{R}^n)} := \sum_{|\alpha|\le[s]} \|\partial^\alpha f\|_{L^\infty(\mathbb{R}^n)} + \sum_{|\alpha|=[s]} \sup_{x\neq y} \frac{|\partial^\alpha f(x) - \partial^\alpha f(y)|}{|x-y|^{s-[s]}} < \infty. \tag{2.124}$$

Here $[\cdot]$ stands for the integer-part function.

Next we turn our attention to Hardy-type spaces in $\mathbb{R}^n$. Fix a function ψ in $C^\infty_c(\mathbb{R}^n)$ with $\operatorname{supp}\psi \subset \{X \in \mathbb{R}^n : |X| < 1\}$ and $\int_{\mathbb{R}^n} \psi(X)\,dX = 1$, and set

$$\psi_t(X) := t^{-n}\psi(X/t) \quad \text{for each } t > 0 \text{ and } X \in \mathbb{R}^n. \tag{2.125}$$

Given a tempered distribution $u \in S'(\mathbb{R}^n)$ we define its radial maximal function and its truncated version, respectively, by setting

$$u^{++} := \sup_{0<t<\infty} |\psi_t * u|, \qquad u^{+} := \sup_{0<t<1} |\psi_t * u|. \tag{2.126}$$

For $p \in (0, \infty)$, the classical `homogeneous` Hardy space $H^p(\mathbb{R}^n)$, and its `local` version, $h^p(\mathbb{R}^n)$, introduced in [52], are then defined as

$$H^p(\mathbb{R}^n) := \{u \in S'(\mathbb{R}^n) : \|u\|_{H^p(\mathbb{R}^n)} := \|u^{++}\|_{L^p(\mathbb{R}^n)} < \infty\}, \tag{2.127}$$

$$h^p(\mathbb{R}^n) := \{u \in S'(\mathbb{R}^n) : \|u\|_{h^p(\mathbb{R}^n)} := \|u^{+}\|_{L^p(\mathbb{R}^n)} < \infty\}. \tag{2.128}$$

Different choices of the function ψ yield equivalent quasi-norms so (2.127), (2.128) viewed as topological spaces, are intrinsically defined. Also,

$$H^p(\mathbb{R}^n) = h^p(\mathbb{R}^n) = L^p(\mathbb{R}^n), \qquad 1 < p < \infty. \tag{2.129}$$

In analogy with (2.122), `Hardy-based Sobolev spaces` $h^p_k(\mathbb{R}^n)$, with $0 < p < \infty$ and $k \in \mathbb{N}_0$, are then defined as

$$h^p_k(\mathbb{R}^n) := \big\{u \in S'(\mathbb{R}^n) : \partial^\gamma u \in h^p(\mathbb{R}^n),\ \forall\, \gamma \in \mathbb{N}_0^n \text{ with } |\gamma| \le k\big\}, \tag{2.130}$$

and are equipped with the quasi-norm $\|u\|_{h^p_k(\mathbb{R}^n)} := \sum_{|\gamma|\le k} \|\partial^\gamma u\|_{h^p(\mathbb{R}^n)}$. For each $0 < p < \infty$ and $k \in \mathbb{N}$ we also set

$$h^p_{-k}(\mathbb{R}^n) := \Big\{u \in S'(\mathbb{R}^n) : u = \sum_{|\gamma|\le k} \partial^\gamma u_\gamma, \text{ where } u_\gamma \in h^p(\mathbb{R}^n)\ \forall\, \gamma \in \mathbb{N}_0^n \text{ with } |\gamma| \le k\Big\} \tag{2.131}$$

which we equip with the natural quasi-norm $\|u\|_{h^p_{-k}(\mathbb{R}^n)} := \inf \sum_{|\gamma|\le k} \|u_\gamma\|_{h^p(\mathbb{R}^n)}$, where the infimum is taken over all representations of u.

Going further, a function $f \in L^2_{\rm loc}(\mathbb{R}^n)$ is said to belong to the space $\mathrm{BMO}(\mathbb{R}^n)$ if

$$\|f\|_{\mathrm{BMO}(\mathbb{R}^n)} := \sup_Q \left(\frac{1}{|Q|}\int_Q |f(X) - f_Q|^2\, dX\right)^{1/2} < \infty, \tag{2.132}$$

where the supremum is taken over all cubes in $\mathbb{R}^n$ and $f_Q := \frac{1}{|Q|}\int_Q f(X)\, dX$. As is well-known, $(H^1(\mathbb{R}^n))^* = \mathrm{BMO}(\mathbb{R}^n)$ (see [43]). The local version of $\mathrm{BMO}(\mathbb{R}^n)$ is defined as follows. A function $f \in L^2_{\rm loc}(\mathbb{R}^n)$ belongs to $\mathrm{bmo}(\mathbb{R}^n)$ if the quantity

$$\|f\|_{\mathrm{bmo}(\mathbb{R}^n)} := \sup\left\{ \sup_{Q:l(Q)\le 1}\left(\frac{1}{|Q|}\int_Q |f(X)-f_Q|^2\,dX\right)^{1/2},\right.$$

$$\left.\sup_{Q:l(Q)>1}\left(\frac{1}{|Q|}\int_Q |f(X)|^2\,dX\right)^{1/2}\right\} \tag{2.133}$$

is finite. Then $(h^1(\mathbb{R}^n))^* = \mathrm{bmo}(\mathbb{R}^n)$ (see [52]).

We now briefly review the classical Besov and Triebel–Lizorkin scales in $\mathbb{R}^n$. Recall that $\mathbb{N}_0 = \mathbb{N}\cup\{0\}$. By $\mathbb{N}_0^n$ we shall then denote the collection of all multi-indices $\alpha = (\alpha_1,\dots,\alpha_n)$ with components in $\mathbb{N}_0$. As is customary, we let $|\alpha| := \sum_{j=1}^n \alpha_j$. One convenient point of view is offered by the classical Littlewood–Paley theory (cf., e.g., [107,124]). More specifically, let Ξ be the collection of all systems $\{\zeta_j\}_{j=0}^\infty \subset S$ with the properties

(a) There exist positive constants $A,\ B,\ C$ such that

$$\begin{cases} \mathrm{supp}\,(\zeta_0)\subset\{X:\ |X|\le A\}, \\ \mathrm{supp}\,(\zeta_j)\subset\{X:\ B2^{j-1}\le|X|\le C2^{j+1}\} \quad \text{if } j=1,2,3\dots. \end{cases} \tag{2.134}$$

(b) For every multi-index α there exists a positive number c_α such that

$$\sup_{x\in\mathbb{R}^n}\sup_{j\in\mathbb{N}} 2^{j|\alpha|}|\partial^\alpha\zeta_j(x)|\le c_\alpha. \tag{2.135}$$

(c)

$$\sum_{j=0}^\infty \zeta_j(x) = 1 \quad \text{for every } x\in\mathbb{R}^n. \tag{2.136}$$

Let $s\in\mathbb{R}$ and $0<q\le\infty$ and fix some family $\{\zeta_j\}_{j=0}^\infty\in\Xi$. Also, let $\mathscr{F}$ denote the Fourier transform in $\mathbb{R}^n$. If $0<p<\infty$ then the Triebel–Lizorkin spaces are defined as

$$F_s^{p,q}(\mathbb{R}^n) := \Big\{ f\in S'(\mathbb{R}^n): \tag{2.137}$$

$$\|f\|_{F_s^{p,q}(\mathbb{R}^n)} := \Big\|\Big(\sum_{j=0}^\infty |2^{sj}\mathscr{F}^{-1}(\zeta_j\mathscr{F}f)|^q\Big)^{1/q}\Big\|_{L^p(\mathbb{R}^n)} < \infty\Big\}.$$

If $0<p\le\infty$ then the Besov spaces are defined as

$$B_s^{p,q}(\mathbb{R}^n) := \Big\{ f\in S'(\mathbb{R}^n): \tag{2.138}$$

$$\|f\|_{B_s^{p,q}(\mathbb{R}^n)} := \Big(\sum_{j=0}^\infty \|2^{sj}\mathscr{F}^{-1}(\zeta_j\mathscr{F}f)\|^q_{L^p(\mathbb{R}^n)}\Big)^{1/q} < \infty\Big\}.$$

A different choice of the system $\{\zeta_j\}_{j=0}^{\infty} \in \Xi$ yields the same spaces (2.137)–(2.138), albeit equipped with equivalent norms. Furthermore, the class of Schwartz functions in $\mathbb{R}^n$ is dense in both $B_s^{p,q}(\mathbb{R}^n)$ and $F_s^{p,q}(\mathbb{R}^n)$ provided the index $s \in \mathbb{R}$ and $0 < p, q < \infty$.

It has long been known that many classical smoothness spaces are encompassed by the Besov and Triebel–Lizorkin scales. For example,

$$C^s(\mathbb{R}^n) = B_s^{\infty,\infty}(\mathbb{R}^n), \quad 0 < s \notin \mathbb{Z}, \tag{2.139}$$

$$L^p(\mathbb{R}^n) = F_0^{p,2}(\mathbb{R}^n), \quad 1 < p < \infty, \tag{2.140}$$

$$L_s^p(\mathbb{R}^n) = F_s^{p,2}(\mathbb{R}^n), \quad 1 < p < \infty, \qquad s \in \mathbb{R}, \tag{2.141}$$

$$W^{k,p}(\mathbb{R}^n) = F_k^{p,2}(\mathbb{R}^n), \quad 1 < p < \infty, \qquad k \in \mathbb{N}, \tag{2.142}$$

$$h^p(\mathbb{R}^n) = F_0^{p,2}(\mathbb{R}^n), \quad 0 < p \leq 1, \tag{2.143}$$

$$\mathrm{bmo}(\mathbb{R}^n) = F_0^{\infty,2}(\mathbb{R}^n), \tag{2.144}$$

$$\mathrm{BMO}(\mathbb{R}^n) = \dot{F}_0^{\infty,2}(\mathbb{R}^n). \tag{2.145}$$

For a more detailed discussion of the Besov and Triebel–Lizorkin scales of spaces in $\mathbb{R}^n$, including the definition and properties of their homogeneous counterparts, $\dot{B}_s^{p,q}(\mathbb{R}^n)$ and $\dot{F}_s^{p,q}(\mathbb{R}^n)$, the interested reader is referred to [46, 47, 107, 124], and the references therein.

A useful lifting result reads as follows

$$F_s^{p,q}(\mathbb{R}^n) = (I - \Delta)^{\mu/2} F_{s+\mu}^{p,q}(\mathbb{R}^n), \qquad \forall\, p, q \in (0, \infty] \text{ and } \forall\, s, \mu \in \mathbb{R}. \tag{2.146}$$

Also, for any $k \in \mathbb{N}$,

$$\begin{aligned} F_s^{p,q}(\mathbb{R}^n) &= \big\{ f \in S'(\mathbb{R}^n) : \partial^\alpha f \in F_{s-k}^{p,q}(\mathbb{R}^n),\ \forall\, \alpha \text{ with } |\alpha| \leq k \big\} \\ &= \big\{ f \in F_{s-k}^{p,q}(\mathbb{R}^n) : \partial^\alpha f \in F_{s-k}^{p,q}(\mathbb{R}^n),\ \forall\, \alpha \text{ with } |\alpha| = k \big\} \end{aligned} \tag{2.147}$$

and

$$\begin{aligned} \|f\|_{F_s^{p,q}(\mathbb{R}^n)} &\approx \sum_{|\alpha| \leq k} \|\partial^\alpha f\|_{F_{s-m}^{p,q}(\mathbb{R}^n)} \\ &\approx \|f\|_{F_{s-k}^{p,q}(\mathbb{R}^n)} + \sum_{|\alpha| = k} \|\partial^\alpha f\|_{F_{s-k}^{p,q}(\mathbb{R}^n)}. \end{aligned} \tag{2.148}$$

In particular, for every $\alpha \in \mathbb{N}_0^n$,

$$\partial^\alpha : F_s^{p,q}(\mathbb{R}^n) \longrightarrow F_{s-|\alpha|}^{p,q}(\mathbb{R}^n) \text{ boundedly, } \forall\, p,q \in (0,\infty],\ s \in \mathbb{R}. \tag{2.149}$$

Similar results are valid for the scale of Besov spaces.

Let us also record here some useful embedding results for Besov and Triebel–Lizorkin spaces in $\mathbb{R}^n$.

Theorem 2.17. *For indices $0 < p_0 \leq p_1 \leq \infty$, $s_0, s_1 \in \mathbb{R}$, $0 < q_0 \leq q_1 \leq \infty$ satisfying $s_0 - \frac{n}{p_0} = s_1 - \frac{n}{p_1}$, the inclusion*

$$B_{s_0}^{p_0,q_0}(\mathbb{R}^n) \hookrightarrow B_{s_1}^{p_1,q_1}(\mathbb{R}^n) \tag{2.150}$$

is continuous with dense range. In addition, the same holds for the inclusion

$$F_{s_0}^{p_0,q_0}(\mathbb{R}^n) \hookrightarrow F_{s_1}^{p_1,q_1}(\mathbb{R}^n), \tag{2.151}$$

if either $0 < p_0 < p_1 < \infty$, $0 < q_0, q_1 \leq \infty$ and $s_0 - \frac{n}{p_0} = s_1 - \frac{n}{p_1}$, or $0 < p_0 = p_1 < \infty$, $0 < q_0 < q_1 \leq \infty$ and $s_0 = s_1$.

Moreover, whenever $0 < p \leq \infty$ and $s \in \mathbb{R}$, one has

$$A_s^{p,q_0}(\mathbb{R}^n) \hookrightarrow A_s^{p,q_1}(\mathbb{R}^n) \text{ whenever } 0 < q_0 \leq q_1 \leq \infty, \tag{2.152}$$

$$A_{s_0}^{p,\infty}(\mathbb{R}^n) \hookrightarrow A_{s_1}^{p,q}(\mathbb{R}^n) \text{ whenever } s_0 > s_1. \tag{2.153}$$

where $A \in \{F, B\}$. Finally,

$$B_s^{p,\min\{p,q\}}(\mathbb{R}^n) \subseteq F_s^{p,q}(\mathbb{R}^n) \subseteq B_s^{p,\max\{p,q\}}(\mathbb{R}^n) \quad \text{for } 0 < p,q \leq \infty,\ s \in \mathbb{R}^n. \tag{2.154}$$

See, e.g., [107]. Later on, we shall also need the following useful membership criterion for certain types of singular functions to Besov and Triebel–Lizorkin spaces. To state our next result recall that $(a)_+ := \max\{a, 0\}$, for every $a \in \mathbb{R}$.

Lemma 2.18. *Assume that $a > 0$, $0 < p,q \leq \infty$, $s > n(1/p - 1)_+$, and fix a function $\psi \in C_c^\infty(\mathbb{R}^n)$ with $\psi \equiv 1$ on $B(0,1)$. Then the following equivalences are true:*

$$\psi(X)|X|^a \in B_s^{p,q}(\mathbb{R}^n) \Longleftrightarrow \text{either } s < \tfrac{n}{p} + a, \text{ or } s = \tfrac{n}{p} + a \text{ and } q = \infty, \tag{2.155}$$

and

$$\psi(X)|X|^a \in F_s^{p,q}(\mathbb{R}^n) \Longleftrightarrow s < \tfrac{n}{p} + a. \tag{2.156}$$

Proof. This is a consequence of Lemma 1 on p. 44 in [107].

In the last part of this section we review some useful mapping properties of pseudodifferential operators acting on Besov and Triebel–Lizorkin scales. To set the stage, given $0 \leq \delta, \rho \leq 1$ and $m \in \mathbb{R}$, let $S^m_{\rho,\delta}$ be the class of symbols consisting of all functions $p \in C^\infty(\mathbb{R}^n \times \mathbb{R}^n)$ such that for each pair of multi-indices $\beta, \gamma \in \mathbb{N}_0^n$ there exists a finite constant $C_{\beta,\gamma} > 0$ such that

$$|\partial_\xi^\beta \partial_X^\gamma p(X, \xi)| \leq C_{\beta,\gamma}(1 + |\xi|)^{m-\rho|\beta|+\delta|\gamma|}, \tag{2.157}$$

uniformly for $(X, \xi) \in \mathbb{R}^n \times \mathbb{R}^n$. For $p \in S^m_{\rho,\delta}$ we define the pseudodifferential operator $p(X, D)$ by (with $\mathscr{F}$ denoting the Fourier transform in $\mathbb{R}^n$)

$$p(X, D)f(X) := (2\pi)^{-n} \int_{\mathbb{R}^n} e^{i\langle X,\xi\rangle} p(X, \xi)\mathscr{F} f(\xi)\, d\xi, \quad f \in S(\mathbb{R}^n), \tag{2.158}$$

and write $p(x, D) \in OPS^m_{\rho,\delta}$. The following is a consequence of [124, Theorem 6.2.2, p. 258] (cf. also [124, Remark. 3, p. 257]).

Theorem 2.19. *Let $m \in \mathbb{R}$, $0 \leq \delta < 1$ and fix $\alpha \in \mathbb{R}$, $0 < q \leq \infty$, arbitrary. Then any $T \in OPS^m_{1,\delta}$ induces a bounded, linear operator*

$$T : F^{p,q}_\alpha(\mathbb{R}^n) \longrightarrow F^{p,q}_{\alpha-m}(\mathbb{R}^n) \tag{2.159}$$

whenever $0 < p < \infty$. Moreover,

$$T : B^{p,q}_\alpha(\mathbb{R}^n) \longrightarrow B^{p,q}_{\alpha-m}(\mathbb{R}^n) \tag{2.160}$$

boundedly, whenever $0 < p \leq \infty$.

We now record two consequences of the above result, which are particularly useful for treating operators akin to the harmonic Newtonian potential operator. The first of these reads as follows (see also [64] for more details).

Corollary 2.20. *Let $a \in S'(\mathbb{R}^n)$ be a tempered distribution for which there exists some $R > 0$ such that a is smooth for $|\xi| > R$ and satisfies*

$$|(\partial^\gamma a)(\xi)| \leq C_\gamma |\xi|^{m-|\gamma|}, \qquad |\xi| > R, \quad \gamma \in \mathbb{N}_0^n \tag{2.161}$$

for some $m \in \mathbb{R}$ such that $m > -n$, and set $T := a(D)$, i.e.

$$Tf(X) := (2\pi)^{-n} \int_{\mathbb{R}^n} e^{i\langle X,\xi\rangle} a(\xi)\mathscr{F} f(\xi)\, d\xi, \qquad \forall\, f \in S(\mathbb{R}^n). \tag{2.162}$$

Finally, fix $0 < q \leq \infty$, $\alpha \in \mathbb{R}$ and $\phi, \psi \in C^\infty_c(\mathbb{R}^n)$ (viewed below as multiplication operators). Then

$$\phi T \psi : F^{p,q}_\alpha(\mathbb{R}^n) \longrightarrow F^{p,q}_{\alpha-m}(\mathbb{R}^n) \tag{2.163}$$

is a bounded operator whenever $0 < p < \infty$*. In fact, so is*

$$\phi T \psi : B_\alpha^{p,q}(\mathbb{R}^n) \longrightarrow B_{\alpha-m}^{p,q}(\mathbb{R}^n) \tag{2.164}$$

if $0 < p \leq \infty$.

To state the second corollary alluded to above, assume that

$$L = \sum_{|\gamma|=2m} A_\gamma \partial^\gamma \tag{2.165}$$

is a homogeneous, (complex, matrix-valued) constant-coefficient, differential operator of order $2m$ (with $m \in \mathbb{N}$) in $\mathbb{R}^n$, which is elliptic, in the sense that there exists a constant $c > 0$ such that

$$(-1)^m \mathrm{Re} \sum_{|\gamma|=2m} A_\gamma \xi^\gamma \geq c|\xi|^{2m}, \qquad \forall\, \xi \in \mathbb{R}^n. \tag{2.166}$$

As a consequence of Theorem 2.19 (cf. also [64] for more details) we then have:

Corollary 2.21. *If L is a homogeneous, constant coefficient, elliptic operator of order m and $\phi, \psi \in C_c^\infty(\mathbb{R}^n)$, then $\phi L^{-1} \psi$ has mapping properties similar to* (2.159)–(2.160).

2.4 Smoothness Spaces in Lipschitz Domains

Given an arbitrary open subset Ω of $\mathbb{R}^n$, denote by $f|_\Omega$ the restriction of a distribution f in $\mathbb{R}^n$ to Ω. For $0 < p, q \leq \infty$ and $s \in \mathbb{R}$ then set

$$A_s^{p,q}(\Omega) := \{f \text{ distribution in } \Omega : \exists\, g \in A_s^{p,q}(\mathbb{R}^n) \text{ such that } g|_\Omega = f\}, \tag{2.167}$$

equipped with the quasi-norm given by

$$\|f\|_{A_s^{p,q}(\Omega)} := \inf\Big\{\|g\|_{A_s^{p,q}(\mathbb{R}^n)} : g \in A_s^{p,q}(\mathbb{R}^n),\ g|_\Omega = f\Big\}, \quad f \in A_s^{p,q}(\Omega), \tag{2.168}$$

where $A \in \{B, F\}$. From the corresponding density result in $\mathbb{R}^n$, it follows that for any bounded Lipschitz domain Ω and any $p, q \in (0, \infty)$, $s \in \mathbb{R}$,

$$C^\infty(\overline{\Omega}) \hookrightarrow B_s^{p,q}(\Omega) \text{ and } C^\infty(\overline{\Omega}) \hookrightarrow F_s^{p,q}(\Omega) \quad \text{densely}. \tag{2.169}$$

Hardy, Sobolev (or Bessel potential), Hölder and bmo spaces are defined analogously, namely (recall that $\mathscr{D}'(\Omega)$ stands for the space of distributions in the open set Ω)

$$L^p_s(\Omega) := \{f \in \mathscr{D}'(\Omega) : \exists g \in L^p_s(\mathbb{R}^n) \text{ such that } g|_\Omega = f\}, \quad 1 < p < \infty,\ s \in \mathbb{R},$$
$$h^p(\Omega) := \{f \in \mathscr{D}'(\Omega) : \exists g \in h^p(\mathbb{R}^n) \text{ such that } g|_\Omega = f\}, \quad 0 < p \leq 1, \tag{2.170}$$
$$\text{bmo}(\Omega) := \{f \in L^2_{\text{loc}}(\Omega) : \exists g \in \text{bmo}(\mathbb{R}^n) \text{ such that } g|_\Omega = f\},$$

equipped, in each case, with the natural, infimum-type, (quasi-)norms. By (2.139)–(2.144), it follows that

$$C^s(\Omega) = B^{\infty,\infty}_s(\Omega), \quad 0 < s \notin \mathbb{Z}, \tag{2.171}$$

$$L^p(\Omega) = F^{p,2}_0(\Omega), \quad 1 < p < \infty, \tag{2.172}$$

$$L^p_s(\Omega) = F^{p,2}_s(\Omega), \quad 1 < p < \infty, \qquad s \in \mathbb{R}, \tag{2.173}$$

$$h^p(\Omega) = F^{p,2}_0(\Omega), \quad 0 < p \leq 1, \tag{2.174}$$

$$\text{bmo}(\Omega) = F^{\infty,2}_0(\Omega), \tag{2.175}$$

where $C^s(\Omega)$ and $L^p(\Omega)$ are, respectively, the standard Hölder and Lebesgue spaces in Ω.

We shall now record a useful characterization of the local Hardy space $h^p(\Omega)$. First, we need some notation. Fix a function $\psi \in C^\infty_c(B(0,1))$ with the property that $\int_{B(0,1)} \psi(X)\,dX = 1$ and set $\psi_t(X) := t^{-n}\psi(X/t)$ for each $t > 0$ and $X \in \mathbb{R}^n$. Then the `radial maximal function` of a distribution u in Ω is defined as

$$u^+(X) := \sup_{0<t<\text{dist}(X,\partial\Omega)} |(\psi_t * u)(X)|, \qquad X \in \Omega. \tag{2.176}$$

For $u \in \mathscr{D}'(\Omega)$, $k \in \mathbb{N}_0$ and $X \in \mathbb{R}^n$ introduce

$$u^*_{k,\Omega}(X) := \sup\{|\langle u, \psi\rangle| : \psi \in \Psi_X\} \tag{2.177}$$

where the class Ψ_X consists of all functions $\psi \in C^\infty_c(\mathbb{R}^n)$ with the property that there exists $r = r_\psi > 0$ with $\text{supp}\,\psi \subset B(X,r) \cap \Omega$ and $\|\partial^\gamma \psi\|_{L^\infty(\mathbb{R}^n)} \leq r^{-n-|\gamma|}$ for each $\gamma \in \mathbb{N}^n_0$, $|\gamma| \leq k$. For a proof of the following theorem, the reader is referred to [79, 80].

Theorem 2.22. *Let Ω be a bounded Lipschitz domain in $\mathbb{R}^n$. Fix ψ as above and define the radial maximal function as in* (2.176)*. Then, for any $0 < p \leq 1$ and any $u \in \mathscr{D}'(\Omega)$*

$$u \in h^p(\Omega) \iff u^+ \in L^p(\Omega), \tag{2.178}$$

with equivalence of quasi-norms. Furthermore, if $k \in \mathbb{N}$ and $\frac{n}{n+k} < p \leq 1$, then

$$\|u^*_{k,\Omega}\|_{L^p(\mathbb{R}^n)} \approx \|u^+\|_{L^p(\Omega)}. \tag{2.179}$$

In particular, a different choice of the function ψ affects the size of u^+ in $L^p(\Omega)$ only up to a fixed multiplicative constant. Finally, similar results are valid in the range $1 < p < \infty$ provided $h^p(\Omega)$ is replaced by $L^p(\Omega)$.

In analogy with the Hardy-based Sobolev spaces in $\mathbb{R}^n$ introduced in (2.130)–(2.131), for an open set $\Omega \subset \mathbb{R}^n$, $k \in \mathbb{N}_0$ and $0 < p \leq 1$, we set

$$h^p_k(\Omega) := \{u \in \mathscr{D}'(\Omega) : \partial^\gamma u \in h^p(\Omega), \ \forall\, \gamma \in \mathbb{N}_0^n \text{ with } |\gamma| \leq k\}, \tag{2.180}$$

equipped with the quasi-norm $\|u\|_{h^p_k(\Omega)} := \sum_{|\gamma|=k} \|\partial^\gamma u\|_{h^p(\Omega)}$, and

$$h^p_{-k}(\Omega) := \Big\{u \in \mathscr{D}'(\Omega) : \ u = \sum_{|\gamma|\leq k} \partial^\gamma u_\gamma, \ \text{ where}$$

$$u_\gamma \in h^p(\Omega), \ \forall\, \gamma \in \mathbb{N}_0^n \text{ with } |\gamma| \leq k\Big\} \tag{2.181}$$

equipped with $\|u\|_{h^p_{-k}(\Omega)} := \inf \sum_{|\gamma|\leq k} \|u_\gamma\|_{h^p(\Omega)}$, where the infimum is taken over all representations of u. Then the following higher-order smoothness version of the identification in (2.174) holds. A proof is found in [64].

Theorem 2.23. *Let Ω be a bounded Lipschitz domain in $\mathbb{R}^n$. Assume that $0 < p \leq 1$ and that $k \in \mathbb{Z}$ is either ≤ 0, or else satisfies $k > n(1/p - 1)$. Then*

$$h^p_k(\Omega) = F^{p,2}_k(\Omega). \tag{2.182}$$

We continue by recording a couple of useful lifting results on Besov and Triebel–Lizorkin spaces on bounded Lipschitz domains. The following has been proved in [83].

Proposition 2.24. *Let $1 < p, q < \infty$, $k \in \mathbb{N}$ and $s \in \mathbb{R}$. Then for any distribution u in the bounded Lipschitz domain $\Omega \subset \mathbb{R}^n$, the following implication holds:*

$$\partial^\alpha u \in A^{p,q}_s(\Omega) \ \textit{for every} \ \alpha \in \mathbb{N}_0^n \ \textit{with} \ |\alpha| = k \implies u \in A^{p,q}_{s+k}(\Omega), \tag{2.183}$$

where, as usual, $A \in \{B, F\}$.

This should be compared to the following result, proven on p. 173 in [58].

Proposition 2.25. *Let Ω be a bounded Lipschitz domain in $\mathbb{R}^n$. Suppose that $\alpha > 0$ and $1 \leq p \leq \infty$. Then*

$$u \in B^{p,p}_{1+\alpha}(\Omega) \iff u \in L^p(\Omega) \ \textit{and} \ \nabla u \in B^{p,p}_\alpha(\Omega). \tag{2.184}$$

As an application of Proposition 2.27 (discussed later) and Proposition 2.25 we record here the following.

Proposition 2.26. *Let Ω be a bounded Lipschitz domain in $\mathbb{R}^n$, $k \in \mathbb{N}_0$, and assume that $0 < s < 1$. Then*

$$B_{k+s}^{\infty,\infty}(\Omega) = C^{k+s}(\Omega), \tag{2.185}$$

where

$$C^{k+s}(\Omega) := \Big\{u \in C^k(\Omega) : \|u\|_{C^{k+s}(\Omega)} < \infty, \text{ where} \tag{2.186}$$

$$\|u\|_{C^{k+s}(\Omega)} := \sum_{j=1}^{k} \|\nabla^j u\|_{L^\infty(\Omega)} + \sum_{|\alpha|=k} \sup_{X \neq Y \in \Omega} \frac{|\partial^\alpha u(X) - \partial^\alpha u(Y)|}{|X-Y|^s}\Big\}.$$

Proof. Fix an integer $k \in \mathbb{N}_0$ and some $s \in (0,1)$. It is well known (see e.g., [107]) that $B_{k+s}^{\infty,\infty}(\mathbb{R}^n) = C^{k+s}(\mathbb{R}^n)$ and, consequently, $B_{k+s}^{\infty,\infty}(\Omega) \subseteq C^{k+s}(\Omega)$. A key observation in showing the opposite inclusion is that

$$B_s^{\infty,\infty}(\Omega) = C^s(\Omega). \tag{2.187}$$

This can be seen from the intrinsic characterization of the space $B_s^{\infty,\infty}(\Omega)$ given by Proposition 2.27 in which we take $M = 1$. Then, the inclusion $C^{k+s}(\Omega) \subseteq B_{k+s}^{\infty,\infty}(\Omega)$, and consequently (2.185), follows from Proposition 2.25. □

Moving on, recall that, in general, $(a)_+ := \max\{a, 0\}$ for each $a \in \mathbb{R}$. An intrinsic characterization of the spaces $B_s^{p,q}(\Omega)$ with $0 < p,q \leq \infty$ and $n\big(\frac{1}{p} - 1\big)_+ < s$ is given in [39] on p. 30 where the following is proved.

Proposition 2.27. *Let Ω be a bounded Lipschitz domain in $\mathbb{R}^n$. Assume that indices p, q, s satisfy $0 < p,q \leq \infty$, $n(\frac{1}{p} - 1)_+ < s$ and suppose that $M \in \mathbb{N}$ is such that $M > s$. Then*

$$u \in B_s^{p,q}(\Omega) \iff \|f\|_{L^p(\Omega)} + \Big\| t^{-s} \sup_{|h| \leq t} \|\Delta_h^M u\|_{L^p(\Omega)} \Big\|_{L^q\left((0,1), \frac{dt}{t}\right)} < \infty, \tag{2.188}$$

with naturally accompanying quasi-norm estimates, where, if $u : \Omega \to \mathbb{C}$, $M \in \mathbb{N}$ and $h \in \mathbb{R}^n$,

$$\Delta_h^M u(X) := \sum_{j=0}^{M} \binom{M}{j} (-1)^{M-j} u(X + jh) \tag{2.189}$$

if $X, X+h, \dots, X+Mh \in \Omega$, and $\Delta_h^M u(X) := 0$ otherwise.

Another, related, intrinsic characterization of the space $B_\alpha^{p,p}(\Omega)$ for the case when $n/(n+1) < p < \infty$ and $n(\frac{1}{p} - 1)_+ < \alpha < 1$ is presented in the proposition below.

Proposition 2.28. *Assume that Ω is an arbitrary nonempty open subset in $\mathbb{R}^n$ and suppose that $n/(n+1) < p < \infty$ and $n(1/p - 1)_+ < \alpha < 1$. Also, fix $k \in \mathbb{N}_0$. Then there exists a finite constant $C > 0$ such that*

$$\sum_{|\gamma|\le k}\|\partial^\gamma u\|_{L^p(\Omega)}+\sum_{|\gamma|=k}\Big(\int_\Omega\int_\Omega\frac{|(\partial^\gamma u)(X)-(\partial^\gamma u)(Y)|^p}{|X-Y|^{n+\alpha p}}\,dX\,dY\Big)^{1/p}$$
$$\le C\|u\|_{B^{p,p}_{\alpha+k}(\Omega)}. \tag{2.190}$$

Moreover, in the case when Ω is a bounded Lipschitz domain in $\mathbb{R}^n$ and $k=0$, the converse inequality also holds. Consequently, in the situation when Ω is a bounded Lipschitz domain in $\mathbb{R}^n$ and under the assumption that $n/(n+1)<p<\infty$ and $n(1/p-1)_+<\alpha<1$, one has

$$\|u\|_{B^{p,p}_\alpha(\Omega)}\approx\|u\|_{L^p(\Omega)}+\Big(\int_\Omega\int_\Omega\frac{|u(X)-u(Y)|^p}{|X-Y|^{n+\alpha p}}\,dX\,dY\Big)^{1/p}, \tag{2.191}$$

uniformly in u.

Finally, if Ω is a bounded Lipschitz domain in $\mathbb{R}^n$, $p\in(1,\infty)$, $\alpha\in(0,1)$ and $k\in\mathbb{N}_0$ then

$$\|u\|_{B^{p,p}_{\alpha+k}(\Omega)}\approx\sum_{|\gamma|\le k}\|\partial^\gamma u\|_{L^p(\Omega)} \tag{2.192}$$
$$+\sum_{|\gamma|=k}\Big(\int_\Omega\int_\Omega\frac{|(\partial^\gamma u)(X)-(\partial^\gamma u)(Y)|^p}{|X-Y|^{n+\alpha p}}\,dX\,dY\Big)^{1/p},$$

uniformly in u.

Proof. Assume that $u\in B^{p,p}_{\alpha+k}(\Omega)$. Then there exists a function $w\in B^{p,p}_{\alpha+k}(\mathbb{R}^n)$ with the property that $w\big|_\Omega=u$ and such that $\|w\|_{B^{p,p}_{\alpha+k}(\mathbb{R}^n)}\le 2\|u\|_{B^{p,p}_{\alpha+k}(\Omega)}$. In concert with a well-known characterization of the Besov quasi-norm in the entire Euclidean space (cf., e.g., [107]) this allows us to estimate

$$\sum_{|\gamma|=k}\int_\Omega\int_\Omega\frac{|(\partial^\gamma u)(X)-(\partial^\gamma u)(Y)|^p}{|X-Y|^{n+\alpha p}}\,dX\,dY$$
$$=\sum_{|\gamma|=k}\int_\Omega\int_\Omega\frac{|(\partial^\gamma w)(X)-(\partial^\gamma w)(Y)|^p}{|X-Y|^{n+\alpha p}}\,dX\,dY$$
$$\le\sum_{|\gamma|=k}\int_{\mathbb{R}^n}\int_{\mathbb{R}^n}\frac{|(\partial^\gamma w)(X)-(\partial^\gamma w)(Y)|^p}{|X-Y|^{n+\alpha p}}\,dX\,dY$$
$$\le C\|\nabla^k w\|^p_{B^{p,p}_\alpha(\mathbb{R}^n)}\le C\|w\|^p_{B^{p,p}_{\alpha+k}(\mathbb{R}^n)}$$
$$\le C\|u\|^p_{B^{p,p}_{\alpha+k}(\Omega)}. \tag{2.193}$$

From this, (2.190) readily follows.

Consider now the opposite inequality in (2.190), under the assumptions that Ω is a bounded Lipschitz domain in $\mathbb{R}^n$ and $k = 0$. In such a scenario, we invoke Corollary 1 from [110] (with $m = 1$ and $r = 1$) which gives

$$\begin{aligned}\|u\|^p_{B^{p,p}_\alpha(\Omega)} &\leq C \int\limits_{X\in\Omega} \left(\int_0^\infty \frac{1}{t^n}\Big(\int\limits_{\substack{|X-Y|\leq t\\ Y\in\Omega}} |u(X)-u(Y)|^p\, dY\Big) \frac{dt}{t^{1+\alpha p}} \right) dX + \|u\|^p_{L^p(\Omega)} \\ &\leq C \int_{X\in\Omega}\int_{Y\in\Omega} |u(X)-u(Y)|^p \left(\int_{|X-Y|}^\infty \frac{dt}{t^{n+1+\alpha p}} \right) dY\, dX + \|u\|^p_{L^p(\Omega)} \\ &\leq C \int_\Omega\int_\Omega \frac{|u(X)-u(Y)|^p}{|X-Y|^{n+\alpha p}}\, dX\, dY + \|u\|^p_{L^p(\Omega)}. \end{aligned} \tag{2.194}$$

In concert with the result from the first part of the proof, this justifies (2.191).

As regards (2.192), assume that u is a function for which

$$\sum_{|\gamma|\leq k} \|\partial^\gamma u\|_{L^p(\Omega)} + \sum_{|\gamma|=k} \Big(\int_\Omega\int_\Omega \frac{|(\partial^\gamma u)(X)-(\partial^\gamma u)(Y)|^p}{|X-Y|^{n+\alpha p}}\, dX\, dY \Big)^{1/p} < \infty. \tag{2.195}$$

Relying on (2.191), we then deduce that $\partial^\gamma u \in B^{p,p}_\alpha(\Omega)$ for every $\gamma \in \mathbb{N}^n_0$ with $|\gamma| = k$. Having established this membership, the lifting result from Proposition 2.24 applies (assuming that $1 < p < \infty$) and gives that $u \in B^{p,p}_{\alpha+k}(\Omega)$. This establishes the right-pointing inequality in (2.192). Together with (2.190), this yields (2.192), thus completing the proof of the proposition. □

The existence of an universal extension operator for Besov and Triebel–Lizorkin spaces in an arbitrary Lipschitz domain $\Omega \subset \mathbb{R}^n$ has been established by V. Rychkov in [108]. To state this result, let $\mathscr{R}_\Omega$ denote the operator of restriction to Ω, which maps distributions from $\mathbb{R}^n$ into distributions in Ω,

$$\mathscr{R}_\Omega(u) := u\Big|_\Omega, \qquad u \text{ distribution in } \mathbb{R}^n. \tag{2.196}$$

Theorem 2.29 ([108]). *Let $\Omega \subset \mathbb{R}^n$ be either a bounded Lipschitz domain, the exterior of a bounded Lipschitz domain, or an unbounded Lipschitz domain. Then there exists a linear, continuous operator E_Ω, mapping distributions in Ω into tempered distributions in $\mathbb{R}^n$, and such that whenever $0 < p, q \leq \infty$, $s \in \mathbb{R}^n$,*

$$\begin{aligned} &E_\Omega : A^{p,q}_s(\Omega) \longrightarrow A^{p,q}_s(\mathbb{R}^n) \text{ boundedly,} \\ &\text{satisfying } \mathscr{R}_\Omega(E_\Omega f) = f,\ \forall\, f \in A^{p,q}_s(\Omega), \end{aligned} \tag{2.197}$$

for $A = B$ or $A = F$, in the latter case assuming $p < \infty$.

Assume that $1 \leq p \leq \infty$ and $k \in \mathbb{N}_0$. Then the standard L^p-based Sobolev space of order k in an open set $\Omega \subseteq \mathbb{R}^n$ is defined as

$$W^{k,p}(\Omega) := \Big\{ f \in L^p(\Omega) : \partial^\gamma f \in L^p(\Omega), \ \forall \, \gamma \in \mathbb{N}_0^n \ \text{ with } \ |\gamma| \leq k \Big\}, \tag{2.198}$$

and is equipped with the norm

$$\|f\|_{W^{k,p}(\Omega)} := \sum_{|\gamma| \leq k} \|\partial^\gamma f\|_{L^p(\Omega)}. \tag{2.199}$$

In view of Theorem 2.29, for any Lipschitz domain Ω we have

$$W^{k,p}(\Omega) = F_k^{p,2}(\Omega), \quad 1 < p < \infty, \quad k \in \mathbb{N}_0. \tag{2.200}$$

It is also useful to define Sobolev spaces of negative smoothness. Specifically, given an open set $\Omega \subset \mathbb{R}^n$, $k \in \mathbb{N}$ and $p \in (1, \infty)$, let us set (with $\mathscr{D}'(\Omega)$ denoting the space of distributions in Ω)

$$\begin{aligned} W^{-k,p}(\Omega) := \Big\{ u \in \mathscr{D}'(\Omega) : \ u = \sum_{|\gamma| \leq k} \partial^\gamma u_\gamma, \ \text{ where} \\ u_\gamma \in L^p(\Omega) \ \ \forall \, \gamma \in \mathbb{N}_0^n \text{ with } |\gamma| \leq k \Big\} \end{aligned} \tag{2.201}$$

equipped with the natural norm $\|u\|_{W^{-k,p}(\Omega)} := \inf \sum_{|\gamma| \leq k} \|u_\gamma\|_{L^p(\Omega)}$, where the infimum is taken over all representations of u as in the right-hand side of (2.201).

Proposition 2.30. *Let Ω be a Lipschitz domain in $\mathbb{R}^n$. Then for every $k \in \mathbb{N}$ and $p \in (1, \infty)$ there holds*

$$W^{-k,p}(\Omega) = F_{-k}^{p,2}(\Omega), \tag{2.202}$$

with equivalent norms.

Proof. Fix $k \in \mathbb{N}$ and $p \in (1, \infty)$. From (2.146), (2.200), and (2.201) it follows that

$$\begin{aligned} F_{-k}^{p,2}(\Omega) &= \{ u\big|_\Omega : u \in F_{-k}^{p,2}(\mathbb{R}^n) \} = \{ u\big|_\Omega : u \in (I - \Delta)^k F_k^{p,2}(\mathbb{R}^n) \} \\ &= \{ (I - \Delta)^k v : v \in W^{k,p}(\Omega) \} \hookrightarrow W^{-k,p}(\Omega). \end{aligned} \tag{2.203}$$

Since $\partial^\gamma : L^p(\Omega) = F_0^{p,2}(\Omega) \to F_{-k}^{p,2}(\Omega)$ is bounded for every $\gamma \in \mathbb{N}_0^n$ with $|\gamma| \leq k$ (cf. (2.149)), it follows from (2.201) that every distribution $u \in W^{-k,p}(\Omega)$ actually belongs to $F_{-k}^{p,2}(\Omega)$. Hence, $W^{-k,p}(\Omega) \hookrightarrow F_{-k}^{p,2}(\Omega)$ and (2.202) follows.

Going further, for $0 < p,q \leq \infty$, $s \in \mathbb{R}$, we set

$$\begin{aligned} A^{p,q}_{s,0}(\Omega) &:= \{f \in A^{p,q}_s(\mathbb{R}^n) : \operatorname{supp} f \subseteq \overline{\Omega}\}, \\ \|f\|_{A^{p,q}_{s,0}(\Omega)} &:= \|f\|_{A^{p,q}_s(\mathbb{R}^n)}, \quad f \in A^{p,q}_{s,0}(\Omega), \end{aligned} \tag{2.204}$$

where, as usual, either $A = F$ and $p < \infty$ or $A = B$. Thus, $B^{p,q}_{s,0}(\Omega)$, $F^{p,q}_{s,0}(\Omega)$ are closed subspaces of $B^{p,q}_{s,0}(\mathbb{R}^n)$ and $F^{p,q}_{s,0}(\mathbb{R}^n)$, respectively. In the same vein, we also define

$$L^p_{s,0}(\Omega) := \{f \in L^p_s(\mathbb{R}^n) : \operatorname{supp} f \subseteq \overline{\Omega}\}, \quad 1 < p < \infty, \ s \in \mathbb{R}, \tag{2.205}$$

with the norms inherited from $L^p_s(\mathbb{R}^n)$. Given $\Omega \subset \mathbb{R}^n$ and a function $f : \Omega \to \mathbb{R}$, define

$$\widetilde{f}(X) := \begin{cases} f(X) & \text{if } X \in \Omega, \\ 0 & \text{if } X \in \mathbb{R}^n \setminus \Omega, \end{cases} \tag{2.206}$$

It follows that if Ω is a bounded Lipschitz domain in $\mathbb{R}^n$ and $0 < p,q < \infty$, $s \in \mathbb{R}$, then

$$\widetilde{C^\infty_c(\Omega)} \hookrightarrow A^{p,q}_{s,0}(\Omega) \ \text{densely}, \tag{2.207}$$

$$C^\infty(\overline{\Omega}) \hookrightarrow A^{p,q}_s(\Omega) \ \text{densely}, \tag{2.208}$$

where, as above, tilde denotes the extension by zero outside Ω, and A stands for either B or F. Indeed, the same proof as in the Remark 2.7 on p. 170 of [58] gives (2.207) and a minor variation of it justifies (2.208) as well.

Next, for $0 < p,q \leq \infty$ and $s \in \mathbb{R}$, we introduce the space

$$A^{p,q}_{s,z}(\Omega) := \{f \text{ distribution in } \Omega : \exists\, g \in A^{p,q}_{s,0}(\Omega) \text{ with } g|_\Omega = f\}, \tag{2.209}$$

equipped with the quasi-norm

$$\|f\|_{A^{p,q}_{s,z}(\Omega)} := \inf\Big\{\|g\|_{A^{p,q}_s(\mathbb{R}^n)} : g \in A^{p,q}_{s,0}(\Omega),\ g|_\Omega = f\Big\}, \quad f \in A^{p,q}_{s,z}(\Omega), \tag{2.210}$$

where, as before, $A = F$ and $p < \infty$ or $A = B$. In this regard, we wish to note that if $\Omega \subset \mathbb{R}^n$ is a bounded Lipschitz domain and $0 < p,q < \infty$, $s \in \mathbb{R}$, then for $A \in \{B, F\}$ there holds

$$C^\infty_c(\Omega) \hookrightarrow A^{p,q}_{s,z}(\Omega) \ \text{densely}. \tag{2.211}$$

This is a consequence of (2.207) and the fact that $\mathscr{R}_\Omega$ maps $A^{p,q}_{s,0}(\Omega)$ continuously onto $A^{p,q}_{s,z}(\Omega)$. In keeping with earlier conventions, if $1 < p < \infty$ and $s \in \mathbb{R}$, let us also consider

$$L^p_{s,z}(\Omega) := F^{p,2}_{s,z}(\Omega) = \{f \in \mathscr{D}'(\Omega) : \exists\, g \in L^p_{s,0}(\Omega) \text{ with } g|_\Omega = f\}, \tag{2.212}$$

where, as before, by $\mathscr{D}'(\Omega)$ we have denoted the set of distributions on Ω.

For further use, we make the simple yet important observation that the operator of restriction to Ω induces linear, bounded and onto mappings in the following settings

$$\mathscr{R}_\Omega : A^{p,q}_s(\mathbb{R}^n) \longrightarrow A^{p,q}_s(\Omega) \text{ and } \mathscr{R}_\Omega : A^{p,q}_{s,0}(\mathbb{R}^n) \longrightarrow A^{p,q}_{s,z}(\Omega) \tag{2.213}$$

for $0 < p, q \le \infty$, $s \in \mathbb{R}$.

Proposition 2.31 ([123]). *Assume that Ω is a bounded Lipschitz domain in $\mathbb{R}^n$, and suppose that*

$$0 < p, q \le \infty \text{ and } s > \max\left\{\tfrac{1}{p} - 1, n\left(\tfrac{1}{p} - 1\right)\right\}. \tag{2.214}$$

Then extension by zero defined in (2.206) *induces a linear and bounded operator from $B^{p,q}_{s,z}(\Omega)$ to $B^{p,q}_{s,0}(\Omega)$ and, if $p < \infty$, from $F^{p,q}_{s,z}(\Omega)$ to $F^{p,q}_{s,0}(\Omega)$. Furthermore, if*

$$0 < p, q < \infty \text{ and } \max\left\{\tfrac{1}{p} - 1, n\left(\tfrac{1}{p} - 1\right)\right\} < s < \tfrac{1}{p}, \tag{2.215}$$

this operator also maps the space $B^{p,q}_s(\Omega)$ to $B^{p,q}_{s,0}(\Omega)$, and the space $F^{p,q}_s(\Omega)$ to $F^{p,q}_{s,0}(\Omega)$ if $\min\{p, 1\} \le q$.

If $1 < p, q < \infty$ and $1/p + 1/p' = 1/q + 1/q' = 1$, then

$$\left(A^{p,q}_{s,z}(\Omega)\right)^* = A^{p',q'}_{-s}(\Omega) \quad \text{if } s > -1 + \tfrac{1}{p}, \tag{2.216}$$

$$\left(A^{p,q}_s(\Omega)\right)^* = A^{p',q'}_{-s,z}(\Omega) \quad \text{if } s < \tfrac{1}{p}. \tag{2.217}$$

Furthermore, for each $s \in \mathbb{R}$ and $1 < p, q < \infty$, the spaces $A^{p,q}_s(\Omega)$ and $A^{p,q}_{s,0}(\Omega)$ are reflexive.

There is yet another type of smoothness space which will play a significant role in this work. Specifically, for $\Omega \subset \mathbb{R}^n$ Lipschitz domain, we set

$$\mathring{A}^{p,q}_s(\Omega) := \text{the closure of } C^\infty_c(\Omega) \text{ in } A^{p,q}_s(\Omega), \quad 0 < p, q \le \infty,\ s \in \mathbb{R}, \tag{2.218}$$

where, as usual, $A = F$ or $A = B$. For every $0 < p, q < \infty$ and $s \in \mathbb{R}$, we then have

$$A^{p,q}_{s,z}(\Omega) \hookrightarrow \mathring{A}^{p,q}_s(\Omega) \hookrightarrow A^{p,q}_s(\Omega), \qquad \text{continuously.} \tag{2.219}$$

The second inclusion is trivial from (2.218), whereas the first can be justified as follows. If $f \in A^{p,q}_{s,z}(\Omega)$ then there exists $u \in A^{p,q}_{s,0}(\Omega)$ such that $\mathscr{R}_\Omega(u) = f$.

By (2.207), there exists a sequence $\{u_j\}_{j\in\mathbb{N}} \subseteq C_c^\infty(\Omega)$ with the property that $\widetilde{u}_j \to u$ in $A_s^{p,q}(\mathbb{R}^n)$ as $j \to \infty$, which then implies $u_j = \mathscr{R}_\Omega(\widetilde{u}_j) \to \mathscr{R}_\Omega(u) = f$ in $A_s^{p,q}(\Omega)$ $j \to \infty$. This proves that $f \in \mathring{A}_s^{p,q}(\Omega)$, and the desired conclusion follows easily from this.

Going further, Proposition 3.1 in [123] ensures that

$$\mathring{A}_s^{p,q}(\Omega) = A_s^{p,q}(\Omega) = A_{s,z}^{p,q}(\Omega), \qquad A \in \{F, B\}, \tag{2.220}$$

if the indices p, q, s satisfy $0 < p, q < \infty$, $\max\Big\{1/p - 1, n(1/p-1)\Big\} < s < 1/p$, and $\min\{p, 1\} \leq q < \infty$ in the case $A = F$. Other cases of interest will be considered later, in Proposition 3.15.

Recall that $(\cdot,\cdot)_{\theta,q}$ and $[\cdot,\cdot]_\theta$ denote, respectively, the interpolation brackets for the real and complex method of interpolation. As regards the latter, special care must be taken when dealing with spaces which are not Banach (which is the setting in which this method has been originally developed by A.P. Calderón in [17]) but merely quasi-Banach. Throughout the monograph, the complex method is understood in the sense described in [63, 64] where Calderón's complex method has been extended to the class of analytically convex quasi-Banach spaces. The reader is referred to [63, 64] for more details.

A proof of the following result can be found in [64].

Theorem 2.32. *Suppose Ω is a bounded Lipschitz domain in $\mathbb{R}^n$. Let $\alpha_0, \alpha_1 \in \mathbb{R}$, $\alpha_0 \neq \alpha_1$, $0 < q_0, q_1, q \leq \infty$, $0 < \theta < 1$, $\alpha = (1-\theta)\alpha_0 + \theta\alpha_1$. Then*

$$(F_{\alpha_0}^{p,q_0}(\Omega), F_{\alpha_1}^{p,q_1}(\Omega))_{\theta,q} = B_\alpha^{p,q}(\Omega), \quad 0 < p < \infty, \tag{2.221}$$

$$(B_{\alpha_0}^{p,q_0}(\Omega), B_{\alpha_1}^{p,q_1}(\Omega))_{\theta,q} = B_\alpha^{p,q}(\Omega), \quad 0 < p \leq \infty. \tag{2.222}$$

Furthermore, if $\alpha_0, \alpha_1 \in \mathbb{R}$, $0 < p_0, p_1 \leq \infty$ and $0 < q_0, q_1 \leq \infty$ are such that

$$\text{either} \quad \max\{p_0, q_0\} < \infty, \quad \text{or} \quad \max\{p_1, q_1\} < \infty, \tag{2.223}$$

then

$$\big[F_{\alpha_0}^{p_0,q_0}(\Omega), F_{\alpha_1}^{p_1,q_1}(\Omega)\big]_\theta = F_\alpha^{p,q}(\Omega), \tag{2.224}$$

where $0 < \theta < 1$, $\alpha = (1-\theta)\alpha_0 + \theta\alpha_1$, $\frac{1}{p} = \frac{1-\theta}{p_0} + \frac{\theta}{p_1}$ and $\frac{1}{q} = \frac{1-\theta}{q_0} + \frac{\theta}{q_1}$.

On the other hand, if $\alpha_0, \alpha_1 \in \mathbb{R}$, $0 < p_0, p_1, q_0, q_1 \leq \infty$ are such that

$$\min\{q_0, q_1\} < \infty, \tag{2.225}$$

then also

$$\big[B_{\alpha_0}^{p_0,q_0}(\Omega), B_{\alpha_1}^{p_1,q_1}(\Omega)\big]_\theta = B_\alpha^{p,q}(\Omega), \tag{2.226}$$

where θ, α, p, q are as above.

Finally, the same interpolation results are valid if the spaces $B^{p,q}_{\alpha}(\Omega)$, $F^{p,q}_{\alpha}(\Omega)$ *are replaced by* $B^{p,q}_{\alpha,0}(\Omega)$ *and* $F^{p,q}_{\alpha,0}(\Omega)$*, respectively.*

Let us also register here a well-known, elementary sufficient condition for membership to the Hölder scale. Specifically, if Ω is a bounded Lipschitz domain and $s \in (0,1)$ then

$$u \in L^{\infty}(\Omega) \cap C^{1}(\Omega) \ \text{ with } \ \rho^{1-s}|\nabla u| \in L^{\infty}(\Omega) \Longrightarrow u \in C^{s}(\Omega), \tag{2.227}$$

plus a natural estimate (where ρ is as in (2.57)).

We continue our discussion concerning interpolation of function spaces in Lipschitz domains by recording a result which essentially asserts that the property of being a null-solution for a fixed elliptic differential operator is preserved under (complex and real) interpolation on Besov and Triebel–Lizorkin scales.

Theorem 2.33. *Consider an elliptic, homogeneous, constant coefficient differential operator* L *and fix a bounded Lipschitz domain* Ω *in* $\mathbb{R}^n$*. Define*

$$\operatorname{Ker} L := \{u \in L^1_{loc}(\Omega) : \ Lu = 0 \ \text{ in } \ \Omega\}. \tag{2.228}$$

In addition, assume that $0 < q_0, q_1, q \le \infty$, $\alpha_0, \alpha_1 \in \mathbb{R}$, $\alpha_0 \neq \alpha_1$, $0 < \theta < 1$, *and suppose that* $\alpha = (1-\theta)\alpha_0 + \theta\alpha_1$*. Then, if* $0 < p < \infty$,

$$\Big(F^{p,q_0}_{\alpha_0}(\Omega) \cap \operatorname{Ker} L\,,\ F^{p,q_1}_{\alpha_1}(\Omega) \cap \operatorname{Ker} L\Big)_{\theta,q} = B^{p,q}_{\alpha}(\Omega) \cap \operatorname{Ker} L, \tag{2.229}$$

and if $0 < p \le \infty$,

$$\Big(B^{p,q_0}_{\alpha_0}(\Omega) \cap \operatorname{Ker} L\,,\ B^{p,q_1}_{\alpha_1}(\Omega) \cap \operatorname{Ker} L\Big)_{\theta,q} = B^{p,q}_{\alpha}(\Omega) \cap \operatorname{Ker} L. \tag{2.230}$$

Let $0 < p_0, p_1 < \infty$, $0 < q_0, q_1 \le \infty$, $\alpha_0, \alpha_1 \in \mathbb{R}$, $0 < \theta < 1$, $\alpha = (1-\theta)\alpha_0 + \theta\alpha_1$, $\frac{1}{p} = \frac{1-\theta}{p_0} + \frac{\theta}{p_1}$ *and* $\frac{1}{q} = \frac{1-\theta}{q_0} + \frac{\theta}{q_1}$*. Then*

$$\Big[F^{p_0,q_0}_{\alpha_0}(\Omega) \cap \operatorname{Ker} L\,,\ F^{p_1,q_1}_{\alpha_1}(\Omega) \cap \operatorname{Ker} L\Big]_{\theta} = F^{p,q}_{\alpha}(\Omega) \cap \operatorname{Ker} L. \tag{2.231}$$

Finally, if $\alpha_0, \alpha_1 \in \mathbb{R}$ *and* $0 < p_0, p_1, q_0, q_1 \le \infty$ *are such that* $\min\{q_0, q_1\} < \infty$*, then*

$$\Big[B^{p_0,q_0}_{\alpha_0}(\Omega) \cap \operatorname{Ker} L\,,\ B^{p_1,q_1}_{\alpha_1}(\Omega) \cap \operatorname{Ker} L\Big]_{\theta} = B^{p,q}_{\alpha}(\Omega) \cap \operatorname{Ker} L, \tag{2.232}$$

where $0 < \theta < 1$, $\alpha = (1-\theta)\alpha_0 + \theta\alpha_1$, $\frac{1}{p} = \frac{1-\theta}{p_0} + \frac{\theta}{p_1}$ *and* $\frac{1}{q} = \frac{1-\theta}{q_0} + \frac{\theta}{q_1}$*.*

See [64, Theorem 1.5].

We conclude this section with a technical result, pertaining to the membership to Besov spaces (which is going to play a significant role later), whose proof makes essential use of tools from interpolation theory.

Lemma 2.34. *Let Ω be a bounded Lipschitz domain in $\mathbb{R}^n$ and assume that*

$$\theta \in (0,1), \quad p \in (1,\infty), \quad \text{and} \quad q \in [p,\infty). \tag{2.233}$$

Also, having fixed a large geometric constant $C > 1$, for each $r > 0$ define

$$\mathcal{O}_r := \{X \in \Omega : \operatorname{dist}(X, \partial\Omega) < r/C\}. \tag{2.234}$$

Finally, pick an atlas (cf. Definition 2.1) with associated Lipschitz maps $\{\varphi_i\}_{1\leq i\leq N}$ (whose graphs therefore cover $\partial\Omega$), and fix some $r_o > 0$ which is sufficiently small relative to the quantitative characteristics of this atlas.

Then for each $r \in (0, r_o)$ there exists a finite constant $C = C(\Omega, r, p, q, \theta) > 0$ with the property that for every function $v \in C^1(\mathcal{O}_{4r})$ one has

$$\begin{aligned} \|v\|_{B^{p,q}_\theta(\mathcal{O}_r)} &\leq C\|\nabla v\|^q_{L^p(\mathcal{O}_{2r}\setminus\mathcal{O}_{r/2})} + C\|\mathcal{N} v\|^q_{L^p(\partial\Omega)} \\ &\quad + C\sum_{i=1}^N \Big[\int_{|x'|<r}\Big(\int_0^r |(\nabla v)(x', \varphi_i(x') + s)|^q\, s^{q(1-\theta+1/p)-1}\,ds\Big)^{p/q} dx'\Big]^{q/p} \end{aligned} \tag{2.235}$$

(ignoring the effect of various systems of coordinates in which the graphs of the φ_i's are considered).

Proof. As in [58], the proof relies on the so-called trace method of interpolation, which we shall briefly recall. Specifically, given a compatible couple of Banach spaces A_0, A_1 and $1 \leq q < \infty$, $\theta \in (0,1)$, set $(A_0, A_1)_{\theta,q}$ for the intermediate space obtained via the standard real interpolation method (cf., e.g., [8, Chap. 3]). Then, if $1 \leq q_0, q_1 < \infty$ are such that $1/q = (1-\theta)/q_0 + \theta/q_1$, we have

$$\begin{aligned} &\|w\|_{(A_0,A_1)_{\theta,q}} \\ &\approx \inf\Big\{\Big(\int_0^\infty \|t^\theta f(t)\|^{q_0}_{A_0}\frac{dt}{t}\Big)^{1/q_0} + \Big(\int_0^\infty \|t^\theta f'(t)\|^{q_1}_{A_1}\frac{dt}{t}\Big)^{1/q_1}\Big\}, \end{aligned} \tag{2.236}$$

uniformly for $w \in (A_0, A_1)_{\theta,q}$, where the infimum is taken over all functions

$$f : (0,\infty) \longrightarrow A_0 + A_1 \tag{2.237}$$

with the property that f is locally A_0-integrable, f' (taken in the sense of distributions) is locally A_1-integrable, and such that $\lim_{t\to 0+} f(t) = w$ in $A_0 + A_1$. See Theorem 3.12.2 on p. 73 in [8].

To proceed in earnest, fix $r > 0$ sufficiently small relative to the atlas with associated Lipschitz maps $\{\varphi_i\}_{1\leq i\leq N}$. In particular, we may assume that $\mathcal{O}_r$ is a Lipschitz domain (cf. [86] for a proof). Assuming that this is the case, for every $p \in (1,\infty)$, $q \in (0,\infty]$, and $\theta \in (0,1)$ we have (cf. (2.200) and (2.221))

$$B^{p,q}_\theta(\mathcal{O}_r) = \big(W^{1,p}(\mathcal{O}_r)\,,\,L^p(\mathcal{O}_r)\big)_{1-\theta,q}. \tag{2.238}$$

Pick an arbitrary function $v \in C^1(\mathcal{O}_{4r})$. We aim to show that the infimum of

$$\int_0^\infty \|t^{1-\theta} f(t)\|^q_{W^{1,p}(\mathcal{O}_r)}\,\frac{dt}{t} + \int_0^\infty \|t^{1-\theta} f'(t)\|^q_{L^p(\mathcal{O}_r)}\,\frac{dt}{t} \tag{2.239}$$

taken over all functions $f : (0,\infty) \to L^p(\mathcal{O}_r) + W^{1,p}(\mathcal{O}_r)$ with the property that f is locally $W^{1,p}(\mathcal{O}_r)$-integrable, f' (taken in the sense of distributions) is locally $L^p(\mathcal{O}_r)$-integrable, and satisfying $\lim\limits_{t\to 0+} f(t) = v$ in $L^p(\mathcal{O}_r) + W^{1,p}(\mathcal{O}_r)$, may be controlled by the right-hand side of (2.235). Note that since v belongs to $W^{1,p}(\mathcal{U})$ for any relatively compact open subset $\mathcal{U}$ of Ω, we only need to prove the corresponding estimate for a small, fixed neighborhood of a point on the boundary. Since the domain Ω is Lipschitz, it suffices to do so in the case when the boundary point is the origin and when $B_r(0) \cap \partial\Omega$ is part of the graph of a Lipschitz function φ with $\varphi(0) = 0$ (where φ is one of the φ_i's considered earlier). In addition, we may assume that

$$\operatorname{dist}\big((x',x_n),\partial\Omega\big) \approx x_n - \varphi(x'), \quad \forall\, X = (x',x_n) \in B_r(0). \tag{2.240}$$

Next, choose

$$\begin{aligned} &\eta \in C^\infty_c(B_r(0)),\ |\eta| \leq 1, && &&\xi \in C^\infty_c((-r,r)),\ |\xi| \leq 1,\\ &\eta(X) = 1 \text{ for } |X| \leq r/2, &&\text{and} &&\xi(t) = 1 \text{ for } |t| < r/2. \end{aligned} \tag{2.241}$$

If we now consider the function f such that

$$(f(t))(X) := \eta(X)v(x',x_n+t)\xi(t) \ \text{ if } \ X = (x',x_n), \tag{2.242}$$

then clearly the function f is locally $W^{1,p}(\mathcal{O}_r)$-integrable, f' is locally $L^p(\mathcal{O}_r)$-integrable, and we have $\lim\limits_{t\to 0+} f(t) = \eta v$. Thus, it is enough to bound (2.239) for this choice of f by the right-hand side of (2.235). To this end, first we note that since $q(1-\theta)-1 > -1$ (and Ω has finite measure),

$$\begin{aligned} I &:= \int_0^r \Big(\int_{B_{2r}(0)\cap\Omega} |v(x',x_n)|^p\,dx'dx_n\Big)^{q/p} t^{q(1-\theta)-1}\,dt\\ &\leq C\,\|v\|^q_{L^p(\Omega)} \leq C\,\|v\|^q_{L^{pn/(n-1)}(\Omega)} \leq C\,\|\mathcal{N}(v)\|^q_{L^p(\partial\Omega)}, \end{aligned} \tag{2.243}$$

where the last step uses (2.42). Given the goal we have in mind, the estimate just derived suits our purposes. Second, we have

$$
\begin{aligned}
II &:= \int_0^\infty \Big(\int_{\mathcal{O}_r} \Big(|\eta(x',x_n)||\nabla v(x',x_n+t)||\xi(t)|t^{1-\theta}\Big)^p\,dx'dx_n\Big)^{q/p}\frac{dt}{t} \\
&\le C\int_0^r t^{q(1-\theta)}\Big(\int_{|x'|<r}\int_t^r |\nabla v(x',\varphi(x')+s)|^p\,dx'\,ds\Big)^{q/p}\frac{dt}{t} \\
&= C\int_0^r \Big(\int_t^r h(s)\,ds\Big)^{q/p} t^{q(1-\theta)-1}\,dt, \qquad (2.244)
\end{aligned}
$$

where we have set

$$
h(s) := \int_{|x'|<r} |\nabla v(x',\varphi(x')+s)|^p\,dx'. \qquad (2.245)
$$

To continue, recall Hardy's inequality (cf. [119, p. 272])

$$
\Big(\int_0^\infty \Big(\int_t^\infty g(s)\,ds\Big)^\beta t^{\alpha-1}\,dt\Big)^{1/\beta} \le \frac{\beta}{\alpha}\Big(\int_0^\infty (sg(s))^\beta s^{\alpha-1}\,ds\Big)^{1/\beta}, \qquad (2.246)
$$

which holds for $g \ge 0$ measurable, $\beta \ge 1$ and $\alpha > 0$. When applied with $g := h\chi_{[0,r]}$, $\alpha := q(1-\theta)$ and $\beta := q/p$, in the case when $q \ge p$ this inequality gives

$$
\begin{aligned}
&\int_0^r \Big(\int_t^r h(s)\,ds\Big)^{q/p} t^{q(1-\theta)-1}\,dt \\
&\le C\int_0^\infty \Big(sh(s)\chi_{[0,r]}(s)\Big)^{q/p} s^{q(1-\theta)-1}\,ds \\
&= C\int_0^r h(s)^{q/p}s^{q-1}\,ds \qquad (2.247) \\
&\le C\Big[\int_{|x'|<r}\Big(\int_0^r |\nabla v(x',\varphi(x')+s)|^q s^{q(1-\theta+1/p)-1}\,ds\Big)^{p/q}dx'\Big]^{q/p},
\end{aligned}
$$

where the last step in (2.247) uses Minkowski's inequality, which once again requires that $q \ge p$. Since the format of the last term above suits our goals, this completes the proof of the lemma. □

2.5 Weighted Sobolev Spaces in Lipschitz Domains

Fix a nonempty, open, proper subset Ω of $\mathbb{R}^n$ and, as in (2.57), denote by ρ the distance function to the boundary of Ω. For each $p \in [1,\infty]$, $a \in (-1/p, 1-1/p)$, and $k \in \mathbb{N}_0$, then set

$$W_a^{k,p}(\Omega) := \Big\{u : \Omega \to \mathbb{R} : u \text{ locally integrable, and} \qquad (2.248)$$
$$\|u\|_{W_a^{k,p}(\Omega)} := \sum_{|\alpha|\le k}\Big(\int_\Omega |\partial^\alpha u(X)|^p \rho(X)^{ap}\,dX\Big)^{1/p} < \infty\Big\}.$$

Let us assume that $\Omega \subset \mathbb{R}^n$ is a bounded Lipschitz domain. Then it is not difficult to check that there exists some open, relatively compact set $\mathscr{O} \subset \Omega$ for which

$$\|u\|_{W_a^{k,p}(\Omega)} \approx \sum_{|\alpha|=k}\Big(\int_\Omega |\partial^\alpha u(X)|^p \rho(X)^{ap}\,dX\Big)^{1/p} + \|u\|_{L^p(\mathscr{O})}. \qquad (2.249)$$

In relation to the standard Besov scale in $\mathbb{R}^n$, we would like to point out that, thanks to Theorem 4.1 in [58] on the one hand, and Theorems 1.4.2.4 and 1.4.4.4 in [53] on the other, for each $p \in (1,\infty)$, $s \in (0,1)$ and $k \in \mathbb{N}_0$ we have:

$$\begin{aligned} a &:= 1 - s - \tfrac{1}{p} \in (0, 1-1/p) \Longrightarrow W_a^{k,p}(\Omega) \hookrightarrow B^{p,p}_{k-1+s+1/p}(\Omega), \\ a &:= 1 - s - \tfrac{1}{p} \in (-1/p, 0) \Longrightarrow B^{p,p}_{k-1+s+1/p}(\Omega) \hookrightarrow W_a^{k,p}(\Omega). \end{aligned} \qquad (2.250)$$

Of course, $W_a^{k,p}(\Omega)$ is just the classical Sobolev space $W^{k,p}(\Omega)$ (cf. (2.198)) when $a = 0$.

Next, let L be a homogeneous, elliptic differential operator of even order with (possibly matrix-valued) constant coefficients, say

$$L := \sum_{|\alpha|=|\beta|=m} \partial^\alpha A_{\alpha\beta}\,\partial^\beta, \qquad (2.251)$$

and fix an open set $\Omega \subset \mathbb{R}^n$. Denote by $\operatorname{Ker} L$ the space of locally integrable functions u satisfying $Lu = 0$ in Ω. Also, recall that given $j \in \mathbb{N}_0$, ∇^j stands for the vector of all mixed-order partial derivatives of order j. Then, for $0 < p \le \infty$ and $s \in \mathbb{R}$, denote by $\mathbb{H}^p_s(\Omega; L)$ the space of functions $u \in \operatorname{Ker} L$ subject to the size/smoothness condition

$$\|u\|_{\mathbb{H}^p_s(\Omega;L)} := \Big\|\rho^{\langle s\rangle - s}|\nabla^{\langle s\rangle}u|\Big\|_{L^p(\Omega)} + \sum_{j=0}^{\langle s\rangle-1}\|\nabla^j u\|_{L^p(\Omega)} < \infty. \qquad (2.252)$$

Here and elsewhere, given $s \in \mathbb{R}$ we set

$$\langle s\rangle := \begin{cases} s \text{ if } s \in \mathbb{N}_0, \\ [s] + 1 \text{ if } s > 0,\ s \notin \mathbb{N}, \\ 0 \text{ if } s < 0, \end{cases} \qquad (2.253)$$

where $[\cdot]$ is the integer-part function. That is, $\langle s\rangle$ is the smallest nonnegative integer greater than or equal to s.

The above weighted Sobolev spaces behave reasonably well under real and complex interpolation. Specifically, with $(\cdot,\cdot)_{\theta,p}$ and $[\cdot,\cdot]_\theta$ standing, respectively, for the real and the complex method of interpolation, we have:

Theorem 2.35. *Let L be a homogeneous, constant-coefficient, elliptic differential operator and suppose that Ω is a bounded Lipschitz domain in $\mathbb{R}^n$, $s_0, s_1 \in \mathbb{R}$, $0 < p < \infty$, $0 < \theta < 1$ and $s = (1-\theta)s_0 + \theta s_1$. Then the following interpolation formulas hold with equivalent norms:*

$$\left(\mathbb{H}^p_{s_0}(\Omega;L), \mathbb{H}^p_{s_1}(\Omega;L)\right)_{\theta,p} = \mathbb{H}^p_s(\Omega;L), \tag{2.254}$$

$$\left[\mathbb{H}^p_{s_0}(\Omega;L), \mathbb{H}^p_{s_1}(\Omega;L)\right]_{\theta} = \mathbb{H}^p_s(\Omega;L). \tag{2.255}$$

When the Lipschitz domain in question is star-like this result has been established in [64] and the slightly more general case recorded above follows from that.

For further reference, let us now record a well-known interior estimate and a reverse Hölder inequality (cf., e.g., [122]). To state this result, we make the convention that a barred integral indicates averaging. One instance, frequently used throughout the paper, is as follows. For a given measurable set $S \subset \partial\Omega$ we define

$$⨍_S f(X)\,d\sigma(X) := \frac{1}{|S|}\int_S f(X)\,d\sigma(X), \qquad \text{where } |S| := \sigma(S). \tag{2.256}$$

Lemma 2.36. *Let L be an elliptic differential operator as in* (2.251) *and assume that Ω is an arbitrary nonempty open set in $\mathbb{R}^n$. Then for each function $u \in \operatorname{Ker} L$, $0 < p < \infty$, $\lambda \in (0,1)$, $k \in \mathbb{N}_0$, $X \in \Omega$, and $0 < r < \rho(X)$,*

$$\sup_{Z\in B(X,\lambda r)} |\nabla^k u(Z)| \le \frac{C}{r^k}\left(⨍_{B(X,r)} |u(Y)|^p\,dY\right)^{1/p}, \tag{2.257}$$

where $C = C(L,p,k,\lambda,n) > 0$ is a finite constant.

Moreover, for any number $q \in (0,\infty)$ there exists a finite constant $C = C(L,p,q,\lambda,n) > 0$ such that, with p, λ, X, r as before, one has

$$\left(⨍_{B(X,\lambda r)} |u(Y)|^q\,dY\right)^{\frac{1}{q}} \le C\left(⨍_{B(X,r)} |u(Y)|^p\,dY\right)^{\frac{1}{p}}, \tag{2.258}$$

for every $u \in \operatorname{Ker} L$.

These local estimates should be contrasted with the following *global* weighted interior estimates and reverse interior estimates, which are going to be relevant shortly (more general results of this type have been obtained in [85]).

Lemma 2.37. *Assume that L is a constant coefficient elliptic differential operator as in* (2.251) *and that Ω is a bounded Lipschitz domain in $\mathbb{R}^n$. Also, fix $k \in \mathbb{N}_0$, $0 < p < \infty$, and $s \in \mathbb{R}$ with $sp > -1$. Then there exists a relatively compact subset $\mathcal{O}$ of Ω and $C > 0$ such that*

$$\Big(\int_\Omega |u(X)|^p \rho(X)^{sp}\,dX\Big)^{1/p} \leq C\Big[\Big(\int_\Omega |\nabla^k u(X)|^p \rho(X)^{(s+k)p}\,dX\Big)^{1/p} + \sup_{X\in\mathcal{O}} |u(X)|\Big], \tag{2.259}$$

uniformly for $u \in \operatorname{Ker} L$.

Lemma 2.38. *Let Ω be a Lipschitz domain in $\mathbb{R}^n$ and assume that L is a constant coefficient elliptic differential operator as in* (2.251)*. Also, fix $0 < p < \infty$, $s \in \mathbb{R}$, and $k \in \mathbb{N}_0$. Then for any $u \in \operatorname{Ker} L$,*

$$\Big(\int_\Omega |\nabla^k u(X)|^p \rho(X)^{(s+k)p}\,dX\Big)^{1/p} \leq C\Big(\int_\Omega |u(X)|^p \rho(X)^{sp}\,dX\Big)^{1/p}, \tag{2.260}$$

where $C = C(L, \Omega, p, s, k) > 0$ is a finite constant.

In particular, the above lemmas are useful in establishing the following result.

Corollary 2.39. *Let Ω be a bounded Lipschitz domain in $\mathbb{R}^n$ and assume that L is a constant coefficient elliptic differential operator as in* (2.251)*. Also, fix $s \in \mathbb{R}$ along with $p \in (0, \infty)$. Then there exists an open, relatively compact set $\mathcal{O}$ in Ω with the following property. If $r \in \mathbb{N}_0$ is such that*

$$s - \tfrac{1}{p} < r \leq \langle s \rangle \tag{2.261}$$

then, with ρ as in (2.57)*, there holds*

$$\|u\|_{\mathbb{H}^p_s(\Omega;L)} \approx \big\|\rho^{r-s}\,|\nabla^r u|\big\|_{L^p(\Omega)} + \|u\|_{L^p(\mathcal{O})}, \tag{2.262}$$

uniformly for $u \in \operatorname{Ker} L$.

Proof. The equivalence in (2.262) follows from (2.252), Lemmas 2.37, and 2.38. □

A related result is contained in the proposition below.

Proposition 2.40. *Suppose that Ω is a bounded Lipschitz domain in $\mathbb{R}^n$ and let L be a constant coefficient elliptic differential operator as in* (2.251)*. Then, for each $1 \leq p \leq \infty$, $a \in (-1/p, 1 - 1/p)$ and $k \in \mathbb{N}_0$,*

$$W^{k,p}_a(\Omega) \cap \operatorname{Ker} L = \mathbb{H}^p_{k-a}(\Omega; L). \tag{2.263}$$

Proof. Consider first the case when $1 \leq p < \infty$, $a \in (-1/p, 1-1/p)$ and $k \in \mathbb{N}_0$. Then the identity (2.263) is a consequence of Corollary 2.39 with $r = k$ and $s = k - a$, in which case (2.261) holds. When $p = \infty$, (2.263) follows directly from definitions. □

The relationship between the spaces $\mathbb{H}^p_s(\Omega; L)$ and the classical Triebel–Lizorkin scale is clarified in the theorem below. This extends earlier work in [58] where the authors have dealt with the case $p, q \in [1, \infty]$, $s > 0$, and $L = \Delta$, and in [2] where the case $p, q \in [1, \infty]$, $s > 0$, and $L = \Delta^2$ has been treated. The present version further builds on, and completes work in, [64].

Theorem 2.41. *Let L be a constant coefficient elliptic differential operator as in (2.251) and let Ω be a bounded Lipschitz domain in $\mathbb{R}^n$. Then for each $s \in \mathbb{R}$ and $p, q \in (0, \infty)$,*

$$\mathbb{H}^p_s(\Omega; L) = F^{p,q}_s(\Omega) \cap \operatorname{Ker} L. \tag{2.264}$$

As a corollary,

$$F^{p,q}_s(\Omega) \cap \operatorname{Ker} L = B^{p,p}_s(\Omega) \cap \operatorname{Ker} L \tag{2.265}$$

whenever $s \in \mathbb{R}$ and $p, q \in (0, \infty)$.

Proof. We divide the proof into a number of steps, starting with the following claim.

Step 1. *For each function u which satisfies $Lu = 0$ in Ω, one has*

$$u \in L^p(\Omega) \Longleftrightarrow u \in h^p(\Omega), \qquad \forall\, p \in (0, 1]. \tag{2.266}$$

Let us deal with the left-to-right implication in (2.266). Since membership to $h^p(\Omega)$ is a local property, it suffices to check that every $X_o \in \partial\Omega$ has an open neighborhood $\mathcal{O} \subset \mathbb{R}^n$ such that $u \in h^p(\mathcal{O} \cap \Omega)$. To see this, identify X_o with the origin in $\mathbb{R}^n$ and consider a Lipschitz function $\varphi : \mathbb{R}^{n-1} \to \mathbb{R}$ such that $\varphi(0) = 0$ and for which there exist $M, R > 0$ with

$$\Sigma_R := \{(x', \varphi(x')) : x' \in \mathbb{R}^{n-1},\ |x'| < R\} \subset \partial\Omega, \tag{2.267}$$

$$D_{R,M} := \{X + te_n : x \in \Sigma_R,\ 0 < t < 2M\} \subset \Omega.$$

As pointed out before, it suffices to check that $u \in h^p(D_{R,M})$. To this end, note that there exists a vertical, circular, truncated cone Γ with vertex at $0 \in \mathbb{R}^n$ such that the cone $\Gamma(X) := X + \Gamma$ is contained in $\overline{\Omega}$ whenever $X \in \overline{D_{R,M}}$. We fix $\psi \in C^\infty_c(B(0,1))$ such that $\int_{B(0,1)} \psi(X)\, dX = 1$ and set $\psi_t(X) := t^{-n}\psi(X/t)$ for each $t > 0$ and each $X \in \mathbb{R}^n$. Furthermore, we shall assume that $\operatorname{supp}\psi \subset B(0,1) \cap (-\Gamma)$. Then, for each $X \in D_{R,M}$ and $0 < t < \operatorname{dist}(X, \partial D_{R,M})$, we may write

$$
\begin{aligned}
|(\psi_t * u)(X)| &= \left| \int_{\Omega} u(X-Y)\psi_t(Y)\,dY \right| \\
&\leq \|u\|_{L^\infty(\Gamma(X)\cap D_{R,M})} \int |\psi_t(Y)|\,dY \leq \mathcal{N}u(X), \qquad (2.268)
\end{aligned}
$$

where $\mathcal{N}$ is a version of the non-tangential maximal operator defined by the formula

$$
(\mathcal{N}u)(X) := \|u\|_{L^\infty(\Gamma(X))}, \qquad X \in \overline{D_{R,M}}. \tag{2.269}
$$

Consequently,

$$
u^+(X) \leq \mathcal{N}u(X) \quad \text{for every} \quad X \in D_{R,M}, \tag{2.270}
$$

where, in the present context,

$$
u^+(X) := \sup_{0<t<\operatorname{dist}(X,\partial D_{R,M})} |(\psi_t * u)(X)|, \qquad X \in D_{R,M}. \tag{2.271}
$$

Next, we propose to show that

$$
\|\mathcal{N}u\|_{L^p(D_{R,M})} \leq C\,\|u\|_{L^p(\Omega)}. \tag{2.272}
$$

In the proof of (2.272) we shall adapt an argument from [120]. Begin by considering the Whitney decomposition of Ω, i.e. a countable family of balls $\{B_j\}_j$ which cover Ω, whose concentric doubles have finite overlap, and such that for each j there holds diam $B_j \approx$ dist $(B_j, \partial\Omega)$ (cf. [119]). Then Fatou's lemma implies

$$
\int_{D_{R,M}} \mathcal{N}(|u|^p)(X)\,dX \leq C \sum_j \int_{D_{R,M}} \mathcal{N}(\chi_j |u|^p)(X)\,dX, \tag{2.273}
$$

where χ_j denotes the characteristic function of the ball B_j. Next, we set

$$
\widehat{B}_j := \{X \in D_{R,M} : \Gamma(X) \cap B_j \neq \emptyset\} \tag{2.274}
$$

and note that $\mathcal{N}(\chi_j|u|^p) = 0$ on $D_{R,M} \setminus \widehat{B}_j$ and that $\|\mathcal{N}(\chi_j|u|^p)\|_{L^\infty(\widehat{B}_j)} \leq \|u\|^p_{L^\infty(B_j)}$. Also, $|\widehat{B}_j| \leq C|B_j|$. Consequently,

$$
\int_{D_{R,M}} \mathcal{N}(\chi_j|u|^p)(X)\,dX \leq C|B_j| \sup_{X\in B_j} |u(X)|^p \leq C \int_{2B_j} |u(X)|^p\,dX, \tag{2.275}
$$

where the last inequality follows from interior estimates for u (cf. Lemma 2.36; here we make use of the fact that $Lu = 0$ in Ω). Recalling that no point of Ω is contained in more than a fixed number of the balls $2B_j$, (2.272) follows from (2.273) and (2.275). Now, (2.270) and (2.272) give that $u^+ \in L^p(D_{R,M})$ so that $u \in h^p(D_{R,M})$, as desired. This proves the left-to-right implication in (2.266).

To see the converse implication, assume that $u \in h^p(\Omega)$ is arbitrary. Fix a function $\psi \in C_c^\infty(B(0,1))$ such that $\int_{B(0,1)} \psi(X)\,dX = 1$ and write

$$|u(X)| = \lim_{t\to 0^+} |(u * \psi_t)(X)| \leq u^+(X), \qquad \text{a.e. } X \in \Omega, \tag{2.276}$$

by Lebesgue's Differentiation Theorem applied to the locally integrable function u. Since $u^+ \in L^p(\Omega)$, this forces $u \in L^p(\Omega)$, as wanted.

`Step 2.` *We claim that for every $p \in (0,\infty)$ and $s \in \mathbb{R}$ there holds*

$$\mathbb{H}_s^p(\Omega; L) \hookrightarrow F_s^{p,2}(\Omega) \cap \operatorname{Ker} L. \tag{2.277}$$

Fix $p \in (0,\infty)$. The plan is to check that (2.277) holds whenever $s \in \mathbb{Z}$ with $|s|$ sufficiently large. Then the fact that (2.277) holds for the larger range $s \in \mathbb{R}$ follows by interpolation, based on Theorem 2.35 and Theorem 2.33. To begin with, we observe that

$$\operatorname{Ker} L \cap L^p(\Omega) \hookrightarrow h^p(\Omega), \qquad 0 < p < \infty. \tag{2.278}$$

This is, of course, trivial if $1 < p < \infty$ and it follows from Step 1 if $0 < p \leq 1$. Consider first the case when $s \in \mathbb{N}$ satisfies $s > n\left(\frac{1}{p} - 1\right)_+$. Now, the fact that the function $u \in \mathbb{H}_s^p(\Omega; L)$ entails $u, \nabla^s u \in L^p(\Omega) \cap \operatorname{Ker} L$ so $u, \nabla^s u \in h^p(\Omega)$ by (2.278). Thus, $u \in h_s^p(\Omega) = F_s^{p,2}(\Omega)$ by Theorem 2.23 (cf. also its proof in [64]) and this justifies the embedding (2.277) in the case we are considering.

Next, we propose to prove (2.277) when $s \in \mathbb{Z}$ is negative. For the sake of clarity, let us deal with the situation when $s := -1$ and then indicate the alterations necessary to treat the general case. Thus, fix $u \in \mathbb{H}_{-1}^p(\Omega; L)$ so that $Lu = 0$ in Ω and $u, \rho \cdot u \in L^p(\Omega)$ (with ρ as in (2.57)). Our goal is to show that $u \in F_{-1}^{p,2}(\Omega)$. Retaining the notation introduced in the discussion in Step 1, it suffices to prove a local version of this claim, i.e., that $u \in F_{-1}^{p,2}(D_{R,M})$. To this end, for each given function w defined in $D_{R,M}$ we set

$$w_1(X) := -\int_0^M w(X + te_n)\,dt, \qquad X \in D_{R,M/2}, \tag{2.279}$$

and, inductively, $w_{k+1} := (w_k)_1$ for each $k \in \mathbb{N}$. Note that in the case when $w \in \operatorname{Ker} L$ it follows that $Lw_k = 0$ in $D_{R,M/2}$ for every $k \in \mathbb{N}$. Returning to the mainstream discussion, observe that

$$\partial_n u_1(X) = u(X) - u(X + Me_n), \qquad X \in D_{R,M/2}. \tag{2.280}$$

In particular, granted the current assumptions on the function u, we may conclude that, on the one hand, $\rho\,\partial_n u_1 \in L^p(D_{R,M/2})$. On the other hand, Lemma 2.37 (with $s := 0$) yields

$$\int_{D_{R,M/2}} |u_1(X)|^p\,dX \leq C\int_{D_{R,M/2}} \rho(X)^p |\partial_n u_1(X)|^p\,dX + \text{l.o.t.} < \infty, \tag{2.281}$$

where "l.o.t." (lower order terms) are expressions which can be dominated in terms of the supremum of $|u|$ on compact subsets of Ω. As such, $v \in L^p(D_{R,M/2})$ and, consequently, $u_1 \in h^p(D_{R,M/2})$ by Step 1. As a result, $\partial_n u_1 \in h^p_{-1}(D_{R,M/2})$ which further implies that $u \in h^p_{-1}(D_{R,M/2})$ as well. Thus, $u \in F^{p,2}_{-1}(D_{R,M/2})$ by Theorem 2.23, as desired.

When $s = -k$, $k = 2, 3, \ldots$, we work with the function u_k which satisfies the identity $\partial_n^k u_k = u + \text{l.o.t.}$ in $D_{R,M}$ and, instead of (2.281), we have

$$\int_{D_{R,M/2}} |u_k(X)|^p\,dX \leq C\int_{D_{R,M/2}} \rho(X)^{kp} |\partial_n^k u_k(X)|^p\,dX + \text{l.o.t.} < \infty, \tag{2.282}$$

which serves our purposes well. This finishes the proof of (2.277) when $s \in \mathbb{R}$ and $0 < p < \infty$.

Step 3. *We claim that if $p \in (0,\infty)$ and $s \in \mathbb{N}$ satisfies $s > n(1/p-1)_+$ then*

$$F^{p,2}_s(\Omega) \cap \operatorname{Ker} L \hookrightarrow \mathbb{H}^p_s(\Omega; L). \tag{2.283}$$

In the case we are currently considering we have $F^{p,2}_s(\Omega) = h^p_s(\Omega)$, thanks to Theorem 2.23. Consider now an arbitrary $u \in h^p_s(\Omega) \cap \operatorname{Ker} L$. Then $u, \partial^s u \in L^p(\Omega)$ by Step 1 and, ultimately, $u \in \mathbb{H}^p_s(\Omega; L)$, given that $s \in \mathbb{N}$. Hence, (2.283) follows.

Step 4. *We claim that* (2.283) *holds whenever $p \in (0,\infty)$ and $s \in \mathbb{Z} \setminus \mathbb{N}_0$.* In this scenario, fix $\xi \in C^\infty_c(\mathbb{R}^n)$ such that $\xi \equiv 1$ in a neighborhood of $\overline{\Omega}$, and select some $N \in \mathbb{N}$ sufficiently large so that $s + 2Nm > n(1/p-1)_+$ where m is as in (2.251). In order to continue, let Π_L be the Newtonian potential operator defined as in (2.162) with $a(\xi) := L(-i\xi)^{-1}$. Also, let E_Ω denote Rychkov's extension operator from Theorem 2.29. Then, given an arbitrary $u \in F^{p,2}_s(\Omega) \cap \operatorname{Ker} L$, it follows that (with ξ viewed as a multiplication operator)

$$v := \Big[(\xi\Pi_L\xi)^N(E_\Omega u)\Big]\Big|_\Omega \tag{2.284}$$

$$\Longrightarrow v \in F^{p,2}_{s+2Nm}(\Omega) \cap \operatorname{Ker} L^{N+1} \hookrightarrow \mathbb{H}^p_{s+2Nm}(\Omega; L^{N+1}),$$

where the inclusion is a consequence of our assumptions on N and Step 3. From the implication in (2.284) we further deduce that $v \in \mathbb{H}^p_{s+2Nm}(\Omega; L^{N+1})$ which, in turn, entails $\nabla^{s+2Nm} v \in L^p(\Omega)$. In concert with Lemma 2.38, this allows us to conclude that $\rho^{-s}\nabla^{2Nm} v \in L^p(\Omega)$. Since $L^N v = u$ and L^N is a constant coefficient differential operator of order $2Nm$, this ultimately yields $\rho^{-s} u \in L^p(\Omega)$. With this in hand, the conclusion that $u \in \mathbb{H}^p_s(\Omega; L)$ readily follows from the definition of $\mathbb{H}^p_s(\Omega; L)$ and this finishes the proof of (2.283) in the case when $p \in (0,\infty)$ and $s \in \mathbb{Z} \setminus \mathbb{N}_0$.

Step 5. *The inclusion* (2.283) *holds whenever* $p \in (0,\infty)$ *and* $s \in \mathbb{R}$. Indeed, this is a consequence of Step 3, Step 4, Theorem 2.33, and Theorem 2.35.

Step 6. *If* $p \in (0,\infty)$ *and* $s \in \mathbb{R}$ *then*

$$\mathbb{H}^p_s(\Omega; L) = F^{p,2}_s(\Omega) \cap \operatorname{Ker} L, \tag{2.285}$$

$$\mathbb{H}^p_s(\Omega; L) = F^{p,p}_s(\Omega) \cap \operatorname{Ker} L. \tag{2.286}$$

Formula (2.285) is an immediate consequence of the inclusions proved in Step 2 and Step 5. Formula (2.286) then follows from (2.285), formula (2.254) in Theorem 2.35, and formula (2.229) (with $p = q$).

Step 7. *We claim that if*

$$k \in \mathbb{N}_0, \quad 0 < p < \infty, \quad 0 < q \le \infty, \quad n\Big(\tfrac{1}{\min\{p,q\}} - 1\Big)_+ - k < \alpha < 1 \tag{2.287}$$

and if $u \in L^1_{loc}(\Omega)$ *satisfies* $Lu = 0$ *in* Ω *as well as*

$$\rho^{1-\alpha}|\nabla^{k+1}u| + |u| \in L^p(\Omega), \tag{2.288}$$

then necessarily $u \in F^{p,q}_{\alpha+k}(\Omega)$, *and a naturally accompanying estimate holds.*

The starting point is the observation that, as a consequence of [110, Corollary 1 p. 398], if

$$0 < p < \infty, \qquad 0 < q \le \infty, \qquad s > n\Big(\tfrac{1}{\min\{p,q\}} - 1\Big)_+ \tag{2.289}$$

and if N is an integer such that $N > s$ then

$$\|u\|_{F^{p,q}_s(\Omega)} \approx \|u\|_{L^p(\Omega)} + \left\| \left(\int_0^{\frac{1}{2}\rho(\cdot)} \big[\operatorname{osc}^{N-1}(u,\cdot,t)\big]^q \,\frac{dt}{t^{1+sq}} \right)^{1/q} \right\|_{L^p(\Omega)}, \tag{2.290}$$

where, generally speaking, for a locally integrable function u in Ω, each integer $M \in \mathbb{N}_0$, each point $X \in \Omega$, and each scale $0 < t < \operatorname{dist}(X, \partial\Omega)$, the M-th order oscillation of u is defined by

$$\operatorname{osc}^M(u, X, t) := \inf_{P \in \mathscr{P}_M} \fint_{B_t(X)} |u(Y) - P(Y)|\, dY, \tag{2.291}$$

where $\mathscr{P}_M$ stands for the space of all polynomials of degree $\le M$ in $\mathbb{R}^n$.

In order to estimate the $(N-1)$-th order oscillation of u, we shall employ a higher-order Poincaré-type inequality, proved in [35, Theorem 3.4, p. 18] to the effect that there exists $C = C(n,N) > 0$ such that for any ball B_R and any $u \in W^{N,1}(B_R)$ one can find $P \in \mathscr{P}_{N-1}$ with the property that

$$\int_{B_R} |u(Y) - P(Y)|\,dY \leq C\,R^N \sum_{|\gamma|=N} \int_{B_R} |\partial^\gamma u(Y)|\,dY. \tag{2.292}$$

In concert with Lemma 2.36, for any $X \in \Omega$ and $0 < t < \frac{1}{2}\rho(X)$, this entails

$$\begin{aligned} \operatorname{osc}^{N-1}(u,X,t) &\leq C t^N \fint_{B(X,t)} |\nabla^N u(Y)|\,dY \\ &\leq C\,t^N\Big(\sup_{B(X,t)} |\nabla^N u|\Big) \leq C\,t^N\Big(\sup_{B(X,\rho(X)/2)} |\nabla^N u|\Big) \\ &\leq C\,t^N\left(\fint_{B(X,7\rho(X)/8)} |\nabla^N u(Y)|^q\,dY\right)^{1/q}. \end{aligned} \tag{2.293}$$

Given p,q,α,k as in the setup to Step 7, choose $N := k+1$, $s := k+\alpha$. Based on (2.290), (2.293), we may then write

$$\begin{aligned} &\left\| \left(\int_0^{\frac{1}{2}\rho(\cdot)} \big[\operatorname{osc}^k(u,\cdot,t)\big]^q \frac{dt}{t^{1+(k+\alpha)q}} \right)^{1/q} \right\|_{L^p(\Omega)} \\ &\quad \leq C \left\| \left(\int_0^{\frac{1}{2}\rho(X)} \left[\fint_{B(X,7\rho(X)/8)} |\nabla^{k+1} u(Y)|^q\,dY \right] \frac{dt}{t^{1+q(\alpha-1)}} \right)^{1/q} \right\|_{L^p_X(\Omega)} \\ &\quad \leq C \left\| \rho(X)^{1-\alpha} \left(\fint_{B(X,7\rho(X)/8)} |\nabla^{k+1} u(Y)|^q\,dY \right)^{1/q} \right\|_{L^p_X(\Omega)}. \end{aligned} \tag{2.294}$$

By the reverse Hölder estimate (2.258), the last expression above may be further dominated by

$$\begin{aligned} &C \left\| \rho(X)^{1-\alpha} \left(\fint_{B(X,8\rho(X)/9)} |\nabla^{k+1} u(Y)|^p\,dY \right)^{1/p} \right\|_{L^p_X(\Omega)} \\ &\quad \leq C \left(\int_\Omega \int_\Omega \rho(X)^{p(1-\alpha)-n} |\nabla^{k+1} u(Y)|^p\,\chi_{|X-Y|\leq 8\rho(X)/9}\,dX\,dY \right)^{1/p}. \end{aligned} \tag{2.295}$$

Next, let $X \in \Omega$ be arbitrary and pick $X^* \in \partial\Omega$ such that $|X^* - X| = \rho(X)$. Then, for each $Y \in \Omega$ with $|X - Y| \leq 8\rho(X)/9$, we have

$$\begin{aligned}\rho(Y) &= \operatorname{dist}(Y, \partial\Omega) \leq |Y - X^*| \leq |X - Y| + |X - X^*| \\ &\leq 8\rho(X)/9 + \rho(X) = 17\rho(X)/9. \end{aligned} \tag{2.296}$$

On the other hand, if given $Y \in \Omega$ with $|X - Y| \leq 8\rho(X)/9$ we pick $Y^* \in \partial\Omega$ such that $\rho(Y) = |Y - Y^*|$ we then have

$$\rho(Y) = |Y - Y^*| \geq |X - Y^*| - |X - Y| \geq \rho(X) - 8\rho(X)/9 = \rho(X)/9. \tag{2.297}$$

Thus, altogether, $|X - Y| \leq 8\rho(X)/9 \Rightarrow 9\rho(Y)/17 \leq \rho(X) \leq 9\rho(Y)$. Availing ourselves of this equivalence back in (2.295) yields

$$\begin{aligned}&\left(\int_\Omega\int_\Omega \rho(X)^{p(1-\alpha)-n}|\nabla^{k+1}u(Y)|^p\,\chi_{|X-Y|\leq 8\rho(X)/9}\,dXdY\right)^{1/p} \\ &\quad\leq C\left(\int_\Omega\int_\Omega \rho(Y)^{p(1-\alpha)-n}|\nabla^{k+1}u(Y)|^p\,\chi_{|X-Y|\leq 8\rho(Y)}\,dXdY\right)^{1/p} \\ &\quad\leq C\left(\int_\Omega \rho(Y)^{p(1-\alpha)}|\nabla^{k+1}u(Y)|^p\,dY\right)^{1/p}. \end{aligned} \tag{2.298}$$

Collectively, (2.290), (2.294), (2.295), and (2.298) imply that

$$\|u\|_{F^{p,q}_{\alpha+k}(\Omega)} \leq C\left\|\rho^{1-\alpha}|\nabla^{k+1}u|\right\|_{L^p(\Omega)} + C\|u\|_{L^p(\Omega)}. \tag{2.299}$$

This finishes the proof of the claim made in Step 7.

`Step 8.` *We claim that if* $0 < p < \infty$, $0 < q \leq \infty$, *and* $\alpha > n\left(\frac{1}{\min\{p,q\}} - 1\right)_+$, *then*

$$\mathbb{H}^p_\alpha(\Omega; L) \subseteq F^{p,q}_\alpha(\Omega) \cap \operatorname{Ker} L. \tag{2.300}$$

When, additionally, $\alpha \notin \mathbb{N}$, this is just a rephrasing of the result proved in Step 7. The larger range described above follows from this, Theorem 2.35, Theorem 2.33, and complex interpolation.

`Step 9.` *If* $0 < p < \infty$, $0 < q \leq \infty$ *and* $\alpha \in \mathbb{R}$ *then* (2.300) *holds true*. To see this, first recall that (2.300) holds if the indices are as in Step 8. The extension to $\alpha \in \mathbb{R}$ can then be obtained by interpolating via the complex method (cf. Theorems 2.33 and 2.35) between $\mathbb{H}^p_{\alpha_1}(\Omega; L) \subseteq F^{p,2}_{\alpha_1}(\Omega) \cap \operatorname{Ker} L$ for $0 < p < \infty$, $\alpha_1 \in \mathbb{R}$ and (2.300) with $\alpha > n\left(\frac{1}{\min\{p,q\}} - 1\right)_+$ and $0 < p, q < \infty$. When $q = \infty$, use the monotonicity of the Triebel–Lizorkin scale in the second integrability exponent (cf. (2.152)).

Step 10. *We claim that*

$$F_{\alpha}^{p,2}(\Omega) \cap \operatorname{Ker} L = F_{\alpha}^{p,q}(\Omega) \cap \operatorname{Ker} L, \quad \text{if } 0 < p < \infty,\ 0 < q \le 2, \text{ and } \alpha \in \mathbb{R}. \tag{2.301}$$

Indeed, the right-to-left inclusion in (2.301) follows from the monotonicity of the Triebel–Lizorkin scale in the second integrability exponent, whereas the opposite inclusion follows from Step 9 and the fact that $F_{\alpha}^{p,2}(\Omega) \cap \operatorname{Ker} L = \mathbb{H}_{\alpha}^{p}(\Omega; L)$, by (2.285) in Step 6.

Step 11. *We claim that*

$$F_{\alpha}^{p,2}(\Omega) \cap \operatorname{Ker} L = F_{\alpha}^{p,q}(\Omega) \cap \operatorname{Ker} L, \quad \text{whenever } 0 < p, q < \infty \text{ and } \alpha \in \mathbb{R}. \tag{2.302}$$

To prove this claim, observe first that, as a consequence of (2.285)–(2.286) we have

$$F_{\alpha}^{p,2}(\Omega) \cap \operatorname{Ker} L = F_{\alpha}^{p,p}(\Omega) \cap \operatorname{Ker} L \quad \text{for any } p \in (0, \infty) \text{ and } \alpha \in \mathbb{R}. \tag{2.303}$$

Then (2.302) follows by interpolating between (2.303) and (2.301), while making use of formula (2.231) in Theorem 2.33.

At this stage, formula (2.264) follows by combining (2.285) in Step 6 with (2.302) in Step 11, while (2.265) follows from this by writing

$$F_{s}^{p,q}(\Omega) \cap \operatorname{Ker} L = \mathbb{H}_{s}^{p}(\Omega; L) = F_{s}^{p,p}(\Omega) \cap \operatorname{Ker} L = B_{s}^{p,p}(\Omega) \cap \operatorname{Ker} L, \tag{2.304}$$

for every $p, q \in (0, \infty)$ and $s \in \mathbb{R}$. This finishes the proof of the theorem. □

Corresponding to $p = \infty$, we also have the following result.

Proposition 2.42. *If L is a constant coefficient elliptic differential operator as in* (2.251) *and if Ω is a bounded Lipschitz domain in $\mathbb{R}^n$, then*

$$\mathbb{H}_{k+s}^{\infty}(\Omega; L) = B_{k+s}^{\infty,\infty}(\Omega) \cap \operatorname{Ker} L \tag{2.305}$$

for each $k \in \mathbb{N}_0$ and $s \in (0, 1)$.

Proof. This formula is a consequence of definitions, standard interior estimates and Proposition 2.26. □

We conclude this section by singling out a useful consequence of Theorem 2.41.

Corollary 2.43. *Assume that Ω is a bounded Lipschitz domain in $\mathbb{R}^n$ and that L is a constant coefficient elliptic differential operator as in* (2.251). *Then, whenever $1 \le p < \infty$, $a \in (-1/p, 1 - 1/p)$, $q \in (0, \infty)$, and $k \in \mathbb{N}_0$,*

$$W_{a}^{k,p}(\Omega) \cap \operatorname{Ker} L = F_{k-a}^{p,q}(\Omega) \cap \operatorname{Ker} L = B_{k-a}^{p,p}(\Omega) \cap \operatorname{Ker} L. \tag{2.306}$$

Proof. This is a direct consequence of Proposition 2.40 and Theorem 2.41. □

2.6 Stein's Extension Operator on Weighted Sobolev Spaces in Lipschitz Domains

Extension results from a domain $\Omega \subseteq \mathbb{R}^n$ to the entire Euclidean space play a fundamental role in analysis. In the class of Lipschitz domains, two results which have historically influenced much of the development of the theory on this topic are Calderón's and Stein's extension theorems (see [16, 119]).

Other pertinent references and a broader perspective may be found in the excellent monographs [119] by E.M. Stein, [74] by V.G. Maz'ya, and [62] by A. Jonsson and H. Wallin, and in the informative survey by V. Burenkov in [15], among many others. The interested reader may also consult the influential work of P. Jones [61], V. Rychkov [108], and D. Jerison and C. Kenig [58].

Two attractive features of the extension operator constructed by E. Stein are: (1) its universality (i.e., its format does not change as the amount of smoothness considered increases), and (2) its apparent simplicity. The latter makes it tempting to consider the possibility that the same operator continues to work for other classes of smoothness spaces than originally intended (i.e., classical Sobolev spaces). A case in point is the article [66] where the author shows that Stein's extension operator preserves smoothness measured on the Triebel–Lizorkin scales.

In this section, the goal is to further explore this issue and study the action of Stein's extension operator on the *weighted Sobolev spaces* in Lipschitz domains considered in the previous section. Our main result in this regard (cf. Theorem 2.46) shows that Stein's extension operator continues to preserve smoothness measured in this weighted fashion. The presentation in this section follows that from [13].

To get started, let $\psi : [1,\infty) \to \mathbb{R}$ be the function given by

$$\psi(\lambda) := \frac{e}{\pi\lambda} \cdot \operatorname{Im}\{e^{-e^{-i\pi/4}\cdot(\lambda-1)^{1/4}}\}, \qquad \forall\,\lambda \geq 1. \tag{2.307}$$

Then according to [119, Lemma 1, p. 182], this function enjoys the following properties:

$$\psi \in C^0([1,\infty)), \tag{2.308}$$

$$\int_1^\infty \psi(\lambda)\,d\lambda = 1, \tag{2.309}$$

$$\int_1^\infty \lambda^k \psi(\lambda)\,d\lambda = 0, \qquad \forall\,k \in \mathbb{N}, \tag{2.310}$$

$$\psi(\lambda) = \mathcal{O}(\lambda^{-N}), \qquad \forall\,N \in \mathbb{N} \text{ as } \lambda \to \infty. \tag{2.311}$$

In particular, (2.311) guarantees that $|\psi|$ decays at infinity faster than the reciprocal of any polynomial. Parenthetically, we wish to point out that a function satisfying similar conditions to (2.308)–(2.311) has been constructed in [22, Lemma 3.4,

p. 309]. The function in question has compact support though one needs to specify a priori the number of vanishing moments.

On a different topic, recall from [119, Theorem 2, p. 171] that for any closed set $F \subseteq \mathbb{R}^n$ there exists a function $\rho_{reg} : \mathbb{R}^n \to [0, +\infty)$ such that

$$\rho_{reg} \in C^\infty(\mathbb{R}^n \setminus F), \quad \rho_{reg} \approx \mathrm{dist}(\cdot, F) \ \text{ on } \ \mathbb{R}^n, \tag{2.312}$$

and, with $\mathbb{N}_0 := \mathbb{N} \cup \{0\}$,

$$|\partial^\alpha \rho_{reg}(X)| \leq C_\alpha \big[\mathrm{dist}(X, F)\big]^{1-|\alpha|}, \quad \forall\, \alpha \in \mathbb{N}_0^n \ \text{ and } \ \forall\, X \in \mathbb{R}^n \setminus F. \tag{2.313}$$

To proceed, let Ω be an unbounded Lipschitz domain in $\mathbb{R}^n$ and denote by $C_b^\infty(\overline{\Omega})$ the vector space of restrictions to Ω of functions from $C_c^\infty(\mathbb{R}^n)$. Also, if ρ_{reg} stands for the regularized distance function associated with $\overline{\Omega}$, we set

$$\delta := C\rho_{reg}, \tag{2.314}$$

where $C > 0$ is a fixed constant chosen large enough so that

$$\delta(Z - se_n) > 2s, \qquad \forall\, Z \in \partial\Omega \ \text{ and } \ \forall\, s > 0, \tag{2.315}$$

where the family $\{e_j\}_{1\leq j\leq n}$ denotes the standard orthonormal basis in $\mathbb{R}^n$ (hence, in particular, $e_n := (0, \dots, 0, 1) \in \mathbb{R}^n$). The above normalization condition on δ ensures that

$$X + \lambda\delta(X)e_n \in \Omega, \qquad \forall\, X \in \mathbb{R}^n \setminus \overline{\Omega} \ \text{ and } \ \forall\, \lambda \geq 1. \tag{2.316}$$

Let us also note that in the current case (i.e., when $F := \overline{\Omega}$ where Ω is an unbounded Lipschitz domain in $\mathbb{R}^n$), there holds[1]

$$\delta \in \mathrm{Lip}\,(\mathbb{R}^n), \tag{2.317}$$

where $\mathrm{Lip}\,(\mathbb{R}^n)$ stands for the set of Lipschitz functions in $\mathbb{R}^n$. Indeed, given that Ω is a Lipschitz domain, for any two points $X, Y \in \mathbb{R}^n \setminus \overline{\Omega}$ it is possible to find a rectifiable curve $\gamma_{X,Y} : \big[0, L_{X,Y}\big] \to \mathbb{R}^n \setminus \overline{\Omega}$, parametrized by arc-length, joining X in Y and whose length $L_{X,Y}$ satisfies

$$L_{X,Y} \leq C|X - Y|, \tag{2.318}$$

[1]A quick inspection reveals that the same result and proof are valid in the more general case of domains whose complement is *regular in the sense of Whitney* (in the terminology of [57, p. 52]; cf. also [101, p. 1372] where the notion of *quasi-Euclideanity* is employed), i.e., subsets of $\mathbb{R}^n$ with the property that any two points X, Y in the complement may be joined with a rectifiable curve $\gamma_{X,Y}$ disjoint from the set in question and which satisfies (2.318).

for some finite constant $C = C(\Omega) > 0$. Using the fact that $\rho_{reg} \in C^\infty(\mathbb{R}^n \setminus \overline{\Omega})$ (cf. (2.312)) we may estimate, keeping in mind (2.313) and the fact that $|\dot{\gamma}| = 1$

$$\begin{aligned} |\rho_{reg}(X) - \rho_{reg}(Y)| &= \Big| \int_0^{L_{X,Y}} \frac{d}{ds}\big[\rho_{reg}(\gamma(s))\big]\,ds \Big| \\ &\leq \int_0^{L_{X,Y}} |\nabla \rho_{reg}(\gamma(s)) \cdot \dot{\gamma}(s)|\,ds \\ &\leq C L_{X,Y}, \qquad \forall\, X, Y \in \mathbb{R}^n \setminus \overline{\Omega}. \end{aligned} \tag{2.319}$$

From (2.318), (2.319) we deduce that there exists a finite constant $C > 0$ such that

$$|\rho_{reg}(X) - \rho_{reg}(Y)| \leq C|X - Y|, \tag{2.320}$$

whenever $X, Y \in \mathbb{R}^n \setminus \overline{\Omega}$. In fact, (2.312) shows that (2.320) continues to hold in the case when $X, Y \in \overline{\Omega}$. When $X \in \overline{\Omega}$ and $Y \in \mathbb{R}^n \setminus \overline{\Omega}$, pick $Y_* \in \partial\Omega$ such that $\operatorname{dist}(Y, \partial\Omega) = |Y - Y_*|$, and note that this forces $|Y - Y_*| \leq |X - Y|$. Then,

$$\begin{aligned} |\rho_{reg}(X) - \rho_{reg}(Y)| &= \rho_{reg}(Y) = |\rho_{reg}(Y_*) - \rho_{reg}(Y)| \\ &\leq C|Y - Y_*| \leq C|X - Y|, \end{aligned} \tag{2.321}$$

where the first inequality above follows from reasoning as in (2.320). Finally, the case when $X \in \mathbb{R}^n \setminus \overline{\Omega}$ and $y \in \overline{\Omega}$ is treated similarly. Hence, $\rho_{reg} \in \operatorname{Lip}(\mathbb{R}^n)$ which immediately yields (2.317).

On a different topic, given $m \in \mathbb{N}$, we shall let $\mathscr{A}_m$ stand for the collection of all ordered m-tuples $A = (\alpha^{(1)}, \dots, \alpha^{(m)})$ consisting of nonzero multi-indices $\alpha^{(1)}, \dots, \alpha^{(m)} \in \mathbb{N}_0^n \setminus \{(0, \dots, 0)\}$. For each such m-tuple $A = (\alpha^{(1)}, \dots, \alpha^{(m)}) \in \mathscr{A}_m$ define

$$\#A := m \quad \text{and} \quad |A| := \sum_{i=1}^{m} |\alpha^{(i)}|. \tag{2.322}$$

Corresponding to the case when $m = 0$ we shall take the set $\mathscr{A}_0$ to be a singleton, namely the null multi-index $\mathbf{0} := (0, \dots, 0) \in \mathbb{N}_0^n$. We also agree that $\#\mathbf{0} := |\mathbf{0}| := 0$. Finally, set

$$\mathscr{A} := \bigcup_{m \in \mathbb{N}_0} \mathscr{A}_m. \tag{2.323}$$

Next, for each $A \in \mathscr{A}$ and each $k \in \mathbb{N}_0$ we introduce the operator $\mathscr{R}_{k,A}$ acting on functions $f \in C_b^\infty(\overline{\Omega})$ at each $X \in \mathbb{R}^n \setminus \overline{\Omega}$ according to

$$(\mathscr{R}_{k,A} f)(X) := \left\{ \prod_{i=1}^{\#A} (\partial^{\alpha^{(i)}} \delta)(X) \right\} \int_1^\infty f\big(X + \lambda\delta(X)e_n\big)\lambda^k \psi(\lambda)\, d\lambda, \quad (2.324)$$

assuming the $\alpha^{(i)}$'s are the components of A, and with the convention that the product is omitted if $A = \mathbf{0}$.

The main role of the family $\{\mathscr{R}_{k,A}\}_{k,A}$ is to facilitate the understanding of the operator

$$E := \mathscr{R}_{0,0} \quad (2.325)$$

which is the main object of interest. This is because Ef, which naturally extends to $\mathbb{R}^n$ by considering

$$(Ef)(X) = \int_1^\infty f\big(X + \lambda\delta(X)e_n\big)\psi(\lambda)\, d\lambda, \qquad \forall\, X \in \mathbb{R}^n, \quad (2.326)$$

is precisely Stein's extension operator (cf. [119, (24), p. 182]) acting on the function $f \in C_b^\infty(\overline{\Omega})$. Incidentally, the fact that

$$Ef \in \mathrm{Lip}\,(\mathbb{R}^n) \quad \text{and} \quad (Ef)\Big|_{\Omega} = f, \quad \forall\, f \in C_b^\infty(\overline{\Omega}), \quad (2.327)$$

is a direct consequence of (2.317), (2.326) and (2.309).

In relation to the family $\{\mathscr{R}_{k,A}\}_{k,A}$, we note that

$$\mathscr{R}_{k,A} f \in C_b^\infty(\mathbb{R}^n \setminus \overline{\Omega}) \ \text{ for each } \ f \in C_b^\infty(\overline{\Omega}). \quad (2.328)$$

To proceed, we claim that

$$\begin{gathered} \mathscr{R}_{k,A} f \text{ vanishes of infinite order on } \partial\Omega, \\ \text{for each } f \in C_b^\infty(\overline{\Omega}), \text{ each } k \in \mathbb{N}, \text{ and each } A \in \mathscr{A}, \end{gathered} \quad (2.329)$$

i.e., for each $N \in \mathbb{N}$ one can find a finite constant $C = C(N, \psi, k, A, f) > 0$ such that

$$\big|(\mathscr{R}_{k,A} f)(X)\big| \leq C\, \mathrm{dist}\,(X, \overline{\Omega})^N, \qquad \forall\, X \in \mathbb{R}^n \setminus \overline{\Omega} \quad (2.330)$$

(of course, this is not the case when $k = 0$). To justify this claim we first note that, for each number $N \in \mathbb{N}$, Taylor's formula gives

$$\begin{aligned} f\big(X + \lambda\delta(X)e_n\big) &= \sum_{j=0}^{N-1} \frac{(\lambda - 1)^j \delta(X)^j}{j!} (\partial_n^j f)\big(X + \delta(X)e_n\big) \quad (2.331) \\ &\quad + \frac{1}{N!} \int_{\delta(X)}^{\lambda\delta(X)} (\lambda\delta(X) - t)^{N-1} (\partial_n^N f)\big(X + te_n\big)\, dt, \end{aligned}$$

for each $X \in \mathbb{R}^n \setminus \overline{\Omega}$ and each $\lambda \in [1, \infty)$. Using this in concert with (2.313) and the fact that $\int_1^\infty \lambda^k (\lambda - 1)^j \psi(\lambda)\, d\lambda = 0$ for any $j \in \mathbb{N}_0$ and any $k \in \mathbb{N}$, we therefore obtain

$$\begin{aligned} \big|(\mathscr{R}_{k,A} f)(X)\big| &\le C_N \delta(X)^{\#A-|A|+N} \|f\|_{C^N(\overline{\Omega})} \Big(\int_1^\infty (\lambda - 1)^N \lambda^k |\psi(\lambda)|\, d\lambda \Big) \\ &\le C(N, \psi, k, A, f) \operatorname{dist}(X, \overline{\Omega})^{\#A-|A|+N}, \quad \forall\, X \in \mathbb{R}^n \setminus \overline{\Omega}. \end{aligned} \tag{2.332}$$

Since $N \in \mathbb{N}$ has been arbitrarily chosen, (2.329) follows.

To formulate the next claim about the operators introduced in (2.324), fix some arbitrary $k \in \mathbb{N}_0$ along with $A \in \mathscr{A}$ and $\alpha \in \mathbb{N}_0^n$. Then there exists a family of finitely many nonzero constant coefficients $\{C_{\ell,\beta,B}\}_{\ell,\beta,B}$ with the property that for every function $f \in C_b^\infty(\overline{\Omega})$ we have

$$\partial^\alpha (\mathscr{R}_{k,A} f) = \mathscr{R}_{k,A}(\partial^\alpha f) + \sum_{\ell=0}^{|\alpha|} \sum_{\beta,B} C_{\ell,\beta,B}\, \mathscr{R}_{k+\ell,B}(\partial^\beta f) \quad \text{in } \mathbb{R}^n \setminus \overline{\Omega}, \tag{2.333}$$

where the second sum is performed over $\beta \in \mathbb{N}_0^n$ and $B \in \mathscr{A}$ such that

$$|\beta| \le |\alpha| \quad \text{and} \quad |B| - \#B + |\beta| = |A| - \#A + |\alpha|. \tag{2.334}$$

Furthermore,

$$\begin{gathered} \text{if } A = \mathbf{0} \text{ then the first sum in the} \\ \text{right-hand side of (2.333) starts from } \ell = 1. \end{gathered} \tag{2.335}$$

This follows by a routine induction argument once we observe that for every $A \in \mathscr{A}$, $k \in \mathbb{N}_0$, $j \in \{1, \dots, n\}$, and every function $f \in C_b^\infty(\overline{\Omega})$ we have

$$\partial_j (\mathscr{R}_{k,A} f) = \mathscr{R}_{k,A}(\partial_j f) + \sum_{i=1}^{\#A} \mathscr{R}_{k,A'_{ij}} f + \mathscr{R}_{k+1,A''_j}(\partial_n f) \quad \text{in } \mathbb{R}^n \setminus \overline{\Omega}, \tag{2.336}$$

(with the natural convention that the sum in the right-hand side of (2.336) is discarded if $\#A = 0$, i.e., if $A = \mathbf{0}$), where, if $m := \#A$ and $A = (\alpha^{(1)}, \dots, \alpha^{(m)})$ then

$$A'_{ij} := (\alpha^{(1)}, \dots, \alpha^{(i-1)}, \alpha^{(i)} + e_j, \alpha^{(i+1)}, \dots, \alpha^{(m)}), \tag{2.337}$$

$$A''_j := (\alpha^{(1)}, \dots, \alpha^{(m)}, e_j) \text{ if } A \neq \mathbf{0}, \text{ and } A''_j := (e_j) \text{ if } A = \mathbf{0}. \tag{2.338}$$

The key features of the newly assigned tuples $A'_{ij}, A''_j \in \mathscr{A}$ to each given $A \in \mathscr{A}$ are

$$\#A'_{ij} - |A'_{ij}| = \#A - |A| - 1 \quad \text{and} \quad \#A''_j - |A''_j| = \#A - |A|. \tag{2.339}$$

Indeed, it is (2.339) which allows us to iterate (2.336) (while keeping proper count of how the various indices change) and obtain (2.333).

Going further, combining (2.333) with (2.329) gives

$$\begin{gathered}\partial^\alpha(\mathscr{R}_{k,A} f)\text{vanishes of infinite order on } \partial\Omega, \\ \text{for each } f \in C_b^\infty(\overline{\Omega}),\ \text{each } k \in \mathbb{N},\ \text{each } \alpha \in \mathbb{N}_0^n \text{ and each } A \in \mathscr{A}.\end{gathered} \tag{2.340}$$

Hence, if for each $k \in \mathbb{N}$, $\alpha \in \mathbb{N}_0^n$, and $A \in \mathscr{A}$, we now introduce

$$\widetilde{\mathscr{R}}_{k,A} f := \begin{cases} \mathscr{R}_{k,A} f & \text{in } \mathbb{R}^n \setminus \overline{\Omega}, \\ 0 & \text{in } \overline{\Omega}, \end{cases} \tag{2.341}$$

it follows from (2.340) and the fact that Ω is a Lipschitz domain (cf. [119, p. 186] for more details) that

$$\begin{gathered}\widetilde{\mathscr{R}}_{k,A} : C_b^\infty(\overline{\Omega}) \longrightarrow C_c^\infty(\mathbb{R}^n) \ \text{ is a well-defined, linear operator} \\ \text{for each } k \in \mathbb{N},\ \text{and each } A \in \mathscr{A}.\end{gathered} \tag{2.342}$$

Specializing (2.333) to the case when $A := \mathbf{0}$ and $k := 0$ yields, on account of (2.325) and (2.335), that for every function $f \in C_b^\infty(\overline{\Omega})$ and every $\alpha \in \mathbb{N}_0^n$,

$$\begin{aligned}\partial^\alpha(Ef) &= E(\partial^\alpha f) \\ &\quad + \sum_{\ell=1}^{|\alpha|} \sum_{|\beta|\le|\alpha|} \sum_{\substack{B\in\mathscr{A} \text{ so that} \\ |B|-\#B=|\alpha|-|\beta|}} C_{\ell,\beta,B}\, \mathscr{R}_{\ell,B}(\partial^\beta f) \quad \text{in } \mathbb{R}^n \setminus \overline{\Omega}.\end{aligned} \tag{2.343}$$

In turn, (2.343), (2.329), and (2.327), allow us to conclude (compare with (2.342)) that

$$E : C_b^\infty(\overline{\Omega}) \longrightarrow C_c^\infty(\mathbb{R}^n) \ \text{ is a well-defined, linear operator,} \tag{2.344}$$

and that

$$\partial^\alpha(Ef) = E(\partial^\alpha f) + \sum_{\ell=1}^{|\alpha|} \sum_{|\beta|\le|\alpha|} \sum_{\substack{B\in\mathscr{A} \text{ so that} \\ |B|-\#B=|\alpha|-|\beta|}} C_{\ell,\beta,B}\, \widetilde{\mathscr{R}}_{\ell,B}(\partial^\beta f) \quad \text{in } \mathbb{R}^n, \tag{2.345}$$

for every function $f \in C_b^\infty(\overline{\Omega})$ and every multi-index $\alpha \in \mathbb{N}_0^n$.

We shall work with the weighted Sobolev spaces $W_a^{m,p}(\Omega)$ which have been considered in the previous section. In addition, we make the following definition.

Definition 2.44. If $p \in [1, \infty]$, $a \in (-1/p, 1 - 1/p)$ and $m \in \mathbb{N}_0$ are given and Ω is a nonempty, proper, open subset of $\mathbb{R}^n$, consider the weighted Sobolev space

$$W_a^{m,p}(\mathbb{R}^n) := \Big\{u \in L^1_{loc}(\mathbb{R}^n) : \partial^\alpha u \in L^1_{loc}(\mathbb{R}^n) \ \text{ for } |\alpha| \le m, \ \text{ and} \tag{2.346}$$

$$\|u\|_{W_a^{m,p}(\mathbb{R}^n)} := \sum_{|\alpha|\le m} \Big(\int_{\mathbb{R}^n} |(\partial^\alpha u)(X)|^p \,\mathrm{dist}\,(X,\partial\Omega)^{ap}\,dX\Big)^{1/p} < \infty\Big\}.$$

We wish to stress that $W_a^{m,p}(\mathbb{R}^n)$ is *not* $W_a^{m,p}(\Omega)$ corresponding to $\Omega = \mathbb{R}^n$ (which, incidentally, is not a permissible choice since Ω is assumed to be a proper subset of $\mathbb{R}^n$). Instead, the named space should always be understood in the sense of (2.346).

Moving on, fix an unbounded Lipschitz domain $\Omega \subset \mathbb{R}^n$ and assume that $p \in [1,\infty]$ and $a \in (-1/p,\infty)$. We claim that for each $k \in \mathbb{N}$ and each $A \in \mathscr{A}$ there exists a finite constant $C = C(\Omega,k,A,a,p) > 0$ with the property that

$$\begin{aligned}\|\mathscr{R}_{k,A} f\|_{L^p(\mathbb{R}^n\setminus\overline{\Omega},\,\mathrm{dist}\,(\cdot,\partial\Omega)^{ap}\,dX)} \\ \le C \sum_{|\alpha|=|A|-\#A} \|\partial^\alpha f\|_{L^p(\Omega,\,\mathrm{dist}\,(\cdot,\partial\Omega)^{ap}\,dX)},\end{aligned} \tag{2.347}$$

for every function $f \in C_b^\infty(\overline{\Omega})$. To justify (2.347), select a positive integer N and a real number $M > 1$. Then, by arguing as in (2.332) we deduce that there exists a finite constant $C = C(N,M,k,A,\psi) > 0$ with the property that for each $f \in C_b^\infty(\overline{\Omega})$ we have

$$\begin{aligned}\big|(\mathscr{R}_{k,A} f)(X)\big| &\le C\delta(X)^{\#A-|A|+N-1}\int_1^\infty \Big(\int_{\delta(X)}^{\lambda\delta(X)} |(\partial_n^N f)(X+te_n)|\,dt\Big)\frac{d\lambda}{\lambda^M} \\ &= C\delta(X)^{\#A-|A|+N-1}\int_{\delta(X)}^\infty |(\partial_n^N f)(X+te_n)|\Big(\int_{t/\delta(X)}^\infty \frac{d\lambda}{\lambda^M}\Big)\,dt \\ &\approx C\delta(X)^{\#A-|A|+N+M-2}\int_{\delta(X)}^\infty |(\partial_n^N f)(X+te_n)|\frac{dt}{t^{M-1}},\end{aligned} \tag{2.348}$$

at each point $X \in \mathbb{R}^n \setminus \overline{\Omega}$. Above, the first inequality is derived much as in (2.332), the subsequent equality is based on Fubini's Theorem, while the last step utilizes $M > 1$. Moreover, a direct argument based on (2.324) and the decay condition for ψ, shows that the estimate

$$\big|(\mathscr{R}_{k,A} f)(X)\big| \le C\delta(X)^{\#A-|A|+N+M-2}\int_{\delta(X)}^\infty |(\partial_n^N f)(X+te_n)|\frac{dt}{t^{M-1}} \tag{2.349}$$

at each point $X \in \mathbb{R}^n \setminus \overline{\Omega}$, is also valid in the case when $N = 0$. Having established this, for each fixed $Z \in \partial\Omega$ and any $p \in [1,\infty]$, $a \in (-1/p,\infty)$ we may estimate, for any $k \in \mathbb{N}_0$, $N \in \mathbb{N}_0$, and $M > 1$,

$$\Big(\int_0^\infty |(\mathscr{R}_{k,A}f)(Z-se_n)|^p\,\mathrm{dist}\,(Z-se_n,\partial\Omega)^{ap}\,ds\Big)^{1/p}$$

$$\leq C\Big(\int_0^\infty\Big(s^{\#A-|A|+N+M-2}\int_{\delta(Z-se_n)}^\infty \frac{|(\partial_n^N f)(Z+(t-s)e_n)|}{t^{M-1}}\,dt\Big)^p s^{ap}\,ds\Big)^{1/p}$$

$$\leq C\Big(\int_0^\infty\Big(s^{\#A-|A|+N+M-2}\int_{\delta(Z-se_n)-s}^\infty \frac{|(\partial_n^N f)(Z+re_n)|}{(r+s)^{M-1}}\,dr\Big)^p s^{ap}\,ds\Big)^{1/p}$$

$$\leq C\Big(\int_0^\infty\Big(s^{\#A-|A|+N+M-2}\int_s^\infty \frac{|(\partial_n^N f)(Z+re_n)|}{r^{M-1}}\,dr\Big)^p s^{ap}\,ds\Big)^{1/p}, \quad (2.350)$$

where the last inequality above makes use of (2.315). In turn, Hardy's inequality (cf., e.g., [119, p. 272]) together with the version of the estimate (2.350) corresponding to the choices $N := |A| - \#A$ and $M := 2$, permits us to estimate (for each $k \in \mathbb{N}_0$ and $A \in \mathscr{A}$)

$$\Big(\int_0^\infty |(\mathscr{R}_{k,A}f)(Z-se_n)|^p\,\mathrm{dist}\,(Z-se_n,\partial\Omega)^{ap}\,ds\Big)^{1/p}$$

$$\leq C\Big(\int_0^\infty |(\partial_n^{|A|-\#A}f)(Z+se_n)|^p s^{ap}\,ds\Big)^{1/p} \quad (2.351)$$

$$\approx C\Big(\int_0^\infty |(\partial_n^{|A|-\#A}f)(Z+se_n)|^p\,\mathrm{dist}\,(Z+se_n,\partial\Omega)^{ap}\,ds\Big)^{1/p},$$

where $C > 0$ is a finite constant independent of $f \in C_b^\infty(\overline{\Omega})$ and $Z \in \partial\Omega$. Let us now recall two useful estimates to the effect that, given $p \in (0,\infty]$ along with two measurable functions $F : \mathbb{R}^n \setminus \overline{\Omega} \to \mathbb{R}$ and $G : \Omega \to \mathbb{R}$, we have

$$\Big(\int_{\mathbb{R}^n\setminus\overline{\Omega}} |F(X)|^p\,dX\Big)^{1/p} \approx \Big(\int_{\partial\Omega}\int_0^\infty |F(Z-se_n)|^p\,ds\,d\sigma(Z)\Big)^{1/p} \quad (2.352)$$

and

$$\Big(\int_{\Omega} |G(X)|^p\,dX\Big)^{1/p} \approx \Big(\int_{\partial\Omega}\int_0^\infty |G(Z+se_n)|^p\,ds\,d\sigma(Z)\Big)^{1/p}. \quad (2.353)$$

Based on (2.351)–(2.353), we then deduce that, given $p \in [1,\infty]$,

$$\Big(\int_{\mathbb{R}^n\setminus\overline{\Omega}} |(\mathscr{R}_{k,A}f)(X)|^p \operatorname{dist}(X,\partial\Omega)^{ap}\,dX\Big)^{1/p}$$

$$\leq C\Big(\int_{\Omega} |(\partial_n^{|A|-\#A}f)(X)|^p \operatorname{dist}(X,\partial\Omega)^{ap}\,dX\Big)^{1/p}, \tag{2.354}$$

where $C>0$ is a finite constant independent of $f\in C_b^\infty(\overline{\Omega})$. This finishes the proof of (2.347).

Let $k\in\mathbb{N}_0$, $A\in\mathscr{A}$, $N\in\mathbb{N}_0$, $p\in[1,\infty]$, and $a\in(-1/p,\infty)$ be given. Having dispensed with (2.347), we may then proceed to estimate for any $f\in C_b^\infty(\overline{\Omega})$

$$\sum_{|\alpha|\leq N}\big\|\partial^\alpha(\mathscr{R}_{k,A}f)\big\|_{L^p(\mathbb{R}^n\setminus\overline{\Omega},\,\operatorname{dist}(\cdot,\partial\Omega)^{ap}dX)}$$

$$\leq \sum_{|\alpha|\leq N}\big\|\mathscr{R}_{k,A}(\partial^\alpha f)\big\|_{L^p(\mathbb{R}^n\setminus\overline{\Omega},\,\operatorname{dist}(\cdot,\partial\Omega)^{ap}dX)} \tag{2.355}$$

$$+\sum_{|\alpha|\leq N}\sum_{\ell=0}^{|\alpha|}\sum_{\beta,B} C_{\ell,\beta,B}\big\|\mathscr{R}_{k+\ell,B}(\partial^\beta f)\big\|_{L^p(\mathbb{R}^n\setminus\overline{\Omega},\,\operatorname{dist}(\cdot,\partial\Omega)^{ap}dX)},$$

and thus

$$\sum_{|\alpha|\leq N}\big\|\partial^\alpha(\mathscr{R}_{k,A}f)\big\|_{L^p(\mathbb{R}^n\setminus\overline{\Omega},\,\operatorname{dist}(\cdot,\partial\Omega)^{ap}dX)}$$

$$\leq C\sum_{|\alpha|\leq N}\ \sum_{|\gamma|=|A|-\#A+|\alpha|}\big\|\partial^\gamma f\big\|_{L^p(\Omega,\,\operatorname{dist}(\cdot,\partial\Omega)^{ap}dX)}$$

$$+C\sum_{|\alpha|\leq N}\sum_{\beta,B}\ \sum_{|\tau|=|B|-\#B+|\beta|}\big\|\partial^\tau f\big\|_{L^p(\Omega,\,\operatorname{dist}(\cdot,\partial\Omega)^{ap}dX)}$$

$$\leq C\sum_{|\gamma|\leq|A|-\#A+N}\big\|\partial^\gamma f\big\|_{L^p(\Omega,\,\operatorname{dist}(\cdot,\partial\Omega)^{ap}dX)}, \tag{2.356}$$

thanks to (2.333), (2.347) and (2.334). In summary, the analysis so far shows that for any $k\in\mathbb{N}_0$, $A\in\mathscr{A}$, $N\in\mathbb{N}_0$, $p\in[1,\infty]$ and $a\in(-1/p,\infty)$ there exists a finite constant $C=C(k,A,N,a,p,\Omega)>0$ with the property that

$$\big\|\mathscr{R}_{k,A}f\big\|_{W_a^{N,p}(\mathbb{R}^n\setminus\overline{\Omega})}\leq C\|f\|_{W_a^{|A|-\#A+N,p}(\Omega)},\qquad \forall\, f\in C_b^\infty(\overline{\Omega}). \tag{2.357}$$

We are now prepared to formally state the following result.

Proposition 2.45. *Let Ω be an unbounded Lipschitz domain in $\mathbb{R}^n$ and recall* (2.341) *and* (2.326)*. Then for any $k \in \mathbb{N}$, $A \in \mathscr{A}$, $N \in \mathbb{N}_0$, $p \in [1, \infty]$, and any $a \in (-1/p, \infty)$, there exists a finite constant $C = C(k, A, N, a, p, \Omega) > 0$ such that*

$$\|\widetilde{\mathscr{R}}_{k,A} f\|_{W_a^{N,p}(\mathbb{R}^n)} \leq C \|f\|_{W_a^{|A|-\#A+N,p}(\Omega)}, \qquad \forall\, f \in C_b^\infty(\overline{\Omega}). \tag{2.358}$$

Moreover, for every $N \in \mathbb{N}_0$, $p \in [1, \infty]$ and $a \in (-1/p, \infty)$ there exists a finite constant $C = C(N, a, p, \Omega) > 0$ for which

$$\|Ef\|_{W_a^{N,p}(\mathbb{R}^n)} \leq C \|f\|_{W_a^{N,p}(\Omega)}, \qquad \forall\, f \in C_b^\infty(\overline{\Omega}). \tag{2.359}$$

Proof. We deduce from (2.357) that (2.358) holds, thanks to (2.342) and (2.341). Finally, (2.359) follows from (2.345), (2.358), (2.357), (2.344), as well as the fact that Ef coincides with f on Ω, for every $f \in C_b^\infty(\overline{\Omega})$. □

Further elaboration on the theme of Proposition 2.45 requires some density results. In this vein, first recall from [1, Theorem 3.22, p. 68] that, since the Lipschitz domain Ω satisfies the so-called segment condition, the inclusion operator

$$C_b^\infty(\overline{\Omega}) \hookrightarrow W^{N,p}(\Omega) \text{ has dense range, if } p \in [1, \infty),\ N \in \mathbb{N}_0. \tag{2.360}$$

On the other hand, in the weighted case, given any Lipschitz domain Ω,

$$\begin{gathered} C_b^\infty(\overline{\Omega}) \hookrightarrow W_a^{N,p}(\Omega) \text{ has dense range,} \\ \text{if } p \in (1, \infty),\ N \in \mathbb{N}_0, \text{ and } a \in (-1/p, 1 - 1/p). \end{gathered} \tag{2.361}$$

This is proved much as in (2.360), the new key technical ingredient being the fact that, given any Lipschitz domain $\Omega \subseteq \mathbb{R}^n$,

$$\begin{gathered} \operatorname{dist}(\cdot, \partial\Omega)^{ap} \text{ is a Muckenhoupt } A_p\text{-weight in } \mathbb{R}^n \\ \text{whenever } p \in (1, \infty) \text{ and } a \in (-1/p, 1 - 1/p). \end{gathered} \tag{2.362}$$

See [93] for more details on this. For the convenience of the reader, let us recall that, if $1 < p < \infty$, a measurable, a.e. positive function w defined in $\mathbb{R}^n$ is said to belong to the Muckenhoupt class A_p provided

$$[w]_{A_p} := \sup_{B \text{ ball in } \mathbb{R}^n} \Big(\fint_B w\, dX\Big)\Big(\fint_B w^{-1/(p-1)}\, dX\Big)^{p-1} < +\infty. \tag{2.363}$$

In fact, (2.362) also permits us to show that

$$\begin{gathered} C_c^\infty(\mathbb{R}^n) \hookrightarrow W_a^{N,p}(\mathbb{R}^n) \text{ has dense range,} \\ \text{if } p \in (1, \infty),\ N \in \mathbb{N}_0, \text{ and } a \in (-1/p, 1 - 1/p). \end{gathered} \tag{2.364}$$

We now arrive at the main result in this section.

Theorem 2.46. *Let Ω be a Lipschitz domain in $\mathbb{R}^n$. Then for any $k \in \mathbb{N}$, $A \in \mathscr{A}$, and $N \in \mathbb{N}_0$, the operator $\mathscr{R}_{k,A}$ initially considered as in (2.342) extends to a bounded mapping*

$$\widetilde{\mathscr{R}}_{k,A} : W_a^{|A|-\#A+N,p}(\Omega) \longrightarrow \{u \in W_a^{N,p}(\mathbb{R}^n) : \operatorname{supp} u \subseteq \mathbb{R}^n \setminus \Omega\}, \tag{2.365}$$

provided

$$\begin{aligned} &\text{either } p \in (1,\infty) \text{ and } a \in (-1/p, 1-1/p),\\ &\text{or } p = 1 \text{ and } a = 0. \end{aligned} \tag{2.366}$$

Moreover, for each $N \in \mathbb{N}_0$ and a, p as in (2.366), the operator E initially defined as in (2.344) may be extended to a bounded mapping

$$E : W_a^{N,p}(\Omega) \longrightarrow W_a^{N,p}(\mathbb{R}^n) \text{ such that } \big(Ef\big)\Big|_{\Omega} = f, \quad \forall\, f \in W_a^{N,p}(\Omega). \tag{2.367}$$

Proof. Consider first the case when Ω is an unbounded Lipschitz domain. Then the first claim in the statement of the theorem follows by combining (2.360)–(2.361) with (2.358), while the second claim is proved in a similar fashion, this time making use of (2.360)–(2.361) and (2.359).

The above results continue to hold in the case when Ω is an arbitrary bounded Lipschitz domain. One way to see this is to glue together results proved for various unbounded Lipschitz domains via arguments very similar to those in [119, § 3.3, p. 189–192]. As far as the extension operator is concerned, another, perhaps more elegant argument, is to change formula (2.326) to

$$(Ef)(X) := \int_1^\infty f\big(X + \lambda\delta(X)h(X)\big)\psi(\lambda)\,d\lambda, \qquad \forall\, X \in \mathbb{R}^n, \tag{2.368}$$

where $h \in C_c^\infty(\mathbb{R}^n, \mathbb{R}^n)$ is a suitably chosen vector field. In particular, it is assumed that h is transversal to $\partial\Omega$ in a uniform fashion, i.e., that for some constant $\kappa > 0$ there holds

$$\nu \cdot h \geq \kappa \quad \sigma\text{-a.e. on} \quad \partial\Omega, \tag{2.369}$$

where ν is the outward unit normal to Ω. The vector field h is a replacement of e_n and this permits us to avoid considering a multitude of special local systems of coordinates. □

A noteworthy consequence of the above theorem is the following corollary.

Corollary 2.47. *Let Ω be a Lipschitz domain in $\mathbb{R}^n$. Given $k \in \mathbb{N}$, $A \in \mathscr{A}$, $s \in [0,\infty)$, and $p \in (1,\infty)$, the operator*

$$\begin{aligned} &\widetilde{\mathscr{R}}_{k,A} : L^p_{|A|-\#A+s}(\Omega) \longrightarrow L^p_{s,0}(\mathbb{R}^n \setminus \overline{\Omega})\\ &\text{is well-defined, linear and bounded.} \end{aligned} \tag{2.370}$$

Also, the operator

$$E : L^p_s(\Omega) \longrightarrow L^p_s(\mathbb{R}^n) \text{ is well-defined, linear, bounded, and satisfies } \big(Ef\big)\Big|_{\Omega} = f, \quad \forall\, f \in L^p_s(\Omega). \tag{2.371}$$

Proof. This is obtained from Theorem 2.46 (used with $a := 0$) via interpolation. Specifically, for each $p \in (1,\infty)$, $\theta \in (0,1)$ and $N \in \mathbb{N}_0$, we have

$$\big[W^{N,p}(\Omega), L^p(\Omega)\big]_\theta = L^p_{N\theta}(\Omega). \tag{2.372}$$

This completes the proof of the corollary. □

In the last part of this section we shall show that our earlier estimates in weighted Sobolev spaces for Stein's extension operator on an unbounded Lipschitz domain continue to hold in the range $0 < p < 1$ with a suitable interpretation of the weighted Sobolev spaces involved. We begin by clarifying the latter issue.

Definition 2.48. Assume that Ω is an unbounded Lipschitz domain in $\mathbb{R}^n$ and consider three numbers, $p \in (0,1)$, $a \in (-1/p,\infty)$, and $m \in \mathbb{N}_0$. Given any $u : \Omega \to \mathbb{R}$, define its associated radial maximal function $u^* : \Omega \to [0,\infty]$ by the formula

$$u^*(X) := \sup_{\eta>0} \big|u(X + \eta\, e_n)\big|, \qquad \forall\, X \in \Omega. \tag{2.373}$$

Also, for any $u \in C^\infty_b(\overline{\Omega})$ set

$$\|u\|_{W^{m,p}_a(\Omega)} := \Big(\sum_{|\alpha|\le m} \int_\Omega |(\partial^\alpha u)^*(X)|^p \operatorname{dist}(X,\partial\Omega)^{ap}\, dX\Big)^{1/p}. \tag{2.374}$$

Then the space $W^{m,p}_a(\Omega)$ is defined as the completion of $C^\infty_b(\overline{\Omega})$ in the quasi-norm (2.374).

Finally, with Ω, p, a, m as above, the space $W^{m,p}_a(\mathbb{R}^n)$ is defined as the completion of $C^\infty_c(\mathbb{R}^n)$ with respect to the quasi-norm

$$\|u\|_{W^{m,p}_a(\mathbb{R}^n)} := \Big(\sum_{|\alpha|\le m} \int_{\mathbb{R}^n} |(\partial^\alpha u)^*(X)|^p \operatorname{dist}(X,\partial\Omega)^{ap}\, dX\Big)^{1/p} \tag{2.375}$$

where, this time, the understanding is that for any $u : \mathbb{R}^n \setminus \partial\Omega \to \mathbb{R}$ the radial maximal function $u^* : \mathbb{R}^n \setminus \partial\Omega \to [0,\infty]$ is defined as

$$u^*(Z \pm \tau e_n) := \sup_{\eta>0} \big|u(Z \pm (\tau+\eta)e_n)\big|, \qquad \forall\, Z \in \partial\Omega \text{ and } \forall\, \tau > 0. \tag{2.376}$$

Our next result is the following companion to Theorem 2.46.

Theorem 2.49. *Let Ω be an unbounded Lipschitz domain in $\mathbb{R}^n$ and assume that $N \in \mathbb{N}_0$, $p \in (0,1)$, and $a \in (-1/p, \infty)$. Then for any $A \in \mathscr{A}$ and $k \in \mathbb{N}$, the operator $\mathscr{R}_{k,A}$ initially considered as in (2.342) extends to a bounded mapping*

$$\widetilde{\mathscr{R}}_{k,A} : W_a^{|A|-\#A+N,p}(\Omega) \longrightarrow W_a^{N,p}(\mathbb{R}^n). \tag{2.377}$$

Moreover, the operator E initially defined as in (2.344) may be extended to a bounded mapping

$$E : W_a^{N,p}(\Omega) \longrightarrow W_a^{N,p}(\mathbb{R}^n). \tag{2.378}$$

Proof. Fix $A \in \mathscr{A}$, $k \in \mathbb{N}$, and let $f \in C_b^\infty(\overline{\Omega})$ be arbitrary. Writing the estimate (2.348) for N and $M := 2$ at the point $X := Z - (s+\eta)e_n$ for some fixed $Z \in \partial\Omega$ and $s, \eta > 0$ yields

$$\big|(\mathscr{R}_{k,A} f)(Z-(s+\eta)e_n)\big| \leq C \int\limits_{\delta(Z-(s+\eta)e_n)}^{\infty} \big|(\partial_n^N f)(Z+(t-(s+\eta))e_n)\big| \frac{dt}{t}. \tag{2.379}$$

Changing variables in the right-hand side of (2.348), we let $\tau := t - (s+2\eta)$ and obtain

$$\begin{aligned} &\big|(\mathscr{R}_{k,A} f)(Z-(s+\eta)e_n)\big| \\ &\quad \leq C \int\limits_{\delta(Z-(s+\eta)e_n)-(s+2\eta)}^{\infty} \big|(\partial_n^N f)(Z+(\tau+\eta))e_n)\big| \frac{d\tau}{\tau+(s+2\eta)}. \end{aligned} \tag{2.380}$$

However, thanks to (2.315) there holds

$$\delta(Z-(s+\eta)e_n) - (s+2\eta) > 2(s+\eta) - (s+2\eta) = s. \tag{2.381}$$

In particular $\tau \geq \delta(Z-(s+\eta)e_n) - (s+2\eta)$ guarantees that $\tau \geq 0$. Since $s, \eta > 0$ this further implies that $1/(\tau+(s+2\eta)) \leq 1/\tau$ whenever

$$\tau \geq \delta(Z-(s+\eta)e_n) - (s+2\eta). \tag{2.382}$$

Combining this with (2.381) and (2.380) yields

$$\begin{aligned} \big|(\mathscr{R}_{k,A} f)(Z-(s+\eta)e_n)\big| &\leq C \int_s^\infty \big|(\partial_n^N f)(Z+(\tau+\eta))e_n)\big| \frac{d\tau}{\tau} \\ &\leq C \int_s^\infty (\partial_n^N f)^*(Z+\tau e_n) \frac{d\tau}{\tau}, \end{aligned} \tag{2.383}$$

where the last inequality follows from the definition of the maximal operator from (2.376). Taking supremum in (2.383) over $\eta > 0$ and using again (2.376) gives

$$(\mathscr{R}_{k,A} f)^*(Z - s e_n) := \sup_{\eta>0} \big|(\mathscr{R}_{k,A} f)(Z - (s+\eta)e_n)\big| \le C \int_s^\infty (\partial_n^N f)^*(Z + \tau e_n)\,\frac{d\tau}{\tau}. \tag{2.384}$$

Now, based on (2.384), we may write

$$\Big(\int_0^\infty \big((\mathscr{R}_{k,A} f)^*(Z - s e_n)\big)^p \operatorname{dist}(Z - s e_n, \partial\Omega)^{ap}\, ds\Big)^{1/p} \le C \Big(\int_0^\infty \Big(\int_s^\infty (\partial_n^N f)^*(Z + r e_n)\frac{dr}{r}\Big)^p s^{ap}\, ds\Big)^{1/p}. \tag{2.385}$$

Having established (2.385), we shall utilize that Hardy's inequality continues to hold even when $p \in (0,1)$ provided that the function in question is nonincreasing (which is the case for $r \mapsto (\partial_n^N f)^*(Z + r e_n)$). Keeping this in mind and proceeding as in the case of (2.347) we arrive at the conclusion that there exists a finite constant $C = C(\Omega, k, A, a, p) > 0$ with the property that

$$\|(\mathscr{R}_{k,A} f)^*\|_{L^p(\mathbb{R}^n \setminus \overline{\Omega}, \operatorname{dist}(\cdot,\partial\Omega)^{ap}\, dX)} \le C \sum_{|\alpha| = |A| - \# A} \|(\partial^\alpha f)^*\|_{L^p(\Omega, \operatorname{dist}(\cdot,\partial\Omega)^{ap}\, dX)}, \tag{2.386}$$

for every $f \in C_b^\infty(\overline{\Omega})$. At this stage, from (2.333)–(2.334), (2.386), and (2.341)–(2.342) we deduce that

$$\|\widetilde{\mathscr{R}}_{k,A} f\|_{W_a^{N,p}(\mathbb{R}^n)} \le C \|f\|_{W_a^{|A| - \# A + N, p}(\Omega)}, \qquad \forall\, f \in C_b^\infty(\overline{\Omega}). \tag{2.387}$$

From this and Definition 2.48, the claim about the operator (2.377) follows. Finally, the claim about the operator (2.378) is handled similarly, and this completes the proof of the theorem. □

In our last result in this section we take up the issue of proving that the weighted Sobolev scale is stable under complex interpolation. Specifically, we shall prove the following theorem.

Theorem 2.50. *Let Ω be a Lipschitz domain in $\mathbb{R}^n$ and assume that $1 < p_i < \infty$ and $-1/p_i < a_i < 1 - 1/p_i$ for $i \in \{0,1\}$. Fix $\theta \in (0,1)$ and suppose that $p \in (0,\infty)$ and $a \in \mathbb{R}$ are such that $1/p = (1-\theta)/p_0 + \theta/p_1$ and $a = (1-\theta)a_0 + \theta a_1$. Then for each $m \in \mathbb{N}_0$ there holds*

$$\big[W^{m,p_0}_{a_0}(\Omega), W^{m,p_1}_{a_1}(\Omega)\big]_\theta = W^{m,p}_{a}(\Omega), \tag{2.388}$$

where $[\cdot,\cdot]_\theta$ *denotes the usual complex interpolation bracket.*

Proof. The proof is organized into a number of steps. Throughout, Ω is a fixed Lipschitz domain in $\mathbb{R}^n$.

Step I. *There holds*

$$\big[L^{p_0}(\mathbb{R}^n, \operatorname{dist}(\cdot,\partial\Omega)^{a_0p_0}\,dX), L^{p_1}(\mathbb{R}^n, \operatorname{dist}(\cdot,\partial\Omega)^{a_1p_1}\,dX)\big]_\theta = L^{p}(\mathbb{R}^n, \operatorname{dist}(\cdot,\partial\Omega)^{ap}\,dX), \tag{2.389}$$

granted that the indices involved are as in the statement of the theorem. This formula follows from well-known interpolation results for Lebesgue spaces with change of measure (cf. [8, Theorem 5.5.3, p. 120]).

Step II. *For each* $\alpha > 0$, *let* G_α *stand for the Bessel kernel of order* α *in* $\mathbb{R}^n$. *That is,* $G_\alpha \in \mathscr{C}^\infty(\mathbb{R}^n \setminus \{0\})$ *is the (positive, radial, decreasing) function whose Fourier transform is given by*

$$\widehat{G}_\alpha(\xi) = (1+|\xi|^2)^{-\alpha/2} \quad \text{for all } \xi \in \mathbb{R}^n. \tag{2.390}$$

Also, for each $p \in (1,\infty)$ *and any Muckenhoupt weight* $w \in A_p(\mathbb{R}^n)$, *define the weighted Sobolev spaces of order* $m \in \mathbb{N}$ in $\mathbb{R}^n$ *as*

$$W^{m,p}(\mathbb{R}^n, w\,dX) := \Big\{ f \in L^1_{loc}(\mathbb{R}^n) : \|f\|_{W^{m,p}(\mathbb{R}^n, w\,dX)} < +\infty \Big\}, \tag{2.391}$$

where (with derivatives taken in the sense of distributions),

$$\|f\|_{W^{m,p}(\mathbb{R}^n, w\,dX)} := \sum_{|\gamma|\le m} \left(\int_{\mathbb{R}^n} |\partial^\gamma f|^p w\,dX \right)^{1/p}. \tag{2.392}$$

Then

$$L^p(\mathbb{R}^n, w\,dX) \ni f \mapsto G_m * f \in W^{m,p}(\mathbb{R}^n, w\,dX) \tag{2.393}$$

is a well-defined, linear, bounded isomorphism. Indeed, this is [81, Theorem 3.3].

Step III. *Assume that* X_0, X_1 *and* Y_0, Y_1 *are two compatible pairs of Banach spaces such that* $\{Y_0, Y_1\}$ *is a retract of* $\{X_0, X_1\}$ *(as above, the "extension" and "restriction" operators are denoted by* E *and* R, *respectively). Then for each parameter* $\theta \in (0,1)$ *one has*

$$[Y_0, Y_1]_\theta = R\Big([X_0, X_1]_\theta\Big). \tag{2.394}$$

This is a folklore interpolation result.

Step VI. *Proof of* (2.388). From (2.362) and Steps I–II we obtain that

$$\left[W^{m,p_0}_{a_0}(\mathbb{R}^n), W^{m,p_1}_{a_1}(\mathbb{R}^n)\right]_\theta = W^{m,p}_a(\mathbb{R}^n), \tag{2.395}$$

provided the indices involved are as in the statement of the theorem. With this in hand, (2.388) follows from (2.367) in Theorem 2.46 and the abstract retract-type result from Step III.

2.7 Other Smoothness Spaces on Lipschitz Boundaries

Recall that for each $a \in \mathbb{R}$ we have set $(a)_+ := \max\{a, 0\}$. Consider three parameters p, q, s subject to

$$0 < p, q \leq \infty, \qquad (n-1)\Big(\tfrac{1}{p} - 1\Big)_+ < s < 1. \tag{2.396}$$

Assume first that $\Omega \subset \mathbb{R}^n$ is the upper-graph of a Lipschitz function $\varphi : \mathbb{R}^{n-1} \to \mathbb{R}$. We then define $B^{p,q}_s(\partial\Omega)$ as the space of locally integrable functions f on $\partial\Omega$ for which the assignment $\mathbb{R}^{n-1} \ni x \mapsto f(x, \varphi(x))$ belongs to $B^{p,q}_s(\mathbb{R}^{n-1})$ (cf. §2.3). This space is equipped with the natural quasi-norm

$$\|f\|_{B^{p,q}_s(\partial\Omega)} := \|f(\cdot, \varphi(\cdot))\|_{B^{p,q}_s(\mathbb{R}^{n-1})}. \tag{2.397}$$

As far as Besov spaces with a negative amount of smoothness are concerned, in the same context as above we set

$$f \in B^{p,q}_{s-1}(\partial\Omega) \Longleftrightarrow f(\cdot, \varphi(\cdot))\sqrt{1 + |\nabla\varphi(\cdot)|^2} \in B^{p,q}_{s-1}(\mathbb{R}^{n-1}), \tag{2.398}$$

$$\|f\|_{B^{p,q}_{s-1}(\partial\Omega)} := \|f(\cdot, \varphi(\cdot))\sqrt{1 + |\nabla\varphi(\cdot)|^2}\|_{B^{p,q}_{s-1}(\mathbb{R}^{n-1})}. \tag{2.399}$$

As is well-known, the case when $p = q = \infty$ corresponds to the usual (non-homogeneous) Hölder spaces $C^s(\partial\Omega)$, defined by the requirement that

$$\|f\|_{C^s(\partial\Omega)} := \|f\|_{L^\infty(\partial\Omega)} + \sup_{\substack{X \neq Y \\ X, Y \in \partial\Omega}} \frac{|f(X) - f(Y)|}{|X - Y|^s} < +\infty. \tag{2.400}$$

All the above definitions then readily extend to the case of (bounded) Lipschitz domains in $\mathbb{R}^n$ via a standard partition of unity argument.

We now recall several properties of the Besov scales just introduced above which are going to be of importance for us later on.

Proposition 2.51. *Let Ω be a Lipschitz domain in $\mathbb{R}^n$. For $(n-1)/n < p \leq \infty$ and $(n-1)(1/p-1)_+ < s < 1$, one then has*

$$\|f\|_{B^{p,p}_s(\partial\Omega)} \approx \|f\|_{L^p(\partial\Omega)} + \left(\int_{\partial\Omega}\int_{\partial\Omega} \frac{|f(X)-f(Y)|^p}{|X-Y|^{n-1+sp}}\,d\sigma(X)d\sigma(Y)\right)^{1/p}, \tag{2.401}$$

uniformly for $f \in L^1_{loc}(\partial\Omega)$.

The equivalence (2.401) is well-known when $\partial\Omega$ is flat and the more general case considered here may be derived from this by (locally) making a suitable bi-Lipschitz change of variables. In this vein, let us also point out that, for each fixed $\kappa > 0$, another quasi-norm equivalent to the expression in the right-hand side of (2.401) is

$$\|f\|_{L^p(\partial\Omega)} + \left(\iint\limits_{\substack{(X,Y)\in\partial\Omega\times\partial\Omega\\ |X-Y|<\kappa}} \frac{|f(X)-f(Y)|^p}{|X-Y|^{n-1+sp}}\,d\sigma(X)d\sigma(Y)\right)^{1/p}. \tag{2.402}$$

Also, if

$$\omega_p(f;t) := \left(\iint\limits_{\substack{(X,Y)\in\partial\Omega\times\partial\Omega\\ |X-Y|<t}} |f(X)-f(Y)|^p\,d\sigma(X)d\sigma(Y)\right)^{1/p}, \tag{2.403}$$

is the L^p-modulus of continuity of f, then for indices p, s satisfying the conditions $(n-1)/n < p \leq \infty$ and $(n-1)(1/p-1)_+ < s < 1$ we also have

$$\begin{aligned}\|f\|_{B^{p,p}_s(\partial\Omega)} &\approx \|f\|_{L^p(\partial\Omega)} + \left(\int_0^\infty \frac{\omega_p(f;t)^p}{t^{n-1+sp}}\,dt\right)^{1/p}\\ &\approx \|f\|_{L^p(\partial\Omega)} + \left(\int_0^1 \frac{\omega_p(f;t)^p}{t^{n-1+sp}}\,dt\right)^{1/p}.\end{aligned} \tag{2.404}$$

To state our next result, given $0 < p \leq \infty$, set

$$p' := \begin{cases} \frac{p}{p-1} & \text{if } 1 < p < \infty,\\ \infty & \text{if } 0 < p \leq 1,\\ 1 & \text{if } p = \infty,\end{cases} \tag{2.405}$$

for the conjugate exponent of p.

Proposition 2.52. *Let Ω be a Lipschitz domain in $\mathbb{R}^n$ and assume that $0 < p, q < \infty$ and $(n-1)(\frac{1}{p}-1)_+ < s < 1$. Then*

$$\left(B^{p,q}_{s-1}(\partial\Omega)\right)^* = B^{p',q'}_{1-s+(n-1)(\frac{1}{p}-1)_+}(\partial\Omega),$$

$$(2.406)$$

$$\left(B^{p,q}_{s}(\partial\Omega)\right)^* = B^{p',q'}_{-s+(n-1)(\frac{1}{p}-1)_+}(\partial\Omega),$$

where the pairing between f in a Besov space and g in its dual is a natural extension of the bilinear form $(f,g) \mapsto \int_{\partial\Omega} fg\, d\sigma$.

Proof. This may be proved by reducing matters (via localization using a smooth partition of unity and flattening the boundary via a bi-Lipschitz map) to classical duality results in the entire Euclidean space (cf., e.g., [107] for the latter setting). □

The following theorem has been proved in [73].

Theorem 2.53. *Let Ω be a Lipschitz domain in $\mathbb{R}^n$ and assume that the indices p, s satisfy $\frac{n-1}{n} < p \leq \infty$ and $(n-1)(\frac{1}{p}-1)_+ < s < 1$. Then the following hold:*

(i) The restriction to the boundary extends to a linear, bounded operator

$$\mathrm{Tr} : B^{p,q}_{s+\frac{1}{p}}(\Omega) \longrightarrow B^{p,q}_{s}(\partial\Omega) \quad for\ 0 < q \leq \infty. \qquad (2.407)$$

Moreover, for this range of indices, Tr is onto and has a bounded right inverse

$$\mathrm{Ex} : B^{p,q}_{s}(\partial\Omega) \longrightarrow B^{p,q}_{s+\frac{1}{p}}(\Omega). \qquad (2.408)$$

(ii) Similar considerations hold for

$$\mathrm{Tr} : F^{p,q}_{s+\frac{1}{p}}(\Omega) \longrightarrow B^{p,p}_{s}(\partial\Omega) \qquad (2.409)$$

with the convention that $q = \infty$ if $p = \infty$. More specifically, Tr in (2.409) is a linear, bounded, operator which has a linear, bounded right inverse

$$\mathrm{Ex} : B^{p,p}_{s}(\partial\Omega) \longrightarrow F^{p,q}_{s+\frac{1}{p}}(\Omega). \qquad (2.410)$$

Remark. (i) Assume that Ω is a bounded Lipschitz domain in $\mathbb{R}^n$. Then a pointwise description of the trace operator from part (i) of Theorem 2.53 is

$$\left(\mathrm{Tr}u\right)(X) = \lim_{r \to 0^+} \fint_{B(X,r)\cap\Omega} u(Y)\, dY, \qquad (2.411)$$

where the limit exists at σ-a.e. $X \in \partial\Omega$.

(ii) Let Ω be a bounded Lipschitz domain in $\mathbb{R}^n$ and fix p, q, s as in the statement of Theorem 2.53. Also, assume that $u \in F^{p,q}_{s+\frac{1}{p}}(\Omega) \cap C^0(\Omega)$ is such that $u\lfloor_{\partial\Omega}$, the non-tangential restriction to the boundary, exists almost everywhere on $\partial\Omega$. Then

$$\operatorname{Tr}(u) = u\Big|_{\partial\Omega} \quad \text{almost everywhere on} \quad \partial\Omega. \tag{2.412}$$

Since the statement (2.412) is local, using a suitable partition of unity matters can be reduced to the case in which Ω is a bounded Lipschitz domain, star-like with respect to the origin. In this case, for each $\varepsilon > 0$ we set $u_\varepsilon(X) = u\big(\frac{X}{1+\varepsilon}\big)$, $X \in \Omega$. It follows that $u_\varepsilon \in F^{p,q}_{s+\frac{1}{p}}(\Omega) \cap C^0(\overline{\Omega})$ and

$$\begin{aligned} &u_\varepsilon\Big|_{\partial\Omega} \to u\Big|_{\partial\Omega} \quad \text{pointwise on } \partial\Omega, \text{as } \varepsilon \to 0^+, \\ &\text{and } u_\varepsilon \to u \text{ in } F^{p,q}_{s+\frac{1}{p}}(\Omega), \text{ as } \varepsilon \to 0^+. \end{aligned} \tag{2.413}$$

Employing Theorem 2.53, the second convergence in (2.413) implies that

$$\operatorname{Tr}(u_\varepsilon) \longrightarrow \operatorname{Tr}(u) \text{ in } B^{p,p}_s(\partial\Omega) \hookrightarrow L^1(\partial\Omega) \text{ as } \varepsilon \to 0^+. \tag{2.414}$$

Now, the claim formulated in (2.412) becomes a consequence of (2.414) and the first convergence in (2.413).

■

We now proceed to discuss Triebel–Lizorkin spaces defined on the boundary of a Lipschitz domain $\Omega \subset \mathbb{R}^n$, denoted in the sequel by $F^{p,q}_s(\partial\Omega)$. Compared with the Besov scale, the most important novel aspect here is the possibility of allowing the endpoint case $s = 1$ as part of the general discussion if $q = 2$. To discuss this in more detail, assume that either

$$0 < p < \infty, \quad 0 < q \le \infty, \quad (n-1)\Big(\frac{1}{\min\{p,q\}} - 1\Big)_+ < s < 1, \tag{2.415}$$

or

$$\frac{n-1}{n} < p < \infty, \quad q = 2, \quad s \in \{0, 1\}. \tag{2.416}$$

In this scenario, the Triebel–Lizorkin scale in $\mathbb{R}^{n-1}$ is invariant under pointwise multiplication by Lipschitz maps as well as composition by Lipschitz diffeomorphisms.

When Ω is a Lipschitz domain as in (2.14), we may therefore define the space $F^{p,q}_s(\partial\Omega)$ as the collection of all locally integrable functions f on $\partial\Omega$ such that

$$f(\cdot, \varphi(\cdot)) \in F^{p,q}_s(\mathbb{R}^{n-1}), \tag{2.417}$$

endowed with the quasi-norm

$$\|f\|_{F^{p,q}_s(\partial\Omega)} := \|f(\cdot, \varphi(\cdot))\|_{F^{p,q}_s(\mathbb{R}^{n-1})}. \tag{2.418}$$

Also, $F^{p,q}_{s-1}(\partial\Omega)$ is defined (for the same range of indices p, q, s as in (2.415)–(2.416)) as the collection of all functionals $f \in (\mathrm{Lip}_c(\partial\Omega))'$ such that

$$f(\cdot,\varphi(\cdot))\sqrt{1+|\nabla\varphi(\cdot)|^2} \in F^{p,q}_{s-1}(\mathbb{R}^{n-1}), \tag{2.419}$$

and we equip this space with the quasi-norm

$$\|f\|_{F^{p,q}_{s-1}(\partial\Omega)} := \|f(\cdot,\varphi(\cdot))\sqrt{1+|\nabla\varphi(\cdot)|^2}\|_{F^{p,q}_{s-1}(\mathbb{R}^{n-1})}. \tag{2.420}$$

Finally, when $\Omega \subset \mathbb{R}^n$ is a *bounded* Lipschitz domain and (s, p, q) are as in (2.415)–(2.416), we define $F^{p,q}_s(\partial\Omega)$ and $F^{p,q}_{s-1}(\partial\Omega)$ via localization (using a smooth, finite partition of unity) and pull-back to $\mathbb{R}^{n-1}$ (in the manner described above, for graph-Lipschitz domains). When equipped with the natural quasi-norms, the Triebel–Lizorkin spaces just introduced are quasi-Banach, and different partitions of unity yield equivalent quasi-norms. Two basic identities, relating Triebel–Lizorkin spaces to Sobolev spaces on $\partial\Omega$ read as follows:

$$F^{p,2}_0(\partial\Omega) = L^p(\partial\Omega), \quad F^{p,2}_1(\partial\Omega) = L^p_1(\partial\Omega), \quad \forall\, p \in (1,\infty), \tag{2.421}$$

The above formulas have natural counterparts for values of $p \leq 1$, more specifically

$$F^{p,2}_0(\partial\Omega) = h^p_{at}(\partial\Omega), \quad F^{p,2}_1(\partial\Omega) = h^{1,p}_{at}(\partial\Omega), \qquad \tfrac{n-1}{n} < p \leq 1, \tag{2.422}$$

where $h^p_{at}(\partial\Omega)$ is the inhomogeneous Hardy space and $h^{1,p}_{at}(\partial\Omega)$ is the inhomogeneous Hardy-based Sobolev space of order one on $\partial\Omega$, considered in greater detail further below. For now, we wish to note that introducing

$$\begin{aligned} h^p(\partial\Omega) &:= \begin{cases} h^p_{at}(\partial\Omega) \text{ if } \frac{n-1}{n} < p \leq 1, \\[1ex] L^p(\partial\Omega) \text{ if } 1 < p < \infty, \end{cases} \\ h^p_1(\partial\Omega) &:= \begin{cases} h^{1,p}_{at}(\partial\Omega) \text{ if } \frac{n-1}{n} < p \leq 1, \\[1ex] L^p_1(\partial\Omega) \text{ if } 1 < p < \infty, \end{cases} \end{aligned} \tag{2.423}$$

allows us to combine (2.421)–(2.422) into

$$F^{p,2}_0(\partial\Omega) = h^p(\partial\Omega) \quad \text{and} \quad F^{p,2}_1(\partial\Omega) = h^p_1(\partial\Omega), \qquad \tfrac{n-1}{n} < p < \infty. \tag{2.424}$$

Since these identifications involve working with a multitude of Hardy-type spaces, we shall devote the next few pages to reviewing them, and to discussing their basic properties.

In a first stage, we shall assume that Ω is as in (2.14), i.e., the *unbounded domain in $\mathbb{R}^n$ lying above the graph of the Lipschitz function $\varphi : \mathbb{R}^{n-1} \to \mathbb{R}$*. For the case when $\frac{n-1}{n} < p \leq 1$, the *homogeneous* Hardy space is then defined by

$$H^p_{at}(\partial\Omega) := \Big\{ f = \sum_j \lambda_j a_j \, : \, a_j \, (p, p_o)\text{-atom}, \ (\lambda_j)_j \in \ell^p \Big\}, \tag{2.425}$$

where the series converges in $\big(\mathrm{Lip}_c(\partial\Omega)\big)'$, the dual of $\mathrm{Lip}_c(\partial\Omega)$, and equip the space in (2.425) with the usual infimum quasi-norm. Here, $1 < p_o \leq \infty$ is a fixed parameter and a measurable function $a : \partial\Omega \to \mathbb{R}$ is called a (p, p_o)-`atom` if there exists a surface ball $S_r \subset \partial\Omega$ such that

$$\operatorname{supp} a \subseteq S_r, \quad \|a\|_{L^{p_o}(\partial\Omega)} \leq r^{-(n-1)\left(\frac{1}{p_o} - \frac{1}{p}\right)} \quad \text{and} \quad \int_{\partial\Omega} a \, d\sigma = 0. \tag{2.426}$$

Corresponding to one unit more on the smoothness scale we have $H^{1,p}_{at}(\partial\Omega)$, defined for $\frac{n-1}{n} < p \leq 1$ as the ℓ^p-span of 'regular' atoms. More specifically, if Ω is as in (2.14) and $[f]$ denotes the class of f modulo constants, define

$$H^{1,p}_{at}(\partial\Omega) := \Big\{ [f] : \, f \in L^1_{loc}(\partial\Omega) \text{ and } \exists \, (\lambda_i)_i \in \ell^p \text{ and } a_i \text{ regular}(p, p_o)\text{-atoms}$$
$$\text{with } \partial_{\tau_{jn}} f = \sum_{i=1}^{\infty} \lambda_i \partial_{\tau_{jn}} a_i \text{ whenever } 1 \leq j \leq n-1 \Big\}, \tag{2.427}$$

where the series converges in $\big(\mathrm{Lip}\,(\partial\Omega)\big)'$. Here, if $(n-1)/n < p \leq 1 < p_o \leq \infty$, a function $a \in L^{p_o}_1(\partial\Omega)$ is called a `regular` (p, p_o)*- atom* if there exists a surface ball S_r so that

$$\operatorname{supp} a \subseteq S_r, \qquad \|\nabla_{tan} a\|_{L^{p_o}(\partial\Omega)} \leq r^{(n-1)\left(\frac{1}{p_o} - \frac{1}{p}\right)}. \tag{2.428}$$

Also, set

$$\|[f]\|_{H^{1,p}_{at}(\partial\Omega)} := \inf \Big(\sum_{i \in \mathbb{N}} |\lambda_i|^p \Big)^{1/p}, \qquad [f] \in H^{1,p}_{at}(\partial\Omega), \tag{2.429}$$

where the infimum is taken over all possible representations of the tangential derivatives of f as in the definition of $H^{1,p}_{at}(\partial\Omega)$. It can be shown that

$$[f] \in H^{1,p}_{at}(\partial\Omega) \Longleftrightarrow [f(\cdot, \varphi(\cdot))] \in \dot{F}^{p,2}_1(\mathbb{R}^{n-1}), \tag{2.430}$$

the homogeneous Triebel–Lizorkin space in $\mathbb{R}^{n-1}$. In particular, this shows that different choices of the parameter p_o in (2.428) yield the same vector space and topology on $H^{1,p}_{at}(\partial\Omega)$. An alternative characterization of this space is

$$H^{1,p}_{at}(\partial\Omega) = \Big\{[f] : f \in L^1_{loc}(\partial\Omega) \text{ and } \partial_{\tau_{jn}} f \in H^p_{at}(\partial\Omega),\ 1 \le j \le n-1\Big\}, \tag{2.431}$$

and, in fact,

$$\|[f]\|_{H^{1,p}_{at}(\partial\Omega)} \approx \sum_{j=1}^{n-1} \|\partial_{\tau_{jn}} f\|_{H^p(\partial\Omega)}, \tag{2.432}$$

whenever $\frac{n-1}{n} < p \le 1$.

The space $H^{1,p}_{at}(\partial\Omega)$ in (2.427) is defined modulo constants. A realization of this as a space of genuine functions is as follows. If $\frac{n-1}{n} < p \le 1$ and $p^* \in (1,\infty)$ is such that

$$\frac{1}{p^*} = \frac{1}{p} - \frac{1}{n-1} \tag{2.433}$$

we set

$$\widetilde{H}^{1,p}_{at}(\partial\Omega) := \Big\{f \in L^{p^*}(\partial\Omega) : f = \sum_{j=1}^{\infty} \lambda_j a_j \text{ in } L^{p^*}(\partial\Omega), \quad (\lambda_j)_j \in \ell^p,\ a_j \text{ regular}(p,p_o)\text{-atom}\Big\}, \tag{2.434}$$

and equip it with the natural infimum quasi-norm. It can then be checked that, for each $p \in (\frac{n-1}{n}, 1]$, the application

$$\widetilde{H}^{1,p}_{at}(\partial\Omega) \ni f \mapsto [f] := f + \mathbb{R} \in H^{1,p}_{at}(\partial\Omega) \tag{2.435}$$

is an isomorphism. As a consequence, the definition of $\widetilde{H}^{1,p}_{at}(\partial\Omega)$ is independent of the particular choice of $p_o \in (1,\infty]$. Let us also point out here that, when used in concert with (2.432), the fact that (2.435) is an isomorphism further entails

$$\|f\|_{\widetilde{H}^{1,p}_{at}(\partial\Omega)} \approx \|[f]\|_{H^{1,p}_{at}(\partial\Omega)} \approx \textstyle\sum_{j=1}^{n-1} \|\partial_{\tau_{jn}} f\|_{H^p_{at}(\partial\Omega)} \quad \text{uniformly for } f \in \widetilde{H}^{1,p}_{at}(\partial\Omega). \tag{2.436}$$

A distinctive feature of $\widetilde{H}^{1,p}_{at}(\partial\Omega)$ is that this space is local, in the sense that if Ω is as in (2.14) and $p \in (\frac{n-1}{n}, 1]$ then

$$f \in \widetilde{H}^{1,p}_{at}(\partial\Omega) \Longrightarrow \psi f \in \widetilde{H}^{1,p}_{at}(\partial\Omega), \tag{2.437}$$

plus a naturally accompanying estimate, for every $\psi \in \mathrm{Lip}_c(\partial\Omega)$.

The spaces $H^p_{at}(\partial\Omega)$ and $H^{1,p}_{at}(\partial\Omega)$ have inhomogeneous counterparts, denoted by $h^p_{at}(\partial\Omega)$ and $h^{1,p}_{at}(\partial\Omega)$, respectively. To be precise, fix a graph Lipschitz domain $\Omega \subset \mathbb{R}^n$ as in (2.14) and assume that $\frac{n-1}{n} < p \le 1 < p_o \le \infty$. Also, fix a threshold

$\eta > 0$. Call a function $a \in L^1_{loc}(\partial\Omega)$ an `inhomogeneous` (p, p_o)-`atom` if for some surface ball $S_r \subseteq \partial\Omega$

$$\operatorname{supp} a \subseteq S_r, \quad \|a\|_{L^{p_o}(\partial\Omega)} \leq r^{(n-1)\left(\frac{1}{p_o}-\frac{1}{p}\right)}, \text{ and} \\ \text{either } r = \eta, \text{ or } r < \eta \text{ and } \int_{\partial\Omega} a\, d\sigma = 0. \tag{2.438}$$

We then define $h^p_{at}(\partial\Omega)$ as the ℓ^p-span of inhomogeneous (p, p_o)-atoms, and equip it with the natural infimum-type quasi-norm. One can check that this is a "local" quasi-Banach space, in the sense that

$$h^p_{at}(\partial\Omega) \text{ is a module over } C^\alpha(\partial\Omega) \text{ for any } \alpha > (n-1)\left(\tfrac{1}{p}-1\right). \tag{2.439}$$

Different choices of the parameters p_o, η lead to equivalent quasi-norms and

$$\left(h^p_{at}(\partial\Omega)\right)^* = C^{(n-1)\left(\frac{1}{p}-1\right)}(\partial\Omega). \tag{2.440}$$

It is also useful to note that

$$H^p_{at}(\partial\Omega) \hookrightarrow h^p_{at}(\partial\Omega), \quad L^q_{comp}(\partial\Omega) \subset h^p_{at}(\partial\Omega), \quad \text{if } \tfrac{n-1}{n} < p \leq 1,\ q > 1. \tag{2.441}$$

Furthermore, for each $p \in (\frac{n-1}{n}, 1]$,

$$f \in h^p_{at}(\partial\Omega) \Longleftrightarrow f(\cdot, \varphi(\cdot))\sqrt{1 + |\nabla\varphi(\cdot)|^2} \in F^{p,2}_0(\mathbb{R}^{n-1}), \tag{2.442}$$

$$f \in H^p_{at}(\partial\Omega) \Longleftrightarrow f(\cdot, \varphi(\cdot))\sqrt{1 + |\nabla\varphi(\cdot)|^2} \in \dot{F}^{p,2}_0(\mathbb{R}^{n-1}), \tag{2.443}$$

These characterizations show that as far as the spaces $h^p_{at}(\partial\Omega)$, $H^p_{at}(\partial\Omega)$ are concerned, the particular value of the parameter p_o (used in the normalization of atoms) is immaterial.

Let us briefly digress for the purpose of recalling the local BMO space. As before, we make the convention that a barred integral indicates averaging. Now, for some fixed $0 < \eta < \operatorname{diam}(\partial\Omega)$, the local BMO space alluded to above is introduced as

$$f \in \operatorname{bmo}(\partial\Omega) \overset{def}{\Longleftrightarrow} f \in L^2(\partial\Omega) \quad \text{and} \quad \sup_{\substack{S_r \text{ surface ball} \\ \text{with } r \leq \eta}} \fint_{S_r} |f - f_{S_r}|\, d\sigma < \infty \tag{2.444}$$

where $f_{S_r} := \fint_{S_r} f\, d\sigma$, and is equipped with the natural norm

$$\|f\|_{\operatorname{bmo}(\partial\Omega)} := \|f\|_{L^2(\partial\Omega)} + \left(\sup_{\substack{S_r \text{ surface ball} \\ \text{with } r \leq \eta}} \fint_{S_r} |f - f_{S_r}|\, d\sigma\right). \tag{2.445}$$

Then (cf. [28]),

$$\Big(h^1_{at}(\partial\Omega)\Big)^* = \text{bmo}\,(\partial\Omega) \quad \text{and} \quad h^1_{at}(\partial\Omega) = \Big(\text{vmo}\,(\partial\Omega)\Big)^*, \tag{2.446}$$

where Sarason's space vmo $(\partial\Omega)$ of functions of vanishing mean oscillation on $\partial\Omega$ is defined by the requirement that $f \in \text{vmo}\,(\partial\Omega)$ if and only if

$$f \in \text{bmo}\,(\partial\Omega) \text{ and } \lim_{R\to 0}\left(\sup_{\substack{S_r \text{ surface ball}\\ \text{with } r\leq R}} \fint_{S_r} |f - f_{S_r}|\,d\sigma\right) = 0. \tag{2.447}$$

Let us point out that, for each $s \in (0,1)$, an alternative characterization of the latter space is

$$\text{vmo}\,(\partial\Omega) = \text{the closure of } C^s_c(\partial\Omega) \text{ in bmo}\,(\partial\Omega). \tag{2.448}$$

Also,

$$\Big(h^{1,1}_{at}(\partial\Omega)\Big)^* = \text{bmo}_{-1}(\partial\Omega), \tag{2.449}$$

where

$$\begin{aligned}\text{bmo}_{-1}(\partial\Omega) := \Big\{ g_* + \sum_{j,k=1}^n \partial_{\tau_{jk}} g_{jk} \,:\, g_* \in L^{n-1}(\partial\Omega) \text{ and}\\ g_{jk} \in \text{bmo}(\partial\Omega) \text{ for } 1 \leq j,k \leq n\Big\}.\end{aligned} \tag{2.450}$$

Above, the term $g_* + \sum_{j,k=1}^n \partial_{\tau_{jk}} g_{jk}$ should be understood in the following sense. Given $f \in h^{1,1}_{at}(\partial\Omega)$ we have

$$\Big\langle f, g_* + \sum_{j,k=1}^n \partial_{\tau_{jk}} g_{jk}\Big\rangle := \langle f, g_*\rangle + \sum_{j,k=1}^n \langle \partial_{\tau_{jk}} f, g_{jk}\rangle, \tag{2.451}$$

where the first pairing $\langle\cdot,\cdot\rangle$ in the right hand-side of (2.451) is understood in the sense of the duality between $L^{\frac{n-1}{n-2}}(\partial\Omega) \hookleftarrow h^{1,1}_{at}(\partial\Omega)$ and $L^{n-1}(\partial\Omega)$ while the second denotes the duality pairing between $h^1_{at}(\partial\Omega)$ and bmo$(\partial\Omega)$.

Moving on, with Ω, p, p_o as before, fix an arbitrary small $\eta > 0$ then define

$$\begin{aligned}h^{1,p}_{at}(\partial\Omega) := \Big\{ f \in \big(\text{Lip}_c(\partial\Omega)\big)' : f = \sum_j \lambda_j a_j,\ (\lambda_j)_j \in \ell^p \text{ and } \forall\, j\\ a_j \text{ regular}(p,p_o)\text{-atom supported in a surface ball of radius } \leq \eta\Big\},\end{aligned} \tag{2.452}$$

where the series converges in $\big(\mathrm{Lip}_c(\partial\Omega)\big)'$, and equip it with the natural infimum quasi-norm. Next, if p^* is as in (2.433) then, by Poincaré's inequality,

$$a \text{ regular } (p, p_o)\text{-atom} \Longrightarrow \|a\|_{L^{p^*}(\partial\Omega)} \le C(\partial\Omega, p, p_o), \tag{2.453}$$

$$\left.\begin{array}{l} a \text{ regular } (p, p_o)\text{-atom supported} \\ \text{in a surface ball of radius } \le \eta \end{array}\right\} \Rightarrow \|a\|_{L^p(\partial\Omega)} \le C(\partial\Omega, \eta, p, p_o). \tag{2.454}$$

Thus, if $f = \sum_{j=1}^{\infty} \lambda_j a_j$ is the atomic decomposition of $f \in h^{1,p}_{at}(\partial\Omega)$, it follows that the series $\sum_{j=1}^{\infty} \lambda_j a_j$ converges both in $L^{p^*}(\partial\Omega)$ and $L^p(\partial\Omega)$. As a consequence,

$$h^{1,p}_{at}(\partial\Omega) \hookrightarrow L^p(\partial\Omega) \cap L^{p^*}(\partial\Omega) \tag{2.455}$$

and, hence,

$$h^{1,p}_{at}(\partial\Omega) \hookrightarrow \widetilde{H}^{1,p}_{at}(\partial\Omega) \hookrightarrow L^{p^*}(\partial\Omega) \tag{2.456}$$

boundedly, for each $p \in (\frac{n-1}{n}, 1]$. Let us also record here the fact that, if $\frac{n-1}{n} < p \le 1$, we have

$$f \in h^{1,p}_{at}(\partial\Omega) \Longleftrightarrow f(\cdot, \varphi(\cdot)) \in F^{p,2}_1(\mathbb{R}^{n-1}). \tag{2.457}$$

In particular, various choices of the parameter p_o in (2.452) yield the same vector space and topology on $h^{1,p}_{at}(\partial\Omega)$. The equivalence (2.457) also shows that the space $h^{1,p}_{at}(\partial\Omega)$, $p \in (\frac{n-1}{n}, 1]$, is local, in the sense that

$$f \in h^{1,p}_{at}(\partial\Omega) \Longrightarrow \psi f \in h^{1,p}_{at}(\partial\Omega), \tag{2.458}$$

plus a natural estimate, for every function $\psi \in \mathrm{Lip}_c(\partial\Omega)$. The fact that

$$F^{p,2}_1(\mathbb{R}^{n-1}) = \Big\{f \in L^p(\mathbb{R}^{n-1}) \cap \mathcal{S}'(\mathbb{R}^{n-1}) : \\ [f] \in \dot{F}^{p,2}_1(\mathbb{R}^{n-1})\Big\}, \quad \tfrac{n-1}{n} < p \le 1, \tag{2.459}$$

yields another alternative characterization of $h^{1,p}_{at}(\partial\Omega)$, namely

$$h^{1,p}_{at}(\partial\Omega) = \Big\{f \in L^1_{loc}(\partial\Omega) : f \in L^p(\partial\Omega) \text{ and } \\ \partial_{\tau_{jn}} f \in H^p_{at}(\partial\Omega),\ 1 \le j \le n-1\Big\} \tag{2.460}$$

and, moreover,

$$\|f\|_{h^{1,p}_{at}(\partial\Omega)} \approx \|f\|_{L^p(\partial\Omega)} + \sum_{j=1}^{n-1} \|\partial_{\tau_{jn}} f\|_{H^p(\partial\Omega)}. \tag{2.461}$$

We shall now review the definitions and some of the most basic properties of Hardy–Sobolev spaces in the setting of *bounded* Lipschitz domains. For a bounded Lipschitz domain and $p \in (\frac{n-1}{n}, 1]$, the spaces $H^p_{at}(\partial\Omega)$, $\widetilde{H}^{1,p}_{at}(\partial\Omega)$, $h^p_{at}(\partial\Omega)$ and $h^{1,p}_{at}(\partial\Omega)$ can be defined as before. Thus, when $\Omega \subset \mathbb{R}^n$ is a bounded Lipschitz domain and $\frac{n-1}{n} < p \le 1$, we have:

$$\begin{aligned} & h^p_{at}(\partial\Omega) = H^p_{at}(\partial\Omega) + \mathbb{R} = H^p_{at}(\partial\Omega) + L^q(\partial\Omega) \quad \text{for each } q > 1, \\ & L^q_1(\partial\Omega) \hookrightarrow h^{1,p}_{at}(\partial\Omega) = \widetilde{H}^{1,p}_{at}(\partial\Omega) \hookrightarrow L^{p^*}(\partial\Omega), \quad \text{for each } q > 1, \\ & h^p_{at}(\partial\Omega), \quad h^{1,p}_{at}(\partial\Omega) \text{ are modules over } \mathrm{Lip}\,(\partial\Omega). \end{aligned} \tag{2.462}$$

Let Ω be Lipschitz domain (bounded or unbounded) in $\mathbb{R}^n$. Also, fix a function $\psi \in \mathrm{Lip}_c\,(\partial\Omega)$ and assume that $\frac{n-1}{n} < p \le 1 < p_o \le \infty$. It is then trivial to check that, for a fixed $0 < \eta \le \operatorname{diam}\partial\Omega$, there exists $C = C(\partial\Omega, \psi, \eta, p, p_o) > 0$ such that

$$\begin{aligned} & A \text{ regular } (p, p_o)\text{-atom supported in a surface ball of radius } \le \eta \\ & \quad \Longrightarrow C^{-1}\psi\, A \text{ is a regular } (p, p_o)\text{-atom on } \partial\Omega. \end{aligned} \tag{2.463}$$

A slightly more refined version of this result (allowing for atoms supported in surface balls of arbitrary radii) is as follows.

Lemma 2.54. *Let Ω be Lipschitz domain (bounded or unbounded) in $\mathbb{R}^n$ and assume that $\frac{n-1}{n} < p \le 1$ and $p^* \le p_o \le q \le \infty$, where p^* is as in* (2.433)*. If $\psi \in \mathrm{Lip}_c\,(\partial\Omega)$ then $\psi\, A$ is, up to a fixed multiplicative constant, a regular (p, p_o)-atom on $\partial\Omega$ whenever A is a regular (p, q)-atom on $\partial\Omega$.*

Proof. To fix ideas, let us assume that $\operatorname{supp}\psi \subseteq S_1$, a surface ball of radius 1, and that $\|\psi\|_{L^\infty(\partial\Omega)} + \|\nabla_{tan}\psi\|_{L^\infty(\partial\Omega)} \le 1$. Fix a regular (p,q)-atom A on $\partial\Omega$, i.e. a function $A \in L^q_1(\partial\Omega)$ satisfying $\operatorname{supp} A \subseteq S_r$, for some $r > 0$, and such that the inequality $\|\nabla_{tan} A\|_{L^q(\partial\Omega)} \le r^{(n-1)\left(\frac{1}{q}-\frac{1}{p}\right)}$ holds. In particular, Poincaré's inequality gives

$$\|A\|_{L^q(\partial\Omega)} \le C r \|\nabla_{tan} A\|_{L^q(\partial\Omega)} \le C\, r^{1+(n-1)\left(\frac{1}{q}-\frac{1}{p}\right)}. \tag{2.464}$$

Next, introduce $\tilde{r} := \min\{r, 1\} > 0$ and note that $\operatorname{supp}(\psi\, A) \subseteq S_{\tilde{r}}$. Going further, write $\nabla_{tan}(\psi\, A) = \psi\, \nabla_{tan} A + (\nabla_{tan}\psi) A =: I + II$, and use Hölder's inequality in order to estimate

$$\begin{aligned}\|I\|_{L^{p_o}(\partial\Omega)} &\le \|\psi\|_{L^\infty(\partial\Omega)}\|\nabla_{tan}A\|_{L^{p_o}(S_{\tilde r})} \le C\tilde r^{(n-1)\left(\frac{1}{p_o}-\frac{1}{q}\right)}\|\nabla_{tan}A\|_{L^q(\partial\Omega)}\\ &\le C\tilde r^{(n-1)\left(\frac{1}{p_o}-\frac{1}{q}\right)} r^{(n-1)\left(\frac{1}{q}-\frac{1}{p}\right)} \le C\tilde r^{(n-1)\left(\frac{1}{p_o}-\frac{1}{p}\right)}\end{aligned} \tag{2.465}$$

and

$$\begin{aligned}\|II\|_{L^{p_o}(\partial\Omega)} &\le \|\nabla_{tan}\psi\|_{L^\infty(\partial\Omega)}\|A\|_{L^{p_o}(S_{\tilde r})} \le C\tilde r^{(n-1)\left(\frac{1}{p_o}-\frac{1}{q}\right)}\|A\|_{L^q(\partial\Omega)}\\ &\le C\tilde r^{(n-1)\left(\frac{1}{p_o}-\frac{1}{q}\right)} r^{1+(n-1)\left(\frac{1}{q}-\frac{1}{p}\right)} \le C\tilde r^{(n-1)\left(\frac{1}{p_o}-\frac{1}{p}\right)}.\end{aligned} \tag{2.466}$$

It is only in the last step above that $p_o \ge p^*$ is needed (when r is large). Altogether, the estimates (2.465)–(2.466) give $\|\nabla_{tan}(\psi\, A)\|_{L^{p_o}(\partial\Omega)} \le C\tilde r^{(n-1)\left(\frac{1}{p_o}-\frac{1}{p}\right)}$, so $C^{-1}\psi\, A$ is a regular (p, p_o)-atom. □

Let $\Omega \subset \mathbb{R}^n$ be a bounded Lipschitz domain and assume that $\frac{n-1}{n} < p \le 1$. Let $\Sigma \subset \mathbb{R}^n$ be the graph of a real-valued Lipschitz function, suitably rotated and translated as to agree with $\partial\Omega$ for an non-empty, open subset of $\partial\Omega$. Finally, let $\xi \in \mathrm{Lip}_c(\Sigma \cap \partial\Omega)$. Then there exists $C > 0$ such that

$$\|\widetilde{\xi f}\|_{h^{1,p}_{at}(\partial\Omega)} \le C\|f\|_{\widetilde{H}^{1,p}_{at}(\Sigma)}, \tag{2.467}$$

$$\|\widetilde{\xi f}\|_{h^{1,p}_{at}(\partial\Omega)} \le C\|f\|_{h^{1,p}_{at}(\Sigma)}, \tag{2.468}$$

$$\|\widetilde{\xi f}\|_{\widetilde{H}^{1,p}_{at}(\Sigma)} \le C\|\widetilde{\xi f}\|_{h^{1,p}_{at}(\Sigma)} \le C\|f\|_{h^{1,p}_{at}(\partial\Omega)}, \tag{2.469}$$

where tilde denotes the extension by zero outside the support (naturally interpreted in each case). Indeed, (2.467) is implied by Lemma 2.54, whereas (2.468) is a direct consequence of (2.463), and (2.469) follows from (2.456) and (2.463). In turn, the estimates (2.467)–(2.469) permit to prove that many of the properties established for the scale $h^{1,p}_{at}(\partial\Omega)$ when Ω is a graph Lipschitz domain have natural counterparts in the setting of bounded Lipschitz domains.

Proposition 2.55. *Let Ω be a bounded Lipschitz domain in $\mathbb{R}^n$. Then the embedding*

$$h^p_{at}(\partial\Omega) \hookrightarrow L^{p^*}_{-1}(\partial\Omega) \tag{2.470}$$

is well-defined and bounded for each $\frac{n-1}{n} < p \le 1$, where p^* is as in (2.433).

Proof. First, let a be (p,∞)-atom in $h^p_{at}(\partial\Omega)$, i.e., a function satisfying $\operatorname{supp} a \subseteq S_r$ and $\|a\|_{L^\infty(\partial\Omega)} \le r^{-\frac{n-1}{p}}$ for some $r > 0$. In addition, if $0 < r < \eta$, it is also required that $\int_{\partial\Omega} a\, d\sigma = 0$. Thus, if $r \in (0,\eta)$, then for each $f \in L^{q^*}_1(\partial\Omega)$ with $1/p^* + 1/q^* = 1$ we have

$$\Big|\int_{\partial\Omega} af\,d\sigma\Big| = \Big|\int_{S_r} a\Big(f - \fint_{S_r} f\,d\sigma\Big)\,d\sigma\Big| \leq r^{-\frac{n-1}{p}} \int_{S_r}\Big|f - \fint_{S_r} f\,d\sigma\Big|\,d\sigma$$

$$\leq C r^{1-\frac{n-1}{p}} \int_{S_r} |\nabla_{tan} f|, \tag{2.471}$$

where the last estimate follows from applying Poincaré's inequality. Going further, (2.471) and Hölder's inequality give

$$\Big|\int_{\partial\Omega} af\,d\sigma\Big| \leq C r^{1-\frac{n-1}{p}+(n-1)(1-\frac{1}{q^*})} \|\nabla_{tan} f\|_{L^{q^*}(\partial\Omega)} \leq C\,\|\nabla_{tan} f\|_{L^{q^*}(\partial\Omega)}. \tag{2.472}$$

Hence, any (p,∞)-atom a with small support satisfies

$$a \in L^{p^*}_{-1}(\partial\Omega) \quad \text{and} \quad \|a\|_{L^{p^*}_{-1}(\partial\Omega)} \leq C = C(\partial\Omega, p) < \infty. \tag{2.473}$$

Consider next the case in which the atom a is supported in S_r with $r > \eta$. In this scenario, we write

$$\Big|\int_{\partial\Omega} af\,d\sigma\Big| \leq r^{-\frac{n-1}{p}} \int_{S_r} |f|\,d\sigma \leq C r^{-1} \|f\|_{L^{q^*}(\partial\Omega)} \leq \frac{C}{\eta}\|f\|_{L^{q^*}(\partial\Omega)}, \tag{2.474}$$

where the second estimate follows from Hölder's inequality and the fact that p^* and q^* are conjugate exponents. Hence, once again, (2.473) holds.

Now, for each $f = \sum_j \lambda_j a_j \in h^p_{at}(\partial\Omega)$, with $\{\lambda_j\}_j \in \ell^p$ and each a_j a (p,∞)-atom, we set

$$\Lambda_f : L^{q^*}_1(\partial\Omega) \to \mathbb{R}, \quad \Lambda_f(g) := \sum_j \lambda_j \int_{\partial\Omega} a_j g\,d\sigma. \tag{2.475}$$

Our goal is to show that the mapping (2.475) is well-defined (i.e., it does not depend on the atomic representation of f), linear and bounded. Concerning the well-definiteness of (2.475) it suffices to show that

$$\sum_j \lambda_j a_j = 0 \text{ in } h^p_{at}(\partial\Omega) \implies \sum_j \lambda_j \int_{\partial\Omega} a_j g\,d\sigma = 0 \quad \forall\, g \in L^{q^*}_1(\partial\Omega), \tag{2.476}$$

where $\{\lambda_j\}_j \in \ell^p$ and a_j, $j \in \mathbb{N}$, are as before. With this goal in mind, fix a function $g \in L^{q^*}_1(\partial\Omega)$ and let $\psi_\alpha \in \mathrm{Lip}(\partial\Omega)$ be such that $\psi_\alpha \to g$ in $L^{q^*}_1(\partial\Omega)$ as $\alpha \to \infty$. Since $\sum_j \lambda_j a_j = 0$ in $h^p_{at}(\partial\Omega)$ we have that, for each α,

$$\lim_{N\to\infty} \int_{\partial\Omega} \Big(\sum_{j=1}^{N} \lambda_j a_j\Big)\psi_\alpha\,d\sigma = 0. \tag{2.477}$$

Fix for the moment $\alpha \in \mathbb{N}$. Using the triangle inequality we have

$$\Big|\sum_{j=1}^{N}\lambda_j\int_{\partial\Omega}a_j g\,d\sigma\Big| \le \Big|\sum_{j=1}^{N}\lambda_j\int_{\partial\Omega}a_j\psi_\alpha\,d\sigma\Big| + \sum_{j=1}^{N}|\lambda_j|\Big|\int_{\partial\Omega}a_j(g-\psi_\alpha)\,d\sigma\Big|$$

$$=: I + II. \tag{2.478}$$

Invoking (2.472) and (2.474) we may write

$$II \le C\|g-\psi_\alpha\|_{L_1^{q^*}(\partial\Omega)}\sum_{j=1}^{N}|\lambda_j|$$

$$\le C\|\{\lambda_j\}_j\|_{\ell^p}\|g-\psi_\alpha\|_{L_1^{q^*}(\partial\Omega)} \to 0 \quad \text{as } \alpha\to\infty. \tag{2.479}$$

Then, by passing to the limit as $\alpha\to\infty$ and $N\to\infty$ in (2.478), using (2.477) and (2.479) we finally obtain (2.476). The boundedness of the mapping Λ_f in (2.475) readily follows from (2.472) and (2.474). This finally gives $f\in L_1^{p^*}(\partial\Omega)$ and the desired estimate and completes the proof of Proposition 2.55. □

We continue by recording the analogue of (2.460) in the case when $\Omega\subset\mathbb{R}^n$ is a *bounded* Lipschitz domain.

Proposition 2.56. *Assume that $\Omega\subset\mathbb{R}^n$ is a bounded Lipschitz domain, fix some exponent $p\in\big(\frac{n-1}{n},1\big]$ and suppose that p^* is as in (2.433). Then*

$$h_{at}^{1,p}(\partial\Omega) = \Big\{f\in L^{p^*}(\partial\Omega) : \partial_{\tau_{jk}}f\in H_{at}^p(\partial\Omega),\ 1\le j,k\le n\Big\}$$

$$= \Big\{f\in L^{p^*}(\partial\Omega) : \partial_{\tau_{jk}}f\in h_{at}^p(\partial\Omega),\ 1\le j,k\le n\Big\} \tag{2.480}$$

and, in addition,

$$\|f\|_{h_{at}^{1,p}(\partial\Omega)} \approx \|f\|_{L^{p^*}(\partial\Omega)} + \sum_{j,k=1}^{n}\|\partial_{\tau_{jk}}f\|_{H^p(\partial\Omega)} \tag{2.481}$$

$$\approx \|f\|_{L^{p^*}(\partial\Omega)} + \sum_{j,k=1}^{n}\|\partial_{\tau_{jk}}f\|_{h^p(\partial\Omega)}.$$

Proof. To get started, we claim that for each $j,k\in\{1,\dots,n\}$ the tangential derivative operator

$$\partial_{\tau_{jk}} : h_{at}^{1,p}(\partial\Omega) \longrightarrow H_{at}^p(\partial\Omega) \tag{2.482}$$

is well-defined, linear and bounded. To prove this, fix $1<p_o\le\infty$ and observe that $\partial_{\tau_{jk}}a$ is a (p,p_o)-atom whenever a is a regular (p,p_o)-atom. It is therefore natural

to try to define the operator in (2.482) as

$$\partial_{\tau_{jk}} f := \sum_i \lambda_i \partial_{\tau_{jk}} a_i \text{ whenever } f = \sum_i \lambda_i a_i \text{ in } h^{1,p}_{at}(\partial\Omega). \tag{2.483}$$

Nonetheless, due to the redundancy in the atomic representations of functions in $h^{1,p}_{at}(\partial\Omega)$ the above observation alone does not guarantee that this operator is well-defined. See, e.g., the discussion in [11]. In order to overcome this difficulty, it suffices to show that if $\{\lambda_j\}_j \in \ell^p$ and a_j, $j \in \mathbb{N}$ are (p, p_o)-regular atoms, then

$$\sum_i \lambda_i a_i = 0 \text{ in } h^{1,p}_{at}(\partial\Omega) \Longrightarrow \sum_i \lambda_i \partial_{\tau_{jk}} a_i = 0 \text{ in } h^p_{at}(\partial\Omega). \tag{2.484}$$

This, however, is a consequence of (2.455), Proposition 2.55 and (2.103). Hence, the operator (2.482) is well-defined and bounded.

Turning to the identity (2.480), let us note that, thanks to (2.462), (2.482), the three spaces are listed in increasing order. Hence, it suffices to show that if $f \in L^{p^*}(\partial\Omega)$ has $\partial_{\tau_{jk}} f \in h^p_{at}(\partial\Omega)$ for $1 \le j, k \le n$, then $f \in h^{1,p}_{at}(\partial\Omega)$. Using a smooth partition of unity, matters can be reduced to the case when $\partial\Omega$ is replaced by $\Sigma \subset \mathbb{R}^n$, the graph of a real-valued Lipschitz function, defined in $\mathbb{R}^{n-1}$, and f is compactly supported on Σ. By further flattening Σ to $\mathbb{R}^{n-1}$ using a bi-Lipschitz change of variables, we arrive at the following question. Prove that if $f \in L^{p^*}_{comp}(\mathbb{R}^{n-1}) \hookrightarrow h^p_{at}(\mathbb{R}^{n-1})$ has $\partial_j f \in h^p_{at}(\mathbb{R}^{n-1})$ for every $j = 1, \dots, n-1$, then $f \in F^{p,2}_1(\mathbb{R}^{n-1})$. However, since $h^p_{at}(\mathbb{R}^{n-1}) = F^{p,2}_0(\mathbb{R}^{n-1})$ for $\frac{n-1}{n} < p \le 1$, this latter claim follows from well-known lifting results for Triebel–Lizorkin spaces. Finally, the equivalences in (2.481) are implicit in the above reasoning. □

As already mentioned, the Besov and Triebel–Lizorkin spaces have been defined in such a way that a number of basic properties from the Euclidean setting carry over to spaces defined on $\partial\Omega$ in a rather direct fashion. We continue by recording an interpolation result which is going to be very useful for us here. To state it, recall that $(\cdot,\cdot)_{\theta,q}$ and $[\cdot,\cdot]_\theta$ stand, respectively, for the real and complex interpolation brackets.

Proposition 2.57. *Suppose that Ω is a bounded Lipschitz domain in $\mathbb{R}^n$. Also, assume that $0 < p, q, q_0, q_1 \le \infty$ and that*

$$\begin{aligned} &\textit{either } (n-1)\Big(\tfrac{1}{p}-1\Big)_+ < s_0 \neq s_1 < 1, \\ &\textit{or } -1 + (n-1)\Big(\tfrac{1}{p}-1\Big)_+ < s_0 \neq s_1 < 0. \end{aligned} \tag{2.485}$$

Then, with $0 < \theta < 1$, $s = (1-\theta)s_0 + \theta s_1$,

$$\big(B^{p,q_0}_{s_0}(\partial\Omega), B^{p,q_1}_{s_1}(\partial\Omega)\big)_{\theta,q} = B^{p,q}_s(\partial\Omega). \tag{2.486}$$

Furthermore, if $s_0 \neq s_1$ *and* $0 < p_i, q_i \leq \infty$, $i = 0, 1$, *satisfy* $\min\{q_0, q_1\} < \infty$ *as well as either of the following two conditions*

$$\begin{aligned} &\textit{either}\ (n-1)\Big(\tfrac{1}{p_i}-1\Big)_+ < s_i < 1,\ i = 0,1, \\ &\textit{or}\ -1+(n-1)\Big(\tfrac{1}{p_i}-1\Big)_+ < s_i < 0,\ i = 0,1, \end{aligned} \tag{2.487}$$

then

$$\big[B^{p_0,q_0}_{s_0}(\partial\Omega), B^{p_1,q_1}_{s_1}(\partial\Omega)\big]_\theta = B^{p,q}_s(\partial\Omega), \tag{2.488}$$

where $0 < \theta < 1$, $s := (1-\theta)s_0 + \theta s_1$, $\frac{1}{p} := \frac{1-\theta}{p_0} + \frac{\theta}{p_1}$ *and* $\frac{1}{q} := \frac{1-\theta}{q_0} + \frac{\theta}{q_1}$.

Next, consider

$$\begin{aligned} \mathscr{O} := \Big\{ &(p,q,s) \in (0,\infty) \times (0,\infty] \times [0,1] : \textit{either}\ (n-1)\big(\tfrac{1}{\min\{p,q\}} - 1\big)_+ < s < 1, \\ &\textit{or}\ \tfrac{n-1}{n} < p,\ q = 2\ \textit{and}\ s \in \{0,1\}\Big\}. \end{aligned} \tag{2.489}$$

Then given any (p, q_0, s_0) *and* (p, q_1, s_1) *in* $\mathscr{O}$ *along with* $\theta \in (0,1)$ *it follows that*

$$\begin{cases} \big(F^{p,q_0}_{s_0}(\partial\Omega), F^{p,q_1}_{s_1}(\partial\Omega)\big)_{\theta,q} = B^{p,q}_s(\partial\Omega), \\ \big(F^{p,q_0}_{s_0-1}(\partial\Omega), F^{p,q_1}_{s_1-1}(\partial\Omega)\big)_{\theta,q} = B^{p,q}_{s-1}(\partial\Omega), \end{cases} \tag{2.490}$$

provided $s_0 \neq s_1$ *and* $s = (1-\theta)s_0 + \theta s_1$.

Finally, with $\mathscr{O}$ *as in (2.489), assume that* $(p_i, q_i, s_i) \in \mathscr{O}$, $i = 0, 1$, *and fix a number* $\theta \in (0,1)$. *Also, set* $s := (1-\theta)s_0 + \theta s_1$, $\frac{1}{p} := \frac{1-\theta}{p_0} + \frac{\theta}{p_1}$ *and* $\frac{1}{q} := \frac{1-\theta}{q_0} + \frac{\theta}{q_1}$. *Then*

$$\big[F^{p_0,q_0}_{s_0}(\partial\Omega), F^{p_1,q_1}_{s_1}(\partial\Omega)\big]_\theta = F^{p,q}_s(\partial\Omega) \tag{2.491}$$

provided $\min\{q_0, q_1\} < \infty$.

As an application of the above proposition, we shall establish the following useful result.

Proposition 2.58. *Let* $\Omega \subset \mathbb{R}^n$ *be a bounded Lipschitz domain and fix* p, q, s *such that* $(n-1)/n < p < \infty$, $0 < q \leq \infty$, *and* $(n-1)(\frac{1}{p}-1)_+ < s < 1$. *Then, for each* $j, k \in \{1, \dots, n\}$, *the tangential derivative operator*

$$\partial_{\tau_{jk}} : B^{p,q}_s(\partial\Omega) \longrightarrow B^{p,q}_{s-1}(\partial\Omega) \tag{2.492}$$

is well-defined, linear and bounded.

Proof. Call a point in the plane, whose coordinates $(s, 1/p)$ satisfy $(n-1)/n < p < \infty$ and $(n-1)(\frac{1}{p}-1)_+ < s < 1$, "good" if

$$\partial_{\tau_{jk}} : F^{p,2}_s(\partial\Omega) \longrightarrow F^{p,2}_{s-1}(\partial\Omega) \tag{2.493}$$

is well-defined, linear and bounded, for each $j, k = 1, \ldots, n$. Likewise, call a subregion of the plane "good" if all its points are so. Then Proposition 2.57 ensures that the collection of all good points is a convex set in the plane. Also, from (2.480) and Corollary 2.12, the (open) segments with endpoints $(1,0)$, $(1, \frac{n}{n-1})$ and $(0,0)$, $(0,1)$ are good. The bottom line is that all points whose coordinates $(s, 1/p)$ satisfy $(n-1)/n < p < \infty$ and $(n-1)(\frac{1}{p}-1)_+ < s < 1$ are good. Having established this, the desired conclusion about the operator (2.492) follows from (2.490). □

Proposition 2.59. *Let $\Omega \subset \mathbb{R}^n$ be an unbounded Lipschitz domain (cf. (2.14)) and assume that $u \in C^1(\Omega)$ is such that $\mathcal{N}(\nabla u) \in L^p(\partial\Omega)$ for some $p \in (\frac{n-1}{n}, 1]$. Then u has a non-tangential limit at almost every boundary point on $\partial\Omega$,*

$$\Big[u\Big\lfloor_{\partial\Omega}\Big] \in H^{1,p}_{at}(\partial\Omega) \quad \text{and} \quad \big\|[u\lfloor_{\partial\Omega}]\big\|_{H^{1,p}_{at}(\partial\Omega)} \leq C\,\|\mathcal{N}(\nabla u)\|_{L^p(\partial\Omega)}. \tag{2.494}$$

Proof. The pointwise existence of the non-tangential boundary trace $u\lfloor_{\partial\Omega}$ almost everywhere on $\partial\Omega$ is established as in the proof of Proposition 2.15. Let us show that

$$u\Big\lfloor_{\partial\Omega} \in L^1_{loc}(\partial\Omega). \tag{2.495}$$

Indeed, this follows from the sequence of implications

$$\begin{aligned}\mathcal{N}(\nabla u) \in L^p(\partial\Omega) &\Longrightarrow \nabla u \in L^{\frac{np}{n-1}}(\Omega) \Longrightarrow u \in W^{1,\frac{np}{n-1}}_{loc}(\overline{\Omega})\\ &\Longrightarrow \operatorname{Tr} u \in L^{p^*}_{loc}(\partial\Omega) \Longrightarrow u\Big\lfloor_{\partial\Omega} \in L^1_{loc}(\partial\Omega).\end{aligned} \tag{2.496}$$

Above, the first implication follows from Proposition 2.3, the second one can be justified using Proposition 2.24, the third one is a consequence of (2.200), Theorem 2.53, and the embedding

$$B^{\frac{np}{n-1},\frac{np}{n-1}}_{1-\frac{n-1}{np}}(\partial\Omega) \hookrightarrow L^{p^*}(\partial\Omega) \tag{2.497}$$

where p^* is as in (2.433), and the fourth one is implied by the remark following the statement of Theorem 2.53.

Having dealt with (2.495), we now proceed to show that if the domain Ω is as in (2.14) and $\theta \in (0, 1/M)$ then there exists a constant $C > 0$ with the property that

$$|u(X) - u(Y)| \leq C|X-Y|\Big(\widetilde{\mathcal{N}}_\theta(\nabla u)(X) + \widetilde{\mathcal{N}}_\theta(\nabla u)(Y)\Big), \quad \forall\, X, Y \in \partial\Omega. \tag{2.498}$$

To prove (2.498), fix a point $X = (x', \varphi(x')) \in \partial\Omega$, $Y = (y', \varphi(y')) \in \partial\Omega$, where $x', y' \in \mathbb{R}^{n-1}$, and set $t := |x' - y'|$. We now claim that there exists $\lambda = \lambda(M, \theta) > 0$ such that

$$X^* := (x', \varphi(x') + \lambda t) \in Y + \Gamma_\theta \quad \text{and} \quad Y^* := (y', \varphi(y') + \lambda t) \in X + \Gamma_\theta. \tag{2.499}$$

We shall only check the first membership in (2.499), as the second one is proved in a very similar fashion. We need $(x' - y', \varphi(x') - \varphi(y') + \lambda t) \in \Gamma_\theta$ or, equivalently, $|x' - y'| < \theta(\varphi(x') - \varphi(y') + \lambda t)$. Since

$$\varphi(x') - \varphi(y') + \lambda t \geq \lambda t - M|x' - y'| = (\lambda - M)t, \tag{2.500}$$

the desired conclusion follows by taking $\lambda > M + 1/\theta$. With (2.499) in hand, we now make repeated use of the Fundamental Theorem of Calculus and the fact that the sets $X + \Gamma_\theta$, $Y + \Gamma_\theta$ and $(X + \Gamma_\theta) \cap (Y + \Gamma_\theta)$ are convex, in order to estimate

$$\begin{aligned} |u(X) - u(Y)| &\leq |u(X) - u(X^*)| + |u(X^*) - u(Y^*)| + |u(Y^*) - u(Y)| \\ &\leq |X - X^*|\widetilde{\mathcal{N}}_\theta(\nabla u)(X) + |X^* - Y^*|\widetilde{\mathcal{N}}_\theta(\nabla u)(X) \\ &\quad + |Y^* - Y|\widetilde{\mathcal{N}}_\theta(\nabla u)(Y). \end{aligned} \tag{2.501}$$

Since $|X - X^*| \approx |X^* - Y^*| \approx |Y^* - Y| \approx r \approx |X - Y|$, (2.501) implies (2.498).

Consider now the functions $f, g : \mathbb{R}^{n-1} \to \mathbb{R}$ given by

$$f(x') := u(x', \varphi(x')) \quad \text{and} \quad g(x') := \widetilde{\mathcal{N}}_\theta(\nabla u)(x', \varphi(x')), \quad \forall\, x' \in \mathbb{R}^{n-1}. \tag{2.502}$$

Then (2.498) can be rephrased as

$$|f(x') - f(y')| \leq C|x' - y'|(g(x') + g(y')), \qquad \forall\, x', y' \in \mathbb{R}^{n-1}. \tag{2.503}$$

Also, $f \in L^1_{loc}(\mathbb{R}^{n-1})$ by (2.495), and $g \in L^p(\mathbb{R}^{n-1})$ from assumptions. In the language of Sobolev spaces on metric spaces, (2.503) expresses the fact that g is a generalized gradient for f. According to Theorem 1 in [71], we then have

$$[f] \in \dot{F}^{p,2}_1(\mathbb{R}^{n-1}) \quad \text{and} \quad \|[f]\|_{\dot{F}^{p,2}_1(\mathbb{R}^{n-1})} \leq C\|g\|_{L^p(\mathbb{R}^{n-1})}. \tag{2.504}$$

Now (2.494) readily follows from (2.504) and (2.430). □

Proposition 2.60. *Let $\Omega \subset \mathbb{R}^n$ be a bounded Lipschitz domain and assume that the function $u \in C^1(\Omega)$ is such that $\mathcal{N}(\nabla u) \in L^p(\partial\Omega)$ for some $p \in (\frac{n-1}{n}, 1]$. Then u has a non-tangential limit at almost every boundary point on $\partial\Omega$ and, for any $q \in (0, \infty]$,*

$$u\Big|_{\partial\Omega} \in h^{1,p}_{at}(\partial\Omega) \quad \text{and} \quad \|u\lfloor_{\partial\Omega}\|_{h^{1,p}_{at}(\partial\Omega)} \leq C\|\mathcal{N}(\nabla u)\|_{L^p(\partial\Omega)} + C\|u\|_{L^q(\mathcal{O})}, \tag{2.505}$$

for some relatively compact subset $\mathcal{O} \subset \Omega$. In particular,

$$\|u\lfloor_{\partial\Omega}\|_{h^{1,p}_{at}(\partial\Omega)} \leq C\,\|\mathcal{N}(\nabla u)\|_{L^p(\partial\Omega)} + C\,\|\mathcal{N}u\|_{L^p(\partial\Omega)}. \tag{2.506}$$

Furthermore,

$$\sum_{j,k=1}^{n} \|\partial_{\tau_{jk}}(u\lfloor_{\partial\Omega})\|_{H^p_{at}(\partial\Omega)} \leq C\,\|\mathcal{N}(\nabla u)\|_{L^p(\partial\Omega)}. \tag{2.507}$$

Proof. First, let us point out that there exists a finite family of balls B_j, $j = 1,\dots,N$, in Ω and $C = C(\Omega) > 0$ such that (recall that a barred integral indicates averaging)

$$\mathcal{N}(u) \leq C\,\mathcal{N}(\nabla u) + C \max_{j=1,\dots,N}\Big(⨍_{B_j} |u|^q\,dX\Big)^{1/q}. \tag{2.508}$$

This can be proved by starting with (2.110), raising all terms to the q-th power, average in Q over a suitable ball, then take supremum in P and, finally, take the q-th root of all terms involved. In particular (2.508) shows that $\mathcal{N}(u) \in L^p(\partial\Omega)$. Next, let us consider a open cover $\{B(X_i, r_i)\}_{i=1,\dots,M}$ of $\partial\Omega$ where for each $1 \leq i \leq M$ we have $X_i \in \partial\Omega$ and $r_i > 0$ is small enough. Next, let $\psi_i \in C^\infty_c(\mathbb{R}^n)$, be a partition of unity subordinated to the cover. Going further, for each $i \in \{1,\dots,M\}$ we consider D_i to be a graph Lipschitz domain as in (2.14) such that $D_i \cap B(X_i, r_i) = \Omega \cap B(X_i, r_i)$. Let $w_i := \widetilde{\psi_i u}\Big|_{D_i} \in C^1(D_i)$, where as before, tilde denotes the extension by zero to $\mathbb{R}^n$. Using (2.508) we obtain that $\mathcal{N}_{D_i}(\nabla w_i) \in L^p(\partial D_i)$ and

$$\|\mathcal{N}_{D_i}(\nabla w_i)\|_{L^p(\partial D_i)} \leq C\Big(\|\mathcal{N}(\nabla u)\|_{L^p(\partial\Omega)} + \|\mathcal{N}(u)\|_{L^p(\partial\Omega)}\Big), \quad 1 \leq i \leq M. \tag{2.509}$$

Above, $\mathcal{N}_{D_i}$ denotes the non-tangential maximal operator for the graph Lipschitz domain D_i. Then, employing Proposition 2.59 together with (2.509), we obtain that $[w_i\lfloor_{\partial D_i}] \in H^{1,p}_{at}(\partial D_i)$ and

$$\|[w_i\lfloor_{\partial D_i}]\|_{H^{1,p}_{at}(\partial D_i)} \leq C\Big(\|\mathcal{N}(\nabla u)\|_{L^p(\partial\Omega)} + \|\mathcal{N}(u)\|_{L^p(\partial\Omega)}\Big). \tag{2.510}$$

In particular, there exists $c_i \in \mathbb{R}$ such that $w_i|_{\partial D_i} + c_i \in \widetilde{H}^{1,p}_{at}(\partial D_i) \hookrightarrow L^{p^*}(\partial D_i)$, where the last inclusion is a consequence of (2.456) with p^* as in (2.433). Since $w_i\lfloor_{\partial D_i}$ is compactly supported this readily gives that $c_i = 0$ and hence the membership $w_i|_{\partial D_i} \in \widetilde{H}^{1,p}_{at}(\partial D_i)$ holds. At this point, (2.467), (2.510) and (2.508) allow us to conclude that (2.505) holds. Then (2.506) follows from this and Proposition 2.3.

There remains to justify (2.507). To this end, pick $q := np/(n-1) \in (1,\infty)$ and for $c := \frac{1}{|\mathcal{O}|}\int_{\mathcal{O}} u(X)\,dX$ write, based on (2.505) for $u - c$ and Poincaré's inequality,

$$\sum_{j,k=1}^{n} \|\partial_{\tau_{jk}}(u\lfloor_{\partial\Omega})\|_{H^p_{at}(\partial\Omega)} \leq C\,\|\mathcal{N}(\nabla u)\|_{L^p(\partial\Omega)} + C\,\|\nabla u\|_{L^q(\mathcal{O})}. \quad (2.511)$$

Employing now Proposition 2.3, applied to ∇u, finishes the proof of (2.507). □

We conclude this section with a boundedness result for the trace operator mapping into Triebel–Lizorkin spaces on the boundary of a Lipschitz domain.

Proposition 2.61. *Let $\Omega \subset \mathbb{R}^n$ be a bounded Lipschitz domain. Then the trace operator from Theorem 2.53 induces a bounded, linear mapping*

$$\mathrm{Tr} : B^{p,1}_{s+1/p}(\Omega) \longrightarrow F^{p,q}_{s}(\partial\Omega) \quad (2.512)$$

whenever $1 < p < \infty$, $0 \leq s \leq 1$ and $2 \leq q \leq \infty$.

Proof. It is enough to show that if $1 < p < \infty$ and $0 \leq s \leq 1$ then

$$\mathrm{Tr} : B^{p,1}_{s+1/p}(\Omega) \longrightarrow L^p_s(\partial\Omega) \quad (2.513)$$

boundedly, since (2.512) is a consequence of (2.513) and standard embedding results.

As far as (2.513) is concerned, the case $s = 0$, i.e., the well-definiteness and boundedness of

$$\mathrm{Tr} : B^{p,1}_{1/p}(\Omega) \longrightarrow L^p(\partial\Omega), \qquad 1 < p < \infty, \quad (2.514)$$

follows from the corresponding result proved in [45] for $\Omega = \mathbb{R}^n_+$, localization and bi-Lipschitz changes of variables. With this in hand, for a function $w \in B^{p,1}_{1+1/p}(\Omega)$ and for $j, k \in \{1, \dots, n\}$ we may write

$$\partial_{\tau_{jk}}[\mathrm{Tr}\, w] = \nu_j \,\mathrm{Tr}\,[\partial_k w] - \nu_k \,\mathrm{Tr}\,[\partial_j w] \in L^p(\partial\Omega). \quad (2.515)$$

To justify this identity, pick a sequence $\{w_\ell\}_{\ell\in\mathbb{N}} \subseteq C^\infty(\overline{\Omega})$ with the property that $w_\ell \to w$ in $B^{p,1}_{1+1/p}(\Omega)$ as $\ell \to \infty$. For every $\varphi \in C^\infty_c(\mathbb{R}^n)$ we may then write

$$\begin{aligned}
-\int_{\partial\Omega} (\mathrm{Tr}\, w)(\partial_{\tau_{jk}}\varphi)\, d\sigma &= -\lim_{\ell\to\infty} \int_{\partial\Omega} (\mathrm{Tr}\, w_\ell)(\partial_{\tau_{jk}}\varphi)\, d\sigma \\
&= -\lim_{\ell\to\infty} \int_{\partial\Omega} w_\ell(\partial_{\tau_{jk}}\varphi)\, d\sigma \\
&= \lim_{\ell\to\infty} \int_{\partial\Omega} \Big[\nu_j(\partial_k w_\ell) - \nu_k(\partial_j w_\ell)\Big]\varphi\, d\sigma \\
&= \lim_{\ell\to\infty} \int_{\partial\Omega} \Big[\nu_j \,\mathrm{Tr}\,(\partial_k w_\ell) - \nu_k \,\mathrm{Tr}\,(\partial_j w_\ell)\Big]\varphi\, d\sigma \\
&= \int_{\partial\Omega} \Big[\nu_j \,\mathrm{Tr}\,(\partial_k w) - \nu_k \,\mathrm{Tr}\,(\partial_j w)\Big]\varphi\, d\sigma, \quad (2.516)
\end{aligned}$$

proving (2.515). It follows from identity (2.515) that $\partial_{\tau_{jk}}[\operatorname{Tr} w] \in L^p(\partial\Omega)$ for every $j, k \in \{1, \dots, n\}$ so that, ultimately, $\operatorname{Tr} w \in L_1^p(\partial\Omega)$, plus a natural estimate. This shows that the operator (2.513) is well-defined and bounded in the case $s = 1$. Then the full claim about the operator (2.513) follows from what we have proved so far and complex interpolation. □

2.8 Calderón–Zygmund Theory in the Scalar-Valued Case

The aim in this section is to discuss those aspects of the Calderón–Zygmund theory of scalar-valued singular integral operators which are most relevant for our subsequent work. We start by reviewing issues concerning the existence of principal-value limits and the boundedness of maximal operator associated with singular integrals considered in the entire Euclidean space. Below and elsewhere, given two quasi-Banach spaces $(\mathscr{X}, \|\cdot\|_{\mathscr{X}})$ and $(\mathscr{Y}, \|\cdot\|_{\mathscr{Y}})$ we denote by $\mathscr{L}(\mathscr{X} \to \mathscr{Y})$ the space of linear and bounded operators from $\mathscr{X}$ into $\mathscr{Y}$, and set

$$\|T\|_{\mathscr{L}(\mathscr{X}\to\mathscr{Y})} := \sup\{\|Tf\|_{\mathscr{Y}} : f \in \mathscr{X} \text{ with } \|f\|_{\mathscr{X}} \le 1\}, \tag{2.517}$$

for every $T \in \mathscr{L}(\mathscr{X} \to \mathscr{Y})$.

For the proof of the following basic result the reader is referred to [78, 119].

Theorem 2.62. *Let $A : \mathbb{R}^n \to \mathbb{R}^m$ be a Lipschitz function with Lipschitz constant M, and assume that $F : \mathbb{R}^m \to \mathbb{R}$, $F \in C^N(\mathbb{R}^m)$, $N \ge 5 + m$, F is an odd function. For $x, y \in \mathbb{R}^n$ with $x \neq y$ set $K(x,y) := \frac{1}{|x-y|^n} F\left(\frac{A(x)-A(y)}{|x-y|}\right)$, and for $\varepsilon > 0$, $f \in \operatorname{Lip}_c(\mathbb{R}^n)$, define the truncated operator*

$$T_\varepsilon f(x) := \int_{|x-y|>\varepsilon} K(x,y) f(y)\, dy. \tag{2.518}$$

Then, for each $1 < p < \infty$, the following assertions hold:

1. *The maximal operator $T_* f(x) := \sup\{|T_\varepsilon f(x)| : \varepsilon > 0\}$ is bounded from $L^p(\mathbb{R}^n)$ into $L^p(\mathbb{R}^n)$. Moreover,*

$$\|T_*\|_{\mathscr{L}(L^p\to L^p)} \le C_p + C_p(1+M^{4+m}) \sup\{|D^\alpha F(z)| : |z| \le M+1,\ |\alpha| \le 5+m\}. \tag{2.519}$$

2. *If $f \in L^p(\mathbb{R}^n)$ then the limit $\lim_{\varepsilon\to 0} T_\varepsilon f(x)$ exists for almost every $x \in \mathbb{R}^n$ and the operator*

$$Tf(x) := \lim_{\varepsilon\to 0^+} T_\varepsilon f(x) \tag{2.520}$$

is bounded from $L^p(\mathbb{R}^n)$ into $L^p(\mathbb{R}^n)$.
Furthermore, if $B : \mathbb{R}^n \to \mathbb{R}^n$ is a bi-Lipschitz function then

$$Tf(x) := \lim_{\varepsilon\to 0+} \int_{|B(x)-B(y)|>\varepsilon} K(x,y)f(y)\,dy \tag{2.521}$$

for all $f \in L^p(\mathbb{R}^n)$ *and for almost every* $x \in \mathbb{R}^n$.

3. *The operator* (2.520) *is bounded from* $L^\infty(\mathbb{R}^n)$ *into* $\mathrm{BMO}(\mathbb{R}^n)$.
4. *The operator* (2.520) *is bounded from* $L^1(\mathbb{R}^n)$ *into* $L^{1,\infty}(\mathbb{R}^n)$.
5. *The operator* (2.520) *is bounded from* $H^1(\mathbb{R}^n)$ *into* $L^1(\mathbb{R}^n)$.

The next result in this section deals with non-tangential maximum function estimates, non-tangential limits, etc., for integral operators defined on Lipschitz surfaces. As such, this is essentially due to Coifman, McIntosh and Meyer [27]. More general results of this type may be found in [33] and [56].

Proposition 2.63. *There exists a positive integer* $N = N(n)$ *with the following significance. Consider a Lipschitz domain* $\Omega \subset \mathbb{R}^n$ *and fix some function*

$$\begin{aligned} &k \in C^N(\mathbb{R}^n \setminus \{0\}) \quad \text{such that} \quad k(-X) = -k(X) \\ &\text{and} \quad k(\lambda X) = \lambda^{-(n-1)} k(X) \quad \forall\, \lambda > 0, \ \forall\, X \in \mathbb{R}^n \setminus \{0\}. \end{aligned} \tag{2.522}$$

Next, with σ *denoting the surface measure on* $\partial\Omega$, *define the singular integral operator*

$$\mathscr{T} f(X) := \int_{\partial\Omega} k(X-Y) f(Y)\, d\sigma(Y), \qquad X \in \Omega, \tag{2.523}$$

as well as

$$T_* f(X) := \sup_{\varepsilon>0} |T_\varepsilon f(X)|, \qquad X \in \partial\Omega, \quad \text{where} \tag{2.524}$$

$$T_\varepsilon f(X) := \int_{\substack{Y\in\partial\Omega \\ |X-Y|>\varepsilon}} k(X-Y) f(Y)\, d\sigma(Y), \qquad X \in \partial\Omega. \tag{2.525}$$

Then for each $p \in (1,\infty)$ *there exists a finite constant* $C = C(\partial\Omega, p, n) > 0$ *such that*

$$\|T_* f\|_{L^p(\partial\Omega)} \le C \|k|_{S^{n-1}}\|_{C^N} \|f\|_{L^p(\partial\Omega)} \tag{2.526}$$

for each $f \in L^p(\partial\Omega)$. *Furthermore, for each parameter* $\kappa > 0$ *there exists a finite constant* $C = C(\partial\Omega, p, \kappa, n,) > 0$ *such that*

$$\|\mathscr{N}_\kappa(\mathscr{T} f)\|_{L^p(\partial\Omega)} \le C \|k|_{S^{n-1}}\|_{C^N} \|f\|_{L^p(\partial\Omega)}, \quad \text{if } 1 < p < \infty, \tag{2.527}$$

and

$$\|\mathscr{N}_\kappa(\mathscr{T} f)\|_{L^p(\partial\Omega)} \le C \|k|_{S^{n-1}}\|_{C^N} \|f\|_{H^p_{at}(\partial\Omega)}, \quad \text{if } \tfrac{n-1}{n} < p \le 1. \tag{2.528}$$

Finally, for each $p \in (1,\infty)$, $f \in L^p(\partial\Omega)$, *the limit*

$$Tf(X) := \lim_{\varepsilon\to 0+} T_\varepsilon f(X) \tag{2.529}$$

exists for σ*-a.e.* $X \in \partial\Omega$, *and the jump-formula*

$$\lim_{\substack{Z\to X\\ Z\in R_\kappa(X)}} \mathscr{T} f(Z) = \frac{1}{2\sqrt{-1}}\widehat{k}(\nu(X))f(X) + Tf(X) \tag{2.530}$$

is valid at σ*-a.e.* $X \in \partial\Omega$, *where* ν *is the outward unit normal to* $\partial\Omega$ *and "hat" stands for the Fourier transform in* $\mathbb{R}^n$.

We continue with a result about the mapping properties of integral operators given by singular integrals from Besov spaces into weighted Sobolev spaces. This is a particular case of a more general result found in [85].

Proposition 2.64. *Let* Ω *be a bounded Lipschitz domain in* $\mathbb{R}^n$ *with surface measure* σ, *and consider the integral operator*

$$Rf(X) := \int_{\partial\Omega} r(X,Y)f(Y)\,d\sigma(Y), \qquad \forall\, X \in \Omega, \tag{2.531}$$

whose kernel satisfies the estimates

$$|\nabla_X^k \nabla_Y^j r(X,Y)| \leq C|X-Y|^{-(n-2+j+k)}, \quad j = 0,1, \quad 1 \leq k \leq N, \tag{2.532}$$

for some positive integer N. *Also, recall that* $\rho(X) := \operatorname{dist}(X,\partial\Omega)$ *for every point* $X \in \mathbb{R}^n$. *Then*

$$\left\|\rho^{k-\frac{1}{p}-s}|\nabla^k Rf|\right\|_{L^p(\Omega)} + \sum_{j=0}^{k-1}\|\nabla^j Rf\|_{L^p(\Omega)} \leq C\|f\|_{B^{p,p}_{s-1}(\partial\Omega)}, \quad k = 1,2,\dots,N, \tag{2.533}$$

granted that $\frac{n-1}{n} < p \leq \infty$ *and* $(n-1)(\frac{1}{p}-1)_+ < s < 1$.

Moving on, recall that $\Omega_j \nearrow \Omega$ and $\Omega_j \searrow \Omega$ as $j \to \infty$ indicate that the family of domains $\{\Omega_j\}_{j\in\mathbb{N}}$ approximate Ω in the manner described in the proof of Proposition 2.15. As usual, we set $\Omega_{j,+} := \Omega_j$ and $\Omega_{j,-} := \mathbb{R}^n \setminus \overline{\Omega_j}$ for each $j \in \mathbb{N}$.

Proposition 2.65. *Fix a sufficiently large* $N \in \mathbb{N}$ *and assume that* k *is a function as in* (2.522). *Also, let* Ω *be a bounded Lipschitz domain in* $\mathbb{R}^n$ *and consider the integral operators*

$$\mathscr{T}_k^{\pm} f(X) := \int_{\partial\Omega} k(X-Y)f(Y)\,d\sigma(Y), \qquad X \in \Omega_\pm, \tag{2.534}$$

where $\Omega_\pm$ are as in (2.11). Analogously, given a family of bounded Lipschitz domains $\{\Omega_j\}_{j\in\mathbb{N}}$, define the integral operators

$$\mathscr{T}^{\pm}_{k,j} f(X) := \int_{\partial\Omega_j} k(X-Y) f(Y)\, d\sigma_j(Y), \qquad X \in \Omega_{j,\pm}, \tag{2.535}$$

for each $j \in \mathbb{N}$ (where σ_j stands for the surface measure on $\partial\Omega_j$). Finally, fix an arbitrary $\Phi \in C^\infty_c(\mathbb{R}^n)$ and some $p \in (1,\infty)$.

Then, if $\Omega_j \nearrow \Omega$, there holds

$$\Big[\mathscr{T}^{-}_{k,j}\Big(\Phi|_{\partial\Omega_j}\Big)\Big]\Big|_{\partial\Omega} \longrightarrow \Big[\mathscr{T}^{-}_{k}\Big(\Phi|_{\partial\Omega}\Big)\Big]\Big\lfloor_{\partial\Omega} \quad \text{in } L^p(\partial\Omega), \quad \text{as } j\to\infty. \tag{2.536}$$

If, on the other hand, $\Omega_j \searrow \Omega$ then

$$\Big[\mathscr{T}^{+}_{k,j}\Big(\Phi|_{\partial\Omega_j}\Big)\Big]\Big|_{\partial\Omega} \longrightarrow \Big[\mathscr{T}^{+}_{k}\Big(\Phi|_{\partial\Omega}\Big)\Big]\Big\lfloor_{\partial\Omega} \quad \text{in } L^p(\partial\Omega), \quad \text{as } j\to\infty. \tag{2.537}$$

Proof. We shall only present the proof of (2.536), as (2.537) can be justified in a similar fashion. With this goal in mind, consider the set of all odd spherical harmonics $\{\Psi_{i\ell} : \ell \in 2\mathbb{N}+1,\ 1 \le i \le H_\ell\}$, where

$$H_1 := n, \qquad \text{and} \qquad H_\ell := \binom{n+\ell-1}{\ell} - \binom{n+\ell-3}{\ell-2} \quad \text{if } \ell \ge 3. \tag{2.538}$$

In particular,

$$H_\ell \le C_n(\ell-1)\cdot\ell\cdots(n+\ell-2)\cdot(n+\ell-3) \le C_n\,\ell^{n-1} \tag{2.539}$$

and, whenever $\ell \in 2\mathbb{N}+1$ and $1 \le i \le H_\ell$,

$$\Delta_{S^{n-1}}\Psi_{i\ell} = -\ell(n+\ell-2)\Psi_{i\ell} \quad \text{and} \quad \Psi_{i\ell}\Big(\frac{X}{|X|}\Big) = \frac{P_{i\ell}(X)}{|X|^\ell} \tag{2.540}$$

for some homogeneous, odd, harmonic polynomial $P_{i\ell}$ of degree ℓ in $\mathbb{R}^n$. Thus, if we now set

$$a_{i\ell} := \int_{S^{n-1}} k(\omega)\Psi_{i\ell}(\omega)\, d\omega, \qquad \ell \in 2\mathbb{N}+1, \quad 1 \le i \le H_\ell, \tag{2.541}$$

it follows that

$$|a_{i\ell}| \le C_n \|\nabla^N k\|_{L^\infty(S^{n-1})}\ell^{-2N}, \qquad \ell \in 2\mathbb{N}+1, \quad 1 \le i \le H_\ell. \tag{2.542}$$

Also, for any $X \in \mathbb{R}^n \setminus \{0\}$ we have

$$k(X) = \frac{1}{|X|^{n-1}} k\Big(\frac{X}{|X|}\Big) = \sum_{\ell \in 2\mathbb{N}+1} \sum_{i=1}^{H_\ell} a_{i\ell} \frac{1}{|X|^{n-1}} \Psi_{i\ell}\Big(\frac{X}{|X|}\Big)$$

$$= \sum_{\ell \in 2\mathbb{N}+1} \sum_{i=1}^{H_\ell} a_{i\ell} k_{i\ell}(X), \tag{2.543}$$

where

$$k_{i\ell}(X) := \frac{1}{|X|^{n-1}} \Psi_{i\ell}\Big(\frac{X}{|X|}\Big) = \frac{P_{i\ell}(X)}{|X|^{n-1+\ell}} \tag{2.544}$$

is a kernel which satisfies (2.522) (with $N = \infty$). Let us also point out here that, once an even integer $d > N + (n-1)/2$ has been fixed, then for each $\ell \in 2\mathbb{N}+1$ and $1 \leq i \leq H_\ell$,

$$\|k_{i\ell}|_{S^{n-1}}\|_{C^N} \leq C_n \|(I - \Delta_{S^{n-1}})^{d/2}(k_{i\ell}|_{S^{n-1}})\|_{L^2(S^{n-1})} \leq C_n \ell^d, \tag{2.545}$$

thanks to (2.540), Sobolev's Embedding Theorem, and the fact that $k_{i\ell} = \Psi_{i\ell}$ on S^{n-1}.

Fix $p \in (1, \infty)$ and $\Omega_j \nearrow \Omega$ as $j \to \infty$. Our claim is that in order to prove (2.536) it suffices to show that for each $\ell \in 2\mathbb{N}+1$, $i \in \{1, \dots, H_\ell\}$, and $\Phi \in C_c^\infty(\mathbb{R}^n)$ we have

$$\Big[\mathscr{T}^-_{k_{i\ell},j}(\Phi|_{\partial\Omega_j})\Big]\Big|_{\partial\Omega} \to \big[\mathscr{T}^-_{k_{i\ell}}(\Phi|_{\partial\Omega})\big]\Big\lfloor_{\partial\Omega} \quad \text{in } L^p(\partial\Omega) \quad \text{as } j \to \infty. \tag{2.546}$$

Indeed, for each $M \in \mathbb{N}$, (2.543) entails

$$\Big[\mathscr{T}^-_{k,j}(\Phi|_{\partial\Omega_j})\Big]\Big|_{\partial\Omega} = \sum_{\substack{\ell \geq M+1 \\ \ell\, odd}} \sum_{i=1}^{H_\ell} a_{i\ell} \Big[\mathscr{T}^-_{k_{i\ell},j}(\Phi|_{\partial\Omega_j})\Big]\Big|_{\partial\Omega} + \big[\mathscr{R}_{M,j}(\Phi|_{\partial\Omega_j})\big]\Big|_{\partial\Omega}, \tag{2.547}$$

where

$$\mathscr{R}_{M,j} f(X) := \sum_{\substack{\ell \geq M+1 \\ \ell\, odd}} \sum_{i=1}^{H_\ell} a_{i\ell} \mathscr{T}^-_{k_{i\ell},j} f(X), \qquad X \in \Omega_{j,-}. \tag{2.548}$$

Now, if $\mathscr{N}_j^-$ denotes the non-tangential maximal function for $\mathbb{R}^n \setminus \overline{\Omega}_j$, it follows from (2.548), (2.545), (2.527), (2.542) and (2.539) that

$$
\begin{aligned}
\left\| \left[\mathscr{R}_{M,j} f \right] \Big|_{\partial\Omega} \right\|_{L^p(\partial\Omega)} &\leq \| \mathscr{N}_j^-(\mathscr{R}_{M,j} f) \|_{L^p(\partial\Omega_j)} \\
&\leq C \sum_{\substack{\ell \geq M+1 \\ \ell\, odd}} \sum_{i=1}^{H_\ell} a_{i\ell} \|k_{i\ell}|_{S^{n-1}}\|_{C^N} \|f\|_{L^p(\partial\Omega_j)} \\
&\leq C \sum_{\substack{\ell \geq M+1 \\ \ell\, odd}} \sum_{i=1}^{H_\ell} a_{i\ell} \|k_{i\ell}|_{S^{n-1}}\|_{C^N} \|f\|_{L^p(\partial\Omega_j)} \\
&\leq C \Big(\sum_{\substack{\ell \geq M+1 \\ \ell\, odd}} \ell^{-N+3(n-1)/2+2} \Big) \|f\|_{L^p(\partial\Omega_j)}. \qquad (2.549)
\end{aligned}
$$

Thus, if N is large enough, the following holds. Given $\varepsilon > 0$, the last term in (2.547) is $\leq C\varepsilon \|\Phi\|_{L^\infty(\mathbb{R}^n)}$, provided M is sufficiently large. Since considerations similar to (2.547)–(2.549) also apply to $\left[\mathscr{T}_k^-\left(\Phi|_{\partial\Omega} \right) \right] \Big|_{\partial\Omega}$, we may conclude that (2.536) holds if (2.546) holds for each fixed $\ell \in 2\mathbb{N}+1$ and $i \in \{1, \dots, H_\ell\}$.

There remains to establish (2.546) for each fixed $\ell \in 2\mathbb{N}+1$ and $i \in \{1, \dots, H_\ell\}$. If we now let σ_j denote the surface measure on $\partial\Omega_j$, then

$$
\begin{aligned}
\mathscr{T}_{k_{i\ell},j}^-(\Phi|_{\partial\Omega_j})(X) &= \int_{\partial\Omega_j} k_{i\ell}(X-Y)\,[\Phi(Y) - \Phi(X)]\, d\sigma_j(Y) \\
&\quad + \Phi(X) \int_{\partial\Omega_j} k_{i\ell}(X-Y)\, d\sigma_j(Y) \\
&=: I_j(X) + II_j(X), \qquad X \in \Omega_{j,-}. \qquad (2.550)
\end{aligned}
$$

Since $\partial\Omega \subset \Omega_{j,-}$, an application of the Lebesgue Dominated Convergence Theorem gives

$$
\lim_{j\to\infty} I_j \Big|_{\partial\Omega} = \Big(\int_{\partial\Omega} k_{i\ell}(\cdot - Y)\,[\Phi(Y) - \Phi(\cdot)]\, d\sigma(Y) \Big) \Big|_{\partial\Omega} \quad \text{in } L^p(\partial\Omega), \qquad (2.551)
$$

and, therefore, in order to prove (2.546) it suffices to show that, for each $\ell \in 2\mathbb{N}+1$ and $i \in \{1, \dots, H_\ell\}$,

$$
\mathscr{T}_{k_{i\ell},j}^-(1) \Big|_{\partial\Omega} \longrightarrow \mathscr{T}_{k_{i\ell}}^-(1) \Big|_{\partial\Omega} \quad \text{in } L^p(\partial\Omega) \text{ as } j \to \infty. \qquad (2.552)
$$

At this point, we digress for the purpose of reviewing some basic facts and terminology from Clifford analysis which we shall employ in our proof. To get started, the Clifford algebra with n imaginary units is the minimal enlargement of

$\mathbb{R}^n$ to a unitary real algebra $(\mathcal{C}\ell_n, +, \odot)$, which is not generated (as an algebra) by any proper subspace of $\mathbb{R}^n$ and such that

$$X \odot X = -|X|^2 \quad \text{for any } X \in \mathbb{R}^n. \tag{2.553}$$

This identity readily implies that, if $\{e_j\}_{j=1}^n$ is the standard orthonormal basis in $\mathbb{R}^n$, then

$$e_j \odot e_j = -1 \quad \text{and} \quad e_j \odot e_k = -e_k \odot e_j \quad \text{for any } 1 \leq j \neq k \leq n. \tag{2.554}$$

In particular, we identify the canonical basis $\{e_j\}_j$ from $\mathbb{R}^n$ with the n imaginary units generating $\mathcal{C}\ell_n$, so that we have the embedding

$$\mathbb{R}^n \hookrightarrow \mathcal{C}\ell_n, \qquad \mathbb{R}^n \ni X = (x_1, \dots, x_n) \equiv \sum_{j=1}^n x_j e_j \in \mathcal{C}\ell_n. \tag{2.555}$$

Also, any element $u \in \mathcal{C}\ell_n$ can be uniquely represented in the form

$$u = \sum_{l=0}^{n+1} {\sum_{|I|=l}}' u_I\, e_I, \quad u_I \in \mathbb{R}. \tag{2.556}$$

Here e_I stands for the product $e_{i_1} \odot e_{i_2} \odot \cdots \odot e_{i_l}$ if $I = (i_1, i_2, \dots, i_l)$ and we have set $e_0 := e_\emptyset := 1$ for the multiplicative unit. Also, $\sum'$ indicates that the sum is performed only over strictly increasing multi-indices, i.e. over multi-indices $I = (i_1, i_2, \dots, i_l)$ with $1 \leq i_1 < i_2 < \cdots < i_l \leq n$. We endow $\mathcal{C}\ell_n$ with the natural Euclidean metric $|u| := \Big[\sum_I |u_I|^2\Big]^{1/2}$, if $u = \sum_I u_I e_I \in \mathcal{C}\ell_n$. Next, recall the *Dirac operator*

$$D := \sum_{j=1}^n e_j \partial_j. \tag{2.557}$$

We shall use D_L and D_R to denote the action of D on a C^1 function $u : \Omega \to \mathcal{C}\ell_n$ (where Ω is an open subset of $\mathbb{R}^n$) from the left and from the right, respectively. For any bounded Lipschitz domain Ω with outward unit normal $\nu = (\nu_1, \dots, \nu_n)$ (identified with the $\mathcal{C}\ell_n$-valued function $\nu = \sum_{j=1}^n \nu_j e_j$) and surface measure σ, and for any $\mathcal{C}\ell_n$-valued function u defined in Ω, the following integration by parts formula holds:

$$\int_{\partial\Omega} (u\big\lfloor_{\partial\Omega})(X) \odot \nu(X)\, d\sigma(X) = \int_\Omega (D_R u)(X)\, dX, \tag{2.558}$$

granted that $\mathcal{N} u \in L^1(\partial\Omega)$, $D_R u \in L^1(\Omega)$, and $u\big\lfloor_{\partial\Omega}$ exists σ-a.e. on $\partial\Omega$. Another simple but useful observation in this context is that, for any $1 \leq p \leq \infty$,

$$\nu\odot : L^p(\partial\Omega) \otimes \mathcal{C}\ell_n \longrightarrow L^p(\partial\Omega) \otimes \mathcal{C}\ell_n \quad \text{is an isomorphism.} \tag{2.559}$$

Indeed, by (2.553), its inverse is $-\nu\odot$. More detailed accounts on these and related matters can be found in, e.g., [12, 90].

Let us now return to the mainstream discussion. The proof of (2.552), from which (2.536) follows, proceeds by induction on ℓ. Now, the initial step in the induction scheme (corresponding to $\ell = 1$, when $k_{i\ell}(X) \in \{x_1/|X|^n, \dots, x_n/|X|^n\}$) has been carried out in [82]. Fix next some $\ell \geq 3$ with the property that, for each odd $\tilde{\ell} \in \mathbb{N}$ with $\tilde{\ell} \leq \ell - 2$ and each $i \in \{1, \dots, H_{\tilde{\ell}}\}$, the convergence (2.552) is satisfied with $\tilde{\ell}$ in place of ℓ. Let $[\cdot]_j$ denote the projection onto the j-th Euclidean coordinate, i.e., $[X]_j := x_j$ if $X = (x_1, \dots, x_n) \in \mathbb{R}^n$. According to a useful result proved in [112], for each ℓ as above and $i \in \{1, \dots, H_\ell\}$, there exist a family $P_{sr}(X)$, $1 \leq s, r \leq n$, of harmonic, homogeneous polynomials of degree $\ell - 2$ in $\mathbb{R}^n$, as well as a family of odd, C^∞ functions $h_{sr} : \mathbb{R}^n \setminus \{0\} \to \mathbb{R}^n \hookrightarrow \mathscr{C}\ell_n$, $1 \leq s, r \leq n$, homogeneous of degree $-(n-1)$, such that

$$k_{i\ell}(X) = C_{n,\ell,i} \sum_{s,r=1}^{n} [h_{sr}(X)]_r \quad \text{and} \tag{2.560}$$

$$(D_R h_{sr})(X) = \frac{\partial}{\partial x_s}\left(\frac{P_{sr}(X)}{|X|^{n+\ell-3}}\right), \quad 1 \leq s, r \leq n. \tag{2.561}$$

As a consequence of (2.561) and (2.558), if we set

$$h^{sr}(X) := \frac{P_{sr}(X)}{|X|^{n+\ell-3}}, \tag{2.562}$$

and denote by $\nu^j = (\nu_1^j, \dots, \nu_n^j)$ the unit normal vector to $\partial\Omega_j$, then

$$\int_{\partial\Omega_j} h_{sr}(X) \odot \nu^j(X)\, d\sigma_j(X) = \int_{\partial\Omega_j} h^{sr}(X)\nu_s^j(X)\, d\sigma_j(X), \quad 1 \leq s, r \leq n. \tag{2.563}$$

In the above notation, using (2.560) we may express

$$\mathscr{T}^-_{k_{i\ell},j}(1) = C_{n,\ell,i} \sum_{r,s=1}^{n} \left[\mathscr{T}^-_{h_{sr},j}(1)\right]_r. \tag{2.564}$$

Hereafter, given a measure space (E, μ), we shall interpret the action of generic integral operator T associated with a Clifford algebra-valued kernel $h(X,Y)$ on a Clifford algebra-valued function f as $Tf(X) = \int_E h(X,Y) \odot f(Y)\, d\mu(Y)$.

Hence, in order to prove (2.552) it suffices to show

$$\mathscr{T}^-_{h_{sr},j}(1)\Big|_{\partial\Omega} \to \mathscr{T}^-_{h_{sr}}(1)\Big\lfloor_{\partial\Omega} \quad \text{in } L^p(\partial\Omega), \quad \text{as } j \to \infty, \tag{2.565}$$

a task to which we now turn. In fact, we will prove that for each $\Phi \in C_c^\infty(\mathbb{R}^n) \otimes \mathscr{C}\ell_n$ we have

$$\Big[\mathscr{T}^-_{h_{sr},j}\big(\nu^j \odot \Phi|_{\partial\Omega_j}\big)\Big]\Big|_{\partial\Omega} \to \big[\mathscr{T}^-_{h_{sr}}\big(\nu \odot \Phi|_{\partial\Omega}\big)\big]\Big\lfloor_{\partial\Omega} \quad \text{in } L^p(\partial\Omega), \quad \text{as } j \to \infty. \tag{2.566}$$

To see why this implies (2.565), we first note that for any $\varepsilon > 0$ there exists a function $\boldsymbol{\xi} \in C_c^\infty(\mathbb{R}^n)$ such that

$$\lim_{j\to\infty} \Big\| \nu^j - \boldsymbol{\xi}|_{\partial\Omega_j} \Big\|_{L^p(\partial\Omega_j)} = \Big\| \nu - \boldsymbol{\xi}|_{\partial\Omega} \Big\|_{L^p(\partial\Omega)} < \varepsilon. \tag{2.567}$$

Utilizing (2.566) with $\Phi := \boldsymbol{\xi}$ yields

$$\limsup_{j\to\infty} \Big\| \Big[\mathscr{T}^-_{h_{sr},j}\big(\nu^j \odot \Phi|_{\partial\Omega_j}\big)\Big]\Big|_{\partial\Omega} - \big[\mathscr{T}^-_{h_{sr}}\big(\nu \odot \Phi|_{\partial\Omega}\big)\big]\Big\lfloor_{\partial\Omega} \Big\|_{L^p(\partial\Omega)} < C\varepsilon, \tag{2.568}$$

from which the desired conclusion follows easily. Thus, it remains to establish (2.566).

With this in mind, we note that by employing an identity similar to (2.550) matters are reduced to proving (2.566) in the particular case when $\Phi \equiv 1$. However, in this situation,

$$\begin{aligned}
\Big[\mathscr{T}^-_{h_{sr},j}\nu^j\Big]\Big|_{\partial\Omega} &= \Big[\int_{\partial\Omega_j} h_{sr}(\cdot - Y) \odot \nu^j(Y)\, d\sigma_j(Y)\Big]\Big|_{\partial\Omega} \\
&= \Big[\int_{\partial\Omega_j} h^{sr}(\cdot - Y)\nu^j_s(Y)\, d\sigma_j(Y)\Big]\Big|_{\partial\Omega},
\end{aligned} \tag{2.569}$$

by (2.563). If we now pick an arbitrary $\varepsilon > 0$ and select $\boldsymbol{\xi} \in C_c^\infty(\mathbb{R}^n)$ as in (2.567), then

$$\begin{aligned}
\lim_{j\to\infty} \Big[\mathscr{T}^-_{h_{sr},j}\nu^j\Big]\Big|_{\partial\Omega} &= \lim_{j\to\infty}\Big[\int_{\partial\Omega_j} h^{sr}(\cdot - Y)\nu^j_s(Y)\, d\sigma_j(Y)\Big]\Big|_{\partial\Omega} \\
&= \lim_{j\to\infty}\Big[\int_{\partial\Omega_j} h^{sr}(\cdot - Y)\xi_s(Y)\, d\sigma_j(Y)\Big]\Big|_{\partial\Omega} + O(\varepsilon) \\
&= \Big[\int_{\partial\Omega} h^{sr}(\cdot - Y)\xi_s(Y)\, d\sigma(Y)\Big]\Big\lfloor_{\partial\Omega} + O(\varepsilon) \\
&= \Big[\int_{\partial\Omega} h^{sr}(\cdot - Y)\nu_s(Y)\, d\sigma(Y)\Big]\Big\lfloor_{\partial\Omega} + O(\varepsilon) \\
&= \Big[\int_{\partial\Omega} h_{sr}(\cdot - Y)\odot\nu(Y)\, d\sigma(Y)\Big]\Big\lfloor_{\partial\Omega} + O(\varepsilon) \\
&= \big[\mathscr{T}^-_{h_{sr}}\nu\big]\Big\lfloor_{\partial\Omega} + O(\varepsilon),
\end{aligned} \tag{2.570}$$

in $L^p(\partial\Omega)$. Above, the first equality is a consequence of (2.569), while the second one follows from (2.567). The third identity in (2.570) (which is the crucial step in the proof) is a consequence of the induction's hypothesis (reviewed just after (2.559)), given that the integral kernel h^{sr} has the form (2.562), with $\deg P_{sr} \leq \ell - 2$. Going further, the fourth identity once again follows from (2.567), while the fifth one is (2.563) (written for Ω in place of Ω_j). Finally, the last identity uses just the definition of $\mathscr{T}^{-}_{h_{sr}}$. Since ε is arbitrary, we further conclude from (2.570) that (2.566) holds. This finishes the proof of (2.536). □

We continue our discussion by proving the following useful result.

Theorem 2.66. *Let Ω be a Lipschitz domain in $\mathbb{R}^n$ and denote by $\rho(X)$ the distance from $X \in \Omega$ to the boundary $\partial\Omega$. Assume that $k \in C^2(\mathbb{R}^n \setminus \{0\})$ is an even function, homogeneous of degree $2-n$ and, for each $j, k \in \{1,\dots,n\}$, consider the operator*

$$\mathscr{T}_{jk} f(X) := \int_{\partial\Omega} \partial_{\tau_{jk}(Y)}[k(X-Y)] f(Y)\, d\sigma(Y), \quad X \in \Omega. \tag{2.571}$$

Then, for every $j, k \in \{1,\dots,n\}$, the following implication holds

$$f \in \mathrm{bmo}(\partial\Omega) \Longrightarrow |\nabla \mathscr{T}_{jk} f|^2 \rho\, dX \quad \text{is a Carleson measure on } \Omega$$
$$\text{with Carleson constant} \leq C \|f\|^2_{\mathrm{bmo}(\partial\Omega)}, \tag{2.572}$$

for some $C \in (0,\infty)$ independent of f.

In the proof of the above theorem, the following result is going to be useful.

Proposition 2.67. *Suppose that k is a real-valued function satisfying*

$$\begin{aligned} &k \in C^2(\mathbb{R}^n \setminus \{0\}), \quad k \text{ is odd, and} \\ &k(\lambda X) = \lambda^{1-n} k(X) \text{ for all } \lambda > 0,\ X \in \mathbb{R}^n \setminus \{0\}. \end{aligned} \tag{2.573}$$

Let Ω be a Lipschitz domain in $\mathbb{R}^n$ with surface measure σ and denote by ρ the distance function to $\partial\Omega$. Finally, define the integral operator $\mathscr{T}$ acting on functions $f \in L^2(\partial\Omega)$ by

$$\mathscr{T} f(X) := \int_{\partial\Omega} k(X-Y) f(Y)\, d\sigma(Y), \qquad \forall\, X \in \mathbb{R}^n \setminus \partial\Omega. \tag{2.574}$$

Then there exists $C \in (0,\infty)$ with the property that for each $f \in L^2(\partial\Omega)$ one has

$$\int_{\mathbb{R}^n \setminus \partial\Omega} |(\nabla \mathscr{T} f)(X)|^2 \rho(X)\, dX \leq C \int_{\partial\Omega} |f(X)|^2\, d\sigma(X). \tag{2.575}$$

This is a particular case of a more general theorem from [55] (which, in turn, extends results due to G. David and S. Semmes [34]). We are ready to present the

Proof of Theorem 2.66. Let us start by fixing a function $f \in \mathrm{bmo}(\partial\Omega)$, a point $X_o \in \partial\Omega$ and a scale $r > 0$. As usual, let $S_r(X_o) := B(X_o, r) \cap \partial\Omega$ be the surface ball of radius r centered at X_o. Consider next a cutoff function $\eta \in C_c^\infty(\mathbb{R}^n)$ satisfying

$$\begin{gathered} 0 \le \eta \le 1, \quad \eta \equiv 1 \text{ on } B(X_o, 2r), \quad \eta \equiv 0 \text{ outside } B(X_o, 4r), \\ |\partial^\alpha \eta(X)| \le C_\alpha r^{-|\alpha|}, \quad \forall\, X \in \mathbb{R}^n \text{ and } \forall\, \alpha \in \mathbb{N}_0^n. \end{gathered} \tag{2.576}$$

Going further, recall (2.256) and set $f_{S_{4r}} := \fint_{S_{4r}(X_o)} f \, d\sigma$. Then,

$$f = \eta(f - f_{S_{4r}}) + (1 - \eta)(f - f_{S_{4r}}) + f_{S_{4r}}. \tag{2.577}$$

Next, fix $j, k \in \{1, \dots, n\}$. Using that $\mathscr{T}_{jk} 1 = 0$ (which immediately follows from (2.571)), along with (2.577), yields

$$\nabla \mathscr{T}_{jk} f = \mathscr{T}_{jk}\big(\eta(f - f_{S_{4r}})\big) + \mathscr{T}_{jk}\big((1 - \eta)(f - f_{S_{4r}})\big). \tag{2.578}$$

Thus, if $T(S_r) := B(X_o, r) \cap \Omega$, we have

$$\begin{aligned} \fint_{T(S_r)} \Big|\nabla \mathscr{T}_{jk} f(X)\Big|^2 \rho(X)\, dX \le{} & C \fint_{T(S_r)} \Big|\nabla \mathscr{T}_{jk}\big(\eta(f - f_{S_{4r}})\big)(X)\Big|^2 \rho(X)\, dX \\ & + C \fint_{T(S_r)} \Big|\nabla \mathscr{T}_{jk}\big((1 - \eta)(f - f_{S_{4r}})\big)(X)\Big|^2 \rho(X)\, dX \\ =:{} & I + II. \end{aligned} \tag{2.579}$$

Let us point out that (2.572) immediately follows as soon as we establish

$$I \le C r^{-1} f^\#(X_o) \quad \text{and} \quad II \le C r^{-1} f^\#(X_o), \tag{2.580}$$

where, if $S_R(X) := B(X, R) \cap \partial\Omega$ for $X \in \partial\Omega$ and $R > 0$, and $f_{S_R(X)} := \fint_{S_R(X)} f \, d\sigma$, we have set

$$f^\#(X) := \sup_{R>0} \Big(\fint_{S_R(X)} |f(Y) - f_{S_R(X)}|^2 \, dY \Big)^{1/2}. \tag{2.581}$$

This is because, due to the definition of $\mathrm{bmo}(\partial\Omega)$ and John–Nirenberg's inequality, the following holds:

$$f \in \mathrm{bmo}(\partial\Omega) \iff f^\# \in L^\infty(\partial\Omega), \tag{2.582}$$

Moreover, there exists a finite geometric constant $C > 0$ such that

$$\|f^{\#}\|_{L^\infty(\partial\Omega)} \leq C\|f\|_{\mathrm{bmo}(\partial\Omega)}. \tag{2.583}$$

Turning our attention to the first part of (2.580), we write

$$\begin{aligned}
I &\leq \frac{C}{r^n}\int_\Omega \Big|\nabla \mathscr{T}_{jk}\big(\eta(f - f_{S_{4r}})\big)(X)\Big|^2 \rho(X)\,dX \\
&\leq \frac{C}{r^n}\int_{\partial\Omega} |\eta(f - f_{S_{4r}})|^2\,d\sigma \leq \frac{C}{r^n}\int_{S_{4r}} |\eta(f - f_{S_{4r}})|^2\,d\sigma \\
&\leq C r^{-1} ⨍_{S_{4r}} |\eta(f - f_{S_{4r}})|^2\,d\sigma \leq C r^{-1} f^{\#}(X_o).
\end{aligned} \tag{2.584}$$

Above, the first inequality follows from the definition of I and the fact that the n-dimensional measure of the Carleson box $T(S_r)$ is $\approx r^n$. The second inequality follows from Proposition 2.67, the third one is a consequence of the support properties of the function η introduced in (2.576), the fourth inequality is due to the fact that $\sigma\big(S_{4r}\big) \approx r^{n-1}$ and the last one is due to the definition of $f^{\#}$ in (2.581).

Next, before estimating the term II, let us make the observation that

$$X \in T(S_r) \quad \text{and} \quad Y \in \partial\Omega \setminus S_{2r} \quad \Longrightarrow \quad |X - Y| \approx |X_o - Y|. \tag{2.585}$$

With this in hand, we may estimate

$$\begin{aligned}
&\Big|\nabla \mathscr{T}_{jk}\big((1-\eta)(f - f_{S_{4r}})\big)(X)\Big| \\
&\qquad \leq C \int_{\partial\Omega\setminus S_{2r}} \frac{1}{|X_o - Y|^n} |f(Y) - f_{S_{4r}}|\,d\sigma(Y) \\
&\qquad \leq C \sum_{j=1}^\infty \int_{S_{2^{j+1}r}\setminus S_{2^j r}} \frac{1}{(2^j r)^n} |f(Y) - f_{S_{4r}}|\,d\sigma(Y) \\
&\qquad \leq C \sum_{j=1}^\infty \frac{1}{2^j r} ⨍_{S_{2^{j+1}r}} |f - f_{S_{4r}}|\,d\sigma(Y) \\
&\qquad \leq C \sum_{j=1}^\infty \frac{1}{2^j r} ⨍_{S_{2^{j+1}r}} \Big[|f - f_{S_{2^{j+1}r}}| + \sum_{k=2}^{j} |f_{S_{2^{k+1}r}} - f_{S_{2^k r}}|\Big]\,d\sigma \\
&\qquad \leq C r^{-1} \sum_{j=1}^\infty \frac{1}{2^j}(1 + j) f^{\#}(X_o) \leq C r^{-1} f^{\#}(X_o).
\end{aligned} \tag{2.586}$$

Indeed, the first inequality follows from the definition of $\mathscr{T}_{jk}\big((1-\eta)(f-f_{S_{4r}})\big)$, the properties of the function $1-\eta$ where η is as in (2.576) and (2.585). The second inequality is a consequence of writing the integral over $\partial\Omega\setminus S_{2r}$ as the telescopic sum over $S_{2^{j+1}r}\setminus S_{2^j r}$, $j\in\mathbb{N}$ and the fact that whenever $Y\in S_{2^{j+1}r}\setminus S_{2^j r}$ there holds $|X_o-Y|\approx 2^j r$. The third inequality is a result of enlarging the domain of integration from $S_{2^{j+1}r}\setminus S_{2^j r}$ to $S_{2^{j+1}r}\setminus S_{2^j r}$ and using that $\sigma\big(S_{2^{j+1}r}\big)\approx (2^j r)^{1-n}$. The fourth inequality follows from the triangle inequality after expanding $f-f_{S_{4r}}$ as the sum $f-f_{S_{2^{j+1}r}}+\sum\limits_{k=2}^{j}(f_{S_{2^{k+1}r}}-f_{S_{2^k r}})$. The fifth inequality is a consequence of (2.581) from which one can easily deduce that $\big|f_{S_{2^{k+1}r}}-f_{S_{2^k r}}\big|\le Cf^{\#}(X_o)$. Finally, the last inequality is a direct consequence of the fact that the series $\sum\limits_{j=1}^{\infty}2^{-j}(1+j)$ is convergent.

Now, (2.586) immediately gives the second part of (2.580) and completes the proof of the theorem. □

Given $a,b\in\mathbb{R}$, set $a\vee b:=\max\{a,b\}$. The following result (of purely real variable nature) extends earlier work in [91, 126].

Proposition 2.68. *Let $\Omega\subset\mathbb{R}^n$ be a bounded Lipschitz domain and assume that*

$$\begin{aligned} &k\in C^N(\mathbb{R}^n\setminus\{0\}) \quad \textit{such that} \quad k(-X)=-k(X)\\ &\textit{and} \quad k(\lambda X)=\lambda^{-(n-1)}k(X) \;\; \forall\,\lambda>0,\;\; \forall\,X\in\mathbb{R}^n\setminus\{0\}, \end{aligned} \tag{2.587}$$

for some sufficiently large integer $N=N(n)$. Associated with this kernel, consider the integral operator

$$\mathscr{T}f(X):=\int_{\partial\Omega}k(X-Y)f(Y)\,d\sigma(Y), \qquad \forall\,X\in\Omega. \tag{2.588}$$

Then for each $p\in(1,\infty)$ there exists a finite constant $C=C(\Omega,k,p)>0$ such that

$$\|\mathscr{T}f\|_{B^{p,p\vee 2}_{1/p}(\Omega)}\le C\|f\|_{L^p(\partial\Omega)}, \qquad \forall\,f\in L^p(\partial\Omega). \tag{2.589}$$

Proof. Assume that $0\in\partial\Omega$ and let $\varphi:\mathbb{R}^{n-1}\to\mathbb{R}$ be a Lipschitz function satisfying $\varphi(0)=0$ and having the property that its graph coincides with $\partial\Omega$ in some surface ball $S_r=S_r(0)$, with $r>0$ small. Given $p,q\in(1,\infty]$, consider the $L^q((0,r),ds/s)$-valued operator T defined as the assignment

$$L^p(S_r)\ni f\mapsto s\nabla(\mathscr{T}\widetilde{f})(x',\varphi(x')+s)\in L^p\big(\{x':|x'|<r\},L^q((0,r),ds/s)\big) \tag{2.590}$$

where $\widetilde{f}$ denotes the extension of f by zero to $\partial\Omega$.

The first order of business is to show that the operator T in (2.590) is bounded whenever $p \in (1,\infty)$ and $q \in [2,\infty]$. Based on well-known interpolation results for Lebesgue spaces of vector-valued functions (cf. [8, Theorem 5.12, p. 107]) it suffices to consider only the end-point cases $q = 2$ and $q = \infty$. Concerning the case when the index $q = 2$, the idea is to use the vector-valued Calderón–Zygmund theory which continues to work the Hilbert space setting. To implement this, consider the kernel

$$k_s(x',y') := s\,(\nabla k)\big(x'-y',\varphi(x')-\varphi(y')+s\big), \quad x',y' \in \mathbb{R}^{n-1} \text{ near } 0, \tag{2.591}$$

regarded as an $L^q((0,r),ds/s)$-valued function in the parameter s. The homogeneity and smoothness of the original kernel k ensures that the estimates

$$\big|(\partial^\alpha k)(X)\big| \leq C_\alpha |X|^{1-|\alpha|-n}, \quad \forall\,\alpha \in \mathbb{N}_0^n \text{ with } |\alpha| \leq 2, \quad \forall\, X \in \mathbb{R}^n \setminus \{0\} \tag{2.592}$$

are valid. Keeping in mind that

$$\begin{gathered}\big|(x'-y',\varphi(x')-\varphi(y')+s)\big| \approx |x'-y'|+s,\\ \text{uniformly for } x',y' \in \mathbb{R}^{n-1} \text{ and } s>0,\end{gathered} \tag{2.593}$$

for each $i,j \in \{0,1\}$ we may then estimate

$$\begin{aligned}\Big(\int_0^r |\nabla^i_{x'}\nabla^j_{y'}k_s(x',y')|^2\frac{ds}{s}\Big)^{1/2} &\leq \left(\int_0^\infty \left(\frac{s}{[\,|x'-y'|+s]^{n+i+j}}\right)^2\frac{ds}{s}\right)^{1/2}\\ &= C|x'-y'|^{-(n-1+i+j)},\end{aligned} \tag{2.594}$$

whenever $x' \neq y'$. In turn, (2.594) ensures that the kernel of the vector-valued integral operator T in (2.590) is standard when $q = 2$.

To prove the boundedness of the operator T when $p = 2$ (while continuing to assume that $q = 2$) choose an arbitrary $f \in L^2(S_r)$ and recall that $\widetilde{f}$ denotes the extension of f by zero to $\partial\Omega$. Then

$$\begin{aligned}\|Tf\|_{L^2(\{x':|x'|<r\},L^2((0,r),ds/s))} &\approx \Big(\int_{|x'|<r}\int_0^r s\big|\nabla(\mathscr{T}\widetilde{f})(x',\varphi(x')+s)\big|^2\,ds\,dx'\Big)^{1/2}\\ &\leq C\Big(\int_\Omega \operatorname{dist}(X,\partial\Omega)|\nabla(\mathscr{T}\widetilde{f})(X)|^2\,dX\Big)^{1/2}\\ &\leq C\|\widetilde{f}\|_{L^2(\partial\Omega)} = C\|f\|_{L^2(S_r)}.\end{aligned} \tag{2.595}$$

The crucial step in (2.595) is the square-function estimate in the last inequality, which follows from Proposition 2.67. At this stage, the vector-valued Calderón–Zygmund theory (in the Hilbert space setting) applies and allows us to conclude that

$$\text{the integral operator } T \text{ in (2.590) is bounded when } q = 2 \text{ and } p \in (1, \infty). \tag{2.596}$$

Consider now the issue of establishing the boundedness of the operator T when $q = \infty$ and $p \in (1, \infty)$. To this end, let $X_0 = (x_0', \varphi(x_0'))$ be an arbitrary point in S_r, and recall the Hardy–Littlewood maximal operator $\mathcal{M}$ on $\partial\Omega$ (cf. (2.9)). Then there exists a finite constant $C = C(\Omega, k) > 0$ with the property that if f is a function on $\partial\Omega$ supported in S_r and $s \in (0, r)$ we have

$$\begin{aligned} & s|\nabla(\mathscr{T} f)(x_0', \varphi(x_0') + s)| \\ & \qquad \leq \int_{|y'|<r} s\big|(\nabla k)\big(x_0' - y', \varphi(x_0') - \varphi(y') + s\big)\big| \cdot \big|f\big(y', \varphi(y')\big)\big|\, dy' \\ & \qquad \leq \sup_{s>0} \int_{|y'|<r} \frac{s}{\big[\,|x_0' - y'| + s\big]^n} \cdot \big|f\big(y', \varphi(y')\big)\big|\, dy' \\ & \qquad \leq C \mathcal{M} f(X_0), \end{aligned} \tag{2.597}$$

where we have used (2.592)–(2.593) and a well-known estimate for the convolution of the Poisson kernel for the upper-half space $\mathbb{R}^n_+ \ni (x_0', s) \mapsto \frac{s}{(|x_0'|+s)^n}$ with functions on the boundary of $\mathbb{R}^n_+$ (cf. [119, Theorem 2, pp. 62–63]). Thus, (2.597) implies that for any f as above

$$\sup_{s\in(0,r)} \big(s|\nabla(\mathscr{T} f)(x_0', \varphi(x_0') + s)|\big) \leq C \mathcal{M} f(X_0), \quad \forall\, X_0 = (x_0', \varphi(x_0')) \in S_r. \tag{2.598}$$

Using this and the boundedness of $\mathcal{M}$ on $L^p(\partial\Omega)$ (recall that $1 < p < \infty$), it follows that

$$\bigg(\int_{|x'|<r} \Big(\sup_{s\in(0,r)} \big|s\nabla(\mathscr{T} f)(x', \varphi(x') + s)\big|\Big)^p\, dx'\bigg)^{1/p} \leq C\|f\|_{L^p(\partial\Omega)}, \tag{2.599}$$

for every function f on $\partial\Omega$ supported in S_r. In turn, estimate (2.599) readily shows that T in (2.590) is bounded when $q = \infty$ and $p \in (1, \infty)$. In concert with (2.596) and the interpolation result mentioned earlier, this proves that

$$\text{the operator } T \text{ in (2.590) is bounded when } q \in [2, \infty] \text{ and } p \in (1, \infty). \tag{2.600}$$

The above result is one of the key ingredients in the proof of (2.589). Another basic ingredient is (a special case of) the estimate stated in the conclusion of Lemma 2.34. In fact, the latter result presents us with a clear strategy for the job at hand. In order to be specific, pick an arbitrary function $f \in L^p(\partial\Omega)$, for some fixed $p \in (1, \infty)$, and set

$$v := \mathscr{T} f \quad \text{in } \Omega. \tag{2.601}$$

Then $\|v\|_{B^{p,p\vee 2}_{1/p}(\Omega)}$ is controlled by several terms which we now begin to analyze. First, the contribution away from the boundary is easily handled given the format of (2.588) (since the integral kernel is no longer singular in this case). Second, the contribution near the boundary is estimated as in the right-hand side of (2.235), used here with $\theta := 1/p$ and $q := p \vee 2 \geq p$. There are two types of terms to be considered in this regard. One of these terms is

$$\|\mathscr{N} v\|_{L^p(\partial\Omega)} = \|\mathscr{N}(\mathscr{T} f)\|_{L^p(\partial\Omega)} \leq C\|f\|_{L^p(\partial\Omega)}, \tag{2.602}$$

where the last step uses estimate (2.527) from Proposition 2.63. This, of course, suits our purposes. The other type of term alluded to earlier is a finite sum of expressions which, given our choice of θ, look like (up to an inessential change in scale)

$$\Big(\int_{|x'|<r/2}\Big(\int_0^{r/2} |s(\nabla v)(x', \varphi(x') + s)|^q \frac{ds}{s}\Big)^{p/q} dx'\Big)^{1/p} \quad \text{where } q := p \vee 2. \tag{2.603}$$

To handle such an expression, we write

$$\begin{aligned}
&\Big(\int_{|x'|<r/2}\Big(\int_0^{r/2} |s(\nabla v)(x', \varphi(x') + s)|^q \frac{ds}{s}\Big)^{p/q} dx'\Big)^{1/p} \\
&\quad = \Big(\int_{|x'|<r/2}\Big(\int_0^{r/2} |s(\nabla \mathscr{T} f)(x', \varphi(x') + s)|^q \frac{ds}{s}\Big)^{p/q} dx'\Big)^{1/p} \\
&\quad \leq A_1 + A_2,
\end{aligned} \tag{2.604}$$

where, with $f_1 := f\chi_{S_r(0)}$ and $f_2 := f\chi_{\partial\Omega\setminus S_r(0)}$, we have set

$$A_i := C\Big(\int_{|x'|<r/2}\Big(\int_0^{r/2} |s(\nabla \mathscr{T} f_i)(x', \varphi(x') + s)|^q \frac{ds}{s}\Big)^{p/q} dx'\Big)^{1/p}, \quad i = 1, 2. \tag{2.605}$$

Observe that

$$\begin{aligned}
A_1 &\leq C\Big(\int_{|x'|<r}\Big(\int_0^{r} |s(\nabla \mathscr{T} f_1)(x', \varphi(x') + s)|^q \frac{ds}{s}\Big)^{p/q} dx'\Big)^{1/p} \\
&\approx \|Tf_1\|_{L^p(\{x' : |x'|<r\}, L^q((0,r), ds/s))} \\
&\leq C\|f_1\|_{L^p(S_r)} \leq C\|f\|_{L^p(\partial\Omega)},
\end{aligned} \tag{2.606}$$

by (2.600) given that $p \in (1, \infty)$ and $q = p \vee 2 \in [2, \infty)$. As regards A_2, we first note that for each x', s with $|x'| < r/2$ and $s \in (0, r/2)$ we have the rough estimate

$$\begin{aligned} \big|\nabla(\mathscr{T} f_2)(x', \varphi(x') + s)\big| &= \Big|\int_{\partial\Omega\setminus S_r(0)} (\nabla k)\big((x', \varphi(x') + s) - Y\big) f(Y)\, d\sigma(Y)\Big| \\ &\leq C \int_{\partial\Omega} |f|\, d\sigma \leq C\|f\|_{L^p(\partial\Omega)}, \end{aligned} \tag{2.607}$$

since the distance from $(x', \varphi(x') + s)$ to $\partial\Omega \setminus S_r(0)$ is bounded away from zero uniformly, and $\partial\Omega$ has finite measure. Based on this we may then readily conclude that

$$A_2 \leq C\|f\|_{L^p(\partial\Omega)}. \tag{2.608}$$

In summary, the above analysis shows that if the function v is as in (2.601), for an arbitrary $f \in L^p(\partial\Omega)$ then $\|v\|_{B^{p,p\vee 2}_{1/p}(\Omega)} \leq C\|f\|_{L^p(\partial\Omega)}$ with $C > 0$ finite constant independent of f. This finishes the proof of (2.589). □

Chapter 3
Function Spaces of Whitney Arrays

Here we discuss how to adapt the traditional ways of measuring smoothness for scalar functions (defined on the boundary of a Lipschitz domain) to the case of Whitney arrays.

3.1 Whitney–Lebesgue and Whitney–Sobolev Spaces

Fix a Lipschitz domain $\Omega \subset \mathbb{R}^n$. Given $m \in \mathbb{N}$, we say that a family

$$\dot{f} := \{f_\alpha : \alpha \in \mathbb{N}_0^n,\ |\alpha| \leq m-1\} \tag{3.1}$$

of functions from $L^1_{1,loc}(\partial\Omega)$ is a `Whitney array` if the components satisfy certain compatibility conditions, henceforth abbreviated as CC. More specifically,

$$\dot{f} \in CC \iff \begin{cases} \partial_{\tau_{jk}} f_\gamma = \nu_j f_{\gamma+e_k} - \nu_k f_{\gamma+e_j} & \sigma\text{-a.e. on } \partial\Omega, \\ \text{whenever } |\gamma| \leq m-2 \text{ and } 1 \leq j,k \leq n, \end{cases} \tag{3.2}$$

where the multi-index $e_j := (0,\dots,1,\dots 0) \in \mathbb{N}_0^n$ has the only nonzero component on the j-th position, $j \in \{1,\dots,n\}$.

For each $p \in [1,\infty]$ we then define the `Whitney--Lebesgue space`

$$\dot{L}^p_{m-1,0}(\partial\Omega) := \Big\{\dot{f} = \{f_\alpha\}_{|\alpha|\leq m-1} : \tag{3.3}$$

$$f_\alpha \in L^p(\partial\Omega) \text{ if } |\alpha| \leq m-1 \text{ and } \dot{f} \in CC\Big\},$$

I. Mitrea and M. Mitrea, *Multi-Layer Potentials and Boundary Problems*, Lecture Notes in Mathematics 2063, DOI 10.1007/978-3-642-32666-0_3,

which we equip with the norm

$$\|\dot{f}\|_{\dot{L}^p_{m-1,0}(\partial\Omega)} := \sum_{|\alpha|\leq m-1} \|f_\alpha\|_{L^p(\partial\Omega)}. \tag{3.4}$$

Note that, in the light of (3.2),

$$\dot{f} \in \dot{L}^p_{m-1,0}(\partial\Omega) \Longrightarrow f_\alpha \in L^p_1(\partial\Omega), \quad \text{whenever } |\alpha| \leq m-2 \tag{3.5}$$

and, hence,

$$\|\dot{f}\|_{\dot{L}^p_{m-1,0}(\partial\Omega)} \approx \sum_{|\alpha|\leq m-2} \|f_\alpha\|_{L^p_1(\partial\Omega)} + \sum_{|\alpha|=m-1} \|f_\alpha\|_{L^p(\partial\Omega)}. \tag{3.6}$$

We shall also work with a more regular version of the space introduced in (3.3), namely the Whitney–Sobolev space

$$\dot{L}^p_{m-1,1}(\partial\Omega) := \Big\{\dot{f} = \{f_\alpha\}_{|\alpha|\leq m-1} : \tag{3.7}$$
$$f_\alpha \in L^p_1(\partial\Omega) \text{ if } |\alpha| \leq m-1 \text{ and } \dot{f} \in CC\Big\},$$

which, for each $1 < p < \infty$, we endow with the norm

$$\|\dot{f}\|_{\dot{L}^p_{m-1,1}(\partial\Omega)} := \sum_{|\alpha|\leq m-1} \|f_\alpha\|_{L^p_1(\partial\Omega)}. \tag{3.8}$$

We next elaborate on the operation of multiplication between scalar functions and Whitney arrays. In this regard, we make the following.

Definition 3.1. Given a Lipschitz domain $\Omega \subset \mathbb{R}^n$ and $\eta \in C^\infty_c(\mathbb{R}^n)$, a scalar-valued function, for each family $\dot{f} = \{f_\alpha\}_{|\alpha|\leq m-1}$ with locally integrable components on $\partial\Omega$, set

$$\eta\dot{f} := \Big\{\sum_{\beta+\gamma=\alpha} \frac{\alpha!}{\beta!\gamma!} f_\gamma\big[\partial^\beta\eta\big]\Big|_{\partial\Omega}\Big\}_{|\alpha|\leq m-1}. \tag{3.9}$$

It is then easy to check that

$$\eta\dot{f} + \xi\dot{f} = (\eta+\xi)\dot{f}, \qquad \forall\, \eta,\xi \in C^\infty_c(\mathbb{R}^n). \tag{3.10}$$

Other properties of interest are summarized below.

Proposition 3.2. *Assume that $\Omega \subset \mathbb{R}^n$ is a Lipschitz domain. Then for each $m \in \mathbb{N}$ and $p \in (1,\infty)$, the spaces $\dot{L}^p_{m-1,0}(\partial\Omega)$, $\dot{L}^p_{m-1,1}(\partial\Omega)$ are modules over $C_c^\infty(\mathbb{R}^n)$, with the multiplication described in Definition 3.1.*

Proof. Fix $\dot{f} \in \dot{L}^p_{m-1,0}(\partial\Omega)$ along with $\eta \in C_c^\infty(\mathbb{R}^n)$. It suffices to show that the components of $\eta\dot{f}$ satisfy the compatibility conditions

$$\partial_{\tau_{jk}}(\eta\dot{f})_\alpha = \nu_j(\eta\dot{f})_{\alpha+e_k} - \nu_k(\eta\dot{f})_{\alpha+e_j}, \tag{3.11}$$

whenever $\alpha \in \mathbb{N}_0^n$ has $|\alpha| \leq m-2$ and $j,k \in \{1,\dots,n\}$. Unraveling definitions, this comes down to verifying that, given $\alpha \in \mathbb{N}_0^n$, then for each $k \in \{1,\dots,n\}$ we have

$$\sum_{\beta+\gamma=\alpha} \frac{\alpha!}{\beta!\gamma!}(f_{\gamma+e_k}\partial^\beta\eta + f_\gamma\partial^{\beta+e_k}\eta) = \sum_{\bar\beta+\bar\gamma=\alpha+e_k} \frac{\alpha!}{\bar\beta!\bar\gamma!} f_{\bar\gamma}\partial^{\bar\beta}\eta. \tag{3.12}$$

However, more generally, one has

$$\sum_{\beta+\gamma=\alpha} \frac{\alpha!}{\beta!\gamma!}\Big[F(\gamma+e_k,\beta) + F(\gamma,\beta+e_k)\Big] = \sum_{\bar\beta+\bar\gamma=\alpha+e_k} \frac{\alpha!}{\bar\beta!\bar\gamma!} F(\bar\gamma,\bar\beta), \tag{3.13}$$

for any function $F : \mathbb{N}_0^n \times \mathbb{N}_0^n \to \mathbb{R}$. Indeed, by linearity, it suffices to check that (3.13) holds in the particular case when $F(\theta,\sigma) = \delta_{\theta a}\delta_{\sigma b}$ for some fixed pair of multi-indices $a = (a_1,\dots,a_n)$, $b = (b_1,\dots,b_n)$. In this scenario, (3.13) amounts to the readily verified identity

$$\frac{\alpha!}{b!(a-e_k)!} + \frac{\alpha!}{(b-e_k)!a!} = \frac{(\alpha+e_k)!}{b!a!}, \tag{3.14}$$

with the convention that the factorial of a negative number is $+\infty$. □

We now present a basic density result involving Whitney array spaces.

Proposition 3.3. *Let Ω be a bounded Lipschitz domain in $\mathbb{R}^n$ and fix $1 < p < \infty$ and $m \in \mathbb{N}$. Then for every $\dot{f} = \{f_\alpha\}_{|\alpha|\leq m-1} \in \dot{L}^p_{m-1,1}(\partial\Omega)$ there exists a sequence of functions $F_j \in C_c^\infty(\mathbb{R}^n)$, $j \in \mathbb{N}$, such that*

$$[\partial^\alpha F_j]\Big|_{\partial\Omega} \longrightarrow f_\alpha \ \text{ in } \ L_1^p(\partial\Omega) \ \text{ as } \ j\to\infty, \qquad \forall\,\alpha\in\mathbb{N}_0^n : |\alpha| \leq m-1. \tag{3.15}$$

In particular,

$$\dot{L}^p_{m-1,1}(\partial\Omega) = \text{the closure of } \Big\{(\partial^\alpha F\big|_{\partial\Omega})_{|\alpha|\leq m-1} : F \in C_c^\infty(\mathbb{R}^n)\Big\}$$
$$\text{in } L_1^p(\partial\Omega)\oplus\cdots\oplus L_1^p(\partial\Omega). \tag{3.16}$$

On the other hand, for every $\dot{f} = \{f_\alpha\}_{|\alpha|\le m-1} \in \dot{L}^p_{m-1,0}(\partial\Omega)$ *there exist functions* $F_j \in C_c^\infty(\mathbb{R}^n)$, $j \in \mathbb{N}$, *such that*

$$\begin{aligned} &[\partial^\alpha F_j]\Big|_{\partial\Omega} \longrightarrow f_\alpha \text{ in } L^p_1(\partial\Omega) \text{ as } j\to\infty,\ \forall\,\alpha\in\mathbb{N}_0^n \text{ with } |\alpha|\le m-2,\\ &[\partial^\alpha F_j]\Big|_{\partial\Omega} \longrightarrow f_\alpha \text{ in } L^p(\partial\Omega) \text{ as } j\to\infty, \text{ if } \alpha\in\mathbb{N}_0^n \text{ has } |\alpha|\le m-1. \end{aligned} \tag{3.17}$$

Consequently,

$$\begin{aligned} \dot{L}^p_{m-1,0}(\partial\Omega) = \text{ the closure of } &\Big\{(\partial^\alpha F\big|_{\partial\Omega})_{|\alpha|\le m-1} : F\in C_c^\infty(\mathbb{R}^n)\Big\}\\ &\text{in } L^p_1(\partial\Omega)\oplus\cdots\oplus L^p_1(\partial\Omega)\oplus L^p(\partial\Omega). \end{aligned} \tag{3.18}$$

Proof. Since in all cases, both the hypotheses and the conclusions are stable under multiplication by a smooth function with compact support as in (3.9), there is no loss of generality in assuming that $\Omega = \{X = (x', x_n) \in \mathbb{R}^{n-1}\times\mathbb{R} : x_n > \varphi(x')\}$ for some Lipschitz function $\varphi : \mathbb{R}^{n-1} \to \mathbb{R}$, and that all the functions f_α have compact support. Consider first the case when the Whitney array $\dot{f} = \{f_\alpha\}_{|\alpha|\le m-1} \in \dot{L}^p_{m-1,1}(\partial\Omega)$. In this setting, Proposition 2.9 gives

$$\|f_\alpha\|_{L^p_1(\partial\Omega)} \approx \|f_\alpha(\cdot,\varphi(\cdot))\|_{L^p_1(\mathbb{R}^{n-1})}, \qquad \forall\,\alpha\in\mathbb{N}_0^n : |\alpha|\le m-1, \tag{3.19}$$

and the compatibility conditions (3.2) can be written in the form

$$\begin{aligned} &\frac{\partial}{\partial x_j}\Big[f_\alpha(x',\varphi(x'))\Big] = f_{\alpha+e_j}(x',\varphi(x')) + \partial_j\varphi(x')\,f_{\alpha+e_n}(x',\varphi(x'))\\ &\quad\text{for a.e. } x'\in\mathbb{R}^{n-1},\quad \forall\,\alpha\in\mathbb{N}_0^n : |\alpha|\le m-2,\quad 1\le j\le n-1. \end{aligned} \tag{3.20}$$

For further reference, let us also fix $R > 0$ with the property that for each multi-index α of length $\le m-1$ there holds $\operatorname{supp} f_\alpha(\cdot,\varphi(\cdot)) \subseteq \{x' \in \mathbb{R}^{n-1} : |x'| \le R\}$.

Next, fix a nonnegative function $\eta \in C^\infty(\mathbb{R}^{n-1})$, which vanishes identically for $|x'| > 1$ and such that $\int_{\mathbb{R}^{n-1}} \eta(x')\,dx' = 1$. Also, for each number $\varepsilon \in (0,1)$, set $\eta_\varepsilon(x') := \varepsilon^{1-n}\eta(x'/\varepsilon)$ for $x' \in \mathbb{R}^{n-1}$. Then, for each $|\alpha| \le m-1$, $\varepsilon > 0$, and each point $X = (x', x_n) \in \mathbb{R}^{n-1}\times\mathbb{R} = \mathbb{R}^n$, consider

$$F^\varepsilon_\alpha(X) := \left(\sum_{\ell=0}^{m-1-|\alpha|} \frac{1}{\ell!}\Big[(x_n-\varphi(\cdot))^\ell f_{\alpha+\ell e_n}(\cdot,\varphi(\cdot))\Big] * \eta_\varepsilon\right)(x'). \tag{3.21}$$

Clearly, for each multi-index α, the function $F^\varepsilon_\alpha(X)$ is C^∞-smooth for $X = (x', x_n)$ in $\mathbb{R}^n$, has compact support in the variable $x' \in \mathbb{R}^{n-1}$ (more precisely, $F^\varepsilon_\alpha(x', x_n) = 0$ if $|x'| > R+1$), and depends polynomially on x_n.

Now, based on the compatibility conditions (3.20), a straightforward calculation gives that whenever $r := m - 1 - |\alpha| \geq 1$ and $1 \leq j \leq n-1$,

$$\partial_j F_\alpha^\varepsilon(X) = F_{\alpha+e_j}^\varepsilon(X) + \frac{1}{(r-1)!}\Big[\Big((x_n - \varphi(\cdot))^{r-1}\partial_j\varphi(\cdot) f_{\alpha+r\,e_n}(\cdot\,,\varphi(\cdot))\Big) * \eta_\varepsilon\Big](x') + \frac{1}{r!}\Big[\Big((x_n - \varphi(\cdot))^r f_{\alpha+r\,e_n}(\cdot\,,\varphi(\cdot))\Big) * (\partial_j\eta_\varepsilon)\Big](x'). \tag{3.22}$$

Thus, if $r := m - 1 - |\alpha| \geq 1$, after moving the derivative off of η_ε in the last term above we arrive at the recurrence formula

$$\partial_j F_\alpha^\varepsilon(X) = \begin{cases} F_{\alpha+e_j}^\varepsilon(X) + \frac{1}{r!}\Big[\Big((x_n - \varphi(\cdot))^r \partial_j(f_{\alpha+r\,e_n}(\cdot\,,\varphi(\cdot)))\Big) * \eta_\varepsilon\Big](x') \text{ if } j < n, \\ F_{\alpha+e_n}^\varepsilon(X) \text{ if } j = n. \end{cases} \tag{3.23}$$

Let us now pick $\psi \in C_c^\infty(\mathbb{R})$ such that $\psi(t) = 1$ if $|t| < \sup\{|\varphi(x')| : |x'| \leq R+1\}$, and define $F^\varepsilon(X) := \psi(x_n)F_{(0,\dots,0)}^\varepsilon(x', x_n)$ for $X = (x', x_n) \in \mathbb{R}^n$. Then, obviously,

$$F_\varepsilon \in C_c^\infty(\mathbb{R}^n) \text{ and } F_\varepsilon(x', x_n) = F_{(0,\dots,0)}^\varepsilon(x', x_n) \quad \text{if } |x'| \leq R+1 \text{ and } x_n \text{ is near } \varphi(x'). \tag{3.24}$$

Thus, an inductive argument based on the formula (3.23) shows that, for any multi-index $\alpha = (\alpha', \alpha_n) \in \mathbb{N}_0^{n-1} \times \mathbb{N}$ of length $|\alpha'| + \alpha_n \leq m - 1$, the difference between $\partial^\alpha F^\varepsilon(X)$ and $F_\alpha^\varepsilon(X)$ can be expressed, when $X = (x', x_n)$ with $|x'| \leq R$ and x_n is near $\varphi(x')$, as a finite, constant coefficient, linear combination of terms of the type

$$\varepsilon^{-|\gamma|}\Big[\Big((x_n - \varphi(\cdot))^{m-1-|\beta|-\alpha_n}\partial_j(f_\delta(\cdot\,,\varphi(\cdot)))\Big) * (\partial^\gamma\eta)_\varepsilon\Big](x'), \tag{3.25}$$

where

$$1 \leq j \leq n-1, \quad \beta, \gamma \in \mathbb{N}_0^{n-1}, \quad \text{are such that } e_j + \beta + \gamma = \alpha', \quad \text{and } \delta \in \mathbb{N}_0^n, \ |\delta| = m-1. \tag{3.26}$$

Consequently, (3.15) will follow from Proposition 2.9 once we establish that for every multi-index $\alpha \in \mathbb{N}_0^n$ of length $\leq m - 1$,

$$F_\alpha^\varepsilon(\cdot, \varphi(\cdot)) \to f_\alpha(\cdot, \varphi(\cdot)) \text{ in } L_1^p(\mathbb{R}^{n-1}) \text{ as } \varepsilon \to 0, \tag{3.27}$$

and that, whenever the indices are as in (3.26) and $x_n = \varphi(x')$, the expression in (3.25), viewed as a function of $x' \in \mathbb{R}^{n-1}$, converges to zero in $L_1^p(\mathbb{R}^{n-1})$ as $\varepsilon \to 0$.

As regards (3.27), we begin by noting that

$$F_\alpha^\varepsilon(x',\varphi(x')) = \sum_{\ell=0}^{m-1-|\alpha|} \frac{1}{\ell!} \int_{\mathbb{R}^{n-1}} \Big(\varphi(x')-\varphi(y')\Big)^\ell f_{\alpha+\ell\, e_n}(y',\varphi(y'))\,\eta_\varepsilon(x'-y')\,dy'. \tag{3.28}$$

Thus, $F_\alpha^\varepsilon(\cdot,\varphi(\cdot)) \to f_\alpha(\cdot,\varphi(\cdot))$ in $L^p(\mathbb{R}^{n-1})$ as $\varepsilon \to 0$ (indeed, based on the Lipschitzianity of φ and the fact that $\operatorname{supp}\eta_\varepsilon \subseteq \{x' \in \mathbb{R}^{n-1} : |x'| \leq \varepsilon\}$, one can easily show that any term with $\ell > 0$ converges to zero in $L^p(\mathbb{R}^{n-1})$ as $\varepsilon \to 0$). Consequently, (3.27) is proved as soon as we show that $\nabla_{x'}[F_\alpha^\varepsilon(\cdot,\varphi(\cdot))] \to \nabla_{x'}[f_\alpha(\cdot,\varphi(\cdot))]$ in $L^p(\mathbb{R}^{n-1})$ as $\varepsilon \to 0$. We remark that this is obviously true if $|\alpha| = m-1$ (cf. (3.28)), so we consider the case when $|\alpha| \leq m-2$. In this situation, for each $k \in \{1,\dots,n-1\}$, we use (3.28) (in which we treat the cases $\ell = 0$, $\ell = 1$ and $\ell \geq 2$ separately) to compute

$$\begin{aligned}\frac{\partial}{\partial x_k}\Big[F_\alpha^\varepsilon(x',\varphi(x'))\Big] &= \int_{\mathbb{R}^{n-1}} \frac{\partial}{\partial y_k}\Big[f_\alpha(y',\varphi(y'))\Big]\,\eta_\varepsilon(x'-y')\,dy' \\ &\quad + \int_{\mathbb{R}^{n-1}} (\partial_k\varphi(x')-\partial_k\varphi(y'))\,f_{\alpha+e_n}(y',\varphi(y'))\,\eta_\varepsilon(x'-y')\,dy' \\ &\quad + \mathscr{R}(x'),\end{aligned} \tag{3.29}$$

where (based on the Lipschitzianity of φ, the fact that the function η_ε satisfies $\operatorname{supp}\eta_\varepsilon \subseteq \{x' \in \mathbb{R}^{n-1} : |x'| \leq \varepsilon\}$ and $\partial_k(\eta_\varepsilon) = \varepsilon^{-1}(\partial_k\eta)_\varepsilon$) it can be shown that the reminder satisfies the pointwise estimate

$$|\mathscr{R}(x')| \leq C\varepsilon \sum_{|\gamma|\leq m-1} \Big(|f_\gamma(\cdot,\varphi(\cdot))| * (|\eta| + |\nabla\eta|)_\varepsilon\Big)(x'), \quad \forall\, x' \in \mathbb{R}^{n-1}. \tag{3.30}$$

In particular, $\|\mathscr{R}\|_{L^p(\mathbb{R}^{n-1})}$ goes to zero as $\varepsilon \to 0$. Also, the expression in the second line of the right-hand side of (3.29) converges to zero in $L^p(\mathbb{R}^{n-1})$ as $\varepsilon \to 0$ (this is most easily seen by naturally splitting the integrand into two pieces, before passing to limit). As a result,

$$\frac{\partial}{\partial x_k}\Big[F_\alpha^\varepsilon(x',\varphi(x'))\Big] \to \frac{\partial}{\partial x_k}\Big[f_\alpha(x',\varphi(x'))\Big] \text{ in } L^p(\mathbb{R}^{n-1}) \text{ as } \varepsilon \to 0, \tag{3.31}$$

finishing the proof of (3.27).

Let us now consider the expression (3.25) when $\alpha \in \mathbb{N}_0^n$ with $|\alpha| \leq m-1$ is fixed and the conditions in (3.26) hold. A direct estimate (based on familiar, by now, support considerations, etc.) shows that, when $x_n = \varphi(x')$, the L^p-norm of this quantity (as a function of the variable x' in $\mathbb{R}^{n-1}$) is $\leq C\varepsilon^{m-|\alpha|} \to 0$ as $\varepsilon \to 0$.

To finish the proof, take $x_n = \varphi(x')$ in (3.25) and, for an arbitrary $j \in \{1, \dots, n-1\}$, apply $\partial/\partial x_j$. Much as we just did, the L^p-norm of this quantity is $\leq C\varepsilon^{m-|\alpha|-1}$. Thus, provided that $|\alpha| \leq m-2$, this converges to zero as $\varepsilon \to 0$.

The most delicate case is when $|\alpha| = m-1$. In this situation, we write out the terms obtained as a result of making $x_n = \varphi(x')$ in (3.25) and then applying $\partial/\partial x_k$ for some fixed $k \in \{1, \dots, n-1\}$. They are

$$\varepsilon^{-r} \int_{\mathbb{R}^{n-1}} \Big(\varphi(x') - \varphi(y')\Big)^r g(y')(\partial^{\gamma+e_k}\eta)_\varepsilon(x'-y')\,dy', \tag{3.32}$$

and

$$r\,\varepsilon^{1-r} \int_{\mathbb{R}^{n-1}} \Big(\varphi(x') - \varphi(y')\Big)^{r-1} \partial_k\varphi(x') g(y')(\partial^{\gamma}\eta)_\varepsilon(x'-y')\,dy', \tag{3.33}$$

where we have set $r := m-1-|\beta|-\alpha_n$ and $g := \partial_j(f_\delta(\cdot, \varphi(\cdot)))$. Above, we have used the fact that $|\alpha| = m-1$ forces $|\gamma| = r-1$. Our goal is to prove that the L^p-norm of the sum between (3.33) and (3.32), viewed as functions in $x' \in \mathbb{R}^{n-1}$, converges to zero as $\varepsilon \to 0$. To this end, write $\varphi(x') - \varphi(y') = \Delta(x', y')\,|x'-y'| + \nabla\varphi(x')\cdot(x'-y')$, where

$$\Delta(x', y') := \frac{\varphi(x') - \varphi(y') - (\nabla\varphi)(x')\cdot(y'-x')}{|x'-y'|}, \tag{3.34}$$

then expand

$$\begin{aligned} \Big(\varphi(x') - \varphi(y')\Big)^r &= \sum_{a+b=r} \frac{r!}{a!b!} \Delta(x', y')^a\, |x'-y'|^a\, (\nabla\varphi(x')\cdot(x'-y'))^b \qquad (3.35) \\ &= \sum_{a+b=r} \sum_{\substack{\sigma \in \mathbb{N}_0^{n-1} \\ |\sigma| = b}} \frac{r!}{a!\sigma!} \Delta(x', y')^a\, |x'-y'|^a\, (\nabla\varphi(x'))^\sigma\, (x'-y')^\sigma. \end{aligned}$$

Plugging this back into (3.32) finally yields the expression

$$\sum_{a+b=r} \sum_{\sigma \in \mathbb{N}_0^{n-1} : |\sigma| = b} \frac{r!}{a!\sigma!} (\nabla\varphi(x'))^\sigma \int_{\mathbb{R}^{n-1}} \Delta(x', y')^a g(y')(\Theta^{\sigma,a}_{\gamma+e_k})_\varepsilon(x'-y')\,dy', \tag{3.36}$$

where we have used the notation

$$\Theta^{\sigma,a}_\tau(x') := (x')^\sigma\,|x'|^a\,(\partial^\tau\eta)(x'), \qquad x' \in \mathbb{R}^{n-1},\ \sigma, \tau \in \mathbb{N}_0^{n-1},\ a \in \mathbb{N}_0. \tag{3.37}$$

Each integral above is pointwise dominated by $C(\|\nabla\varphi\|_{L^\infty(\mathbb{R}^{n-1})})(\mathscr{M}g)(x')$ uniformly with respect to $\varepsilon > 0$ (where $\mathscr{M}$ denotes the Hardy–Littlewood maximal operator in $\mathbb{R}^{n-1}$), and converges to zero as $\varepsilon \to 0$ whenever $a > 0$ and x' is a differentiability point for the function φ. Thus, since φ is almost everywhere differentiable, by a well-known theorem of H. Rademacher, and since $\mathscr{M}$ is bounded on $L^p(\mathbb{R}^{n-1})$ if $1 < p < \infty$, Lebesgue's Dominated Convergence Theorem gives that all integrals in (3.36) corresponding to $a > 0$ converge to zero in $L^p(\mathbb{R}^{n-1})$ as $\varepsilon \to 0$.

On the other hand, in the context of (3.36), $a = 0$ forces $|\sigma| = r = |\gamma + e_k|$. Note that in general, if $a = 0$ and $|\sigma| = |\tau|$, definition (3.37) and repeated integrations by parts yield

$$\int_{\mathbb{R}^{n-1}} \Theta_\tau^{\sigma,0}(x')\,dx' = \int_{\mathbb{R}^{n-1}} (x')^\sigma (\partial^\tau \eta)(x')\,dx' = (-1)^{|\tau|}\sigma!\,\delta_{\sigma\tau}, \tag{3.38}$$

where $\delta_{\sigma\tau}$ is the Kronecker symbol. Consequently, as $\varepsilon \to 0$, the portion of (3.36) corresponding to $a = 0$ and, hence, the entire expression in (3.36), converges in $L^p(\mathbb{R}^{n-1})$ to

$$(-1)^r\, r!\,(\nabla\varphi)^{\gamma+e_k} g. \tag{3.39}$$

The analysis of (3.33) closely parallels that of (3.32). In fact, given the close analogy between (3.33) and (3.32), in order to compute the limit of the former in $L^p(\mathbb{R}^{n-1})$ as $\varepsilon \to 0$, we only need to make the following changes in (3.39): replace $\gamma + e_k$ by γ, r by $r-1$ and then multiply the result by $r\,\partial_k\varphi$. The resulting expression is precisely the opposite of (3.39). In summary, the above reasoning shows that the sum of the expressions in (3.32)–(3.33) is convergent to zero in $L^p(\mathbb{R}^{n-1})$ as $\varepsilon \to 0$, and this finishes the proof of (3.16).

The proof of (3.18) proceeds largely as before with the most notable differences being that we will necessarily use (3.22) in place of (3.23), since this time we only have $f_{\alpha+re_n} \in L^p(\partial\Omega)$ for $r = m - 1 - |\alpha|$. As a result, in the present case we find ourselves in a situation when, instead of treating residual terms as in (3.25), we need to show that whenever $\gamma \in \mathbb{N}_0^{n-1}$, $r := |\gamma| + 1$ and $1 \le k \le n-1$, we have

$$T_\varepsilon g \to 0 \text{ in } L^p(\mathbb{R}^{n-1}) \text{ as } \varepsilon \to 0, \qquad \text{for every } g \in L^p(\mathbb{R}^{n-1}), \tag{3.40}$$

where

$$\begin{aligned}(T_\varepsilon g)(x') := {} & r\varepsilon^{1-r}\Big[\Big((\varphi(x') - \varphi(\cdot))^{r-1}\partial_k\varphi(\cdot)g(\cdot)\Big) * (\partial^\gamma\eta)_\varepsilon\Big](x') \\ & + \varepsilon^{-r}\Big[\Big((\varphi(x') - \varphi(\cdot))^r g(\cdot)\Big) * (\partial^{\gamma+e_k}\eta)_\varepsilon\Big](x'). \end{aligned} \tag{3.41}$$

This, in turn, is proved much as we have treated (3.32)–(3.33) (the fact that $\partial_k\varphi$ is evaluated at x' in (3.33) as opposed to being evaluated at y' in (3.41) is inconsequential). This justifies (3.18) and finishes the proof of Proposition 3.3. □

Our next theorem sheds further light on the nature of the compatibility conditions introduced in (3.2).

Theorem 3.4. *Assume that Ω is a bounded Lipschitz domain in $\mathbb{R}^n$, and fix two numbers $p \in (1,\infty)$ and $m \in \mathbb{N}$. Given $(g_0, g_1, \dots, g_{m-1}) \in L^p(\partial\Omega) \oplus \cdots \oplus L^p(\partial\Omega)$, a necessary and sufficient condition for the existence of a Whitney–Lebesgue array $\dot{h} := \{h_\alpha\}_{|\alpha|\le m-1} \in \dot{L}^p_{m-1,0}(\partial\Omega)$ such that*

$$g_k = \sum_{|\alpha|=k} \frac{k!}{\alpha!} \nu^\alpha h_\alpha, \qquad 0 \le k \le m-1, \tag{3.42}$$

is that

$$\begin{cases} f_\alpha \in L^p_1(\partial\Omega), & \forall\, \alpha \in \mathbb{N}_0^n \text{ with } |\alpha| \le m-2, \\ f_\alpha \in L^p(\partial\Omega), & \forall\, \alpha \in \mathbb{N}_0^n \text{ with } |\alpha| = m-1, \end{cases} \tag{3.43}$$

where the family $\{f_\alpha\}_{|\alpha|\le m-1}$ is defined as follows. Set

$$f_{(0,\dots,0)} := g_0 \tag{3.44}$$

and, inductively, if $\{f_\gamma\}_{|\gamma|\le \ell-1}$ have already been defined for some $\ell \in \{1,\dots, m-1\}$, set

$$f_\alpha = \nu^\alpha g_\ell + \frac{\alpha!}{\ell!} \sum_{\substack{\mu+\delta+e_j=\alpha \\ |\theta|=|\delta|}} \frac{|\delta|!}{\delta!}\frac{|\mu|!}{\mu!}\frac{|\theta|!}{\theta!} \nu^{\delta+\theta} (\nabla_{tan} f_{\mu+\theta})_j, \qquad \forall\, \alpha \in \mathbb{N}_0^n : |\alpha| = \ell, \tag{3.45}$$

where $(\cdot)_j$ is the j-th component.

Furthermore, if for some m-tuple of functions

$$(g_0, g_1, \dots, g_{m-1}) \in L^p(\partial\Omega) \oplus \cdots \oplus L^p(\partial\Omega) \tag{3.46}$$

the array $\dot{f} = \{f_\alpha\}_{|\alpha|\le m-1}$ defined as in formulas (3.44)–(3.45) *satisfies* (3.43) *then $\dot{f} := \{f_\alpha\}_{|\alpha|\le m-1}$ belongs to $\dot{L}^p_{m-1,0}(\partial\Omega)$ and*

$$g_k = \sum_{|\alpha|=k} \frac{k!}{\alpha!} \nu^\alpha f_\alpha, \qquad 0 \le k \le m-1. \tag{3.47}$$

Moreover, $\dot{f} = \{f_\alpha\}_{|\alpha|\le m-1}$ is the unique array in $\dot{L}^p_{m-1,0}(\partial\Omega)$ which satisfies (3.47).

Proof. Assume that $g_k \in L^p(\partial\Omega)$, $0 \le k \le m-1$, are such that the functions f_α defined as in formulas (3.44)–(3.45) satisfy (3.43). The claim that we make is that $\dot{f} := \{f_\alpha\}_{|\alpha|\le m-1} \in \dot{L}^p_{m-1,0}(\partial\Omega)$ and (3.47) holds. Of course, the first part of the claim is proved as soon as we verify (3.2) which we aim to prove by reasoning by induction on $\ell := |\alpha| \in \{0,\dots,m-2\}$. Based on (3.44)–(3.45), we compute

$$f_{e_j} = \nu_j g_1 + \sum_{k=1}^{n} \nu_k \partial_{\tau_{kj}} g_0 \qquad \forall\, j \in \{1,\dots,n\}, \tag{3.48}$$

from which we deduce, with the help of

$$(\nabla_{tan} f)_j = \sum_{k=1}^{n} \nu_k \partial_{\tau_{kj}} f, \qquad \forall\, j \in \{1,\ldots,n\}, \tag{3.49}$$

that

$$\begin{aligned}\nu_k f_{e_j} - \nu_j f_{e_k} &= (\nu_k \nu_j g_1 + \nu_k \nu_r \partial_{\tau_{rj}} g_0) - (\nu_j \nu_k g_1 + \nu_j \nu_r \partial_{\tau_{rk}} g_0)\\ &= \partial_{\tau_{kj}} f_{(0,\ldots,0)}, \qquad \forall\, j,k \in \{1,\ldots,n\},\end{aligned} \tag{3.50}$$

i.e., the version of (3.2) with $\alpha = (0,\ldots,0)$. To prove the induction step, assume that (3.2) holds whenever $|\alpha| \leq \ell - 1$. In particular, $\{f_\gamma\}_{|\gamma|\leq\ell} \in \dot{L}^p_{\ell,1}(\partial\Omega)$. In concert with the induction hypothesis, Proposition 3.3 then proves that there exists a sequence of functions $F_\varepsilon \in C^\infty_c(\mathbb{R}^n)$, $\varepsilon > 0$, such that

$$|\gamma| \leq \ell \Longrightarrow [\partial^\gamma F_\varepsilon]\Big|_{\partial\Omega} \to f_\gamma \ \text{ in } \ L^p_1(\partial\Omega) \ \text{ as } \ \varepsilon \to 0. \tag{3.51}$$

Let us now digress in order to note an important algebraic identity. Specifically, for any two multi-indices $\alpha, \beta \in \mathbb{N}^n_0$ of length $\ell + 1$ written as

$$\alpha = e_{j_1} + \cdots e_{j_{\ell+1}}, \qquad \beta = e_{k_1} + \cdots e_{k_{\ell+1}}, \tag{3.52}$$

a direct calculation yields

$$\begin{aligned}&\nu^\beta \partial^\alpha - \nu^\alpha \partial^\beta\\ &= \sum_{r=0}^{\ell} \nu_{k_1} \cdots \nu_{k_{\ell-r}} \nu_{j_{\ell-r+2}} \cdots \nu_{j_{\ell+1}} \partial_{\tau_{k_{\ell-r+1} j_{\ell-r+1}}} \partial_{j_1} \cdots \partial_{j_{\ell-r}} \partial_{k_{\ell-r+2}} \cdots \partial_{k_{\ell+1}}\end{aligned} \tag{3.53}$$

with the convention that products (of components of ν, and of partial derivatives) taken over void sets of indices are discarded. In order to be able to re-write (3.53) in multi-index notation, it is convenient to symmetrize the right-hand side of this identity by adding up all its versions obtained by permuting the indices $j_1,\ldots,j_{\ell+1}$ and $k_1,\ldots,k_{\ell+1}$ in (3.52). In this fashion, we obtain

$$\begin{aligned}\nu^\beta \partial^\alpha - \nu^\alpha \partial^\beta &= \frac{\alpha!}{(\ell+1)!} \frac{\beta!}{(\ell+1)!} \times\\ &\times \sum_{r=0}^{\ell} \sum_{\substack{\mu+\delta+e_j=\alpha,\, |\delta|=r\\ \gamma+\theta+e_k=\beta,\, |\theta|=r}} \frac{(\ell-r)!}{\mu!} \frac{r!}{\delta!} \frac{(\ell-r)!}{\gamma!} \frac{r!}{\theta!} \nu^{\gamma+\delta} \partial_{\tau_{kj}} \partial^{\mu+\theta}.\end{aligned} \tag{3.54}$$

As a corollary, for every $\alpha \in \mathbb{N}_0^n$ of length $\ell + 1$, from (3.54) and the fact that

$$\sum_{|\gamma|=r} \frac{r!}{\gamma!} \nu^{2\gamma} = 1, \qquad \forall r \in \mathbb{N}_0, \tag{3.55}$$

we obtain

$$\sum_{|\beta|=\ell+1} \frac{(\ell+1)!}{\beta!} \nu^{\beta} (\nu^{\beta} \partial^{\alpha} - \nu^{\alpha} \partial^{\beta})$$

$$= \frac{\alpha!}{(\ell+1)!} \sum_{\substack{\mu+\delta+e_j=\alpha \\ |\delta|=|\theta|}} \sum_{k=1}^{n} \frac{|\mu|!}{\mu!} \frac{|\delta|!}{\delta!} \frac{|\gamma|!}{\gamma!} \frac{|\theta|!}{\theta!} \nu^{\gamma+\delta+e_k} \partial_{\tau_{kj}} \partial^{\mu+\theta}. \tag{3.56}$$

Returning to the mainstream discussion, for each $\alpha \in \mathbb{N}_0^n$ with $|\alpha| = \ell + 1$ we now write

$$f_\alpha = \nu^\alpha g_{\ell+1} + \frac{\alpha!}{(\ell+1)!} \sum_{\substack{\mu+\delta+e_j=\alpha \\ |\theta|=|\delta|}} \frac{|\delta|!}{\delta!} \frac{|\mu|!}{\mu!} \frac{|\theta|!}{\theta!} \nu^{\delta+\theta} (\nabla_{tan} f_{\mu+\theta})_j$$

$$= \nu^\alpha g_{\ell+1} + \lim_{\varepsilon\to 0} \frac{\alpha!}{(\ell+1)!} \sum_{\substack{\mu+\delta+e_j=\alpha \\ |\theta|=|\delta|}} \frac{|\delta|!}{\delta!} \frac{|\mu|!}{\mu!} \frac{|\theta|!}{\theta!} \nu^{\delta+\theta} \Big(\nabla_{tan} [\partial^{\mu+\theta} F_\varepsilon]\Big|_{\partial\Omega}\Big)_j$$

$$= \nu^\alpha g_{\ell+1} + \lim_{\varepsilon\to 0} \frac{\alpha!}{(\ell+1)!} \sum_{\substack{\mu+\delta+e_j=\alpha \\ |\theta|=|\delta|}} \sum_{k=1}^{n} \frac{|\delta|!}{\delta!} \frac{|\mu|!}{\mu!} \frac{|\theta|!}{\theta!} \nu^{\delta+\theta+e_k} \partial_{\tau_{kj}} \Big([\partial^{\mu+\theta} F_\varepsilon]\Big|_{\partial\Omega}\Big)$$

$$= \nu^\alpha g_{\ell+1} + \lim_{\varepsilon\to 0} \sum_{|\beta|=\ell+1} \frac{(\ell+1)!}{\beta!} \nu^\beta \Big(\nu^\beta [\partial^\alpha F_\varepsilon]\Big|_{\partial\Omega} - \nu^\alpha [\partial^\beta F_\varepsilon]\Big|_{\partial\Omega}\Big) \tag{3.57}$$

in $L^p(\partial\Omega)$, thanks to (3.45), (3.51), (3.49) and (3.56).

Next, fix an arbitrary $\alpha \in \mathbb{N}_0^n$ with $|\alpha| = \ell$, choose $j, k \in \{1, \ldots, n\}$, and consider the identity (3.57) written twice, with $\alpha + e_k$ and $\alpha + e_j$, respectively, in place of α. If we multiply the first such identity by ν_j, the second one by ν_k and then subtract them from one another, we arrive at

$$\nu_j f_{\alpha+e_k} - \nu_k f_{\alpha+e_j} = \lim_{\varepsilon\to 0} \sum_{|\beta|=\ell+1} \frac{(\ell+1)!}{\beta!} \nu^{2\beta} \partial_{\tau_{jk}} \Big([\partial^\alpha F_\varepsilon]\Big|_{\partial\Omega}\Big)$$

$$= \lim_{\varepsilon\to 0} \partial_{\tau_{jk}} \Big([\partial^\alpha F_\varepsilon]\Big|_{\partial\Omega}\Big)$$

$$= \partial_{\tau_{jk}} f_\alpha, \tag{3.58}$$

by (3.55) and (3.51). This finishes the proof of the induction step. Thus (3.2) holds and, as a result, $\dot{f} := \{f_\alpha\}_{|\alpha|\leq m-1} \in \dot{L}^p_{m-1,0}(\partial\Omega)$, as desired.

As for (3.47), it follows from (3.57) that for each multi-index α with $|\alpha| = \ell$

$$\begin{aligned} f_\alpha &= \nu^\alpha g_\ell + \lim_{\varepsilon\to 0} \sum_{|\beta|=\ell} \frac{\ell!}{\beta!} \nu^\beta \Big(\nu^\beta [\partial^\alpha F_\varepsilon]\Big|_{\partial\Omega} - \nu^\alpha [\partial^\beta F_\varepsilon]\Big|_{\partial\Omega}\Big) \\ &= \nu^\alpha g_\ell + \sum_{|\beta|=\ell} \frac{\ell!}{\beta!} \nu^\beta (\nu^\beta f_\alpha - \nu^\alpha f_\beta). \end{aligned} \tag{3.59}$$

From this, we further deduce that

$$\nu^\alpha g_\ell = \nu^\alpha \sum_{|\beta|=\ell} \frac{\ell!}{\beta!} \nu^\beta f_\beta \tag{3.60}$$

for each multi-index α of length ℓ. Multiplying both sides of (3.60) by $\frac{\ell!}{\alpha!}\nu^\alpha$ and summing over all $\alpha \in \mathbb{N}_0^n$ with $|\alpha| = \ell$ finally yields (3.47), on account of (3.55).

There remains to prove the uniqueness claim made in the last part of the theorem. To this end, assume that $\dot{h} = \{h_\alpha\}_{|\alpha|\leq m-1}$, $\dot{f} = \{f_\alpha\}_{|\alpha|\leq m-1} \in \dot{L}^p_{m-1,0}(\partial\Omega)$ are such that (3.42) and (3.47) hold. Of course, our goal is to show that

$$f_\gamma = h_\gamma \qquad \forall\, \gamma \in \mathbb{N}_0^n, \;\; |\gamma| \leq m-1. \tag{3.61}$$

In the proof of (3.61) we shall proceed by induction on $\ell := |\alpha| \in \{0, \dots, m-1\}$. From definitions, $f_{(0,\dots,0)} = g_0 = h_{(0,\dots,0)}$, proving case $\ell = 0$.

Assume next that (3.61) holds whenever $|\gamma| \leq \ell - 1$, and fix an arbitrary multi-index α of length ℓ. Based on (3.45) (which can be seen to hold in this case by reverse engineering (3.57)) and the induction hypothesis, we may then write

$$\begin{aligned} f_\alpha &= \nu^\alpha g_\ell + \frac{\alpha!}{\ell!} \sum_{\substack{\mu+\delta+e_j=\alpha \\ |\theta|=|\delta|}} \frac{|\delta|!}{\delta!} \frac{|\mu|!}{\mu!} \frac{|\theta|!}{\theta!} \nu^{\delta+\theta} (\nabla_{tan} f_{\mu+\theta})_j \\ &= \nu^\alpha g_\ell + \frac{\alpha!}{\ell!} \sum_{\substack{\mu+\delta+e_j=\alpha \\ |\theta|=|\delta|}} \frac{|\delta|!}{\delta!} \frac{|\mu|!}{\mu!} \frac{|\theta|!}{\theta!} \nu^{\delta+\theta} (\nabla_{tan} h_{\mu+\theta})_j \\ &= \nu^\alpha g_\ell + \frac{\alpha!}{\ell!} \sum_{\substack{\mu+\delta+e_j=\alpha \\ |\theta|=|\delta|}} \frac{|\delta|!}{\delta!} \frac{|\mu|!}{\mu!} \frac{|\theta|!}{\theta!} \sum_{k=1}^{n} \nu^{\delta+\theta+e_k} \partial_{\tau_{kj}} h_{\mu+\theta} \end{aligned}$$

$$= \nu^\alpha g_\ell + \frac{\alpha!}{\ell!} \sum_{\substack{\mu+\delta+e_j=\alpha \\ |\theta|=|\delta|}} \frac{|\delta|!}{\delta!} \frac{|\mu|!}{\mu!} \frac{|\theta|!}{\theta!} \sum_{k=1}^{n} \nu^{\delta+\theta+e_k} (\nu_k h_{\mu+\theta+e_j} - \nu_j h_{\mu+\theta+e_k})$$

$$= \nu^\alpha g_\ell + \sum_{|\beta|=\ell} \frac{\ell!}{\beta!} \nu^\beta \left(\nu^\beta h_\alpha - \nu^\alpha h_\beta \right). \tag{3.62}$$

The last step above can be justified by invoking (3.56) and a limiting argument based on Proposition 3.3. Hence, on account of (3.62), (3.42) and (3.55),

$$\begin{aligned} f_\alpha &= \nu^\alpha g_\ell + \sum_{|\beta|=\ell} \frac{\ell!}{\beta!} \nu^\beta \left(\nu^\beta h_\alpha - \nu^\alpha h_\beta \right) \\ &= \nu^\alpha \sum_{|\beta|=\ell} \frac{\ell!}{\beta!} \nu^\beta h_\beta + h_\alpha - \sum_{|\beta|=\ell} \frac{\ell!}{\beta!} \nu^{\alpha+\beta} h_\beta \\ &= h_\alpha, \end{aligned} \tag{3.63}$$

as wanted. This justifies (3.61) and finishes the proof of the theorem. □

We continue to study the nature of the Whitney–Lebesgue and Whitney–Sobolev spaces introduced in (3.3) and (3.7), respectively.

Proposition 3.5. *Assume that $\Omega \subset \mathbb{R}^n$ is a Lipschitz domain, and fix $m \in \mathbb{N}$, $m \geq 2$, and $1 < p < \infty$. Then the mapping*

$$\Psi : \dot{L}^p_{m-1,0}(\partial\Omega) \longrightarrow \dot{L}^p_{m-2,1}(\partial\Omega) \times L^p(\partial\Omega) \tag{3.64}$$

defined for each $\{h_\alpha\}_{|\alpha|\leq m-1} \in \dot{L}^p_{m-1,0}(\partial\Omega)$ by the formula

$$\Psi\Big(\{h_\alpha\}_{|\alpha|\leq m-1}\Big) := \Big(\{h_\alpha\}_{|\alpha|\leq m-2}\,,\ \sum_{|\alpha|=m-1} \frac{(m-1)!}{\alpha!} \nu^\alpha h_\alpha\Big) \tag{3.65}$$

is an isomorphism.

Proof. To show that the map (3.64)–(3.65) is onto, fix

$$\Big(\{h_\alpha\}_{|\alpha|\leq m-2}\,,\ g\Big) \in \dot{L}^p_{m-2,1}(\partial\Omega) \times L^p(\partial\Omega) \tag{3.66}$$

and set

$$g_k := \sum_{|\alpha|=k} \frac{k!}{\alpha!} \nu^\alpha h_\alpha \quad \text{if} \quad 0 \leq k \leq m-2, \quad \text{and} \quad g_{m-1} := g. \tag{3.67}$$

Obviously, we have $(g_0, g_1, \dots, g_{m-1}) \in L^p(\partial\Omega) \oplus \cdots \oplus L^p(\partial\Omega)$ and we claim that if the family $\{f_\alpha\}_{|\alpha|\leq m-1}$ is defined according to (3.44)–(3.45), then $f_\gamma = h_\gamma$ for every multi-index γ of length $\leq m-2$. This claim can be justified by reasoning much as in the last part of the proof of Theorem 3.4. In turn, this ensures that (3.43) holds, hence

$$\dot{f} := \{f_\alpha\}_{|\alpha|\leq m-1} \in \dot{L}^p_{m-1,0}(\partial\Omega) \text{ and (3.47) is valid,} \tag{3.68}$$

by Theorem 3.4. In turn, (3.47) and (3.61) give that $\Psi(\dot{f}) = (\{h_\alpha\}_{|\alpha|\leq m-2}, g)$, i.e. Ψ is onto.

There remains to show that the map (3.64)–(3.65) is one-to-one. For this it suffices to prove that the assignment

$$\dot{L}^p_{m-1,0}(\partial\Omega) \ni \dot{f} = \{f_\alpha\}_{|\alpha|\leq m-1} \mapsto \Big\{\sum_{|\alpha|=k} \frac{k!}{\alpha!} \nu^\alpha f_\alpha\Big\}_{0\leq k\leq m-1} \in L^p(\partial\Omega), \tag{3.69}$$

is one-to-one. In turn, this follows directly from the uniqueness claim made in the last part of the statement of Theorem 3.4. □

3.2 Whitney–Besov Spaces

We shall now consider Besov spaces exhibiting higher-order smoothness by adopting a point of view similar to (3.3)–(3.7). Concretely, fix a Lipschitz domain $\Omega \subset \mathbb{R}^n$ and, for $m \in \mathbb{N}$ and p, q, s as in (2.396), define the `Whitney-Besov space` $\dot{B}^{p,q}_{m-1,s}(\partial\Omega)$ via the requirement

$$\dot{f} \in \dot{B}^{p,q}_{m-1,s}(\partial\Omega) \Longleftrightarrow f_\alpha \in B^{p,q}_s(\partial\Omega) \text{ if } |\alpha| \leq m-1, \text{ and } \dot{f} \in CC. \tag{3.70}$$

For each $\dot{f} \in \dot{B}^{p,q}_{m-1,s}(\partial\Omega)$ we then set

$$\|\dot{f}\|_{\dot{B}^{p,q}_{m-1,s}(\partial\Omega)} := \sum_{|\alpha|\leq m-1} \|f_\alpha\|_{B^{p,q}_s(\partial\Omega)}. \tag{3.71}$$

The case $1 \leq p = q \leq \infty$ has been studied in the literature in [2, 76] (cf. also [62] for related results).

It can then be shown that the Besov space $\dot{B}^{p,q}_{m-1,s}(\partial\Omega)$ is a module over $C^\infty_c(\mathbb{R}^n)$ (where the multiplication is in the sense of (3.9)). When $1 < p < \infty$, we clearly have

$$\|\dot{f}\|_{\dot{B}^{p,p}_{m-1,s}(\partial\Omega)} \approx \sum_{|\alpha|\leq m-2} \|f_\alpha\|_{L^p_1(\partial\Omega)} + \sum_{|\alpha|=m-1} \|f_\alpha\|_{B^{p,p}_s(\partial\Omega)}. \tag{3.72}$$

Other properties of interest are summarized below. First, we record an alternative description of the Whitney–Besov space (3.70) in the case $p = q \in (1, \infty)$ which has been proved in [76].

Proposition 3.6. *Let $\Omega \subset \mathbb{R}^n$ be a Lipschitz domain with surface measure σ. Assume that $m \in \mathbb{N}$, $1 \leq p \leq \infty$, $s \in (0,1)$ and, for an arbitrary family $\dot{f} = \{f_\alpha\}_{|\alpha| \leq m-1}$ of real-valued, σ-measurable functions on $\partial\Omega$, set*

$$R_\alpha(X,Y) := f_\alpha(X) - \sum_{|\beta| \leq m-1-|\alpha|} \frac{1}{\beta!} f_{\alpha+\beta}(Y)\,(X-Y)^\beta, \qquad X, Y \in \partial\Omega, \tag{3.73}$$

for each multi-index α of length $\leq m-1$. Then

$$\|\dot{f}\|_{\dot{B}^{p,p}_{m-1,s}(\partial\Omega)} \approx \sum_{|\alpha| \leq m-1} \|f_\alpha\|_{L^p(\partial\Omega)} + \sum_{|\alpha| \leq m-1} \Big(\int_{\partial\Omega}\int_{\partial\Omega} \frac{|R_\alpha(X,Y)|^p}{|X-Y|^{p(m-1+s-|\alpha|)+n-1}}\,d\sigma(X)d\sigma(Y)\Big)^{1/p}, \tag{3.74}$$

if $p < \infty$ and, corresponding to $p = \infty$,

$$\|\dot{f}\|_{\dot{B}^{\infty,\infty}_{m-1,s}(\partial\Omega)} \approx \sum_{|\alpha| \leq m-1} \|f_\alpha\|_{L^\infty(\partial\Omega)} + \sum_{|\alpha| \leq m-1} \sup_{X \neq Y \in \partial\Omega} \frac{|R_\alpha(X,Y)|}{|X-Y|^{m-1+s-|\alpha|}}. \tag{3.75}$$

Let us point out that, with the help of the L^p-modulus of continuity

$$r_\alpha(t) := \Big(\iint_{\substack{(X,Y)\in\partial\Omega\times\partial\Omega \\ |X-Y|<t}} |R_\alpha(X,Y)|^p\,d\sigma(X)d\sigma(Y)\Big)^{1/p}, \quad |\alpha| \leq m-1,\ t > 0, \tag{3.76}$$

we have, for $\frac{n-1}{n} < p < \infty$ and $(n-1)\big(\frac{1}{p}-1\big)_+ < s < 1$,

$$\|\dot{f}\|_{\dot{B}^{p,p}_{m-1,s}(\partial\Omega)} \approx \sum_{|\alpha| \leq m-1} \|f_\alpha\|_{L^p(\partial\Omega)} + \sum_{|\alpha| \leq m-1} \Big(\int_0^\infty \frac{r_\alpha(t)^p}{t^{p(m-1+s-|\alpha|)+n-1}}\,dt\Big)^{1/p}. \tag{3.77}$$

We further elaborate on the nature of the remainder (3.73). Given a family of functions $\dot{f} = \{f_\alpha\}_{|\alpha|\le m-1}$ on $\partial\Omega$ and $X \in \mathbb{R}^n$, $Y, Z \in \partial\Omega$, set

$$P_\alpha(X,Y) := \sum_{|\beta|\le m-1-|\alpha|} \frac{1}{\beta!} f_{\alpha+\beta}(Y)\,(X-Y)^\beta, \quad \forall\,\alpha : |\alpha| \le m-1, \tag{3.78}$$

$$P(X,Y) := P_{(0,\dots,0)}(X,Y). \tag{3.79}$$

Then

$$R_\alpha(Y,Z) = f_\alpha(Y) - P_\alpha(Y,Z), \qquad \forall\,\alpha : |\alpha| \le m-1, \tag{3.80}$$

and the following elementary identities hold for each multi-index $\alpha \in \mathbb{N}_0^n$ of length $\le m-1$:

$$\partial_X^\beta P_\alpha(X,Y) = P_{\alpha+\beta}(X,Y), \qquad |\beta| \le m-1-|\alpha|, \tag{3.81}$$

$$P_\alpha(X,Y) - P_\alpha(X,Z) = \sum_{|\beta|\le m-1-|\alpha|} \frac{1}{\beta!} R_{\alpha+\beta}(Y,Z)(X-Y)^\beta. \tag{3.82}$$

See, e.g., [119, p. 177] for the last formula.

Moving on, we now state and prove a useful density result (a topic we shall return to later on, and prove a more general result in Corollary 3.10).

Proposition 3.7. *Fix an integer $m \in \mathbb{N}$ along with three numbers p, q, s satisfying $1 < p < \infty$, $0 < q < \infty$ and $s \in (0,1)$. Also, suppose that $\Omega \subset \mathbb{R}^n$ is a bounded Lipschitz domain. Then for every Whitney array $\dot{f} = \{f_\alpha\}_{|\alpha|\le m-1} \in \dot{B}^{p,q}_{m-1,s}(\partial\Omega)$ there exists a sequence of functions $F_j \in C_c^\infty(\mathbb{R}^n)$, $j \in \mathbb{N}$, such that*

$$\begin{aligned} &[\partial^\alpha F_j]\Big|_{\partial\Omega} \longrightarrow f_\alpha \ \text{ in } \ L_1^p(\partial\Omega) \ \text{ as } \ j \to \infty, \ \forall\,\alpha \in \mathbb{N}_0^n \ \text{ with } |\alpha| \le m-2, \\ &[\partial^\alpha F_j]\Big|_{\partial\Omega} \longrightarrow f_\alpha \ \text{ in } \ B_s^{p,q}(\partial\Omega) \ \text{ as } \ j \to \infty, \ \text{ if } \alpha \in \mathbb{N}_0^n \ \text{ has } |\alpha| \le m-1. \end{aligned} \tag{3.83}$$

Thus,

$$\begin{aligned} \dot{B}^{p,q}_{m-1,s}(\partial\Omega) = \ &\text{the closure of } \Big\{(\partial^\alpha F\big|_{\partial\Omega})_{|\alpha|\le m-1} : F \in C_c^\infty(\mathbb{R}^n)\Big\} \\ &\text{in } L_1^p(\partial\Omega) \oplus \cdots \oplus L_1^p(\partial\Omega) \oplus B_s^{p,q}(\partial\Omega). \end{aligned} \tag{3.84}$$

In particular,

$$\Big\{(\partial^\alpha F|_{\partial\Omega})_{|\alpha|\le m-1} : F \in C_c^\infty(\mathbb{R}^n)\Big\} \ \text{ is dense in } \ \dot{B}^{p,q}_{m-1,s}(\partial\Omega), \tag{3.85}$$

and, if $q \in [1,\infty)$,

$$\dot{B}^{p,q}_{m-1,s}(\partial\Omega) \ \text{ is a reflexive Banach space.} \tag{3.86}$$

Proof. We largely follow the approach described in proof of Proposition 3.3. More specifically, recall the operators (3.41). The density result we seek will follow as soon as we show that

$$T_\varepsilon g \to 0 \ \text{ in } \ B^{p,q}_s(\mathbb{R}^{n-1}) \ \text{ as } \ \varepsilon \to 0, \qquad \text{for every } g \in B^{p,q}_s(\mathbb{R}^{n-1}), \tag{3.87}$$

whenever $p \in (1,\infty)$, $0 < q < \infty$ and $s \in (0,1)$. To this end we note that, in addition to (3.40), we have $\sup_{\varepsilon>0} \|T_\varepsilon\|_{\mathscr{L}(L^p \to L^p)} < \infty$. Next, we make the claim that the family of operators (3.41) also satisfies

$$T_\varepsilon g \to 0 \ \text{ in } \ L^p_1(\mathbb{R}^{n-1}) \ \text{ as } \ \varepsilon \to 0, \qquad \text{for every } g \in L^p_1(\mathbb{R}^{n-1}) \tag{3.88}$$

and

$$\sup_{\varepsilon>0} \|T_\varepsilon\|_{\mathscr{L}(L^p_1 \to L^p_1)} < \infty. \tag{3.89}$$

To prove (3.88), first observe that if $g \in L^p_1(\mathbb{R}^{n-1})$ then $T_\varepsilon g \to 0$ in $L^p(\mathbb{R}^{n-1})$ as $\varepsilon \to 0$ by (3.40). Next, much as we have passed from (3.22) to (3.23), we have

$$(T_\varepsilon g)(x') = \varepsilon^{1-r}\Big[\Big((\varphi(x') - \varphi(\cdot))^r (\partial_k g)(\cdot)\Big) * (\partial^\gamma \eta)_\varepsilon\Big](x'), \tag{3.90}$$

so that

$$\begin{aligned} \partial_j (T_\varepsilon g)(x') = {} & r\varepsilon^{1-r}\Big[\Big((\varphi(x') - \varphi(\cdot))^{r-1} (\partial_j \varphi)(x')(\partial_k g)(\cdot)\Big) * (\partial^\gamma \eta)_\varepsilon\Big](x') \\ & + \varepsilon^{-r}\Big[\Big((\varphi(x') - \varphi(\cdot))^r (\partial_k g)(\cdot)\Big) * (\partial^{\gamma+e_j} \eta)_\varepsilon\Big](x'). \end{aligned} \tag{3.91}$$

The key observation is that the assignment sending $\partial_k g$ into the expression in the right-hand side of (3.91) is of the same type as T_ε in (3.41). Hence, the same type of reasoning that led to (3.40) yields

$$\partial_j (T_\varepsilon g) \to 0 \ \text{ in } \ L^p(\mathbb{R}^{n-1}) \ \text{ as } \ \varepsilon \to 0, \qquad \text{for every } g \in L^p_1(\mathbb{R}^{n-1}). \tag{3.92}$$

This justifies (3.88). Also, the proof of (3.89) is implicit in the above reasoning. Consequently, real interpolation with the counterpart of this result on the Lebesgue scale yields

$$\sup_{\varepsilon>0} \|T_\varepsilon\|_{\mathscr{L}(B^{p,q}_s \to B^{p,q}_s)} < \infty, \tag{3.93}$$

for each $1 < p < \infty$, $0 < q < \infty$ and $s \in (0,1)$. Thus, if $g \in B^{p,q}_s(\mathbb{R}^{n-1})$ is given, along with a positive threshold ϵ_o, choose $h \in L^p_1(\mathbb{R}^{n-1})$ so that $\|g - h\|_{B^{p,q}_s(\mathbb{R}^{n-1})} < \epsilon_o$ and estimate

$$\begin{aligned} \|T_\varepsilon g\|_{B^{p,q}_s(\mathbb{R}^{n-1})} & \leq \|T_\varepsilon (g-h)\|_{B^{p,q}_s(\mathbb{R}^{n-1})} + \|T_\varepsilon h\|_{B^{p,q}_s(\mathbb{R}^{n-1})} \\ & \leq C\|g-h\|_{B^{p,q}_s(\mathbb{R}^{n-1})} + C\|T_\epsilon h\|_{L^p_1(\mathbb{R}^{n-1})} \\ & \leq C\epsilon_o + C\|T_\varepsilon h\|_{L^p_1(\mathbb{R}^{n-1})}, \end{aligned} \tag{3.94}$$

for some $C > 0$ independent of ε. Thus, $\limsup_{\varepsilon\to 0} \|T_\varepsilon g\|_{B_s^{p,q}(\mathbb{R}^{n-1})} \leq C\epsilon_o$ by (3.88). Since ϵ_0 is arbitrary, this finally gives (3.87).

Lastly, when $q \in [1,\infty)$, the claim made in (3.86) is a consequence of (3.84) and the fact that a closed subspace of a reflexive Banach space is itself reflexive. This finishes the proof of the proposition. □

We conclude this section with some useful embedding result for Whitney–Besov spaces. To state it, define

$$\widetilde{\mathscr{O}} := \Big\{(p,q,s) \in (0,\infty] \times (0,\infty] \times (0,1) : (n-1)\big(\tfrac{1}{p} - 1\big)_+ < s < 1\Big\}. \tag{3.95}$$

Proposition 3.8. *Let Ω be a bounded Lipschitz domain in $\mathbb{R}^n$ and fix $m \in \mathbb{N}$. Also, with $\widetilde{\mathscr{O}}$ as in (3.95), consider two triplets $(p_0,q_0,s_0),(p_1,q_1,s_1) \in \widetilde{\mathscr{O}}$. Then the inclusion*

$$\dot{B}^{p_0,q_0}_{m-1,s_0}(\partial\Omega) \hookrightarrow \dot{B}^{p_1,q_1}_{m-1,s_1}(\partial\Omega) \tag{3.96}$$

is continuous with dense range if either

$$p_0 \leq p_1, \quad q_0 \leq q_1 \ \text{ and } \ \tfrac{1}{p_0} - \tfrac{s_0}{n-1} = \tfrac{1}{p_1} - \tfrac{s_1}{n-1}, \tag{3.97}$$

or

$$p_0 = p_1, \quad q_0 \leq q_1 \ \text{ and } \ s_0 = s_1. \tag{3.98}$$

Moreover, if $p_0 = p_1 =: p$ and $s_0 > s_1$ then also

$$\dot{B}^{p,q_0}_{m-1,s_0}(\partial\Omega) \hookrightarrow \dot{B}^{p,q_1}_{m-1,s_1}(\partial\Omega) \ \text{ continuously with dense range.} \tag{3.99}$$

Proof. This is an immediate consequence of Theorem 2.17. □

3.3 Multi-Trace Theory

Let $\Omega \subset \mathbb{R}^n$ be an open, nonempty, proper subset of $\mathbb{R}^n$, and fix some $m \in \mathbb{N}$. In this context, define the `higher-order trace` tr_{m-1} (relative to the given set Ω and order m) as the operator mapping $C^\infty(\overline{\Omega})$ into Whitney arrays of length m according to the formula

$$\operatorname{tr}_{m-1} F := \Big\{[\partial^\alpha F]\Big|_{\partial\Omega}\Big\}_{|\alpha|\leq m-1}. \tag{3.100}$$

We are interested in extending the action of this operator to more general spaces.

Theorem 3.9. *Assume that $\Omega \subset \mathbb{R}^n$ is a bounded Lipschitz domain and suppose that*

$$0 < p,q \leq \infty \ \text{ and } \ (n-1)\Big(\tfrac{1}{p} - 1\Big)_+ < s < 1. \tag{3.101}$$

Also, fix $m \in \mathbb{N}$. Then the higher-order trace operator introduced above extends by continuity to a well-defined, linear and bounded operator

$$\mathrm{tr}_{m-1} : B^{p,q}_{m-1+s+1/p}(\Omega) \longrightarrow \dot{B}^{p,q}_{m-1,s}(\partial\Omega). \tag{3.102}$$

In addition, (3.100) *is onto and, in fact, has a bounded, linear right-inverse. That is, there exists a linear, continuous operator*

$$\mathscr{E} : \dot{B}^{p,q}_{m-1,s}(\partial\Omega) \longrightarrow B^{p,q}_{m-1+s+1/p}(\Omega) \tag{3.103}$$

such that, with Tr *as in Theorem 2.53,*

$$\dot{f} = \{f_\alpha\}_{|\alpha|\leq m-1} \in \dot{B}^{p,q}_{m-1,s}(\partial\Omega) \Longrightarrow \mathrm{Tr}\,[\partial^\alpha(\mathscr{E}\,\dot{f})] = f_\alpha, \quad \forall\,\alpha : |\alpha| \leq m-1. \tag{3.104}$$

Similar results are valid on the Triebel–Lizorkin scale. Concretely, with the convention that $q = \infty$ if $p = \infty$, the operator

$$\mathrm{tr}_{m-1} : F^{p,q}_{m-1+s+1/p}(\Omega) \longrightarrow \dot{B}^{p,p}_{m-1,s}(\partial\Omega) \tag{3.105}$$

is also well-defined, linear and bounded. Moreover, this operator has a linear, bounded right inverse

$$\mathscr{E} : \dot{B}^{p,p}_{m-1,s}(\partial\Omega) \longrightarrow F^{p,q}_{m-1+s+1/p}(\Omega) \tag{3.106}$$

which is compatible with (3.103)*.*

Proof. We shall first establish the well-definiteness and boundedness of the higher-order trace operator. The extension of (3.100) in the context of (3.102) is given by

$$\mathrm{tr}_{m-1}(F) := \Big\{\mathrm{Tr}\,[\partial^\alpha F] : |\alpha| \leq m-1\Big\}, \quad \forall\,F \in B^{p,q}_{m-1+s+1/p}(\Omega), \tag{3.107}$$

where the traces in the right-hand side are taken in the sense of Theorem 2.53. That for each $F \in B^{p,q}_{m-1+s+1/p}(\Omega)$ the family $\dot{f} := \{\mathrm{Tr}\,[\partial^\alpha F]\}_{|\alpha|\leq m-1}$ satisfies the compatibility conditions (3.2) is a consequence of the identity

$$\partial_{\tau_{jk}}[\mathrm{Tr}\,u] = \nu_j\,\mathrm{Tr}\,[\partial_k u] - \nu_k\,\mathrm{Tr}\,[\partial_j u], \qquad j,k = 1,..,n, \tag{3.108}$$

valid for each $u \in B^{p,q}_{1+s+1/p}(\Omega)$. In turn, (3.108) is easily checked using (2.208).

Next, we turn our attention to the issue of a linear extension operator, i.e. a linear right-inverse for the higher-order trace operator. This segment in the proof utilizes results and notation which will be discussed later. Let $\dot{\mathscr{D}}^{\pm}$ be a double multi-layer associated with, say, the polyharmonic operator Δ^m in $\Omega_+ := \Omega$ and $\Omega_- := \mathbb{R}^n \setminus \bar{\Omega}$, respectively (see § 4 for relevant definitions). Also, fix some $\psi \in C^\infty_c(\mathbb{R}^n)$ which is identically one in a neighborhood of $\overline{\Omega}$. What we need is that, whenever s satisfies

$0 < (n-1)(\frac{1}{p}-1)_+ < s < 1$ and $0 < q \le \infty$,

$$\psi\dot{\mathscr{D}}^{\pm} : \dot{B}^{p,q}_{m-1,s}(\partial\Omega) \longrightarrow B^{p,q}_{m-1+s+1/p}(\Omega_{\pm}) \tag{3.109}$$

are well-defined, linear and bounded, and

$$\mathrm{tr}_{m-1} \circ \psi\dot{\mathscr{D}}^{+} - \mathrm{tr}_{m-1} \circ \psi\dot{\mathscr{D}}^{-} = I \quad \text{in } \dot{B}^{p,q}_{m-1,s}(\partial\Omega). \tag{3.110}$$

These are consequences of Theorem 4.19, proved later independently of the current considerations.

To proceed, recall that $\mathscr{R}_{\Omega_\pm}$ stands for the operator of restriction to $\Omega_\pm$ (cf. (2.196)), and that

$$E_{\Omega_\pm} : B^{p,q}_s(\Omega_\pm) \longrightarrow B^{p,q}_s(\mathbb{R}^n), \qquad 0 < p,q \le \infty, \;\; s \in \mathbb{R}, \tag{3.111}$$

stands for Rychkov's universal extension operator (see Theorem 2.29). This satisfies

$$\mathscr{R}_{\Omega_\pm} \circ E_{\Omega_\pm} = I \quad \text{on } B^{p,q}_s(\Omega_\pm). \tag{3.112}$$

If we now set

$$\mathscr{E} := \psi\dot{\mathscr{D}}^{+} - \mathscr{R}_{\Omega_+} \circ E_{\Omega_-} \circ \psi\dot{\mathscr{D}}^{-} \tag{3.113}$$

then, thanks to (3.109), (3.113), (3.111), and (2.213), for p,q,s as in (3.101), the operators

$$\mathscr{E} : \dot{B}^{p,q}_{m-1,s}(\partial\Omega) \longrightarrow B^{p,q}_{m-1+s+\frac{1}{p}}(\Omega), \tag{3.114}$$

$$\mathscr{E} : \dot{B}^{p,p}_{m-1,s}(\partial\Omega) \longrightarrow F^{p,q}_{m-1+s+\frac{1}{p}}(\Omega), \tag{3.115}$$

are well-defined, linear and bounded. Now, generally speaking, if $u \in B^{p,q}_{s+1/p}(\mathbb{R}^n)$ where $0 < p,q \le \infty$ and $(n-1)(\frac{1}{p}-1)_+ < s < 1$, then $\mathrm{Tr}\,(\mathscr{R}_{\Omega_+}(u)) = \mathrm{Tr}\,(\mathscr{R}_{\Omega_-}(u))$, since this obviously holds in the dense subspace of $B^{p,q}_{s+1/p}(\mathbb{R}^n)$ consisting of smooth, compactly supported, functions. Hence, from this, (3.112) and the identity (3.110),

$$\begin{aligned}
\mathrm{tr}_{m-1} \circ \mathscr{E} &= \mathrm{tr}_{m-1} \circ \psi\dot{\mathscr{D}}^{+} - \mathrm{tr}_{m-1} \circ \mathscr{R}_{\Omega_+} \circ E_{\Omega_-} \circ \psi\dot{\mathscr{D}}^{-} \\
&= \mathrm{tr}_{m-1} \circ \psi\dot{\mathscr{D}}^{+} - \mathrm{tr}_{m-1} \circ \mathscr{R}_{\Omega_-} \circ E_{\Omega_-} \circ \psi\dot{\mathscr{D}}^{-} \\
&= \mathrm{tr}_{m-1} \circ \psi\dot{\mathscr{D}}^{+} - \mathrm{tr}_{m-1} \circ \psi\dot{\mathscr{D}}^{-} = I,
\end{aligned} \tag{3.116}$$

i.e., $\mathscr{E}$ is a linear, bounded, right-inverse for the trace operator tr_{m-1} on Besov and Triebel–Lizorkin spaces.

This completes the proof of the portion of Theorem 3.9 dealing with Besov spaces, and the case of (3.105) is handled similarly. □

One useful consequence of Theorem 3.9 is singled out below.

Corollary 3.10. *Fix an integer $m \in \mathbb{N}$ and assume that $\frac{n-1}{n} < p < \infty$, $0 < q < \infty$ and $(n-1)\big(\frac{1}{p}-1\big)_+ < s < 1$. Also, suppose that $\Omega \subset \mathbb{R}^n$ is a bounded Lipschitz domain. Then for every $\dot{f} = \{f_\alpha\}_{|\alpha|\le m-1} \in \dot{B}^{p,q}_{m-1,s}(\partial\Omega)$ there exists a sequence of functions $F_j \in C^\infty_c(\mathbb{R}^n)$, $j \in \mathbb{N}$, such that*

$$\mathrm{Tr}\,[\partial^\alpha F_j] \longrightarrow f_\alpha \ \ in\ \ B^{p,q}_s(\partial\Omega)\ \ as\ \ j \to \infty, \quad \forall\,\alpha \in \mathbb{N}^n_0 : |\alpha| \le m-1. \tag{3.117}$$

Hence,

$$\Big\{(\partial^\alpha F|_{\partial\Omega})_{|\alpha|\le m-1} : F \in C^\infty_c(\mathbb{R}^n)\Big\} \ \ is\ dense\ in\ \ \dot{B}^{p,q}_{m-1,s}(\partial\Omega). \tag{3.118}$$

Proof. This is an immediate consequence of (2.208) and the fact that the trace map (3.102) is bounded and onto, for the range of indices specified in the statement of the corollary. □

The following is a companion result for Theorem 3.9 in the case in which the smoothness inside the Lipschitz domain is measured on the scale of weighted Sobolev spaces.

Theorem 3.11. *Let $\Omega \subset \mathbb{R}^n$ be a bounded Lipschitz domain and suppose that*

$$1 < p < \infty \ \ and\ \ 0 < s < 1. \tag{3.119}$$

Also, set

$$a := 1 - s - \tfrac{1}{p} \in (0,1), \tag{3.120}$$

and fix $m \in \mathbb{N}$. Then the higher-order trace operator originally considered as in (3.100) *extends by continuity to a well-defined, linear and bounded operator*

$$\mathrm{tr}_{m-1} : W^{m,p}_a(\Omega) \longrightarrow \dot{B}^{p,p}_{m-1,s}(\partial\Omega). \tag{3.121}$$

In addition, tr_{m-1} considered in the above context is onto. In fact, the operator $\mathscr{E}$ defined as in (3.113) *with*

$$E_{\Omega_-} : W^{m,p}_a(\mathbb{R}^n \setminus \overline{\Omega}) \longrightarrow W^{m,p}_a(\mathbb{R}^n) \tag{3.122}$$

denoting Stein's extension operator from Theorem 2.46, is a bounded, linear right-inverse for the higher-order trace operator from (3.121). *That is,*

$$\mathscr{E} : \dot{B}^{p,p}_{m-1,s}(\partial\Omega) \longrightarrow W^{m,p}_a(\Omega) \tag{3.123}$$

is a linear, continuous operator and has the property that, if Tr *is as in Theorem 2.53, then*

$$\dot{f} = \{f_\alpha\}_{|\alpha|\leq m-1} \in \dot{B}^{p,p}_{m-1,s}(\partial\Omega) \Longrightarrow \mathrm{Tr}\,[\partial^\alpha(\mathscr{E}\,\dot{f})] = f_\alpha, \quad \forall\,\alpha\,:\,|\alpha|\leq m-1. \tag{3.124}$$

Proof. The fact that the higher-order trace operator from (3.121) is well-defined, linear and bounded has been established in [76]. In [76] the authors have also constructed an extension operator in the context weighted Sobolev spaces which serves as a right-inverse for (3.121). The novelty here is that our old extension operator $\mathscr{E}$ from (3.113) also does the job, when E_{Ω_-} is now viewed as Stein's extension operator discussed in § 2.6. Indeed, if this is the case then Theorem 2.46 guarantees that the operator (3.122) is well-defined and bounded. Based on this, the density result from (2.364) (which continues to hold for arbitrary Lipschitz domains), and the fact that (with p, s as in (3.119), and any cutoff function $\psi \in C^\infty_c(\mathbb{R}^n)$)

$$\psi\dot{\mathscr{D}}^\pm\,:\,\dot{B}^{p,p}_{m-1,s}(\partial\Omega) \longrightarrow W^{m,p}_a(\Omega_\pm) \tag{3.125}$$

are well-defined, linear and bounded operators (cf. Theorem 4.20 proved later, independently of the current considerations), we may then conclude, much as in the proof of Theorem 3.9, that the operator $\mathscr{E}$ from (3.123) is well-defined, linear, bounded, and satisfies (3.124). □

It is significant to note that the extension operators given by (3.103) and (3.123) are *universal*, i.e., they do not depend on the regularity and integrability indices used to define the function spaces on which they act.

Moving on, the goal is to characterize the null-space of the higher-order trace operator (3.100) acting of Besov and Triebel–Lizorkin spaces in Lipschitz domains, as well as to prove certain basic density results for the space of test functions in this setting.

Theorem 3.12. *Let Ω be a bounded Lipschitz domain in $\mathbb{R}^n$ and fix $m \in \mathbb{N}$. Also, suppose that $\frac{n-1}{n} < p < \infty$, $(n-1)(1/p-1)_+ < s < 1$ and $\min\{1,p\} \leq q < \infty$. Then*

$$F^{p,q}_{m-1+s+1/p,z}(\Omega) = \Big\{u \in F^{p,q}_{m-1+s+1/p}(\Omega)\,:\,\mathrm{tr}_{m-1}\,u = 0\Big\} \tag{3.126}$$

and

$$C^\infty_c(\Omega) \hookrightarrow \Big\{u \in F^{p,q}_{m-1+s+1/p}(\Omega)\,:\,\mathrm{tr}_{m-1}\,u = 0\Big\}\ \textit{densely}. \tag{3.127}$$

Furthermore, a similar result is valid for the scale of Besov spaces. Specifically, if $m \in \mathbb{N}$, $\frac{n-1}{n} < p < \infty$, $(n-1)(1/p-1)_+ < s < 1$ and $0 < q < \infty$, then

$$B^{p,q}_{m-1+s+1/p,z}(\Omega) = \Big\{u \in B^{p,q}_{m-1+s+1/p}(\Omega) : \operatorname{tr}_{m-1} u = 0\Big\} \tag{3.128}$$

and

$$C^\infty_c(\Omega) \hookrightarrow \Big\{u \in B^{p,q}_{m-1+s+1/p}(\Omega) : \operatorname{tr}_{m-1} u = 0\Big\} \text{ densely.} \tag{3.129}$$

Proof. By (2.218)–(2.219), if $u \in F^{p,q}_{m-1+s+1/p,z}(\Omega)$ then $u \in F^{p,q}_{m-1+s+1/p}(\Omega)$ and there exists a sequence $u_j \in C^\infty_c(\Omega)$, $j \in \mathbb{N}$, such that $u_j \to u$ in $F^{p,q}_{m-1+s+1/p}(\Omega)$ as $j \to \infty$. It follows that $0 = \operatorname{tr}_{m-1}(u_j) \to \operatorname{tr}_{m-1}(u)$ in $\dot{B}^{p,p}_{m-1,s}(\partial\Omega)$ as $j \to \infty$, hence u belongs to the space in the right-hand side of (3.126). This proves the left-to-right inclusion in (3.126).

Conversely, assume that $u \in F^{p,q}_{m-1+s+1/p}(\Omega)$ has $\operatorname{tr}_{m-1}(u) = 0$. Since, generally speaking, $\operatorname{tr}_{m-1}(\psi\, w) = \psi \operatorname{tr}_{m-1}(w)$ for any nice cut-off function ψ and any reasonable w (cf. (3.9)), it follows that the problem is local in character. Hence, matters can be readily reduced to the case when Ω is the domain in $\mathbb{R}^n$ lying above the graph of a Lipschitz function and

$$\operatorname{supp} u \subset \mathcal{O} \tag{3.130}$$

for some open, relatively compact set $\mathcal{O} \subset \mathbb{R}^n$. In particular, there exists an infinite, upright circular cone Γ, with vertex at the origin in $\mathbb{R}^n$ such that $X + \Gamma \subset \Omega$ for every $X \in \overline{\Omega}$. Also, fix $\psi \in C^\infty_c(\mathbb{R}^n)$ such that $\psi \equiv 1$ on $\mathcal{O}$.

In this context, we shall work with Calderón's extension operator of order m, which we denote by $\mathscr{E}_m$ and whose construction we briefly review. The starting point is the so-called Sobolev's representation formula to the effect that if

$$\phi \in C^\infty(S^{n-1}) \text{ is supported in } \Gamma \cap S^{n-1} \tag{3.131}$$

and satisfies

$$\int_{S^{n-1}} \phi(\omega)\, d\omega = \frac{(-1)^m}{(m-1)!} \tag{3.132}$$

then for every $w \in C^m(\overline{\Gamma})$ with compact support,

$$w(0) = \sum_{|\alpha|=m} \int_\Gamma \frac{m!}{\alpha!} \phi\Big(\frac{Y}{|Y|}\Big) \frac{Y^\alpha}{|Y|^n} (\partial^\alpha w)(Y)\, dY. \tag{3.133}$$

Indeed, for such a function w and a fixed $\omega \in S^{n-1}$, repeated integrations by parts give

$$\begin{aligned} w(0) &= \frac{(-1)^m}{(m-1)!} \int_0^\infty t^{m-1} \frac{d^m}{dt^m}[w(t\omega)]\, dt \\ &= \frac{(-1)^m}{(m-1)!} \sum_{|\alpha|=m} \frac{m!}{\alpha!} \int_0^\infty t^{m-1} \omega^\alpha (\partial^\alpha w)(t\omega)\, dt. \end{aligned} \tag{3.134}$$

Multiplying (3.134) by $\phi(\omega)$ and integrating for $\omega \in S^{n-1}$ then readily yields (3.133), given (3.131)–(3.132).

Now, for every $w \in C^m(\overline{\Omega})$ with bounded support we define

$$\mathscr{E}_m(w)(X) := \sum_{|\alpha|=m} \psi(X) \int_\Omega \phi_\alpha(X-Y)(\partial^\alpha w)(Y)\, dY, \qquad X \in \mathbb{R}^n, \quad (3.135)$$

where, for each $\alpha \in \mathbb{N}_0^n$ with $|\alpha| = m$, we have set

$$\phi_\alpha(X) := (-1)^m \frac{m!}{\alpha!} \phi\Big(-\frac{X}{|X|}\Big) \frac{X^\alpha}{|X|^n}, \qquad X \in \mathbb{R}^n \setminus \{0\}. \quad (3.136)$$

(As is customary, $X^\alpha := x_1^{\alpha_1} \cdots x_n^{\alpha_n}$ if $X = (x_1, \dots, x_n) \in \mathbb{R}^n$ where the multi-index $\alpha = (\alpha_1, \dots, \alpha_n) \in \mathbb{N}_0^n$.)

Observe that, for each multi-index α of length m,

$$\operatorname{supp} \phi_\alpha \subset -\Gamma, \quad \phi_\alpha \in C^\infty(\mathbb{R}^n \setminus \{0\}), \quad \phi_\alpha(\lambda X) = \lambda^{m-n} \phi_\alpha(X), \ \forall\, \lambda \in \mathbb{R}. \quad (3.137)$$

It is also useful to note that, since $\operatorname{supp}[\phi(Y/|Y|)] \subset \overline{\Gamma}$, the identity (3.133) yields

$$\sum_{|\alpha|=m} \partial^\alpha \phi_\alpha = \delta, \ \text{ Dirac's distribution, in } \mathbb{R}^n. \quad (3.138)$$

The operator (3.135) enjoys the following properties. First, (3.135) and (3.138) entail that

$$w \in C_c^\infty(\Omega) \Longrightarrow \mathscr{E}_m(w) \equiv 0 \ \text{ in } \mathbb{R}^n \setminus \overline{\Omega}. \quad (3.139)$$

Second,

$$w \in C^\infty(\overline{\Omega}), \text{ with bounded support} \Longrightarrow \mathscr{E}_m(w)\Big|_\Omega = (\psi|_\Omega) w \ \text{ in } \Omega. \quad (3.140)$$

Indeed, given the support condition on ϕ_α and since $-\Gamma \subset X - \Omega$ for every $X \in \Omega$, the change of variables $Z := X - Y$ in the right-hand side of (3.135) yields

$$\mathscr{E}_m(w)(X) = \sum_{|\alpha|=m} \psi(X) \int_{-\Gamma} \phi_\alpha(Z)(\partial^\alpha w)(X-Z)\, dZ, \qquad X \in \Omega. \quad (3.141)$$

Now (3.140) follows readily from (3.141) and (3.133), after changing $Z \mapsto -Z$.

Third, we note that (3.141) can be rephrased as

$$\mathscr{E}_m(w) = \sum_{|\alpha|=m} \psi\, (\phi_\alpha * \widetilde{(\partial^\alpha w)}), \quad (3.142)$$

for every $w \in C^\infty(\overline{\Omega})$ with bounded support, where tilde denotes extension of functions defined in Ω to $\mathbb{R}^n$ by setting them to be identically zero outside Ω.

To proceed, we remark that, as a consequence of (3.137) and Theorem 6.2.2 on p. 258 of [124] (cf. also Remark 3 on p. 257 of [124] and the discussion in the last part of § 2.3), the convolution operators $f \mapsto \phi_\alpha * f$, for $|\alpha| = m$, are locally smoothing of order m on the scale of Triebel–Lizorkin spaces in $\mathbb{R}^n$. Consequently, for each function $w \in C^\infty(\overline{\Omega})$ with bounded support, we may estimate

$$\begin{aligned}\|\mathscr{E}_m(w)\|_{F^{p,q}_{m-1+s+1/p}(\mathbb{R}^n)} &\leq C \sum_{|\alpha|=m} \left\| \psi\, \phi_\alpha * (\widetilde{\partial^\alpha w}) \right\|_{F^{p,q}_{m-1+s+1/p}(\mathbb{R}^n)} \\ &\leq C \sum_{|\alpha|=m} \|\widetilde{\partial^\alpha w}\|_{F^{p,q}_{-1+s+1/p}(\mathbb{R}^n)} \\ &\leq C \sum_{|\alpha|=m} \|\partial^\alpha w\|_{F^{p,q}_{-1+s+1/p}(\Omega)} \\ &\leq C \|w\|_{F^{p,q}_{m-1+s+1/p}(\Omega)}. \end{aligned} \tag{3.143}$$

The third inequality uses the fact that $\max\left(\frac{1}{p} - 1, n(\frac{1}{p} - 1)\right) < -1 + s + \frac{1}{p} < \frac{1}{p}$ and $0 < \min\{1, p\} \leq q < \infty$ which, in turn, ensure that the extension by zero from Ω to $\mathbb{R}^n$ is a bounded operator with preservation of class; cf. Proposition 2.31. Also, the last inequality in (3.143) follows from the boundedness of the differential operator $\partial^\alpha : F^{p,q}_{m-1+s+1/p}(\Omega) \to F^{p,q}_{-1+s+1/p}(\Omega)$ if $|\alpha| = m$ which, in turn, is a direct consequence of (2.149). Thus, by density (cf. (2.208)),

$$\mathscr{E}_m : F^{p,q}_{m-1+s+1/p}(\Omega) \longrightarrow F^{p,q}_{m-1+s+1/p}(\mathbb{R}^n) \tag{3.144}$$

is a well-defined, linear and bounded operator, whenever the indices are as in the statement of the theorem.

Returning now to the study of the function u (considered in the second paragraph of the proof), we pick a sequence $u_j \in C^\infty(\overline{\Omega})$, $j \in \mathbb{N}$, such that

$$u_j \longrightarrow u \quad \text{in} \quad F^{p,q}_{m-1+s+1/p}(\Omega), \quad \text{as} \quad j \to \infty. \tag{3.145}$$

Thus, on the one hand,

$$\operatorname{tr}_{m-1}(u_j) \longrightarrow \operatorname{tr}_{m-1}(u) = 0 \quad \text{in} \quad \dot{B}^{p,p}_{m-1,s}(\partial\Omega), \quad \text{as} \quad j \to \infty. \tag{3.146}$$

On the other hand, from the continuity of (3.144) and (3.145) we have

$$\mathscr{E}_m(u_j) \longrightarrow \mathscr{E}_m(u) \quad \text{in} \quad F^{p,q}_{m-1+s+1/p}(\mathbb{R}^n), \quad \text{as} \quad j \to \infty. \tag{3.147}$$

As a consequence,

$$\mathscr{R}_\Omega(\mathscr{E}_m(u)) = \lim_{j\to\infty} (\mathscr{R}_\Omega \circ \mathscr{E}_m)(u_j) = \lim_{j\to\infty} \psi\, u_j = \psi\, u = u. \tag{3.148}$$

Formula (3.147) also implies that at almost every point $X \in \mathbb{R}^n \setminus \overline{\Omega}$ we have (with $\nu = (\nu_1, \dots, \nu_n)$ denoting the outward unit normal and σ denoting the surface measure on $\partial\Omega$):

$$
\begin{aligned}
\mathscr{E}_m(u)(X) &= \lim_{j\to\infty} \mathscr{E}_m(u_j)(X) = \lim_{j\to\infty} \sum_{|\alpha|=m} \psi(X) \int_{\Omega} \phi_\alpha(X-Y)(\partial^\alpha u_j)(Y)\,dY \\
&= \lim_{j\to\infty} \sum_{|\alpha|=m} \sum_{\beta+\gamma+e_j=\alpha} C_{\beta,\gamma,j}\psi(X) \\
&\quad \times \int_{\partial\Omega} \nu_j(Y)(\partial^\beta\phi_\alpha)(X-Y)(\partial^\gamma u_j)(Y)\,d\sigma(Y) \\
&\quad + \lim_{j\to\infty} \sum_{|\alpha|=m} \psi(X) \int_{\Omega} (-1)^m (\partial^\alpha\phi_\alpha)(X-Y) u_j(Y)\,dY \\
&= \lim_{j\to\infty} \sum_{|\alpha|=m} \sum_{\beta+\gamma+e_j=\alpha} C_{\beta,\gamma,j}\psi(X) \\
&\quad \times \int_{\partial\Omega} \nu_j(Y)(\partial^\beta\phi_\alpha)(X-Y)[\mathrm{tr}_{m-1}(u_j)]_\gamma(Y)\,d\sigma(Y) \\
&\quad + \lim_{j\to\infty} (-1)^m \psi(X)\widetilde{u}_j(X) \\
&= 0.
\end{aligned} \tag{3.149}
$$

In the third step above, we have used repeated integrations by parts (note that there is no singularity in the integrands since we are assuming that $X \notin \overline{\Omega}$). In the fourth step we have used the definition of tr_{m-1} and formula (3.138). Finally, the last step is a consequence of the fact that $X \notin \overline{\Omega}$ as well as (3.146) and (3.71) (in this connection, it is useful to observe that, under the current assumptions on the indices, there exists $r > 1$ such that $B_s^{p,p}(\partial\Omega) \hookrightarrow L^r_{loc}(\partial\Omega)$).

In summary, the above argument shows that $\mathscr{E}_m(u) \in F^{p,q}_{m-1+s+1/p}(\mathbb{R}^n)$ is supported in $\overline{\Omega}$, i.e., $\mathscr{E}_m(u) \in F^{p,q}_{m-1+s+1/p,0}(\Omega)$. Based on this, (3.148) and (2.213) we may therefore conclude that $u \in F^{p,q}_{m-1+s+1/p,z}(\Omega)$. This justifies the right-to-left inclusion in (3.126) and, hence, finishes the proof of this identity.

At this stage, (3.127) follows from (3.126) and (2.211). The case of Besov spaces is treated analogously and this finishes the proof of the theorem. □

To state a consequence of Theorem 3.12 of functional analytic nature, we first recall some relevant terminology. A closed subspace $\mathscr{Y}$ of a Hausdorff topological vector space $\mathscr{X}$ is said to be `complemented in` $\mathscr{X}$ provided there exists a closed subspace $\mathscr{Z}$ of $\mathscr{X}$ with the property that $\mathscr{Y} \oplus \mathscr{Z} = \mathscr{X}$ (where the direct sum is both algebraic and topologic).

Corollary 3.13. *Let Ω be a bounded Lipschitz domain in $\mathbb{R}^n$ and fix $m \in \mathbb{N}$. Then*

$$F^{p,q}_{m-1+s+1/p,z}(\Omega) \text{ is complemented in } F^{p,q}_{m-1+s+1/p}(\Omega) \tag{3.150}$$

whenever $\frac{n-1}{n} < p < \infty$, $(n-1)(1/p-1)_+ < s < 1$ *and* $\min\{1,p\} \leq q < \infty$, *while*

$$B^{p,q}_{m-1+s+1/p,z}(\Omega) \text{ is complemented in } B^{p,q}_{m-1+s+1/p}(\Omega) \tag{3.151}$$

whenever $\frac{n-1}{n} < p < \infty$, $(n-1)(1/p-1)_+ < s < 1$ *and* $0 < q < \infty$.

Proof. All claims follow directly from Theorem 3.12, Theorem 3.9 and Lemma 3.14, stated below.

Here is the abstract functional analytic result invoked above.

Lemma 3.14. *Let $\mathscr{X}$, $\mathscr{Y}$ be two quasi-Banach spaces and assume that the operator $T : \mathscr{X} \to \mathscr{Y}$ is a linear and continuous operator. Then the following two statements are equivalent:*

(i) T has a right-inverse (i.e., there exists a linear and bounded operator $S : \mathscr{Y} \to \mathscr{X}$ with the property that $T \circ S = I$, the identity on $\mathscr{Y}$);
(ii) T is surjective and its null-space, $\operatorname{Ker} T$, *is complemented in $\mathscr{X}$.*

Proof. Assume first that T has a right-inverse $S : \mathscr{Y} \to \mathscr{X}$. We claim that

$$\operatorname{Im} S \text{ is closed in } \mathscr{X} \text{ and } \operatorname{Im} S \cap \operatorname{Ker} T = \{0\}. \tag{3.152}$$

To justify this claim, assume that $\{y_j\}_{j\in\mathbb{N}}$ is a sequence in $\mathscr{Y}$ with the property that $\{Sy_j\}_{j\in\mathbb{N}}$ converges in $\mathscr{X}$ to some vector $x \in \mathscr{X}$. Given that T is continuous and S is a right-inverse for T, it follows that $\{y_j\}_{j\in\mathbb{N}} = \{TSy_j\}_{j\in\mathbb{N}}$ converges in $\mathscr{Y}$ to Tx. Hence, since S is continuous, $\{Sy_j\}_{j\in\mathbb{N}}$ converges in $\mathscr{X}$ to $S(Tx)$. Consequently, $x = S(Tx) \in \operatorname{Im} S$ since the topology on $\mathscr{X}$ is Hausdorff. Given that the topology on $\mathscr{X}$ is also first-countable, this reasoning shows that $\operatorname{Im} S$ is closed in $\mathscr{X}$. Assume next that $x \in \operatorname{Im} S \cap \operatorname{Ker} T$. Then there exists $y \in \mathscr{Y}$ such that $x = Sy$ which further implies that $0 = Tx = T(Sy) = y$. Hence $y = 0$ which ultimately forces $x = Sy$ to be $0 \in \mathscr{X}$. This concludes the proof of (3.152). Granted this, in order to conclude that $\operatorname{Ker} T$ is complemented in $\mathscr{X}$ it suffices to show that

$$\operatorname{Im} S \oplus \operatorname{Ker} T = \mathscr{X}. \tag{3.153}$$

To see that this is the case, observe that for each $x \in \mathscr{X}$

$$x = \big(x - S(Tx)\big) + S(Tx) \text{ and } \big(x - S(Tx)\big) \in \operatorname{Ker} T, \quad S(Tx) \in \operatorname{Im} S. \tag{3.154}$$

This takes care of the algebraic aspect of (3.153). On the topological side, the mappings $\mathscr{X} \ni x \mapsto \big(x - S(Tx)\big) \in \mathscr{X}$ and $\mathscr{X} \ni x \mapsto S(Tx) \in \mathscr{X}$ are continuous, since both T and S are continuous. Thus, (3.153) holds. Given that the existence of a right-inverse implies ontoness, this finishes the proof of the implication $(i) \Rightarrow (ii)$.

As regards the implication $(ii) \Rightarrow (i)$, assume that T is surjective and that there exists a closed subspace $\mathscr{Z}$ of $\mathscr{X}$ with the property that

$$\mathscr{Z} \oplus \operatorname{Ker} T = \mathscr{X}. \tag{3.155}$$

This makes $\mathscr{Z}$ a quasi-Banach space and forces $T\big|_{\mathscr{Z}}$, the restriction of T to $\mathscr{Z}$, to be bijective. Of course, $T\big|_{\mathscr{Z}} : \mathscr{Z} \to \mathscr{Y}$ remains continuous. As a consequence of the Open Mapping Theorem (which is still valid in the context of quasi-Banach spaces; see [65]),

$$\big(T\big|_{\mathscr{Z}}\big)^{-1} : \mathscr{Y} \longrightarrow \mathscr{Z} \text{ is continuous.} \tag{3.156}$$

Denote by ι the embedding of $\mathscr{Z}$ into $\mathscr{X}$ and set

$$S := \big(T\big|_{\mathscr{Z}}\big)^{-1} \circ \iota : \mathscr{Y} \longrightarrow \mathscr{X}. \tag{3.157}$$

Then S is linear and since $\iota : \mathscr{Z} \to \mathscr{X}$ is continuous, it follows from (3.156)–(3.157) that $S : \mathscr{Y} \to \mathscr{X}$ is continuous. We now claim that

$$T \circ S = I, \text{ the identity on } \mathscr{Y}. \tag{3.158}$$

To prove this, assume that $y \in \mathscr{Y}$ is arbitrary. Since $T : \mathscr{X} \to \mathscr{Y}$ is onto, there exists $x \in \mathscr{X}$ such that $Tx = y$. Use now (3.155) in order to decompose $x = x_1 + x_2$ with $x_1 \in \mathscr{Z}$ and $x_2 \in \operatorname{Ker} T$. In particular, $y = Tx = T(x_1 + x_2) = Tx_1$, and thus we may conclude that $x_1 = \big(T\big|_{\mathscr{Z}}\big)^{-1} y$. In turn, this and (3.157) imply that $Sy = x_1$ hence, ultimately, $T(Sy) = Tx_1 = y$. Since $y \in \mathscr{Y}$ has been arbitrarily chosen, this establishes (3.158), thus finishing the proof of the implication $(ii) \Rightarrow (i)$. □

We now extend formula (2.220) by proving the following result. Recall that, generally speaking, $\{x\} := x - [x]$ denotes the fractional part of a real number x. Also, recall the spaces introduced in (2.218) with $A := F$.

Proposition 3.15. *Let Ω be a bounded Lipschitz domain in $\mathbb{R}^n$. Then*

$$\mathring{F}^{p,q}_s(\Omega) = F^{p,q}_{s,z}(\Omega) \tag{3.159}$$

provided

$$\begin{gathered} 0 < p < \infty, \quad \min\{1, p\} \leq q < \infty, \quad \textit{and} \\ s - \tfrac{1}{p} > -1 \ \textit{and} \ \big\{s - \tfrac{1}{p}\big\} > (n-1)\big(\tfrac{1}{p} - 1\big)_+ . \end{gathered} \tag{3.160}$$

Furthermore,

$$\mathring{B}^{p,q}_s(\Omega) = B^{p,q}_{s,z}(\Omega) \tag{3.161}$$

whenever

$$0 < p,q < \infty, \quad s - \tfrac{1}{p} > -1 \ \textit{and} \ \left\{s - \tfrac{1}{p}\right\} > (n-1)\left(\tfrac{1}{p} - 1\right)_+. \tag{3.162}$$

Proof. Both in the context of (3.160) and (3.162), the conditions on the smoothness parameter s amount to

$$\text{there exists } m \in \mathbb{N}_0 \text{ such that } \max\left(\tfrac{1}{p} - 1, n\left(\tfrac{1}{p} - 1\right)\right) < s - m < \tfrac{1}{p}. \tag{3.163}$$

Indeed, $m := \left[s - \frac{1}{p}\right] + 1$ will do. For the remainder of the proof m will retain this significance.

To begin in earnest, observe first that, since by (2.207) $\widetilde{C^\infty_c(\Omega)}$ is dense in $F^{p,q}_{s,0}(\Omega)$, and since the operator of restriction to Ω maps the latter space boundedly onto $F^{p,q}_{s,z}(\Omega)$, it follows that

$$F^{p,q}_{s,z}(\Omega) \hookrightarrow \mathring{F}^{p,q}_s(\Omega), \qquad 0 < p,q < \infty, \ s \in \mathbb{R}. \tag{3.164}$$

The opposite inclusion is contained in (2.220) when $m = 0$ so it suffices to work under the assumption that $m \in \mathbb{N}$. Also, there is no loss of generality in assuming that Ω is the domain in $\mathbb{R}^n$ lying above the graph of a Lipschitz function. Our goal is to show that if $u \in \mathring{F}^{p,q}_s(\Omega)$ then $u \in F^{p,q}_{s,z}(\Omega)$, plus a natural estimate. For the rest of the proof we will follow largely the approach in the proof of Theorem 3.12, and we retain the notation introduced in that setting. In particular, it can be assumed that u satisfies (3.130).

Recall Calderón's extension operator $\mathscr{E}_m$ introduced in (3.135). This satisfies (3.139), (3.140), and (3.142). In particular, the latter formula and a similar string of inequalities as in (3.143) yield

$$\|\mathscr{E}_m f\|_{F^{p,q}_s(\mathbb{R}^n)} \le C\|f\|_{F^{p,q}_s(\Omega)}, \tag{3.165}$$

so that, analogously to (3.144),

$$\mathscr{E}_m : F^{p,q}_s(\Omega) \longrightarrow F^{p,q}_s(\mathbb{R}^n) \tag{3.166}$$

is a bounded operator whenever the indices are as in (3.160). Next, (3.139)–(3.140) entail

$$w \in C^\infty_c(\Omega) \Longrightarrow \mathscr{E}_m(w) = \psi\,\widetilde{w} \ \text{ in } \mathbb{R}^n. \tag{3.167}$$

Consequently, we also have

$$\mathscr{E}_m : \mathring{F}^{p,q}_s(\Omega) \longrightarrow F^{p,q}_{s,0}(\Omega), \quad p,q,s \text{ as in (3.160)}. \tag{3.168}$$

Now, if R_Ω is the operator of restriction of tempered distributions to the open set Ω, it follows from (3.140) that $R_\Omega \circ \mathscr{E}_m : F_s^{p,q}(\Omega) \to F_s^{p,q}(\Omega)$ is simply the operator of multiplication by ψ. Hence, $u = \psi u = (R_\Omega \circ \mathscr{E}_m)u = R_\Omega(\mathscr{E}_m u) \in F_{s,z}^{p,q}(\Omega)$ by (3.168) and (2.213), plus a natural estimate. This readily shows that

$$\mathring{F}_s^{p,q}(\Omega) \hookrightarrow F_{s,z}^{p,q}(\Omega), \qquad p, q, s \text{ as in (3.160)}. \tag{3.169}$$

Along with (3.164), this proves (3.159), as desired. The case of Besov spaces is treated analogously and this finishes the proof of the proposition. □

Let us also record the following description of the null-space of the multi-trace operator.

Corollary 3.16. *Let Ω be a bounded Lipschitz domain in $\mathbb{R}^n$ and pick $m \in \mathbb{N}$. Then*

$$\mathring{F}^{p,q}_{m-1+s+1/p}(\Omega) = \Big\{u \in F^{p,q}_{m-1+s+1/p}(\Omega) : \operatorname{tr}_{m-1} u = 0\Big\}, \tag{3.170}$$

if $\frac{n-1}{n} < p < \infty$, $(n-1)(1/p-1)_+ < s < 1$ *and* $\min\{1, p\} \leq q < \infty$. *Furthermore,*

$$\mathring{B}^{p,q}_{m-1+s+1/p}(\Omega) = \Big\{u \in B^{p,q}_{m-1+s+1/p}(\Omega) : \operatorname{tr}_{m-1} u = 0\Big\}, \tag{3.171}$$

provided $\frac{n-1}{n} < p < \infty$, $(n-1)(1/p-1)_+ < s < 1$ *and* $0 < q < \infty$.

Proof. This is a direct consequence of Theorem 3.12 and Proposition 3.15. □

As in the case of Besov and Triebel–Lizorkin spaces, it is also useful to characterize the null-space of the higher-order trace operator acting from weighted Sobolev spaces. In the process, we also establish a very useful density result.

Theorem 3.17. *Let Ω be a bounded Lipschitz domain in $\mathbb{R}^n$ and fix $m \in \mathbb{N}$. Also, suppose that* $1 < p < \infty$, $0 < s < 1$ *and set* $a := 1 - s - 1/p$. *Then*

$$\Big\{u \in W_a^{m,p}(\Omega) : \operatorname{tr}_{m-1} u = 0\Big\} = \Big\{u \in W_a^{m,p}(\Omega) : \widetilde{u} \in W_a^{m,p}(\mathbb{R}^n)\Big\} \tag{3.172}$$

and

$$C_c^\infty(\Omega) \hookrightarrow \Big\{u \in W_a^{m,p}(\Omega) : \operatorname{tr}_{m-1} u = 0\Big\} \text{ densely}. \tag{3.173}$$

Finally, the same results are true in the case when Ω is replaced by $\mathbb{R}^n \setminus \overline{\Omega}$.

Proof. We begin by establishing (3.172). To this end, consider $u \in W_a^{m,p}(\Omega)$ such that $\widetilde{u} \in W_a^{m,p}(\mathbb{R}^n)$ and write

$$\operatorname{tr}_{m-1} u = \operatorname{tr}^+_{m-1} \widetilde{u} = \operatorname{tr}^-_{m-1} \widetilde{u} = 0. \tag{3.174}$$

Above, $\mathrm{tr}^{\pm}_{m-1}$ denote the higher-order trace operators associated with the domains $\Omega_+ := \Omega$ and $\Omega_- := \mathbb{R}^n \setminus \overline{\Omega}$, respectively. The second equality in (3.174) is a consequence of the boundedness of $\mathrm{tr}^{\pm}_{m-1}$ (cf. Theorem 3.11) as well as a density result as the one described in (2.361) corresponding to the entire Euclidean space. Finally, the last equality in (3.174) is clear since $\widetilde{u}$ vanishes in Ω_-. This completes the proof of the right-to-left inclusion in (3.172).

Consider next the left-to-right inclusion in (3.172). By arguing as in the proof of Theorem 3.12, there is no loss of generality in assuming that Ω is the domain in $\mathbb{R}^n$ lying above the graph of a Lipschitz function and that u has bounded support. In this setting, consider Calderón's extension operator $\mathscr{E}_m$ of order m, whose construction has been reviewed in the proof of Theorem 3.12. We claim that

$$\mathscr{E}_m : W_a^{m,p}(\Omega) \longrightarrow W_a^{m,p}(\mathbb{R}^n) \tag{3.175}$$

is a well-defined, linear and bounded operator. A result from Harmonic Analysis which is a useful ingredient in the proof of this claim is that, for any multi-index α with $|\alpha| = m$, the convolution operator $f \mapsto \phi_\alpha * f$ maps functions from $L^p_{comp}(\mathbb{R}^n, \rho^{ap}\, dX)$ into functions which locally belong to $W_a^{m,p}(\mathbb{R}^n)$ (with quantitative control). The key aspect here is that, as pointed out in (2.362), the function ρ^{ap} is an A_p-weight, granted that $p \in (1,\infty)$ and $a \in (-1/p, 1-1/p)$ (cf., e.g., [81] for general results of this nature, involving arbitrary Muckenhoupt A_p-weights). Keeping this in mind, and making use of the formula (3.142), for each function $w \in C^\infty(\overline{\Omega})$ with bounded support we may write

$$\begin{aligned}
\|\mathscr{E}_m(w)\|_{W_a^{m,p}(\mathbb{R}^n)} &\leq C \sum_{|\alpha|=m} \left\| \psi\, \phi_\alpha * (\widetilde{\partial^\alpha w}) \right\|_{W_a^{m,p}(\mathbb{R}^n)} \\
&\leq C \sum_{|\alpha|=m} \left\| \widetilde{\partial^\alpha w} \right\|_{L^p(\mathbb{R}^n,\, \rho^{ap}\, dX)} \\
&= C \sum_{|\alpha|=m} \|\partial^\alpha w\|_{L^p(\Omega,\, \rho^{ap}\, dX)} \\
&\leq C\, \|w\|_{W_a^{m,p}(\Omega)}.
\end{aligned} \tag{3.176}$$

At this stage, the claim about (3.175) follows from (3.176) with the help of the density result mentioned in (2.361). With this in hand, the same argument as the one following (3.144) ultimately shows that if $u \in W_a^{m,p}(\Omega)$ satisfies $\mathrm{tr}_{m-1}(u) = 0$ then $\mathscr{E}_m(u)$ is supported in $\overline{\Omega}$, which further entails $\widetilde{u} = \mathscr{E}_m(u) \in W_a^{m,p}(\mathbb{R}^n)$. Hence, $\widetilde{u} \in W_a^{m,p}(\mathbb{R}^n)$ and this finishes the proof of (3.172).

As far as (3.173) is concerned, the goal is to approximate a given $u \in W_a^{m,p}(\Omega)$ with $\mathrm{tr}_{m-1}(u) = 0$ by functions from $C_c^\infty(\Omega)$ in the norm of $W_a^{m,p}(\Omega)$. This problem may be localized using a smooth partition of unity. As such, there is no loss of generality in assuming that Ω is an unbounded Lipschitz domain and u has bounded support. Moreover, from what we have proved so far, $\widetilde{u} \in W_a^{m,p}(\mathbb{R}^n)$. The idea is then to consider the vertical shifts $v_j(X) := \widetilde{u}(X + j^{-1}e_n)$, $X \in \Omega$, for $j \in \mathbb{N}$, which converge to u in $W_a^{m,p}(\Omega)$. The upshot of this artifice is that the distance from the support of each function v_j to the complement of Ω is strictly positive. For each fixed j there remains to mollify each v_j at scale $\varepsilon > 0$ in order to obtain $v_{j,\varepsilon}$ satisfying

$$\begin{aligned} &v_{j,\varepsilon} \in C_c^\infty(\Omega) \text{ whenever } \varepsilon > 0 \text{ is small enough,} \\ &v_{j,\varepsilon} \to v_j, \text{ as } \varepsilon \to 0^+, \text{ in the } W_a^{m,p}(\Omega)\text{-norm.} \end{aligned} \tag{3.177}$$

The second property above follows by controlling the mollifiers by the Hardy–Littlewood maximal operator, making use of the A_p-weight condition (2.362), and Muckenhoupt's theorem. Of course, (3.177) suffices to end the proof of (3.173).

Finally, in the case when Ω is replaced by $\mathbb{R}^n \setminus \overline{\Omega}$, the idea is to work with a suitable truncation of $u \in W_a^{m,p}(\mathbb{R}^n \setminus \overline{\Omega})$ (specifically, ψu with $\psi \in C_c^\infty(B_R(0))$ satisfying $\psi \equiv 1$ near $\overline{\Omega}$, where $R > 0$ is large and fixed, in the case of (3.172), and with the sequence $\psi(X/j)u(X)$ where $j \in \mathbb{N}$ in the case of (3.173)) and then apply the results proved in bounded Lipschitz domains. □

Let us now discuss a related version of the trace map (3.100), namely

$$\mathrm{Tr}_{m-1}(u) := \left\{ \frac{\partial^k u}{\partial \nu^k} \right\}_{0 \le k \le m-1}. \tag{3.178}$$

In the case of a Lipschitz domain, the unit normal ν has only bounded, measurable components, hence taking iterated normal derivatives requires attention. It is however natural to define for each $k \in \{0, \dots, m-1\}$

$$\frac{\partial^k}{\partial \nu^k} := \Big(\sum_{j=1}^n \xi_j \partial/\partial x_j \Big)^k \Big|_{\xi = \nu} \tag{3.179}$$

i.e., set

$$\frac{\partial^k u}{\partial \nu^k} := \sum_{|\alpha| = k} \frac{k!}{\alpha!} \nu^\alpha \, \mathrm{Tr}\,[\partial^\alpha u], \qquad 0 \le k \le m-1, \tag{3.180}$$

where Tr is the boundary trace operator from Theorem 2.53.

Compared to (3.100), a distinguished feature of (3.178) is that the latter has fewer components. More specifically, while $\mathrm{tr}_{m-1}(u)$ has

$$\sum_{k=0}^{m-1} \binom{n+k-1}{n-1} \tag{3.181}$$

components, $\mathrm{Tr}_{m-1}(u)$ has only m components. It is then remarkable that the two trace mappings, tr_{m-1} and Tr_{m-1}, have the same null-space. This, as well as other properties of the trace operator (3.178) are established in the theorem below (which should be compared with Theorem 3.4).

Theorem 3.18. *Suppose that Ω is a bounded Lipschitz domain in $\mathbb{R}^n$, fix $m \in \mathbb{N}$, and assume that $0 < p, q \le \infty$ and $(n-1)(\frac{1}{p}-1)_+ < s < 1$. Then the following properties hold.*

(i) The null-space of the mapping (3.178) acting on the Besov scale is

$$\Big\{u \in B^{p,q}_{m-1+s+1/p}(\Omega) : \mathrm{Tr}_{m-1}(u) = 0\Big\} = B^{p,q}_{m-1+s+1/p,z}(\Omega). \tag{3.182}$$

(ii) The image of the mapping (3.178), $\mathrm{Tr}_{m-1}\big[B^{p,q}_{m-1+s+1/p}(\Omega)\big]$, may be characterized as follows:

Fix $1 < p^* \le \big(\frac{1}{p} - \frac{s}{n-1}\big)^{-1}$ if $p \le 1$, and set $p^* := p$ in the case when $p > 1$. For a family $(g_0, g_1, \dots, g_{m-1})$ of functions from $L^1(\partial\Omega)$ set $f_{(0,\dots,0)} := g_0$ and, inductively, if $\{f_\gamma\}_{|\gamma| \le \ell-1}$ have already been defined for some $\ell \in \{1, \dots, m-1\}$, consider

$$f_\alpha := \nu^\alpha g_\ell + \frac{\alpha!}{\ell!} \sum_{\substack{\mu+\delta+e_j=\alpha \\ |\theta|=|\delta|}} \frac{|\delta|!}{\delta!} \frac{|\mu|!}{\mu!} \frac{|\theta|!}{\theta!} \nu^{\delta+\theta} (\nabla_{tan} f_{\mu+\theta})_j, \quad \forall\, \alpha \in \mathbb{N}_0^n : |\alpha| = \ell, \tag{3.183}$$

where the subscript j stands for the j-th component. Then there exists a function $u \in B^{p,q}_{m-1+s+1/p}(\Omega)$ such that $\mathrm{Tr}_{m-1}(u) = \{g_k\}_{0 \le k \le m-1}$ if and only if

$$f_\alpha \in L^{p^*}_1(\partial\Omega) \text{ for } |\alpha| \le m-2, \text{ and } f_\alpha \in B^{p,q}_s(\partial\Omega) \text{ for } |\alpha| = m-1. \tag{3.184}$$

(iii) When equipped with the quasi-norm

$$\{g_k\}_{0\le k\le m-1} \mapsto \sum_{|\alpha|\le m-2} \|f_\alpha\|_{L^{p^*}_1(\partial\Omega)} + \sum_{|\alpha|=m-1} \|f_\alpha\|_{B^{p,q}_s(\partial\Omega)}, \tag{3.185}$$

where the f_α's are constructed from the g_k's using the algorithm described in part (ii), the space $\mathrm{Tr}_{m-1}\big[B^{p,q}_{m-1+s+1/p}(\Omega)\big]$ becomes quasi-Banach, and there exists a linear and bounded operator

$$E : \mathrm{Tr}_{m-1}\big[B^{p,q}_{m-1+s+1/p}(\Omega)\big] \longrightarrow B^{p,q}_{m-1+s+1/p}(\Omega) \tag{3.186}$$

which is universal (in the sense that it does not depend on p, q, s) and with the property that

$$\mathrm{Tr}_{m-1} \circ E = I, \quad \text{the identity on } \mathrm{Tr}_{m-1}\big[B^{p,q}_{m-1+s+1/p}(\Omega)\big]. \tag{3.187}$$

(iv) Similar results to those described in parts (i)–(iii) are valid for Tr_{m-1} *acting on Triebel–Lizorkin spaces, if*

$$0 < p,q < \infty \text{ and } (n-1)\Big(\tfrac{1}{p}-1\Big)_+ < s < 1. \tag{3.188}$$

Specifically, the following three properties hold. First, a family of functions $(g_0, g_1, \dots, g_{m-1})$ *from* $L^1(\partial\Omega)$ *belongs to the image of* Tr_{m-1}*, acting on the space* $F^{p,q}_{m-1+s+1/p}(\Omega)$*, if and only if*

$$f_\alpha \in L_1^{p^*}(\partial\Omega) \text{ for } |\alpha| \le m-2, \text{ and } f_\alpha \in B_s^{p,p}(\partial\Omega) \text{ for } |\alpha| = m-1, \tag{3.189}$$

where the f_α*'s are constructed from* $g_0, g_1, \dots, g_{m-1}$ *using the same algorithm as before.*

Second,

$$\Big\{u \in F^{p,q}_{m-1+s+1/p}(\Omega) : \mathrm{Tr}_{m-1}(u) = 0\Big\} = F^{p,q}_{m-1+s+1/p,z}(\Omega), \tag{3.190}$$

if, in addition to (3.188)*, one also requires that* $\min\{1,p\} \le q < \infty$.

Third, the space $\mathrm{Tr}_{m-1}\big[F^{p,q}_{m-1+s+1/p}(\Omega)\big]$ *becomes quasi-Banach when equipped with the quasi-norm*

$$\{g_k\}_{0\le k\le m-1} \mapsto \sum_{|\alpha|\le m-2} \|f_\alpha\|_{L_1^{p^*}(\partial\Omega)} + \sum_{|\alpha|=m-1} \|f_\alpha\|_{B_s^{p,p}(\partial\Omega)}, \tag{3.191}$$

where the f_α*'s are constructed from the* g_k*'s using the algorithm described in part (ii), and there exists a linear and bounded operator*

$$E : \mathrm{Tr}_{m-1}\big[F^{p,q}_{m-1+s+1/p}(\Omega)\big] \longrightarrow F^{p,q}_{m-1+s+1/p}(\Omega) \tag{3.192}$$

which is universal (in the sense that it does not depend on p,q,s*) and with the property that*

$$\mathrm{Tr}_{m-1} \circ E = I, \text{ the identity on } \mathrm{Tr}_{m-1}\big[F^{p,q}_{m-1+s+1/p}(\Omega)\big]. \tag{3.193}$$

Proof. We claim that, for p,q,s as in the background hypotheses in the statement of the theorem, there holds

$$\Big\{u \in B^{p,q}_{m-1+s+1/p}(\Omega) : \mathrm{Tr}_{m-1}(u) = 0\Big\} \tag{3.194}$$

$$= \Big\{u \in B^{p,q}_{m-1+s+1/p}(\Omega) : \mathrm{tr}_{m-1}(u) = 0\Big\}.$$

Indeed, this is a consequence of the fact (seen from Proposition 3.5) that the assignment

$$\dot{B}^{p,q}_{m-1,s}(\partial\Omega) \ni \dot{f} = \{f_\alpha\}_{|\alpha|\leq m-1} \mapsto \Big\{\sum_{|\alpha|=k} \frac{k!}{\alpha!}\,\nu^\alpha\, f_\alpha\Big\}_{0\leq k\leq m-1} \in \big[L^1(\partial\Omega)\big]^m \tag{3.195}$$

is one-to-one. In turn, (3.194) and (3.128) prove (3.182).

Let us now turn to the characterization of the image of the operator (3.178) acting on $B^{p,q}_{m-1+s+1/p}(\Omega)$. To this end, assume first that $g_k \in L^1(\partial\Omega)$, $0 \leq k \leq m-1$, are such that the functions $\{f_\alpha\}_{|\alpha|\leq m-1}$ defined by setting $f_{(0,\dots,0)} := g_0$ and then inductively proceeding as in (3.183), satisfy the conditions listed in (3.184). Then by proceeding as in the proof of Theorem 3.4 we obtain that

$$\dot{f} := \{f_\alpha\}_{|\alpha|\leq m-1} \in \dot{B}^{p,q}_{m-1,s}(\partial\Omega) \quad \text{and} \quad g_k = \sum_{|\alpha|=k} \frac{k!}{\alpha!}\,\nu^\alpha\, f_\alpha, \quad 0 \leq k \leq m-1. \tag{3.196}$$

Hence, if with $\mathscr{E}$ as in Theorem 3.9 we set

$$u := \mathscr{E}\dot{f} \in B^{p,q}_{m-1+s+1/p}(\Omega), \tag{3.197}$$

it follows from (3.104), (3.178), and (3.196) that

$$\{g_k\}_{0\leq k\leq m-1} = \operatorname{Tr}_{m-1}(u). \tag{3.198}$$

This proves that the family $\{g_k\}_{0\leq k\leq m-1}$ belongs to the image of the mapping (3.178).

Conversely, assume that $\{g_k\}_{0\leq k\leq m-1} = \operatorname{Tr}_{m-1}(u)$ for some $u \in B^{p,q}_{m-1+s+1/p}(\Omega)$. Then much as in (3.57) we obtain that, for each $\alpha \in \mathbb{N}_0^n$ with $|\alpha| = \ell \leq m-1$,

$$\begin{aligned} f_\alpha &= \nu^\alpha g_\ell + \sum_{|\beta|=\ell} \frac{\ell!}{\beta!}\,\nu^\beta \Big(\nu^\beta \operatorname{Tr}(\partial^\alpha u) - \nu^\alpha \operatorname{Tr}(\partial^\beta u)\Big) \\ &= \nu^\alpha g_\ell + \operatorname{Tr}(\partial^\alpha u) - \nu^\alpha \Big(\sum_{|\beta|=\ell} \frac{\ell!}{\beta!}\,\nu^\beta \operatorname{Tr}(\partial^\beta u)\Big) \\ &= \operatorname{Tr}(\partial^\alpha u), \end{aligned} \tag{3.199}$$

by virtue of (3.178) and the multinomial identity

$$\sum_{|\beta|=\ell} \frac{\ell!}{\beta!}\nu^{2\beta} = |\nu|^{2\ell} = 1. \tag{3.200}$$

Consequently, $f_\alpha \in B^{p,q}_s(\partial\Omega)$ if $|\alpha| = m-1$ and $f_\alpha \in L^{p^*}_1(\partial\Omega)$ for $|\alpha| \leq m-2$, thanks to (3.199), Theorem 2.53 and embeddings. This finishes the proof of the fact that (3.184) characterizes the image of the operator (3.178).

Moving on, it is clear that $\mathrm{Tr}_{m-1}\big[B^{p,q}_{m-1+s+1/p}(\Omega)\big]$ becomes a quasi-Banach space when endowed with the quasi-norm (3.185). To construct a universal linear, bounded right-inverse E for Tr_{m-1} as in (3.186)–(3.187), we set

$$E\big(\{g_k\}_{0\leq k\leq m-1}\big) := \mathscr{E}\dot{f}, \tag{3.201}$$

where $\dot{f} := \{f_\alpha\}_{|\alpha|\leq m-1}$ with f_α's constructed from the g_k's using the algorithm described in part (ii), and $\mathscr{E}$ as in Theorem 3.9. That this does the intended job then follows from (3.197)–(3.198).

Finally, the arguments in case of the Triebel–Lizorkin scale are similar. For example,

$$\Big\{u \in F^{p,q}_{m-1+s+1/p}(\Omega) : \mathrm{Tr}_{m-1}(u) = 0\Big\} \tag{3.202}$$

$$= \Big\{u \in F^{p,q}_{m-1+s+1/p}(\Omega) : \mathrm{tr}_{m-1}(u) = 0\Big\}$$

whenever the indices p, q, s are as in (3.188) and, in addition, $\min\{1, p\} \leq q < \infty$. Thus, (3.190) follows from (3.202) by invoking (3.126). This finishes the proof of the theorem. □

The special case $m = 2$ of Theorem 3.18 (in which scenario Tr_{m-1} becomes Tr_1) deserves to be stated separately.

Corollary 3.19. *Assume that $\Omega \subset \mathbb{R}^n$ is a bounded Lipschitz domain and denote by ν the outward unit normal to $\partial\Omega$. Also, suppose that the numbers p, q, s satisfy $0 < p, q \leq \infty$, $(n-1)\big(\frac{1}{p} - 1\big)_+ < s < 1$. Fix $1 < p^* \leq \big(\frac{1}{p} - \frac{s}{n-1}\big)^{-1}$ if $p \leq 1$, and set $p^* := p$ if $p > 1$. Then the operator*

$$\mathrm{Tr}_1 : B^{p,q}_{1+s+\frac{1}{p}}(\Omega) \to \mathscr{X} \quad \textit{where}$$

$$\mathscr{X} := \big\{(g_0, g_1) \in L^{p^*}_1(\partial\Omega) \oplus L^{p^*}(\partial\Omega) : \nabla_{tan} g_0 + g_1\nu \in \big(B^{p,q}_s(\partial\Omega)\big)^n\big\} \tag{3.203}$$

$$\textit{and} \quad \mathrm{Tr}_1(u) := \big(\mathrm{Tr}\, u, \nu\cdot\mathrm{Tr}\,(\nabla u)\big), \quad \forall\, u \in B^{p,q}_{1+s+\frac{1}{p}}(\Omega),$$

is well-defined, linear, bounded, onto, and has a linear, bounded right-inverse. Above, the space $\big\{(g_0, g_1) \in L^{p^*}_1(\partial\Omega) \oplus L^{p^*}(\partial\Omega) : \nabla_{tan} g_0 + g_1\nu \in \big(B^{p,q}_s(\partial\Omega)\big)^n\big\}$ is considered equipped with the natural quasi-norm

$$(g_0, g_1) \mapsto \|g_0\|_{L^{p^*}_1(\partial\Omega)} + \|g_1\|_{L^{p^*}(\partial\Omega)} + \|\nabla_{tan} g_0 + g_1\nu\|_{(B^{p,q}_s(\partial\Omega))^n}, \tag{3.204}$$

with respect to which the space in question becomes quasi-Banach. Furthermore, the null-space of the operator (3.203) *is given by*

$$\mathrm{Ker}\,\mathrm{Tr}_1 := \big\{u \in B^{p,q}_{1+s+\frac{1}{p}}(\Omega) : \mathrm{Tr}\, u = \nu\cdot\mathrm{Tr}\,(\nabla u) = 0\big\} = B^{p,q}_{1+s+\frac{1}{p},z}(\Omega). \tag{3.205}$$

Finally, similar results are valid for the Triebel–Lizorkin scale, i.e., for

$$\mathrm{Tr}_1 : F^{p,q}_{1+s+\frac{1}{p}}(\Omega) \to \mathscr{Y} \ \textit{where}$$

$$\mathscr{Y} := \big\{(g_0, g_1) \in L^{p^*}_1(\partial\Omega) \oplus L^{p^*}(\partial\Omega) : \nabla_{tan} g_0 + g_1\nu \in \big(B^{p,p}_s(\partial\Omega)\big)^n\big\} \tag{3.206}$$

$$\textit{and} \ \mathrm{Tr}_1(u) := \big(\mathrm{Tr}\, u,\ \nu \cdot \mathrm{Tr}\,(\nabla u)\big), \quad \forall\, u \in F^{p,q}_{1+s+\frac{1}{p}}(\Omega),$$

provided $p, q < \infty$. *In this scenario, the null-space of* (3.206) *is given by*

$$\mathrm{Ker}\,\mathrm{Tr}_1 := \big\{u \in F^{p,q}_{1+s+\frac{1}{p}}(\Omega) : \mathrm{Tr}\, u = \nu \cdot \mathrm{Tr}\,(\nabla u) = 0\big\} = F^{p,q}_{1+s+\frac{1}{p},z}(\Omega) \tag{3.207}$$

granted that also $\min\{1, p\} \le q < \infty$.

Proof. This follows from Theorem 3.18 specialized to the case $m = 2$, upon observing that the f_α's from part (ii) of the statement of Theorem 3.18 are given by

$$f_{(0,\dots,0)} := g_0 \ \text{ and } \ f_{e_j} := \nu_j g_1 + (\nabla_{tan} g_0)_j \ \text{ for each } \ j \in \{1, \dots, n\} \tag{3.208}$$

(cf. (3.183) with $\ell = 1$). □

Remark 3.20. Regarding the nature of the space appearing in (3.203), let us note that since multiplication with functions from $B^{(n-1)/s,q}_s(\partial\Omega) \cap L^\infty(\partial\Omega)$ leaves $B^{p,q}_s(\partial\Omega)$ invariant (cf. § 4.9.2 in [107]), we have

$$\begin{aligned}\big\{(g_0, g_1) \in L^{p^*}_1(\partial\Omega) \oplus L^{p^*}(\partial\Omega) : \nabla_{tan} g_0 + g_1\nu \in \big(B^{p,q}_s(\partial\Omega)\big)^n\big\}& \\ = B^{p,q}_{1+s}(\partial\Omega) \oplus B^{p,q}_s(\partial\Omega)&\end{aligned} \tag{3.209}$$

$$\text{whenever } \ \partial\Omega \in C^r \ \text{ with } \ r > 1 + s.$$

Let us now discuss the following result which, in the particular case of unweighted Sobolev spaces (i.e., when $a = 0$), answers the question raised by J. Nečas in Problem 4.1 on p. 91 of [98].

Corollary 3.21. *Let Ω be a bounded Lipschitz domain, and fix $m \in \mathbb{N}$. For every $p \in (1, \infty)$ and $a \in (-1/p, 1 - 1/p)$ set*

$$\mathring{W}^{m,p}_a(\Omega) := \textit{the closure of } C^\infty_c(\Omega) \textit{ in } \ W^{m,p}_a(\Omega). \tag{3.210}$$

Then, corresponding to $a = 0$,

$$\mathring{W}^{m,p}(\Omega) = \mathring{F}^{p,2}_{m}(\Omega), \tag{3.211}$$

and

$$\begin{aligned} &u \in W_a^{m,p}(\Omega) \text{ and } \frac{\partial^k u}{\partial \nu^k} = 0 \text{ on } \partial\Omega \text{ for } 0 \le k \le m-1 \\ &\Longleftrightarrow u \in \mathring{W}_a^{m,p}(\Omega) \Longleftrightarrow u \in W_a^{m,p}(\Omega) \text{ and } \mathrm{tr}_{m-1}(u) = 0. \end{aligned} \tag{3.212}$$

Proof. The equality in (3.211) is a consequence of (3.190), (3.159), (2.200), and (3.202). In a similar manner to (3.202), we also have

$$\Big\{u \in W_a^{m,p}(\Omega) : \mathrm{Tr}_{m-1}(u) = 0\Big\} = \Big\{u \in W_a^{m,p}(\Omega) : \mathrm{tr}_{m-1}(u) = 0\Big\}. \tag{3.213}$$

Based on this and Theorem 3.17, the equivalences in (3.212) follow, and this finishes the proof. □

In turn, Corollary 3.21 is the key ingredient in the proof of the following result.

Theorem 3.22. *Let Ω be a bounded Lipschitz domain in $\mathbb{R}^n$. Then for each index $p \in (1,\infty)$, each $a \in (-1/p, 1-1/p)$, and each $k, m \in \mathbb{N}_0$,*

$$\mathring{W}_a^{k+m,p}(\Omega) = \Big\{u \in W_a^{k+m,p}(\Omega) \cap \mathring{W}_a^{k,p}(\Omega) : \partial^\gamma u \in \mathring{W}_a^{m,p}(\Omega), \ \forall\, \gamma \in \mathbb{N}_0^n, \ |\gamma| = k\Big\}. \tag{3.214}$$

Proof. Let $u \in \mathring{W}_a^{k+m,p}(\Omega)$. Then clearly

$$u \in W_a^{k+m,p}(\Omega) \cap \mathring{W}_a^{k,p}(\Omega). \tag{3.215}$$

In addition, using the definition of the space $\mathring{W}_a^{k+m,p}(\Omega)$ it follows that there exists a sequence of functions $\{u_j\}_{j\in\mathbb{N}} \subseteq C_c^\infty(\Omega)$ such that $u_j \to u$ in $W_a^{k+m,p}(\Omega)$, as $j \to \infty$. In particular, for every $\gamma \in \mathbb{N}_0^n$ with $|\gamma| = k$ there holds

$$\partial^\gamma u_j \longrightarrow \partial^\gamma u \ \text{ in } W_a^{m,p}(\Omega) \text{ as } \ j \to \infty. \tag{3.216}$$

Since for each $j \in \mathbb{N}$ and each $\gamma \in \mathbb{N}_0^n$ we have $\partial^\gamma u_j \in C_c^\infty(\Omega)$, (3.216) and the definition of $\mathring{W}_a^{m,p}(\Omega)$ guarantee that

$$\partial^\gamma u \in \mathring{W}_a^{m,p}(\Omega), \qquad \forall\, \gamma \in \mathbb{N}_0^n \ \text{ such that } |\gamma| = k. \tag{3.217}$$

Combining (3.217) and (3.215) we obtain that the left-to-right inclusion in (3.214) holds.

There remains to establish the right-to-left inclusion in (3.214). To this end, pick a function u such that

$$u \in W_a^{m+k,p}(\Omega) \cap \mathring{W}_a^{k,p}(\Omega) \text{ and } \partial^\gamma u \in \mathring{W}_a^{m,p}(\Omega), \ \forall \gamma \in \mathbb{N}_0^n, \ |\gamma| = k. \tag{3.218}$$

Keeping in mind that $u \in W_a^{k+m,p}(\Omega)$, it follows from (3.212) that the membership of u to $\mathring{W}_a^{k+m,p}(\Omega)$ is equivalent to

$$\mathrm{Tr}\big[\partial^\alpha u\big] = 0, \qquad \forall \alpha \in \mathbb{N}_0^n \text{ such that } |\alpha| \leq m+k-1. \tag{3.219}$$

With the goal of proving (3.219), first notice that, on the one hand, the last condition in (3.218) implies (thanks to (3.212)) that

$$\mathrm{Tr}\big[\partial^\beta(\partial^\gamma u)\big] = 0, \quad \forall \beta, \gamma \in \mathbb{N}_0^n \text{ such that } |\beta| \leq m-1 \text{ and } |\gamma| = k, \tag{3.220}$$

hence

$$\mathrm{Tr}\big[\partial^\alpha u\big] = 0, \qquad \forall \alpha \in \mathbb{N}_0^n \text{ such that } |\alpha| \in \{k, \dots, m+k-1\}. \tag{3.221}$$

On the other hand, the first condition in (3.218) ensures that $u \in \mathring{W}_a^{k,p}(\Omega)$, and thus, by once again appealing to (3.212),

$$\mathrm{Tr}\big[\partial^\alpha u\big] = 0, \qquad \forall \alpha \in \mathbb{N}_0^n \text{ such that } |\alpha| \in \{0, \dots, k-1\}. \tag{3.222}$$

Altogether, (3.221) and (3.222) give that

$$\mathrm{Tr}\big[\partial^\alpha u\big] = 0, \qquad \forall \alpha \in \mathbb{N}_0^n \text{ such that } |\alpha| \leq m+k-1, \tag{3.223}$$

which, in the light of (3.212), proves that (3.219) holds, as desired. This shows that if u is as in (3.218) then $u \in \mathring{W}_a^{k+m,p}(\Omega)$. Thus, the right-to-left inclusion in (3.214) holds, and this finishes the proof of the theorem. □

Moving on, we shall now state and prove a lifting result which is going to be of importance later on.

Proposition 3.23. *Let Ω be a bounded Lipschitz domain in $\mathbb{R}^n$. If $m \in \mathbb{N}$, $0 < p, q < \infty$ and $\max\Big(1/p - 1, n(1/p-1)\Big) < s < 1/p$, then*

$$\begin{aligned} &\partial^\alpha u \in B_s^{p,q}(\Omega) \text{ whenever } |\alpha| \leq m \Longrightarrow u \in B_{s+m}^{p,q}(\Omega) \\ &\text{and } \|u\|_{B_{s+m}^{p,q}(\Omega)} \leq C(\Omega, p, q, s, m) \sum_{|\alpha| \leq m} \|\partial^\alpha u\|_{B_s^{p,q}(\Omega)}. \end{aligned} \tag{3.224}$$

Furthermore, a similar result holds for the scale of Triebel–Lizorkin spaces in Ω *if, in addition,* $\min\{p,1\} \le q$.

Proof. To prove (3.224), assume that $u \in B_s^{p,q}(\Omega)$ has a suitably small support and satisfies $\partial^\alpha u \in B_s^{p,q}(\Omega)$ for each multi-index α of length $\le m$. Via a localization, (outward) translation and mollification, one can construct a sequence of functions $u_j \in C^\infty(\overline{\Omega})$, $j \in \mathbb{N}$, such that

$$\partial^\alpha u_j \longrightarrow \partial^\alpha u \quad \text{in } B_s^{p,q}(\Omega) \text{ as } j \to \infty, \tag{3.225}$$

for all $\alpha \in \mathbb{N}_0^n$ with $|\alpha| \le m$. Upon recalling Calderón's extension operator $\mathscr{E}_m$ from (3.135), we may then estimate

$$\begin{aligned} \|u_j - u_k\|_{B^{p,q}_{m+s}(\Omega)} &\le \|\mathscr{E}_m(u_j - u_k)\|_{B^{p,q}_{m+s}(\mathbb{R}^n)} \\ &\le C \sum_{|\alpha|=m} \|\partial^\alpha u_j - \partial^\alpha u_k\|_{B_s^{p,q}(\Omega)}. \end{aligned} \tag{3.226}$$

Above, the first inequality follows from (3.140) and (2.213), whereas the second is derived much as in (3.143). It is here that the smallness assumption on smoothness index s is used (cf. Proposition 2.31). Since, from (3.225), $\{\partial^\alpha u_j\}_j$ is a Cauchy sequence in $B_s^{p,q}(\Omega)$ for every α with $|\alpha| \le m$, we see from (3.226) that $\{u_j\}_j$ is also a Cauchy sequence in $B^{p,q}_{m+s}(\Omega)$. This forces $u \in B^{p,q}_{m+s}(\Omega)$, as desired. The estimate in (3.224) is also implicit in what we have proved. □

Moving on, we record a useful result, modeling in abstract the effect of constraints on the real and complex interpolation methods at the level of quasi-Banach spaces. The reader is referred to [85] for a proof. Similar results for the complex method on Banach spaces may me found in [72, Theorem 14.3 on p. 97] (cf. also [58]).

Proposition 3.24. *Let* X_i, Y_i, Z_i, $i = 1, 2$, *be quasi-Banach spaces such that* $X_1 \cap X_2$ *is dense in both* X_1 *and* X_2, *and that* $Z_1 \cap Z_2$ *is dense in both* Z_1 *and* Z_2. *In addition, suppose that*

$$Y_i \hookrightarrow Z_i \quad \textit{continuously, for } \ i = 1, 2, \tag{3.227}$$

and that there exists a linear operator D *such that*

$$D : X_i \longrightarrow Z_i \quad \textit{boundedly, for } \ i = 1, 2. \tag{3.228}$$

Define the spaces

$$X_i(D) := \{u \in X_i : Du \in Y_i\}, \quad i = 1, 2, \tag{3.229}$$

and equip them with the natural graph quasi-norm, i.e.,

$$\|u\|_{X_i(D)} := \|u\|_{X_i} + \|Du\|_{Y_i}, \qquad i = 1, 2. \tag{3.230}$$

Finally, suppose that there exist continuous linear mappings

$$G : Z_i \longrightarrow X_i \quad \text{and} \quad K : Z_i \longrightarrow Y_i, \qquad i = 1, 2, \tag{3.231}$$

with the property that (with I denoting the identity)

$$D \circ G = I + K \quad \text{on} \quad Z_i, \quad \text{for} \ i = 1, 2. \tag{3.232}$$

Then, for each $0 < \theta < 1$ and $0 < q \leq \infty$, there holds

$$\big(X_1(D), X_2(D)\big)_{\theta,q} = \big\{u \in (X_1, X_2)_{\theta,q} : \ Du \in (Y_1, Y_2)_{\theta,q}\big\}. \tag{3.233}$$

Furthermore, if the spaces $X_1 + X_2$ and $Y_1 + Y_2$ are analytically convex, then also

$$\big[X_1(D), X_2(D)\big]_\theta = \big\{u \in [X_1, X_2]_\theta : \ Du \in [Y_1, Y_2]_\theta\big\}, \qquad \theta \in (0, 1). \tag{3.234}$$

The above proposition is useful in establishing the interpolation result described in our next theorem.

Theorem 3.25. *Let Ω be a bounded Lipschitz domain in $\mathbb{R}^n$ and pick $m \in \mathbb{N}$. If*

$$\begin{gathered} \tfrac{n-1}{n} < p < \infty, \quad (n-1)\big(\tfrac{1}{p} - 1\big)_+ < s_1, s_2 < 1, \quad s_1 \neq s_2, \\ \min\{1, p\} \leq q, q_1, q_2 < \infty, \quad \theta \in (0, 1), \quad s = (1-\theta)s_1 + \theta s_2, \end{gathered} \tag{3.235}$$

then

$$\Big(\mathring{F}^{p,q_1}_{m-1+s_1+1/p}(\Omega),\ \mathring{F}^{p,q_2}_{m-1+s_2+1/p}(\Omega)\Big)_{\theta,q} = \mathring{B}^{p,q}_{m-1+s+1/p}(\Omega). \tag{3.236}$$

Also,

$$\Big(\mathring{B}^{p,q_1}_{m-1+s_1+1/p}(\Omega),\ \mathring{B}^{p,q_2}_{m-1+s_2+1/p}(\Omega)\Big)_{\theta,q} = \mathring{B}^{p,q}_{m-1+s+1/p}(\Omega), \tag{3.237}$$

provided

$$\begin{gathered} \tfrac{n-1}{n} < p < \infty, \quad (n-1)\big(\tfrac{1}{p} - 1\big)_+ < s_1, s_2 < 1, \quad s_1 \neq s_2, \\ 0 < q, q_1, q_2 < \infty, \quad \theta \in (0, 1), \quad s = (1-\theta)s_1 + \theta s_2. \end{gathered} \tag{3.238}$$

Furthermore,

$$\Big[\mathring{F}^{p_1,q_1}_{m-1+s_1+1/p_1}(\Omega),\ \mathring{F}^{p_2,q_2}_{m-1+s_2+1/p_2}(\Omega)\Big]_\theta = \mathring{F}^{p,q}_{m-1+s+1/p}(\Omega), \tag{3.239}$$

granted that

$$\tfrac{n-1}{n} < p_1, p_2 < \infty, \quad (n-1)\big(\tfrac{1}{p_i} - 1\big)_+ < s_i < 1 \textit{ for } i = 1, 2,$$
$$\min\{1, p_i\} \le q_i < \infty \textit{ for } i = 1, 2, \quad \theta \in (0,1), \quad s = (1-\theta)s_1 + \theta s_2,$$
$$\textit{and } \tfrac{1}{p} = \tfrac{1-\theta}{p_1} + \tfrac{\theta}{p_2}, \quad \tfrac{1}{q} = \tfrac{1-\theta}{q_1} + \tfrac{\theta}{q_2}. \tag{3.240}$$

Finally,

$$\Big[\mathring{B}^{p_1,q_1}_{m-1+s_1+1/p_1}(\Omega), \mathring{B}^{p_2,q_2}_{m-1+s_2+1/p_2}(\Omega)\Big]_\theta = \mathring{B}^{p,q}_{m-1+s+1/p}(\Omega), \tag{3.241}$$

in the case when

$$\tfrac{n-1}{n} < p_1, p_2 < \infty, \quad (n-1)\big(\tfrac{1}{p_i} - 1\big)_+ < s_i < 1 \textit{ for } i = 1, 2,$$
$$0 < q_1, q_2 < \infty, \quad \theta \in (0,1), \quad s = (1-\theta)s_1 + \theta s_2, \tag{3.242}$$
$$\textit{and } \tfrac{1}{p} = \tfrac{1-\theta}{p_1} + \tfrac{\theta}{p_2}, \quad \tfrac{1}{q} = \tfrac{1-\theta}{q_1} + \tfrac{\theta}{q_2}.$$

Proof. Consider (3.236). The idea is to implement the abstract interpolation result from Proposition 3.24 in a suitable setting. Concretely, for $i = 1, 2$, define

$$X_i := F^{p,q_i}_{m-1+s_i+1/p}(\Omega), \quad Y_i := \{0\}, \quad Z_i := \dot{B}^{p,p}_{m-1,s_i}(\partial\Omega), \tag{3.243}$$

set $K := 0$ and, in the notation employed in Theorem 3.9, take

$$D := \mathrm{tr}_{m-1} \quad \text{and} \quad G := \mathscr{E}. \tag{3.244}$$

Theorem 3.9 guarantees that conditions (3.231)–(3.232) hold for these choices. Furthermore, the spaces defined abstractly in (3.229) now become

$$X_i(D) = \Big\{u \in F^{p,q_i}_{m-1+s_i+1/p}(\Omega) : \mathrm{tr}_{m-1} u = 0\Big\}$$
$$= \mathring{F}^{p,q_i}_{m-1+s_i+1/p}(\Omega), \quad i = 1, 2, \tag{3.245}$$

by Corollary 3.16, whereas

$$\{u \in (X_1, X_2)_{\theta,q} : Du \in (Y_1, Y_2)_{\theta,q}\}$$
$$= \Big\{u \in \big(F^{p,q_1}_{m-1+s_1+1/p}(\Omega), F^{p,q_2}_{m-1+s_2+1/p}(\Omega)\big)_{\theta,q} : \mathrm{tr}_{m-1} u = 0\Big\}$$

$$= \left\{u \in B^{p,q}_{m-1+s+1/p}(\Omega) : \ \mathrm{tr}_{m-1}\, u = 0\right\}$$

$$= \mathring{B}^{p,q}_{m-1+s+1/p}(\Omega), \tag{3.246}$$

by (2.221) and (3.171). Having established these, (3.236) now becomes a consequence of (3.233), (3.245), and (3.246).

Finally, all other interpolation formulas in the statement of the theorem are established analogously. □

In the last part of this section we include a result to the effect that gluing equally smooth functions defined on both sides of a Lipschitz interface and which have matching traces on this surface produces a globally smooth function in $\mathbb{R}^n$.

Proposition 3.26. *Given a bounded Lipschitz domain $\Omega \subset \mathbb{R}^n$, consider the sets $\Omega_+ := \Omega$ and $\Omega_- := \mathbb{R}^n \setminus \overline{\Omega}$. Fix $m \in \mathbb{N}$ and assume that the numbers p, q, s satisfy $\frac{n-1}{n} < p < \infty$, $(n-1)(1/p-1)_+ < s < 1$ and $\min\{1, p\} \le q < \infty$. Also, let the functions $u^\pm \in F^{p,q}_{m-1+s+1/p}(\Omega_\pm)$ be such that*

$$\mathrm{tr}_{m-1}(u^+) = \mathrm{tr}_{m-1}(u^-) \quad in \quad \dot{B}^{p,p}_{m-1+s}(\partial\Omega). \tag{3.247}$$

Then, if

$$u(x) := \begin{cases} u^+(x) & for\ x \in \Omega_+, \\ u^-(x) & for\ x \in \Omega_-, \end{cases} \tag{3.248}$$

it follows that $u \in F^{p,q}_{m-1+s+1/p}(\mathbb{R}^n)$ and

$$\|u\|_{F^{p,q}_{m-1+s+1/p}(\mathbb{R}^n)} \approx \|u^+\|_{F^{p,q}_{m-1+s+1/p}(\Omega_+)} + \|u^-\|_{F^{p,q}_{m-1+s+1/p}(\Omega_-)}. \tag{3.249}$$

Furthermore, a similar result is valid for the scale of Besov spaces in $\Omega_\pm$, granted that the indices p, q, s satisfy $\frac{n-1}{n} < p < \infty$, $(n-1)(1/p-1)_+ < s < 1$ and $0 < q < \infty$.

Proof. Since by hypothesis $u^+ \in F^{p,q}_{m-1+s+1/p}(\Omega_+)$, Theorem 2.29 ensures that there exists a function $U^+ \in F^{p,q}_{m-1+s+1/p}(\mathbb{R}^n)$ such that $U^+|_{\Omega_+} = u^+$ and the inequality $\|U^+\|_{F^{p,q}_{m-1+s+1/p}(\mathbb{R}^n)} \le 2\,\|u^+\|_{F^{p,q}_{m-1+s+1/p}(\Omega)}$ holds. If we now define

$$v := U^+|_{\Omega_-} - u^- \quad \text{in} \quad \Omega_-, \tag{3.250}$$

then

$$v \in F^{p,q}_{m-1+s+1/p}(\Omega_-), \tag{3.251}$$

$$\|v\|_{F^{p,q}_{m-1+s+1/p}(\Omega_-)} \le C\Big(\|u^+\|_{F^{p,q}_{m-1+s+1/p}(\Omega_+)} + \|u^-\|_{F^{p,q}_{m-1+s+1/p}(\Omega_-)}\Big), \tag{3.252}$$

and

$$\mathrm{tr}_{m-1}\, v = \mathrm{tr}_{m-1}(U^+|_{\Omega_+}) - \mathrm{tr}_{m-1}(u^-) = \mathrm{tr}_{m-1}(u^+) - \mathrm{tr}_{m-1}(u^-) = 0. \tag{3.253}$$

According to Theorem 3.12, if

$$\tilde{v}(x) := \begin{cases} v(x), & \text{if } x \in \Omega_-, \\ 0, & \text{if } x \in \mathbb{R}^n \setminus \overline{\Omega_-}, \end{cases} \tag{3.254}$$

then $\tilde{v} \in F^{p,q}_{m-1+s+1/p}(\mathbb{R}^n)$ and $\|\tilde{v}\|_{F^{p,q}_{m-1+s+1/p}(\mathbb{R}^n)} \le C\,\|v\|_{F^{p,q}_{m-1+s+1/p}(\Omega_-)}$. Note that

$$U^+(x) - \tilde{v}(x) = \begin{cases} u^+(x) & \text{for } x \in \Omega_+, \\ U^+|_{\Omega_-}(x) - v(x) & \text{for } x \in \Omega_-, \end{cases} \tag{3.255}$$

or, equivalently, $U^+ - \tilde{v} = u$. Since both U^+ and $\tilde{v}$ belong to $F^{p,q}_{m-1+s+1/p}(\mathbb{R}^n)$, we also have that $u \in F^{p,q}_{m-1+s+1/p}(\mathbb{R}^n)$. Moreover,

$$\begin{aligned} \|u\|_{F^{p,q}_{m-1+s+1/p}(\mathbb{R}^n)} &= \|U^+ - \tilde{v}\|_{F^{p,q}_{m-1+s+1/p}(\mathbb{R}^n)} \\ &\le \|U^+\|_{F^{p,q}_{m-1+s+1/p}(\mathbb{R}^n)} + \|\tilde{v}\|_{F^{p,q}_{m-1+s+1/p}(\mathbb{R}^n)} \\ &\le 2\|u^+\|_{F^{p,q}_{m-1+s+1/p}(\Omega_+)} + \|v\|_{F^{p,q}_{m-1+s+1/p}(\Omega_+)} \\ &\le C\,\|u^+\|_{F^{p,q}_{m-1+s+1/p}(\Omega_+)} + C\,\|u^-\|_{F^{p,q}_{m-1+s+1/p}(\Omega_-)}. \end{aligned} \tag{3.256}$$

This establishes one direction in the equivalence (3.249). Conversely, by the continuity of the restriction operator $\cdot|_{\Omega_\pm}$, we have that

$$\|u^+\|_{F^{p,q}_{m-1+s+1/p}(\Omega_+)} + \|u^-\|_{F^{p,q}_{m-1+s+1/p}(\Omega_-)} \le C\,\|u\|_{F^{p,q}_{m-1+s+1/p}(\mathbb{R}^n)}. \tag{3.257}$$

In concert with (3.256), this ultimately yields (3.249). The proof for the scale of Besov spaces is carried out analogously, and this completes the proof of the proposition. □

We conclude by recording the following consequence of Proposition 3.26.

Corollary 3.27. *Assume that $\Omega \subset \mathbb{R}^n$ is a bounded Lipschitz domain. Fix a bounded neighborhood $\mathcal{O}$ of $\overline{\Omega}$ and assume that $m \in \mathbb{N}$, and that the numbers p, q, s satisfy $\frac{n-1}{n} < p < \infty$, $(n-1)(1/p-1)_+ < s < 1$ and $\min\{1, p\} \le q < \infty$. Then there exists a linear, continuous operator*

$$\mathscr{E} : \dot{B}^{p,p}_{m-1+s}(\partial\Omega) \longrightarrow F^{p,q}_{m-1+s+1/p}(\mathbb{R}^n) \tag{3.258}$$

such that, if $\Omega_+ := \Omega$ and $\Omega_- := \mathbb{R}^n \setminus \overline{\Omega}$, then for every $\dot{f} \in \dot{B}^{p,p}_{m-1+s}(\partial\Omega)$ one has

$$\operatorname{supp}(\mathscr{E}\dot{f}) \subseteq \mathscr{O} \quad \textit{and} \quad \operatorname{tr}_{m-1}(\mathscr{E}\dot{f}|_{\Omega_+}) = \dot{f} = \operatorname{tr}_{m-1}(\mathscr{E}\dot{f}|_{\Omega_-}). \tag{3.259}$$

Moreover, a similar result is valid for the scale of Besov spaces in $\mathbb{R}^n$, granted that the indices p, q, s satisfy $\frac{n-1}{n} < p < \infty$, $(n-1)(1/p-1)_+ < s < 1$ and $0 < q < \infty$.

Proof. This is proved by gluing together, with the help of Proposition 3.26, the two extension operators from Theorem 3.9 corresponding to Ω_+ and Ω_- (with the latter further truncated by some $\psi \in C^\infty_c(\mathscr{O})$ with $\psi \equiv 1$ near $\overline{\Omega}$). □

3.4 Whitney–Hardy and Whitney–Triebel–Lizorkin Spaces

The goal of this section is to systematically develop a function theory for Whitney–Hardy and Whitney–Triebel–Lizorkin spaces on Lipschitz surfaces. To set the stage, consider a bounded Lipschitz domain Ω in $\mathbb{R}^n$ and, for $1 < p < \infty$ and $0 \leq s \leq 1$, abbreviate

$$L^p_s(\partial\Omega) := F^{p,2}_s(\partial\Omega), \tag{3.260}$$

which, by (2.421), is in agreement with the notation already employed for $s \in \{0, 1\}$. Going further, we also define

$$\dot{L}^p_{m-1,s}(\partial\Omega) := \Big\{\dot{f} = \{f_\alpha\}_{|\alpha|\leq m-1} \in CC : \\ f_\alpha \in L^p_s(\partial\Omega) \text{ for all } |\alpha| \leq m-1\Big\}, \tag{3.261}$$

and equip this space with the norm

$$\|\dot{f}\|_{\dot{L}^p_{m-1,s}(\partial\Omega)} := \sum_{|\alpha|\leq m-1} \|f_\alpha\|_{L^p_s(\partial\Omega)}. \tag{3.262}$$

Once again, the notation for $\dot{L}^p_{m-1,s}(\partial\Omega)$ is consistent with that used for the spaces (3.3) and (3.7), which occur for $s = 0$ and $s = 1$, respectively.

A natural continuation of the Whitney-Sobolev scale $\dot{L}^p_{m-1,1}(\partial\Omega)$, originally defined for $1 < p < \infty$, to the range $p \in (\frac{n-1}{n}, 1]$ can be constructed using Hardy–Sobolev spaces. More specifically, if $\frac{n-1}{n} < p < \infty$, set

$$\dot{h}^p_{m-1,1}(\partial\Omega) := \Big\{\dot{f} = \{f_\alpha\}_{|\alpha|\leq m-1} \in CC : \\ f_\alpha \in h^p_1(\partial\Omega) \text{ for all } |\alpha| \leq m-1\Big\}, \tag{3.263}$$

endowed with the quasi-norm

$$\|\dot{f}\|_{\dot{h}^p_{m-1,1}(\partial\Omega)} := \sum_{|\alpha|\le m-1} \|f_\alpha\|_{h^p_1(\partial\Omega)}, \tag{3.264}$$

where $h^p_1(\partial\Omega)$ is as in (2.423). Clearly,

$$\dot{h}^p_{m-1,1}(\partial\Omega) = \dot{L}^p_{m-1,1}(\partial\Omega) \quad \text{if } 1 < p < \infty. \tag{3.265}$$

Finally, it is convenient to introduce an even larger scale of Whitney arrays, based on Triebel–Lizorkin spaces. Concretely, for p, q, s as in (2.415)–(2.416), we define the space

$$\dot{F}^{p,q}_{m-1,s}(\partial\Omega) := \Big\{\dot{f} = \{f_\alpha\}_{|\alpha|\le m-1} \in CC : \\ f_\alpha \in F^{p,q}_s(\partial\Omega) \text{ for all } |\alpha| \le m-1\Big\}, \tag{3.266}$$

equipped with the quasi-norm $\|\dot{f}\|_{\dot{F}^{p,q}_{m-1,s}(\partial\Omega)} := \sum_{|\alpha|\le m-1} \|f_\alpha\|_{F^{p,q}_s(\partial\Omega)}$.

Various useful characteristics of the ordinary (scalar) Triebel–Lizorkin spaces are inherited by their Whitney array versions, and several properties of this nature are collected in the proposition below.

Proposition 3.28. *Let Ω be a bounded Lipschitz domain in $\mathbb{R}^n$ and fix $m \in \mathbb{N}$. Also, recall the set $\mathcal{O}$ introduced in* (2.489)*. Then the following claims are true.*

(i) One has

$$\dot{F}^{p,2}_{m-1,s}(\partial\Omega) = \dot{L}^{p,2}_{m-1,s}(\partial\Omega) \text{ whenever } 1 < p < \infty \text{ and } s \in \{0,1\}. \tag{3.267}$$

(ii) For every p, q, s as in (2.415)–(2.416) *one has*

$$\dot{B}^{p,\min\{p,q\}}_{m-1,s}(\partial\Omega) \hookrightarrow \dot{F}^{p,q}_{m-1,s}(\partial\Omega) \hookrightarrow \dot{B}^{p,\max\{p,q\}}_{m-1,s}(\partial\Omega). \tag{3.268}$$

(iii) For any two triplets (p_0, q_0, s_0), (p_1, q_1, s_1) belonging to the set $\mathcal{O}$ one has

$$\dot{F}^{p_0,q_0}_{m-1,s_0}(\partial\Omega) \hookrightarrow \dot{F}^{p_1,q_1}_{m-1,s_1}(\partial\Omega) \text{ continuously, with dense range}, \tag{3.269}$$

whenever

$$\text{either } p_0 < p_1 \text{ and } \tfrac{1}{p_0} - \tfrac{s_0}{n-1} = \tfrac{1}{p_1} - \tfrac{s_1}{n-1}, \\ \text{or } p_0 = p_1, \quad q_0 \le q_1 \text{ and } s_0 = s_1. \tag{3.270}$$

(iv) For any two triplets, (p, q_0, s_0) and (p, q_1, s_1), belonging to the set $\mathcal{O}$ one has

$$\dot{F}^{p,q_0}_{m-1,s_0}(\partial\Omega) \hookrightarrow \dot{F}^{p,q_1}_{m-1,s_1}(\partial\Omega) \text{ whenever } s_0 > s_1. \tag{3.271}$$

Proof. The identification in (3.267) is a consequence of (2.421), (3.266), (3.3), and (3.7). Also, (2.154) readily translates into (3.268), while the claims made in parts (iii)–(iv) are immediate consequences of Theorem 2.17 and (2.153). □

In the next segment we shall discuss certain trace results involving some of the spaces just introduced. To set the stage, for a sufficiently nice function u defined in Ω set

$$u\Big|_{\partial\Omega}^{m-1}(X) := \Big\{(\partial^\alpha u)\Big|_{\partial\Omega} : \alpha \in \mathbb{N}_0^n \text{ with } |\alpha| \leq m-1\Big\}, \tag{3.272}$$

where the boundary traces in the right-hand side are taken in the sense of (2.59).

Proposition 3.29. *Let Ω be a bounded Lipschitz domain in $\mathbb{R}^n$ and $m \in \mathbb{N}$. Then whenever a function $u \in C^\infty(\Omega)$ is such that $\mathcal{N}(\nabla^{m-1}u) \in L^p(\partial\Omega)$ for some $p \in (0,\infty)$, and*

$$u\Big|_{\partial\Omega}^{m-1} \textit{ exists}, \tag{3.273}$$

it follows that

$$\Big\|u\Big|_{\partial\Omega}^{m-1}\Big\|_{L^p(\partial\Omega)} \leq C\sum_{j=0}^{m-1}\|\mathcal{N}(\nabla^j u)\|_{L^p(\partial\Omega)}. \tag{3.274}$$

In addition,

$$u\Big|_{\partial\Omega}^{m-1} \in \dot{L}^p_{m-1,0}(\partial\Omega) \textit{ whenever } 1 < p < \infty. \tag{3.275}$$

A sufficient condition that guarantees (3.273) *is that the function $u \in C^\infty(\Omega)$ is such that $Lu = 0$ in Ω and $\mathcal{N}(\nabla^{m-1}u) \in L^p(\partial\Omega)$ for some $p \in (0,\infty)$, where L is a homogeneous, constant (real) coefficient, symmetric, strongly elliptic differential operator of order $2m$.*

Proof. First, (3.274) is a simple consequence of (3.273) and definitions. Next, assume that $1 < p < \infty$. In order to prove (3.275) it suffices to show that, for each $u \in C^\infty(\Omega)$ such that $\mathcal{N}(\nabla^{m-1}u) \in L^p(\partial\Omega)$ and (3.273) holds, we have

$$\dot{f} := u\Big|_{\partial\Omega}^{m-1} \in CC, \tag{3.276}$$

i.e., that the components of $u\Big|_{\partial\Omega}^{m-1}$ satisfy the compatibility conditions (3.2). From (3.274) we know that

$$f_\alpha := (\partial^\alpha u)\Big|_{\partial\Omega} \in L^p(\partial\Omega), \quad \text{for any } \alpha \in \mathbb{N}_0^n \text{ with } |\alpha| \leq m-1, \tag{3.277}$$

and, for each $1 \leq j,k \leq n$ and $\gamma \in \mathbb{N}_0^n$, $|\gamma| \leq m-2$, we may write

$$\begin{aligned}\partial_{\tau_{jk}} f_\gamma &= \partial_{\tau_{jk}} \Big[(\partial^\gamma u)\Big|_{\partial\Omega} \Big] = \nu_j (\partial^{\gamma+e_k} u)\Big|_{\partial\Omega} - \nu_k (\partial^{\gamma+e_j} u)\Big|_{\partial\Omega} \\ &= \nu_j f_{\gamma+e_k} - \nu_k f_{\gamma+e_j}. \end{aligned} \tag{3.278}$$

Above, for the first and third equality we used (3.277) while the second one follows from (2.109). Thus, (3.276) holds.

Finally, assume that L is a differential operator as in the statement of the proposition, and that $u \in C^\infty(\Omega)$ has $\mathcal{N}(\nabla^{m-1}u) \in L^p(\partial\Omega)$ for some $p \in (0,\infty)$. Then the almost everywhere existence of $u\Big|_{\partial\Omega}^{m-1}$ is a consequence of the unique solvability of the L^2-Dirichlet problem for L in Ω, and a well-known approximation scheme. The aforementioned well-posedness result can be found in [128] which, in turn, further builds on the work in [104]. □

Proposition 3.30. *Let Ω be a bounded Lipschitz domain in $\mathbb{R}^n$. If $u \in C^\infty(\Omega)$ is such that $\mathcal{N}(\nabla^m u) \in L^p(\partial\Omega)$ for some $p \in (\frac{n-1}{n},\infty)$ and $m \in \mathbb{N}$, then*

$$u\Big|_{\partial\Omega}^{m-1} \in \dot{h}^p_{m-1,1}(\partial\Omega) \quad \textit{and} \quad \Big\| u\Big|_{\partial\Omega}^{m-1} \Big\|_{\dot{h}^p_{m-1,1}(\partial\Omega)} \leq C \sum_{j=0}^{m} \|\mathcal{N}(\nabla^j u)\|_{L^p(\partial\Omega)}. \tag{3.279}$$

Proof. We start by fixing $p \in (\frac{n-1}{n},\infty)$, $m \in \mathbb{N}$ and a function $u \in C^\infty(\Omega)$ such that $\mathcal{N}(\nabla^m u) \in L^p(\partial\Omega)$. Set $\dot{f} := u\Big|_{\partial\Omega}^{m-1}$, i.e., $f_\alpha := (\partial^\alpha u)\Big|_{\partial\Omega}$ for each multi-index $\alpha \in \mathbb{N}_0^n$ with $|\alpha| \leq m-1$. If $1 < p < \infty$, Proposition 2.15 applies and gives that $f_\alpha \in L_1^p(\partial\Omega)$ for each multi-index $\alpha \in \mathbb{N}_0^n$ of length $\leq m-1$. Much as in the proof of Proposition 3.29, $\dot{f} \in CC$ and, consequently, (3.279) holds whenever $1 < p < \infty$.

In the case in which $p \in (\frac{n-1}{n}, 1]$, Proposition 2.60 yields

$$f_\alpha \in h^{1,p}_{at}(\partial\Omega), \quad \forall\, \alpha \in \mathbb{N}_0^n,\ |\alpha| \leq m-1. \tag{3.280}$$

Therefore, in order to conclude that (3.279) holds in the current case, we are left with establishing that $\dot{f} \in CC$. Since (2.109) requires $p > 1$, we can no longer argue as in (3.278). However, in this case, applying Proposition 2.3 we may conclude that

$$u \in W^{m,\frac{np}{n-1}}(\Omega) = F^{q,2}_{m-1+s+1/q}(\Omega), \quad \text{where } q := \tfrac{np}{n-1}, \quad s := 1 - \tfrac{n-1}{pn}. \tag{3.281}$$

Hence, using (3.105) from Theorem 3.9, we obtain that $\mathrm{tr}_{m-1}(u) \in \dot{B}^{q,q}_{m-1,s}(\partial\Omega)$. In particular, $\mathrm{tr}_{m-1}u \in CC$. Finally, in the light of the remark following the statement of Theorem 2.53, $\dot{f} = u\Big|_{\partial\Omega}^{m-1} = \mathrm{tr}_{m-1}(u)$ and, hence, $\dot{f} \in CC$ as desired. □

The limiting cases $s \in \{0, 1\}$ in Theorem 3.9 fail in the general setting of all bounded Lipschitz domains. Nonetheless, if suitable addition information is available, then certain versions of (3.105) do hold, and we next wish to elaborate on this issue. As a preamble, we shall prove a result relating the membership of a function to the Triebel–Lizorkin scale to the integrability properties of its non-tangential maximal function.

Proposition 3.31. *Let Ω be a bounded Lipschitz domain in $\mathbb{R}^n$ and assume that L is a homogeneous, constant (real) coefficient, symmetric, strongly elliptic differential operator of order $2m$. Also, fix $k \in \mathbb{N}_0$, $\frac{n-1}{n} < p \leq 2$ and $0 < q < \infty$. Then there exists $C > 0$ with the property that*

$$\sum_{j=0}^{m-1+k} \|\mathcal{N}(\nabla^j u)\|_{L^p(\partial\Omega)} \leq C\,\|u\|_{F^{p,q}_{m-1+k+1/p}(\Omega)} \tag{3.282}$$

for each $u \in F^{p,q}_{m-1+k+1/p}(\Omega) \cap \operatorname{Ker} L$.

Proof. Using estimates similar to (2.508), it suffices to prove the following seemingly weaker version of (3.282):

$$\begin{aligned} \|\mathcal{N}(\nabla^{m-1+k} u)\|_{L^p(\partial\Omega)} &\leq C\,\|u\|_{F^{p,q}_{m-1+k+1/p}(\Omega)} \\ &\forall\, u \in F^{p,q}_{m-1+k+1/p}(\Omega) \cap \operatorname{Ker} L. \end{aligned} \tag{3.283}$$

With this goal in mind, recall the area function

$$\mathcal{A}(u)(X) := \Big(\int_{R_\kappa(X)} |\nabla u(Y)|^2 \rho(Y)^{2-n}\,dY\Big)^{\frac{1}{2}}, \quad X \in \partial\Omega, \tag{3.284}$$

where $\kappa > 0$ is some fixed parameter, and ρ is the distance to the boundary. As proved in [31], for every $0 < p < \infty$ there exists a finite geometric constant $C > 0$ such that

$$\|\mathcal{N}(\nabla^{m-1+k} u)\|_{L^p(\partial\Omega)} \leq C\,\|\mathcal{A}(\nabla^{m-1+k} u)\|_{L^p(\partial\Omega)} + C \sum_{j=0}^{m-1} \|\nabla^{j+k} u\|_{L^1(\Omega)}. \tag{3.285}$$

If $\{Q_j\}_j$ is a Whitney decomposition of Ω into Euclidean cubes Q_j of side-length $l(Q_j)$, we may then estimate

$$\int_{\partial\Omega} \Big(\mathcal{A}(\nabla^{m-1+k} u)(X)\Big)^p\,d\sigma(X) \tag{3.286}$$

$$= \int_{\partial\Omega} \Big(\int_{\Omega} \rho(Y)^{2-n} |\nabla^{m+k} u(Y)|^2 \chi_{\{Y \in R_\kappa(X)\}} dY\Big)^{\frac{p}{2}} d\sigma(X)$$

$$= \int_{\partial\Omega} \Big(\sum_j \int_{Q_j} \rho(Y)^{2-n} |\nabla^{m+k} u(Y)|^2 \chi_{\{Y \in R_\kappa(X)\}} dY\Big)^{\frac{p}{2}} d\sigma(X) =: I.$$

If the points $Y \in Q_j$ and $X \in \partial\Omega$ are such that $Y \in R_\kappa(X)$, then $X \in \Delta_j$, where Δ_j is the "cone shadow" of Q_j on $\partial\Omega$, i.e., $\Delta_j := \{X \in \partial\Omega : R_\kappa(X) \cap Q_j \neq \emptyset\}$. In particular, $\sigma(\Delta_j) \approx l(Q_j)^{n-1}$, uniformly in j.

Assume that $0 < p \leq 2$. Then (with barred integrals indicating averaging) we have

$$
\begin{aligned}
I &\leq \int_{\partial\Omega} \sum_j \Big(\int_{Q_j} \rho(Y)^{2-n} |\nabla^{m+k} u(Y)|^2 \chi_{\{Y \in R_\kappa(X)\}} dY \Big)^{\frac{p}{2}} d\sigma(X) \\
&\leq \sum_j \int_{\Delta_j} \Big[l(Q_j) \Big(\fint_{Q_j} |\nabla^{m+k} u|^2 \Big)^{\frac{1}{2}} \Big]^p \, d\sigma \\
&\leq C \sum_j l(Q_j)^{n-1+p} \fint_{Q_j^*} |\nabla^{m+k} u|^p \leq C \int_\Omega \Big[\rho^{1-\frac{1}{p}} |\nabla^{m+k} u| \Big]^p \\
&\leq C \|u\|^p_{\mathbb{H}^p_{m-1+k+1/p}(\Omega;L)} \leq C \|u\|_{F^{p,q}_{m-1+k+1/p}(\Omega)},
\end{aligned}
\tag{3.287}
$$

provided $\frac{1}{p} > n(\frac{1}{p} - 1)$ (or, equivalently, $p > \frac{n-1}{n}$). For the second inequality in (3.287) we have used the fact that the function $\nabla^{m+k} u \in \operatorname{Ker} L$ satisfies the reverse Hölder inequality

$$
\Big(\fint_{Q_j} |\nabla^{m+k} u|^2 \Big)^{\frac{1}{2}} \leq C \Big(\fint_{Q_j^*} |\nabla^{m+k} u|^p \Big)^{\frac{1}{p}},
\tag{3.288}
$$

where Q_j^* is the concentric double of Q_j. Moreover, the fourth inequality relies on the fact that the Q_j^*'s have bounded overlap. Let us also point out that the next-to-last estimate in (3.287) follows straight from definitions when $1 \leq p \leq 2$, and is a consequence of Lemma 2.37 when $\frac{n-1}{n} < p < 1$. Finally, the last estimate in (3.287) is implied by Theorem 2.41.

The above argument shows that $\|\mathscr{A}(\nabla^{m-1+k} u)\|_{L^p(\partial\Omega)} \leq C \|u\|_{F^{p,q}_{m-1+k+1/p}(\Omega)}$, for some finite constant C independent of u. Since we also have the embedding $F^{p,q}_{1/p}(\Omega) \hookrightarrow L^{np/(n-1)}(\Omega)$, the estimate (3.283) now follows from Theorem 2.41. □

Our next result addresses the converse direction discussed in Proposition 3.31, i.e., passing from non-tangential maximal function estimates to membership to Besov spaces.

Proposition 3.32. *Let Ω be a bounded Lipschitz domain in $\mathbb{R}^n$ and fix $p \in (1, 2]$. Also, let L be a homogeneous, constant (real) coefficient, symmetric, strongly elliptic system of differential operators of order $2m$. If $k \in \mathbb{N}_0$ and the function u is such that $Lu = 0$ in Ω and $\mathcal{N}(\nabla^{m-1+k} u) \in L^p(\partial\Omega)$, then $u \in B^{p,2}_{m-1+k+1/p}(\Omega)$, plus a natural estimate.*

Proof. Working with $\nabla^k u$ in place of u (and relying on the lifting result from Proposition 2.24), it suffices to consider the case $k = 0$ only. Assuming that this is the case, let u be as in the hypotheses of the theorem and, for an arbitrary multi-index γ with $|\gamma| = m - 1$, set $v := \partial^\gamma u$. Our aim is to show that $v \in B^{p,2}_{1/p}(\Omega)$ since, granted this, Proposition 2.24 yields the desired conclusion. The strategy is to combine the result established in Lemma 2.34 with the area-function estimate from [31]. Specifically, we shall employ the result from Lemma 2.34 in the case when $q = 2$ and $\theta := 1 - 1/p$ (where $p \in (1, 2]$ is given). As such, the focus becomes estimating, in terms of the non-tangential maximal function of $\nabla^{m-1}u$, an integral of the form

$$\Big[\int_{|x'|<r}\Big(\int_0^r |\nabla v(x',\varphi(x')+s)|^2 s\,ds\Big)^{p/2} dx'\Big]^{2/p}, \tag{3.289}$$

where $r > 0$ is a small number with the property that $B_r(0) \cap \partial\Omega$ is part of the graph of some Lipschitz function $\varphi : \mathbb{R}^{n-1} \to \mathbb{R}$, assumed to satisfy $\varphi(0) = 0$. To this end, we first claim that

$$\int_0^r |\nabla v(x',\varphi(x')+s)|^2 s\,ds \le C \int_{R_\kappa(x',\varphi(x'))} |\nabla v(Z)|^2 \operatorname{dist}(Z,\partial\Omega)^{2-n}\,dZ, \tag{3.290}$$

uniformly, for $(x',\varphi(x')) \in B_r(0) \cap \partial\Omega$. To justify the inequality (3.290), fix $(x',\varphi(x')) \in B_r(0) \cap \partial\Omega$ and note that there exists a number $\lambda > 0$ such that $B_{\lambda s}(x',\varphi(x')+s) \subset R_\kappa(x',\varphi(x'))$ for all $s \in (0,r)$. Using the fact that v is a null-solution for the elliptic operator L in Ω, interior estimates give that

$$|\nabla v(x',\varphi(x')+s)|^2 \le C \fint_{B_{\lambda s}(x',\varphi(x')+s)} |\nabla v(Z)|^2\,dZ, \quad \forall\, s \in (0,r). \tag{3.291}$$

Hence, by choosing λ small enough (relative to the Lipschitz character of Ω), we may write (for some $0 < c_0 < c_1 < \infty$)

$$\begin{aligned}
\int_0^r |\nabla v(x',\varphi(x')+s)|^2 s\,ds &\le C \int_0^r s^{1-n} \int_{R_\kappa(x',\varphi(x'))} |\nabla v(Z)|^2 \chi_{|Z-(x',\varphi(x')+s)|<\lambda s}\,dZ\,ds \\
&\le C \int_{R_\kappa(x',\varphi(x'))} |\nabla v(Z)|^2 \Big(\int_0^\infty s^{1-n} \chi_{|Z-(x',\varphi(x')+s)|<\lambda s}\,ds\Big)\,dZ \\
&\le C \int_{R_\kappa(x',\varphi(x'))} |\nabla v(Z)|^2 \Big(\int_{c_0 \operatorname{dist}(Z,\partial\Omega)}^{c_1 \operatorname{dist}(Z,\partial\Omega)} s^{1-n}\,ds\Big)\,dZ \\
&= C \int_{R_\kappa(x',\varphi(x'))} |\nabla v(Z)|^2 \operatorname{dist}(Z,\partial\Omega)^{2-n}\,dZ.
\end{aligned} \tag{3.292}$$

This concludes the justification of (3.290).

Moving on, recall that, generally speaking,

$$(\mathscr{A}w)(X) := \left(\int_{R_\kappa(X)} |\nabla w(Y)|^2 \operatorname{dist}(Y,\partial\Omega)^{2-n}\, dY\right)^{\frac{1}{2}}, \qquad X \in \partial\Omega, \quad (3.293)$$

is the area-function of w. Making use of (3.290) we may therefore estimate

$$\begin{aligned}
\int_{|x'|<r} &\left(\int_0^r |\nabla v(x',\varphi(x')+s)|^2 s\, ds\right)^{p/2} dx' \\
&\le C \int_{|x'|<r} \left(\int_{R_\kappa(x',\varphi(x'))} |\nabla v(Y)|^2 \operatorname{dist}(Y,\partial\Omega)^{2-n}\, dY\right)^{p/2} dx' \\
&\le C \|\mathscr{A}(\nabla^{m-1}u)\|^p_{L^p(B_{cr}(0)\cap\partial\Omega)} \\
&\le C \|\mathscr{N}(\nabla^{m-1}u)\|^2_{L^p(\partial\Omega)}, \qquad (3.294)
\end{aligned}$$

where the last inequality in (3.294) is due to [31]. Since the last quantity in (3.294) is finite, Lemma 2.34 applies (in the manner indicated earlier) and yields that u belongs to the Besov space $B^{p,2}_{m-1+k+1/p}$ in a one-sided neighborhood of $\partial\Omega$. Given that u is a null-solution of L, this suffices to ultimately conclude that $u \in B^{p,2}_{m-1+k+1/p}(\Omega)$ (with quantitative control), as wanted. □

The result below can be viewed as the limiting case $s \in \{0,1\}$ of (3.105) in Theorem 3.9.

Theorem 3.33. *Let Ω be a bounded Lipschitz domain in $\mathbb{R}^n$ and assume that L is a homogeneous, constant (real) coefficient, symmetric, strongly elliptic differential operator of order $2m$, where $m \in \mathbb{N}$. Then the higher-order trace operator* tr_{m-1} *introduced in* (3.100) *extends to a bounded mapping in each of the following cases:*

$$\mathrm{tr}_{m-1} : F^{p,q}_{m-1+1/p}(\Omega) \cap \operatorname{Ker} L \longrightarrow \dot{L}^p_{m-1,0}(\partial\Omega), \quad \textit{if } 1 < p \le 2,\ 0 < q < \infty, \qquad (3.295)$$

and

$$\mathrm{tr}_{m-1} : F^{p,q}_{m+1/p}(\Omega) \cap \operatorname{Ker} L \longrightarrow \dot{h}^p_{m-1,1}(\partial\Omega), \quad \textit{if } \tfrac{n-1}{n} < p \le 2,\ 0 < q < \infty. \qquad (3.296)$$

As a consequence,

$$\begin{aligned}
&\mathrm{tr}_{m-1} : F^{p,q}_{m+1/p}(\Omega) \cap \operatorname{Ker} L \longrightarrow \dot{L}^p_{m-1,1}(\partial\Omega) \\
&\textit{boundedly, if } 1 < p \le 2,\ 0 < q < \infty.
\end{aligned} \qquad (3.297)$$

Proof. The claim about (3.295) is a consequence of (3.282) with $k = 0$ and Proposition 3.29. On the other hand, the claim about (3.296) follows from (3.282) with $k = 1$ along with Proposition 3.30. Finally, (3.297) is a particular case of (3.296). □

A related result is discussed below.

Theorem 3.34. *Let $\Omega \subset \mathbb{R}^n$ be a bounded Lipschitz domain satisfying a uniform exterior ball condition. Then, for each $m \in \mathbb{N}$, the higher-order trace operator*

$$\mathrm{tr}_{m-1} : F^{p,2}_{m+1/p}(\Omega) \longrightarrow \dot{L}^p_{m-1,1}(\partial\Omega) \tag{3.298}$$

is well-defined and bounded whenever $1 < p \leq 2$.

Proof. Since $F^{p,2}_{m+1/p}(\Omega) \hookrightarrow F^{p,2}_{m-1+s+1/p}(\Omega)$ for every $s \in (0,1)$ and $p \in (1,\infty)$, Theorem 3.9 implies that $\mathrm{tr}_{m-1}u$ exists in $\dot{B}^{p,p}_{m-1,s}(\partial\Omega)$, $0 < s < 1$, whenever the function $u \in F^{p,2}_{m+1/p}(\Omega)$. In particular $\mathrm{tr}_{m-1}u \in CC$ for every $u \in F^{p,2}_{m+1/p}(\Omega)$. Assuming that the bounded Lipschitz domain Ω satisfies a uniform exterior ball condition and that $1 < p \leq 2$, there remains to show that if $u \in F^{p,2}_{m+1/p}(\Omega)$ then $\mathrm{Tr}\,(\partial^\alpha u) \in L^p_1(\partial\Omega)$ for every multi-index $\alpha \in \mathbb{N}^n_0$ of length $\leq m-1$.

Of course, as far as the latter claim is concerned, it suffices to treat the case when $m = 1$, that is, show that the ordinary trace operator

$$\mathrm{Tr} : F^{p,2}_{1+1/p}(\Omega) \longrightarrow L^p_1(\partial\Omega) \tag{3.299}$$

is well-defined and bounded in the current geometrical setting, whenever $1 < p \leq 2$. Now, on the one hand, it has been established in Corollary 2 on p. 232 of [49] that, if Ω is as in the statement of the theorem and $1 < p < \infty$, then

$$\mathrm{Tr}\Big[F^{p,2}_{1+1/p}(\Omega)\Big] = \mathrm{Tr}\Big[F^{p,2}_{1+1/p}(\Omega) \cap \mathrm{Ker}\,\Delta\Big]. \tag{3.300}$$

On the other hand, by specializing (3.297) further to the case when $L = \Delta$ (hence, $m = 1$) we obtain

$$\mathrm{Tr}\Big[F^{p,2}_{1+1/p}(\Omega) \cap \mathrm{Ker}\,\Delta\Big] = L^p_1(\partial\Omega) \quad \text{if } 1 < p \leq 2. \tag{3.301}$$

Altogether, (3.300)–(3.301) yield the desired conclusion about (3.299). □

Theorem 3.34 should be compared with Proposition 3.2 on p. 176 in [58], according to which the following is true. If $1 < p \leq 2$ then there exists a bounded C^1 domain Ω in the plane and $u \in F^{p,2}_{1+1/p}(\Omega)$ such that $\mathrm{Tr}\,u \notin L^p_1(\partial\Omega)$. On the other hand, any bounded $C^{1,1}$ domain is Lipschitz and satisfies a uniform exterior ball condition. Hence, the conclusion in Theorem 3.34 holds for such domains.

Other classes of domains for which the conclusion in Theorem 3.34 is valid is the family of *bounded, convex domains*, and the family of *polygons in the plane*.

The issue of characterizing the space $\mathrm{Tr}\Big[F^{p,2}_{1+1/p}(\Omega)\Big]$, when $1 < p < \infty$ and Ω is an *arbitrary bounded Lipschitz domain* and, more generally, the trace of $F^{p,2}_{m+1/p}(\Omega)$ for $m \in \mathbb{N}$, is listed as Problem 3.2.20 on p. 121 of [68]. Related results are in [62]. Here we are able to prove the following.

Proposition 3.35. *Let Ω be a bounded Lipschitz domain in $\mathbb{R}^n$ and fix $m \in \mathbb{N}$. Then the extension operator (from* (3.113)*) induces a bounded mapping*

$$\begin{aligned} &\mathscr{E} : \dot{L}^p_{m-1,s}(\partial\Omega) \longrightarrow F^{p,q}_{m-1+s+1/p}(\Omega) \\ &\text{if } 2 \leq p < \infty,\ 0 < q \leq \infty,\ s \in \{0,1\}. \end{aligned} \tag{3.302}$$

As a corollary,

$$\dot{L}^p_{m-1,1}(\partial\Omega) \subseteq \mathrm{tr}_{m-1}\Big[F^{p,q}_{m+1/p}(\Omega)\Big] \ \ \text{if } 2 \leq p < \infty \ \text{and}\ 0 < q \leq \infty. \tag{3.303}$$

In particular,

$$L^p_1(\partial\Omega) \subseteq \mathrm{Tr}\Big[F^{p,2}_{1+1/p}(\Omega)\Big] \ \ \text{if } 2 \leq p < \infty. \tag{3.304}$$

Proof. That (3.302) is bounded, follows from (3.113) and Theorem 4.24 (employed both for $\Omega_+ := \Omega$ and $\Omega_- := \mathbb{R}^n \setminus \overline{\Omega}$). □

The following proposition augments results established earlier, in Theorem 3.33 and Theorem 3.34.

Proposition 3.36. *Let $\Omega \subset \mathbb{R}^n$ be a bounded Lipschitz domain and fix $m \in \mathbb{N}$. Then the higher-order trace operator from* (3.100) *induces a bounded, linear mapping*

$$\mathrm{tr}_{m-1} : B^{p,1}_{m-1+s+1/p}(\Omega) \longrightarrow \dot{F}^{p,q}_{m-1,s}(\partial\Omega) \tag{3.305}$$

whenever $1 < p < \infty$, $0 \leq s \leq 1$ and $2 \leq q \leq \infty$.

In particular,

$$\mathrm{tr}_{m-1} : B^{p,1}_{m-1+1/p}(\Omega) \longrightarrow \dot{L}^p_{m-1,0}(\partial\Omega), \tag{3.306}$$

$$\mathrm{tr}_{m-1} : B^{p,1}_{m+1/p}(\Omega) \longrightarrow \dot{L}^p_{m-1,1}(\partial\Omega), \tag{3.307}$$

are well-defined, linear and bounded mappings for each $p \in (1,\infty)$.

Proof. The claim about (3.305) is a direct consequence of (3.100), Proposition 2.61, and (3.266). In turn, the fact that the higher-order trace operator is well-defined, linear and bounded in the context of (3.306) and (3.307) follows by specializing (3.305) to the case when $s \in \{0,1\}$ and $q = 2$ (keeping in mind (3.267)).

For further reference, we conclude this section with a brief discussion of the operation of multiplication between scalar functions and Whitney arrays belonging to various smoothness spaces, in the sense of Definition 3.1. Our goal is to augment Proposition 3.2 with the following.

Proposition 3.37. *Assume that $\Omega \subset \mathbb{R}^n$ is a bounded Lipschitz domain. Then, for each $m \in \mathbb{N}$, the spaces*

$$\begin{aligned} &\dot{h}^p_{m-1,1}(\partial\Omega), && \tfrac{n-1}{n} < p < \infty, \\ &\dot{B}^{p,q}_{m-1,s}(\partial\Omega), && \tfrac{n-1}{n} < p \leq \infty,\ 0 < q \leq \infty,\ (n-1)\big(\tfrac{1}{p} - 1\big)_+ < s < 1, \\ &\dot{F}^{p,q}_{m-1,s}(\partial\Omega), && \tfrac{n-1}{n} < p \leq \infty,\ 0 < q \leq \infty,\ (n-1)\big(\tfrac{1}{\min\{p,q\}} - 1\big)_+ < s < 1, \end{aligned} \tag{3.308}$$

are all modules over $C^\infty_c(\mathbb{R}^n)$, in the sense described in Definition 3.1.

Furthermore, if $\dot{f} := \mathrm{tr}_{m-1}(u)$ where $u \in A^{p,q}_s(\Omega)$, with $A \in \{B, F\}$ and the indices p, q, s as in Theorem 2.53, then

$$\eta\dot{f} = \mathrm{tr}_{m-1}(\eta u). \tag{3.309}$$

If, in addition u has the property that $u\Big|^{m-1}_{\partial\Omega}$ exists, then also

$$\eta\Big(u\Big|^{m-1}_{\partial\Omega}\Big) = (\eta u)\Big|^{m-1}_{\partial\Omega}. \tag{3.310}$$

Proof. The claim made about the spaces (3.308) is a consequence of the compatibility condition (3.11), definition (3.9), and the fact that the scalar spaces $h^p_1(\partial\Omega)$ (cf. (2.423)), $B^{p,q}_s(\partial\Omega)$ and $F^{p,q}_s(\partial\Omega)$ are, for the indicated ranges of indices, modules over $C^\infty_c(\mathbb{R}^n)|_{\partial\Omega}$.

Next, (3.309) is going to be a consequence of the fact that multiplication of arrays by a cutoff function (3.9) commutes with the multi-trace (3.100). More specifically, we claim that if $u \in A^{p,q}_s(\Omega)$ with $A \in \{B, F\}$ and p, q, s as in Theorem 2.53, then

$$\eta\,\mathrm{tr}_{m-1}(u) = \mathrm{tr}_{m-1}(\eta u). \tag{3.311}$$

Indeed, for each $\alpha \in \mathbb{N}^n_0$ with $|\alpha| \leq m-1$, we have

$$\begin{aligned} \big(\mathrm{tr}_{m-1}(\eta u)\big)_\alpha &= \mathrm{Tr}\left[\partial^\alpha(\eta u)\right] \\ &= \sum_{\beta+\gamma=\alpha} \frac{\alpha!}{\beta!\gamma!}\mathrm{Tr}\left[(\partial^\beta\eta)\right]\mathrm{Tr}\left[(\partial^\gamma u)\right] \\ &= (\eta\,\mathrm{tr}_{m-1}u)_\alpha, \end{aligned} \tag{3.312}$$

as wanted. Finally, (3.310) is proved in a similar fashion. □

3.5 Interpolation

The first order of business is to prove a higher-order version of Proposition 2.57. More specifically, for the Whitney–Besov scale we have the following.

Theorem 3.38. *Let Ω be a bounded Lipschitz domain in $\mathbb{R}^n$ and fix some $m \in \mathbb{N}$. Also, assume that $s_0, s_1 \in \mathbb{R}$ and $0 < p \leq \infty$ satisfy*

$$(n-1)\Big(\tfrac{1}{p}-1\Big)_+ < s_0 \neq s_1 < 1, \tag{3.313}$$

fix $0 < \theta < 1$ and set $s := (1-\theta)s_0 + \theta s_1$. Then for every $q, q_0, q_1 \in (0, \infty]$ there holds

$$\Big(\dot{B}^{p,q_0}_{m-1,s_0}(\partial\Omega)\,,\, \dot{B}^{p,q_1}_{m-1,s_1}(\partial\Omega)\Big)_{\theta,q} = \dot{B}^{p,q}_{m-1,s}(\partial\Omega). \tag{3.314}$$

Furthermore, if $0 < p_i, q_i \leq \infty$ with $\min\{q_0, q_1\} < \infty$ and $s_0, s_1 \in \mathbb{R}$ satisfy

$$(n-1)\Big(\tfrac{1}{p_i}-1\Big)_+ < s_i < 1, \quad i = 0, 1, \tag{3.315}$$

then for each $\theta \in (0,1)$ one has

$$\Big[\dot{B}^{p_0,q_0}_{m-1,s_0}(\partial\Omega)\,,\, \dot{B}^{p_1,q_1}_{m-1,s_1}(\partial\Omega)\Big]_{\theta} = \dot{B}^{p,q}_{m-1,s}(\partial\Omega), \tag{3.316}$$

where $s := (1-\theta)s_0 + \theta s_1$, $\frac{1}{p} := \frac{1-\theta}{p_0} + \frac{\theta}{p_1}$ and $\frac{1}{q} := \frac{1-\theta}{q_0} + \frac{\theta}{q_1}$.

Proof. The starting point is the observation that if $0 < p, p_0, p_1, q, q_0, q_1 \leq \infty$, and $\theta \in (0,1)$, $s_0, s_1 \in \mathbb{R}$ are such that $s_0 \neq s_1$, and we set $s := (1-\theta)s_0 + \theta s_1$, then (cf. Theorem 2.32)

$$(B^{p,q_0}_{s_0}(\Omega), B^{p,q_1}_{s_1}(\Omega))_{\theta,q} = B^{p,q}_{s}(\Omega). \tag{3.317}$$

On the other hand, if $s_0, s_1 \in \mathbb{R}$, $\theta \in (0,1)$, and $0 < p_0, p_1, q_0, q_1 \leq \infty$ are such that $\min\{q_0, q_1\} < \infty$, then

$$[B^{p_0,q_0}_{s_0}(\Omega)\,,\, B^{p_1,q_1}_{s_1}(\Omega)]_{\theta} = B^{p^*,q^*}_{s}(\Omega), \tag{3.318}$$

where $s := (1-\theta)s_0 + \theta s_1$ and indices p^*, q^* satisfy $1/p^* = (1-\theta)/p_0 + \theta/p_1$ and $1/q^* = (1-\theta)/q_0 + \theta/q_1$. Again, see Theorem 2.32.

Then the identities (3.314)–(3.316) are direct consequences of (3.317), Theorem 3.9 and general properties of retracts. See [64] for a related discussion which, in particular, implies that $B^{p,q}_{\alpha}(\Omega)$ for $0 < p, q \leq \infty$ and $\alpha \in \mathbb{R}$, as well as $\dot{B}^{p,q}_{m-1,s}(\partial\Omega)$ for $0 < p, q \leq \infty$ and $(n-1)\Big(\frac{1}{p}-1\Big)_+ < s < 1$, are analytically convex. In particular, the complex method discussed in [64] applies. □

The companion result formulated for Whitney arrays associated with boundary Triebel–Lizorkin spaces is formulated next.

Theorem 3.39. *Let Ω be a bounded Lipschitz domain in $\mathbb{R}^n$ and fix some number $m \in \mathbb{N}$. Assume that $1 < p < \infty$, $0 < q \leq \infty$, $\theta \in (0,1)$, $s_0, s_1 \in \mathbb{R}$, $s_0 \neq s_1$ and set $s := (1-\theta)s_0 + \theta s_1$. Then*

$$\Big(\dot{F}^{p,q_0}_{m-1,s_0}(\partial\Omega)\,,\,\dot{F}^{p,q_1}_{m-1,s_1}(\partial\Omega)\Big)_{\theta,q} = \dot{B}^{p,q}_{m-1,s}(\partial\Omega) \tag{3.319}$$

if q_0, q_1, s_0, s_1 satisfy $2 \leq q_0, q_1 \leq \infty$ and $0 \leq s_0, s_1 \leq 1$, or if they verify the inequalities $\min\{p,2\} \leq q_0, q_1 \leq \infty$ *and* $0 < s_0, s_1 < 1$.

Proof. Let us first verify (3.319), by working under the assumption that q_0, q_1 satisfy $2 \leq q_0, q_1 \leq \infty$ and $0 \leq s_0, s_1 \leq 1$. Thanks to (3.314) and the reiteration theorem for the real method of interpolation (cf., e.g., the discussion in [7]), it suffices to prove (3.319) in the case when $1 < p = q < \infty$, i.e.,

$$\Big(\dot{F}^{p,q_0}_{m-1,s_0}(\partial\Omega)\,,\,\dot{F}^{p,q_1}_{m-1,s_1}(\partial\Omega)\Big)_{\theta,p} = \dot{B}^{p,p}_{m-1,s}(\partial\Omega), \quad s := (1-\theta)s_0 + \theta s_1. \tag{3.320}$$

Fix $m \in \mathbb{N}$, $1 < p < \infty$, $0 \leq s_0 < s_1 \leq 1$, $\theta \in (0,1)$, and set $s := (1-\theta)s_0 + \theta s_1$. Our first goal is to prove the estimate

$$\|\dot{f}\|_{\big(\dot{F}^{p,q_0}_{m-1,s_0}(\partial\Omega)\,,\,\dot{F}^{p,q_1}_{m-1,s_1}(\partial\Omega)\big)_{\theta,p}} \leq C\|\dot{f}\|_{\dot{B}^{p,p}_{m-1,s}(\partial\Omega)}, \tag{3.321}$$

for some finite constant $C = C(\Omega, n, p, q_0, q_1, s_0, s_1, \theta) > 0$. To set the stage, using the properties of the J-functional we may write

$$\|\dot{f}\|_{\big(\dot{F}^{p,q_0}_{m-1,s_0}(\partial\Omega)\,,\,\dot{F}^{p,q_1}_{m-1,s_1}(\partial\Omega)\big)_{\theta,p}} \tag{3.322}$$

$$= \inf\Big\{\Big(\int_0^\infty [t^{-\theta}J(t,\dot{u}(t))]^p\,\tfrac{dt}{t}\Big)^{1/p} : \dot{u}(t) \in CC \text{ and } \dot{f} = \int_0^\infty \dot{u}(t)\,\tfrac{dt}{t}\Big\},$$

where

$$J(t,\dot{u}(t)) := \max\Big\{\|\dot{u}(t)\|_{\dot{F}^{p,q_0}_{m-1,s_0}(\partial\Omega)}\,,\,t\,\|\dot{u}(t)\|_{\dot{F}^{p,q_1}_{m-1,s_1}(\partial\Omega)}\Big\}. \tag{3.323}$$

Thus, given a fixed, arbitrary $\dot{f} = \{f_\alpha\}_{|\alpha|\leq m-1} \in \dot{B}^{p,p}_{m-1,s}(\partial\Omega)$, it suffices to construct a family of functions $\dot{u}(t;X) = \{u_\alpha(t;X)\}_{|\alpha|\leq m-1}$, for $t > 0$ and $X \in \partial\Omega$, with the following properties (with $\nu_1, \ldots, \nu_n$ denoting the components of the outward unit normal to $\partial\Omega$):

$$(\partial_{\tau_{jk}} u_\alpha)(t;X) = \nu_j(X)u_{\alpha+e_k}(t;X) - \nu_k(X)u_{\alpha+e_j}(t;X) \tag{3.324}$$

$$\text{whenever } 1 \leq j,k \leq n,\ \ |\alpha| \leq m-2, \text{ and } X \in \partial\Omega,\ \ t > 0,$$

$$f_\alpha(X) = \int_0^\infty u_\alpha(t;X)\,\tfrac{dt}{t}, \qquad X \in \partial\Omega, \quad |\alpha| \leq m-1, \tag{3.325}$$

$$\Big(\int_0^\infty t^{-p\theta}\sum_{|\alpha|\le m-1}\|u_\alpha(t;\cdot)\|^p_{F^{p,q_0}_{s_0}(\partial\Omega)}\frac{dt}{t}\Big)^{1/p}\le C\|\dot f\|_{\dot B^{p,p}_{m-1,s}(\partial\Omega)},\tag{3.326}$$

$$\Big(\int_0^\infty t^{p(1-\theta)}\sum_{|\alpha|\le m-1}\|u_\alpha(t;\cdot)\|^p_{F^{p,q_1}_{s_1}(\partial\Omega)}\frac{dt}{t}\Big)^{1/p}\le C\|\dot f\|_{\dot B^{p,p}_{m-1,s}(\partial\Omega)}.\tag{3.327}$$

To this end, recall the extension operators $\mathscr{E}$, E_Ω from (2.197) and (3.103) and define the function $F := E_\Omega(\mathscr{E}\dot f)$ (note that matters can be arranged so that F is compactly supported). It follows that

$$\begin{aligned}&F\in B^{p,p}_{m-1+s+1/p}(\mathbb{R}^n),\quad \|F\|_{B^{p,p}_{m-1+s+1/p}(\mathbb{R}^n)}\approx\|\dot f\|_{\dot B^{p,p}_{m-1,s}(\partial\Omega)},\\ &\text{and }\ \mathrm{Tr}\,[\partial^\alpha F]=f_\alpha \text{ whenever } \alpha\in\mathbb{N}_0^n \text{ has } |\alpha|\le m-1.\end{aligned}\tag{3.328}$$

Let now φ be a Schwartz function in $\mathbb{R}^n$ such that

$$\begin{aligned}&\operatorname{supp}\varphi=\{\xi\in\mathbb{R}^n: 2^{-1}\le|\xi|\le 2\},\quad \varphi(\xi)>0 \text{ if } 2^{-1}<|\xi|<2,\\ &\text{and }\ \textstyle\sum_{k=-\infty}^{\infty}\varphi(2^{-k}\xi)=1 \text{ for each } \xi\in\mathbb{R}^n\setminus\{0\}.\end{aligned}\tag{3.329}$$

The existence of such a function is well-known. See, e.g., Lemma 6.1.7 on p. 135 in [8]. Also, consider the Schwartz functions ψ and φ_k, $k\in\mathbb{N}$, such that

$$\widehat{\varphi_k}(\xi)=\varphi(2^{-k}\xi)\ \text{ for } k\in\mathbb{N}\ \text{ and }\ \widehat{\psi}(\xi)=1-\sum_{k=1}^\infty\widehat{\varphi_k}(\xi),\tag{3.330}$$

with the "hat" denoting here the Fourier transform in $\mathbb{R}^n$. Thus, since for each $k\in\mathbb{N}$ there holds $\operatorname{supp}\varphi_k=\{\xi\in\mathbb{R}^n: 2^{k-1}\le|\xi|\le 2^{k+1}\}$, it follows that $\widehat{\varphi_k}\widehat{\varphi_j}=0$ if $j,k\in\mathbb{N}$, $|j-k|\ge 2$ and, ultimately,

$$\varphi_j*\varphi_k=0\ \text{ in } \mathbb{R}^n\ \text{ if }\ j,k\in\mathbb{N},\ |j-k|\ge 2.\tag{3.331}$$

Also, from definitions,

$$\psi+\sum_{k=1}^\infty\varphi_k=\delta,\ \text{ Dirac's distribution in } \mathbb{R}^n,\tag{3.332}$$

and from Lemma 6.2.1 (and its proof) on p. 140 of [8], for every $1<p<\infty$, $r\in\mathbb{R}$, $k\in\mathbb{N}$,

$$\|\varphi_j*\varphi_k*F\|_{L^p_r(\mathbb{R}^n)}\le C2^{kr}\|\varphi_k*F\|_{L^p(\mathbb{R}^n)}\ \text{ if }\ |j-k|\le 2,\tag{3.333}$$

where the constant C is independent of k and p.

For the uniformity of notation, set $\varphi_0 := \psi$. It follows that

$$F = \sum_{k=0}^{\infty} \varphi_k * F \quad \text{in } \mathbb{R}^n, \qquad \text{and} \tag{3.334}$$

$$\|F\|_{B^{p,p}_{m-1+s+1/p}(\mathbb{R}^n)} \approx \left(\sum_{k=0}^{\infty} \Big[2^{(m-1+s+1/p)k} \|\varphi_k * F\|_{L^p(\mathbb{R}^n)} \Big]^p \right)^{1/p}. \tag{3.335}$$

Let us also note here that, as a consequence of (3.334), we also have

$$f_\alpha = \sum_{k=0}^{\infty} \mathrm{Tr}\,[\varphi_k * (\partial^\alpha F)], \text{ whenever } \alpha \in \mathbb{N}_0^n \text{ has } |\alpha| \le m-1. \tag{3.336}$$

After this preamble, for each $\alpha \in \mathbb{N}_0^n$ of length $\le m-1$ we now define

$$u_\alpha(t) := \begin{cases} 0 & \text{if } t > 1, \\ \frac{1}{(s_1-s_0)\ln 2} \mathrm{Tr}\,[\varphi_k * (\partial^\alpha F)] & \text{if } 2^{-(k+1)(s_1-s_0)} < t < 2^{-k(s_1-s_0)},\ k \in \mathbb{N}_0. \end{cases} \tag{3.337}$$

Since the conditions (3.324)–(3.325) are readily verified from (3.337), there remains to prove that the estimates (3.326)–(3.327) hold as well. To check the validity of (3.326), for each multi-index $\alpha \in \mathbb{N}_0^n$ with $|\alpha| \le m-1$ we write

$$\begin{aligned} \int_0^\infty & t^{-p\theta} \|u_\alpha(t;\cdot)\|^p_{F^{p,q_0}_{s_0}(\partial\Omega)} \frac{dt}{t} \\ & \le C \sum_{k=0}^{\infty} \int_{2^{-(k+1)(s_1-s_0)}}^{2^{-k(s_0-s_1)}} t^{-p\theta-1} \|\varphi_k * (\partial^\alpha F)\|^p_{B^{p,1}_{s_0+1/p}(\mathbb{R}^n)} \, dt \\ & \le C \sum_{k=0}^{\infty} 2^{kp\theta(s_1-s_0)} \Big(\sum_{j=0}^{\infty} 2^{j(s_0+1/p)} \|\varphi_j * \varphi_k * (\partial^\alpha F)\|_{L^p(\mathbb{R}^n)} \Big)^p \\ & \le C \sum_{k=0}^{\infty} 2^{k(ps_0+1+p\theta(s_1-s_0))} \|\varphi_k * (\partial^\alpha F)\|^p_{L^p(\mathbb{R}^n)} \qquad (3.338) \\ & \le C \sum_{k=0}^{\infty} 2^{k(ps+1)} \|\varphi_k * F\|^p_{L^p_{m-1}(\mathbb{R}^n)} \end{aligned}$$

$$
\begin{aligned}
&\le C\sum_{k=0}^{\infty} 2^{kp(s+1/p+m-1)}\|\varphi_k * F\|^p_{L^p(\mathbb{R}^n)}\\
&\le C\|F\|^p_{B^{p,p}_{m-1+s+1/p}(\mathbb{R}^n)}\\
&\le C\|\dot f\|^p_{\dot B^{p,p}_{m-1,s}(\partial\Omega)}.
\end{aligned}
$$

Above, the first estimate follows from the boundedness of (2.512), the second one uses the definition of the norm in the Besov space $B^{p,1}_{s_0+1/p}(\mathbb{R}^n)$ in $\mathbb{R}^n$, the third one is based on (3.331) and (3.333) with $r = 0$, the fourth one is a consequence of the fact that the operator $\partial^\alpha : L^p_{m-1}(\mathbb{R}^n) \to L^p(\mathbb{R}^n)$ is bounded if $|\alpha| \le m-1$, the fifth inequality is implied by (3.333) with $r = m-1$, the next-to-the-last one is just the definition of the norm in the Besov space $B^{p,p}_{m-1+s+1/p}(\mathbb{R}^n)$ and, finally, the last inequality is justified by the equivalence in (3.328).

Similarly,

$$
\begin{aligned}
&\int_0^\infty t^{p-p\theta}\|u_\alpha(t;\cdot)\|^p_{F^{p,q_1}_{s_1}(\partial\Omega)}\frac{dt}{t}\\
&\le C\sum_{k=0}^{\infty}\int_{2^{-(k+1)(s_1-s_0)}}^{2^{-k(s_1-s_0)}} t^{p-p\theta-1}\|\varphi_k * (\partial^\alpha F)\|^p_{B^{p,1}_{s_1+1/p}(\mathbb{R}^n)}\,dt\\
&\le C\sum_{k=0}^{\infty} 2^{kp(\theta-1)(s_1-s_0)}\Big(\sum_{j=0}^{\infty} 2^{j(s_1+1/p)}\|\varphi_j * \varphi_k * (\partial^\alpha F)\|_{L^p(\mathbb{R}^n)}\Big)^p\\
&\le C\sum_{k=0}^{\infty} 2^{kp(s_1+1/p+(\theta-1)(s_1-s_0))}\|\varphi_k * (\partial^\alpha F)\|^p_{L^p(\mathbb{R}^n)} \qquad (3.339)\\
&\le C\sum_{k=0}^{\infty} 2^{kp(s+1/p)}\|\varphi_k * F\|^p_{L^p_{m-1}(\mathbb{R}^n)}\\
&\le C\sum_{k=0}^{\infty} 2^{kp(s+1/p+m-1)}\|\varphi_k * F\|^p_{L^p(\mathbb{R}^n)}\\
&\le C\|F\|^p_{B^{p,p}_{m-1+s+1/p}(\mathbb{R}^n)}\\
&\le C\|\dot f\|^p_{\dot B^{p,p}_{m-1,s}(\partial\Omega)}.
\end{aligned}
$$

Thus (3.321) holds, which proves that

$$
\dot B^{p,p}_{m-1,s}(\partial\Omega) \hookrightarrow \Big(\dot F^{p,q_0}_{m-1,s_0}(\partial\Omega),\, \dot F^{p,q_1}_{m-1,s_1}(\partial\Omega)\Big)_{\theta,p}. \qquad (3.340)
$$

To prove the opposite inclusion, we start with the observation that, thanks to (2.490),

$$\Big(\dot{F}^{p,q_0}_{m-1,s_0}(\partial\Omega)\,,\,\dot{F}^{p,q_1}_{m-1,s_1}(\partial\Omega)\Big)_{\theta,p}\hookrightarrow\Big(\oplus_{i=1}^{N_{n,m}}F^{p,q_0}_{s_0}(\partial\Omega)\,,\,\oplus_{i=1}^{N_{n,m}}F^{p,q_1}_{s_1}(\partial\Omega)\Big)_{\theta,p}$$
$$\hookrightarrow\oplus_{i=1}^{N_{n,m}}B^{p,p}_{s}(\partial\Omega),\tag{3.341}$$

where $N_{n,m}$ has been defined in (3.181). Furthermore, if for each $\gamma\in\mathbb{N}_0^n$, $|\gamma|\leq m-2$, and $1\leq j,k\leq n$ we define the operator

$$T^{\gamma}_{j,k}\Big(\{f_\alpha\}_{|\alpha|\leq m-1}\Big):=\partial_{\tau_{jk}}f_\gamma-\nu_j f_{\gamma+e_k}+\nu_k f_{\gamma+e_j},\tag{3.342}$$

where $\{f_\alpha\}_{|\alpha|\leq m-1}$ is an arbitrary family in $\oplus_{i=1}^{N_{n,m}}B^{p,p}_s(\partial\Omega)$, then

$$T^{\gamma}_{j,k}:\dot{F}^{p,q_0}_{m-1,s_0}(\partial\Omega)\longrightarrow 0,\qquad T^{\gamma}_{j,k}:\dot{F}^{p,q_1}_{m-1,s_1}(\partial\Omega)\longrightarrow 0,\tag{3.343}$$

so that, by interpolation,

$$T^{\gamma}_{j,k}:\Big(\dot{F}^{p,q_0}_{m-1,s_0}(\partial\Omega)\,,\,\dot{F}^{p,q_1}_{m-1,s_1}(\partial\Omega)\Big)_{\theta,p}\longrightarrow 0.\tag{3.344}$$

We may then conclude from this, (3.341) and (3.70), that

$$\Big(\dot{F}^{p,q_0}_{m-1,s_0}(\partial\Omega)\,,\,\dot{F}^{p,q_1}_{m-1,s_1}(\partial\Omega)\Big)_{\theta,p}$$
$$\hookrightarrow\Big\{\{f_\alpha\}_{|\alpha|\leq m-1}\in\oplus_{i=1}^{N_{n,m}}B^{p,p}_s(\partial\Omega):\,\partial_{\tau_{jk}}f_\alpha-\nu_j f_{\alpha+e_k}+\nu_k f_{\alpha+e_j}=0$$
$$\forall\,\alpha\in\mathbb{N}_0^n\text{ with }|\alpha|\leq m-2\text{ and }\forall\,j,k\in\{1,\dots,n\}\Big\}$$
$$=\dot{B}^{p,p}_{m-1,s}(\partial\Omega).\tag{3.345}$$

This finishes the proof of the theorem in the case when the indices q_0,q,s_0,s_1 satisfy $2\leq q_0,q_1\leq\infty$ and $0\leq s_0,s_1\leq 1$.

The argument when $\min\{p,2\}\leq q_0,q_1\leq\infty$ and $0<s_0,s_1<1$ is very similar except that, this time, we use the fact that the operator (2.512) is also bounded when $1<p<\infty$, $0<s<1$ and $\min\{p,2\}\leq q\leq\infty$ (as one can readily see from Theorem 2.53). □

We continue by recording a couple of useful consequences of Theorem 3.39 and its proof.

Corollary 3.40. *Let Ω be a bounded Lipschitz domain in $\mathbb{R}^n$ and fix some $m\in\mathbb{N}$. Also, assume that $1<p<\infty$ and $0<q\leq\infty$. Then*

$$\Big(\dot{L}^p_{m-1,0}(\partial\Omega)\,,\,\dot{L}^p_{m-1,1}(\partial\Omega)\Big)_{\theta,q} = \dot{B}^{p,q}_{m-1,\theta}(\partial\Omega), \qquad \forall\,\theta\in(0,1). \tag{3.346}$$

Proof. This follows directly from (3.267) and Theorem 3.39. □

Corollary 3.41. *Let Ω be a bounded Lipschitz domain in $\mathbb{R}^n$ and fix some $m\in\mathbb{N}$. Also, recall the set $\mathcal{O}$ from* (2.489). *Then*

$$\Big[\dot{F}^{p_0,q_0}_{m-1,s_0}(\partial\Omega)\,,\,\dot{F}^{p_1,q_1}_{m-1,s_1}(\partial\Omega)\Big]_\theta \hookrightarrow \dot{F}^{p,q}_{m-1,s}(\partial\Omega) \tag{3.347}$$

provided

$$\begin{aligned}&(p_0,q_0,s_0),(p_1,q_1,s_1)\in\mathcal{O}, \quad \theta\in(0,1), \quad s:=(1-\theta)s_0+\theta s_1,\\ &\min\{q_0,q_1\}<\infty, \qquad \tfrac{1}{p}:=\tfrac{1-\theta}{p_0}+\tfrac{\theta}{p_1} \text{ and } \tfrac{1}{q}:=\tfrac{1-\theta}{q_0}+\tfrac{\theta}{q_1}.\end{aligned} \tag{3.348}$$

Furthermore, for any two triplets (p_0,q_0,s_0), (p_1,q_1,s_1), belonging to the set $\mathcal{O}$ and any $\theta\in(0,1)$, one has

$$\Big[\dot{F}^{p_0,q_0}_{m-1,s_0}(\partial\Omega)\,,\,\dot{F}^{p_1,q_1}_{m-1,s_1}(\partial\Omega)\Big]_\theta \hookrightarrow \dot{B}^{p,\max\{p,q\}}_{m-1,s}(\partial\Omega) \tag{3.349}$$

whenever

$$\begin{aligned}&\min\{q_0,q_1\}<\infty, \quad s=(1-\theta)s_0+\theta s_1,\\ &\tfrac{1}{p}=\tfrac{1-\theta}{p_0}+\tfrac{\theta}{p_1} \text{ and } \tfrac{1}{q}=\tfrac{1-\theta}{q_0}+\tfrac{\theta}{q_1}.\end{aligned} \tag{3.350}$$

Proof. The embedding in (3.347) is established with the help of (2.491) by reasoning in a similar manner to (3.341)–(3.345). Finally, the embedding in (3.349) follows from (3.347) and (3.268). □

The remaining portion of this section is devoted to proving the boundedness criterion formulated in Theorem 3.48 (stated a little later). This requires a number of preparations and we begin by establishing certain embedding results involving the intermediate spaces obtained by interpolating Whitney–Lebesgue with Whitney–Sobolev spaces via the complex method, on the one hand, and Whitney–Besov spaces, on the other hand.

Proposition 3.42. *Suppose that Ω is a bounded Lipschitz domain in $\mathbb{R}^n$ and fix some number $m\in\mathbb{N}$. Then*

$$\dot{B}^{p,1}_{m-1,\theta}(\partial\Omega) \hookrightarrow \Big[\dot{L}^{p_0}_{m-1,0}(\partial\Omega)\,,\,\dot{L}^{p_1}_{m-1,1}(\partial\Omega)\Big]_\theta \hookrightarrow \dot{B}^{p,p\vee 2}_{m-1,\theta}(\partial\Omega) \tag{3.351}$$

continuously, whenever

$$\theta\in(0,1), \quad 1<p_0,p_1<\infty, \quad \text{and } \tfrac{1}{p}:=\tfrac{1-\theta}{p_0}+\tfrac{\theta}{p_1}. \tag{3.352}$$

Proof. Let θ, p_0, p_1, p be as in (3.352). Then the second inclusion in (3.351) follows from (3.349)–(3.350), employed with $q_0 = q_1 = 2$, $s_0 = 0$, $s_1 = 1$, and (3.267). As such, there remains to prove the first inclusion in (3.351). To this end, observe first that from (2.226) and assumptions,

$$\Big[B^{p_0,1}_{m-1+\frac{1}{p_0}}(\Omega), B^{p_1,1}_{m+\frac{1}{p_1}}(\Omega)\Big]_\theta = B^{p,1}_{m-1+\frac{1-\theta}{p_0}+(1+\frac{1}{p_1})\theta}(\Omega), \tag{3.353}$$

with equivalent norms. Based on this, (3.306)–(3.307) and complex interpolation, it follows that

$$\mathrm{tr}_{m-1} : B^{p,1}_{m-1+\frac{1-\theta}{p_0}+(1+\frac{1}{p_1})\theta}(\Omega) \longrightarrow \Big[\dot{L}^{p_0}_{m-1,0}(\partial\Omega)\,,\,\dot{L}^{p_1}_{m-1,1}(\partial\Omega)\Big]_\theta \tag{3.354}$$

linearly and boundedly. Next, pick an arbitrary $\dot{f} \in \dot{B}^{p,1}_{m-1,\theta}(\partial\Omega)$ and note that there exists

$$u \in B^{p,1}_{m-1+\theta+\frac{1}{p}}(\Omega) = B^{p,1}_{m-1+\frac{1-\theta}{p_0}+(1+\frac{1}{p_1})\theta}(\Omega) \tag{3.355}$$

satisfying (for some finite constant $C = C(\Omega, p_0, p_1, m, n, \theta) > 0$, independent of $\dot{f}$)

$$\|u\|_{B^{p,1}_{m-1+\frac{1-\theta}{p_0}+(1+\frac{1}{p_1})\theta}(\Omega)} \leq C\|\dot{f}\|_{\dot{B}^{p,1}_{m-1,\theta}(\partial\Omega)} \quad \text{and} \quad \mathrm{tr}_{m-1}u = \dot{f}. \tag{3.356}$$

Indeed, this may be seen from Theorem 3.9 taking $u := \mathscr{E}\dot{f}$ with $\mathscr{E}$ as in (3.103) with $q = 1$ and $s = \theta$.

At this stage, from (3.353)–(3.356) we deduce that the Whitney array $\dot{f}$ satisfies $\dot{f} \in \Big[\dot{L}^{p_0}_{m-1,0}(\partial\Omega)\,,\,\dot{L}^{p_1}_{m-1,1}(\partial\Omega)\Big]_\theta$ and

$$\begin{aligned}\|\dot{f}\|_{[\dot{L}^{p_0}_{m-1,0}(\partial\Omega),\dot{L}^{p_1}_{m-1,1}(\partial\Omega)]_\theta} &= \|\mathrm{tr}_{m-1}u\|_{[\dot{L}^{p_0}_{m-1,0}(\partial\Omega),\dot{L}^{p_1}_{m-1,1}(\partial\Omega)]_\theta}\\ &\leq C\|u\|_{B^{p,1}_{m-1+\frac{1-\theta}{p_0}+(1+\frac{1}{p_1})\theta}(\Omega)}\\ &\leq C\|\dot{f}\|_{\dot{B}^{p,1}_{m-1,\theta}(\partial\Omega)}.\end{aligned} \tag{3.357}$$

This proves that the first inclusion in (3.351) is well-defined, linear and bounded, thus completing the proof of the proposition. □

The nature of the double inclusion in (3.351) suggests making the following definition.

Definition 3.43. Assume that Ω is a bounded Lipschitz domain in $\mathbb{R}^n$ and fix some number $m \in \mathbb{N}$. In this context, we say that a Banach space $\mathscr{X}$ of Whitney arrays with components from $L^1(\partial\Omega)$ is of `Besov type` (p, s), for some $p \in (1, \infty)$ and $s \in (0, 1)$ provided there exist two indices q_0, q_1 satisfying $0 < q_0 \leq q_1 < \infty$ and such that

$$\dot{B}^{p,q_0}_{m-1,s}(\partial\Omega) \hookrightarrow \mathscr{X} \hookrightarrow \dot{B}^{p,q_1}_{m-1,s}(\partial\Omega) \quad \text{continuously.} \tag{3.358}$$

We now proceed to show that the Besov type property transforms naturally under complex interpolation.

Proposition 3.44. *Assume that Ω is a bounded Lipschitz domain in $\mathbb{R}^n$ and fix some number $m \in \mathbb{N}$. In this context, suppose that two Banach spaces $\mathscr{X}_j$ of Besov type (p_j, s_j), for $p_j \in (1,\infty)$ and $s_j \in (0,1)$, $j = 0, 1$, have been given. Then for each $\theta \in (0,1)$ it follows that*

$$\big[\mathscr{X}_0, \mathscr{X}_1\big]_\theta \text{ is of Besov type } (p, s), \tag{3.359}$$

where

$$s := (1-\theta)s_0 + \theta s_1 \quad \text{and} \quad \tfrac{1}{p} := \tfrac{1-\theta}{p_0} + \tfrac{\theta}{p_1}. \tag{3.360}$$

Proof. By definition, there exist indices $0 < q_j \leq r_j < \infty$, $j = 0, 1$, such that

$$\dot{B}^{p_j,q_j}_{m-1,s_j}(\partial\Omega) \hookrightarrow \mathscr{X}_j \hookrightarrow \dot{B}^{p_j,r_j}_{m-1,s_j}(\partial\Omega) \quad \text{continuously, } j = 0, 1. \tag{3.361}$$

Consequently, for every $\theta \in (0,1)$,

$$\begin{aligned} &\Big[\dot{B}^{p_0,q_0}_{m-1,s_0}(\partial\Omega), \dot{B}^{p_1,q_1}_{m-1,s_1}(\partial\Omega)\Big]_\theta \\ &\quad \hookrightarrow \big[\mathscr{X}_0, \mathscr{X}_1\big]_\theta \hookrightarrow \Big[\dot{B}^{p_0,r_0}_{m-1,s_0}(\partial\Omega), \dot{B}^{p_1,r_1}_{m-1,s_1}(\partial\Omega)\Big]_\theta \end{aligned} \tag{3.362}$$

continuously. Assume now that indices s, p are as in (3.360) and select q_*, r_* such that $\frac{1}{q_*} := \frac{1-\theta}{q_0} + \frac{\theta}{q_1}$ and $\frac{1}{r_*} := \frac{1-\theta}{r_0} + \frac{\theta}{r_1}$. Then $0 < q_* \leq r_* < \infty$ and from (3.362), (3.316) it follows that

$$\dot{B}^{p,q_*}_{m-1,s}(\partial\Omega) \hookrightarrow \mathscr{X} \hookrightarrow \dot{B}^{p,r*}_{m-1,s}(\partial\Omega) \quad \text{continuously.} \tag{3.363}$$

In light of Definition 3.43, (3.359) is now implied by (3.363). □

Another significant property of spaces of Besov type is the fact that they interpolate like genuine Whitney–Besov spaces when the real method of interpolation is employed. This is made precise in the proposition below.

Proposition 3.45. *Assume that Ω is a bounded Lipschitz domain in $\mathbb{R}^n$ and fix two numbers, $m \in \mathbb{N}$ and $p \in (1,\infty)$. In this context, suppose that two Banach spaces $\mathscr{X}_j$ of Besov type (p, s_j), $j = 0, 1$, with $s_0, s_1 \in (0,1)$ satisfying $s_0 \neq s_1$, have been given. Then for each $\theta \in (0,1)$ and $q \in (0,\infty]$ it follows that*

$$\big(\mathscr{X}_0, \mathscr{X}_1\big)_{\theta,q} = \dot{B}^{p,q}_{m-1,s}(\partial\Omega) \tag{3.364}$$

provided

$$s := (1-\theta)s_0 + \theta s_1. \tag{3.365}$$

Proof. By definition, there exist indices $0 < q_j \le r_j < \infty$, $j = 0, 1$, such that

$$\dot{B}^{p,q_j}_{m-1,s_j}(\partial\Omega) \hookrightarrow \mathscr{X}_j \hookrightarrow \dot{B}^{p,r_j}_{m-1,s_j}(\partial\Omega) \quad \text{continuously,} \ j = 0, 1. \tag{3.366}$$

As such, for every $\theta \in (0, 1)$ and $q \in (0, \infty]$ we have

$$\Big(\dot{B}^{p,q_0}_{m-1,s_0}(\partial\Omega), \dot{B}^{p,q_1}_{m-1,s_1}(\partial\Omega)\Big)_{\theta,q} \hookrightarrow (\mathscr{X}_0, \mathscr{X}_1)_{\theta,q} \tag{3.367}$$
$$\hookrightarrow \Big(\dot{B}^{p,r_0}_{m-1,s_0}(\partial\Omega), \dot{B}^{p,r_1}_{m-1,s_1}(\partial\Omega)\Big)_{\theta,q}$$

continuously. Hence, if s is as in (3.365), if follows from (3.367) and (3.314) that (3.364) holds. □

We are now ready to state the following important interpolation result, of mixed character, pertaining to the juxtaposition of the complex and real methods for Whitney–Lebesgue and Whitney–Sobolev spaces.

Corollary 3.46. *Assume that Ω is a bounded Lipschitz domain in $\mathbb{R}^n$ and fix a number $m \in \mathbb{N}$. Then*

$$\Big([\dot{L}^{p_0}_{m-1,0}(\partial\Omega), \dot{L}^{p_1}_{m-1,1}(\partial\Omega)]_\theta, [\dot{L}^{\widetilde{p_0}}_{m-1,0}(\partial\Omega), \dot{L}^{\widetilde{p_1}}_{m-1,1}(\partial\Omega)]_{\widetilde{\theta}}\Big)_{\eta,q} = \dot{B}^{p,q}_{m-1,s}(\partial\Omega) \tag{3.368}$$

whenever

$$\theta, \widetilde{\theta} \in (0,1) \ \textit{with} \ \theta \ne \widetilde{\theta}, \quad p_0, p_1, \widetilde{p}_0, \widetilde{p}_1 \in (1, \infty),$$
$$p \in (1, \infty) \ \textit{with} \ \tfrac{1}{p} = \tfrac{1-\theta}{p_0} + \tfrac{\theta}{p_1} = \tfrac{1-\widetilde{\theta}}{\widetilde{p}_0} + \tfrac{\widetilde{\theta}}{\widetilde{p}_1}, \tag{3.369}$$
$$\eta \in (0,1), \ q \in (0, \infty], \ \textit{and} \ s = (1-\eta)\theta + \eta\widetilde{\theta}.$$

Proof. From Proposition 3.42 and Definition 3.43 it follows that

$$\Big[\dot{L}^{p_0}_{m-1,0}(\partial\Omega), \dot{L}^{p_1}_{m-1,1}(\partial\Omega)\Big]_\theta \ \text{is of Besov type} \ (p, \theta)$$
$$\text{whenever} \ \theta \in (0,1), \ p_0, p_1 \in (1,\infty) \ \text{and} \ \tfrac{1}{p} := \tfrac{1-\theta}{p_0} + \tfrac{\theta}{p_1}. \tag{3.370}$$

With this in hand, formula (3.368) follows by appealing to Proposition 3.45 (whose applicability is ensured by the conditions in (3.369)). □

In tandem, Propositions 3.44 and 3.45 also yield the following versatile criterion for boundedness of linear operators on Whitney–Besov spaces.

Lemma 3.47. *Assume that Ω is a bounded Lipschitz domain in $\mathbb{R}^n$ and fix a number $m \in \mathbb{N}$. In this context, suppose that three Banach spaces $\mathscr{X}_j$ of Besov type (p_j, s_j) with $s_j \in (0, 1)$ and $p_j \in (1, \infty)$, $j = 0, 1, 2$, are given. In addition, suppose that T is a linear operator with the property that*

$$T : \mathscr{X}_j \longrightarrow \mathscr{X}_j \text{ boundedly, for } j = 0, 1, 2. \tag{3.371}$$

Assume that the points in the plane with coordinates $(s_j, 1/p_j)$, $j = 0, 1, 2$, are not collinear and denote by Δ the open triangular region in the plane whose vertices have coordinates $(s_j, 1/p_j)$ for $j = 0, 1, 2$. Then

$$\begin{aligned} &T : \dot{B}^{p,q}_{m-1,s}(\partial\Omega) \longrightarrow \dot{B}^{p,q}_{m-1,s}(\partial\Omega) \text{ boundedly} \\ &\text{whenever } q \in (0, \infty] \text{ and } (s, 1/p) \in \Delta. \end{aligned} \tag{3.372}$$

Proof. Fix $q \in (0, \infty]$ along with $(1/p, s) \in \Delta$. For the sake of specificity, assume that the horizontal line in the plane passing through $(s, 1/p)$ intersects the side of Δ with vertices at $(s_0, 1/p_0)$ and $(s_1, 1/p_1)$ as well as the side of Δ with vertices at $(s_0, 1/p_0)$ and $(s_2, 1/p_2)$. In this case, it follows that

$$\theta_1 := \tfrac{1/p_0 - 1/p}{1/p_0 - 1/p_1} \in (0, 1) \text{ and } \theta_2 := \tfrac{1/p_0 - 1/p}{1/p_0 - 1/p_2} \in (0, 1). \tag{3.373}$$

Moreover,

$$\begin{aligned} s_{01} &:= s_0 + \Big(\tfrac{1/p_0 - 1/p}{1/p_0 - 1/p_1}\Big)(s_1 - s_0) \in (0, 1), \\ s_{02} &:= s_0 + \Big(\tfrac{1/p_0 - 1/p}{1/p_0 - 1/p_2}\Big)(s_2 - s_0) \in (0, 1), \end{aligned} \tag{3.374}$$

and

$$s_{01} \neq s_{02}. \tag{3.375}$$

Finally,

$$\text{there exists } \theta \in (0, 1) \text{ such that } s = (1 - \theta)s_{01} + \theta s_{02}. \tag{3.376}$$

Using (3.373), (3.374), (3.371) and Proposition 3.44 we may then conclude that

$$\begin{aligned} &\big[\mathscr{X}_0, \mathscr{X}_j\big]_{\theta_j} \text{ is of Besov type } (p, s_{0j}) \text{ for } j = 1, 2, \\ &T : \big[\mathscr{X}_0, \mathscr{X}_j\big]_{\theta_j} \longrightarrow \big[\mathscr{X}_0, \mathscr{X}_j\big]_{\theta_j} \text{ boundedly, for } j = 1, 2. \end{aligned} \tag{3.377}$$

Hence, with $\theta \in (0, 1)$ as in (3.376), for any $q \in (0, \infty]$ it follows that

$$T : \Big(\big[\mathscr{X}_0, \mathscr{X}_1\big]_{\theta_1}, \big[\mathscr{X}_0, \mathscr{X}_2\big]_{\theta_2}\Big)_{\theta,q} \longrightarrow \Big(\big[\mathscr{X}_0, \mathscr{X}_1\big]_{\theta_1}, \big[\mathscr{X}_0, \mathscr{X}_2\big]_{\theta_2}\Big)_{\theta,q} \text{ boundedly.} \tag{3.378}$$

In concert with Proposition 3.45 and (3.375), the choice of θ in (3.376) ensures that

$$\Big(\big[\mathscr{X}_0, \mathscr{X}_1\big]_{\theta_1}, \big[\mathscr{X}_0, \mathscr{X}_2\big]_{\theta_2}\Big)_{\theta,q} = \dot{B}^{p,q}_{m-1,s}(\partial\Omega). \tag{3.379}$$

Collectively, (3.378) and (3.379) establish the boundedness of the operator T on the Whitney–Besov space $\dot{B}^{p,q}_{m-1,s}(\partial\Omega)$, as desired. The case when the horizontal line in the plane passing through $(s, 1/p)$ intersects the boundary of Δ at a vertex is similar (and simpler), and this completes the proof of the lemma. □

We conclude the presentation in this section with the following useful boundedness result.

Theorem 3.48. *Let Ω be a bounded Lipschitz domain in $\mathbb{R}^n$ and fix $m \in \mathbb{N}$. Assume that I_0 and I_1 are two (nonempty) subsets of $(1, \infty)$ and that T is a linear mapping with the property that*

$$\begin{aligned} &T : \dot{L}^p_{m-1,0}(\partial\Omega) \longrightarrow \dot{L}^p_{m-1,0}(\partial\Omega) \ \ \textit{boundedly, for each} \ \ p \in I_0, \\ &T : \dot{L}^p_{m-1,1}(\partial\Omega) \longrightarrow \dot{L}^p_{m-1,1}(\partial\Omega) \ \ \textit{boundedly, for each} \ \ p \in I_1. \end{aligned} \tag{3.380}$$

Denote by $\mathscr{U}$ the convex hull of the region in the plane consisting of points with coordinates

$$\Big\{(0, 1/p) : p \in I_0\Big\} \cup \Big\{(1, 1/p) : p \in I_1\Big\}. \tag{3.381}$$

Then, with $\mathring{\mathscr{U}}$ denoting the interior of the region $\mathscr{U}$, there holds

$$\begin{aligned} &T : \dot{B}^{p,q}_{m-1,s}(\partial\Omega) \longrightarrow \dot{B}^{p,q}_{m-1,s}(\partial\Omega) \ \ \textit{boundedly} \\ &\textit{whenever} \ \ q \in (0, \infty] \ \ \textit{and} \ \ (s, 1/p) \in \mathring{\mathscr{U}}. \end{aligned} \tag{3.382}$$

Proof. The claim made in (3.382) follows by making use of (3.370) and repeated applications of Lemma 3.47. □

3.6 Whitney–BMO and Whitney–VMO Spaces

Here we discuss Whitney–BMO and Whitney–VMO spaces on Lipschitz surfaces. Recall that, given a bounded Lipschitz domain Ω in $\mathbb{R}^n$ along with an integer $m \in \mathbb{N}$, we denote by tr_{m-1} the multi-trace operator onto the boundary of Ω (cf. (3.100)). We also introduce

$$\mathscr{P}_{m-1} := \text{the set of polynomials in } \mathbb{R}^n \text{ of degree} \leq m-1, \tag{3.383}$$

$$\dot{\mathscr{P}}_{m-1}(E) := \Big\{\big[\mathrm{tr}_{m-1} P\big]\Big|_E : P \in \mathscr{P}_{m-1}\Big\}, \qquad E \subseteq \partial\Omega. \tag{3.384}$$

To proceed, recall (2.256). Whitney–BMO spaces are then defined as follows.

Definition 3.49. Let $\Omega \subset \mathbb{R}^n$ be a bounded Lipschitz domain with surface measure σ, and assume that $m \in \mathbb{N}$. For each $\dot{f} \in \dot{L}^2_{m-1,0}(\partial\Omega)$ and $X \in \partial\Omega$ consider the following version of the `sharp maximal function` (originally introduced by Fefferman and Stein)

$$\dot{f}^{\#}(X) := \sup_{r>0} \left\{ \inf_{\dot{P} \in \dot{\mathscr{P}}_{m-1}(\partial\Omega)} \left(⨍_{S_r(X)} |\dot{f}(Q) - \dot{P}(Q)|^2 \, d\sigma(Q) \right)^{\frac{1}{2}} \right\}, \tag{3.385}$$

where, as before, $S_r(X) := \partial\Omega \cap B(X, r)$ denotes the surface ball of radius $r > 0$ centered at $X \in \partial\Omega$. The higher-order version of the space of bounded mean oscillations is then defined as

$$\dot{\mathrm{BMO}}_{m-1}(\partial\Omega) := \{\dot{f} \in L^2_{m-1,0}(\partial\Omega) : \dot{f}^{\#} \in L^\infty(\partial\Omega)\}, \tag{3.386}$$

equipped with the norm

$$\|\dot{f}\|_{\dot{\mathrm{BMO}}_{m-1}(\partial\Omega)} = \|\dot{f}\|_{L^2_{m-1,0}(\partial\Omega)} + \|\dot{f}^{\#}\|_{L^\infty(\partial\Omega)}. \tag{3.387}$$

Recall (2.444)–(2.445).

Proposition 3.50. *Let Ω be a bounded Lipschitz domain in $\mathbb{R}^n$ and assume that $m \in \mathbb{N}$. Then for each Whitney array $\dot{f} = (f_\alpha)_{|\alpha| \le m-1} \in \dot{L}^2_{m-1,0}(\partial\Omega)$ the following equivalence holds*

$$\dot{f} \in \dot{\mathrm{BMO}}_{m-1}(\partial\Omega) \iff f_\alpha \in \mathrm{bmo}(\partial\Omega) \quad \text{if } |\alpha| = m-1. \tag{3.388}$$

Proof. Let us start with the left to right implication in (3.388). A direct consequence of the definition of the sharp maximal function in (3.385) is that for each $X \in \Omega$,

$$\dot{f}^{\#}(X) \ge \sup_{r>0} \left\{ \inf_{P \in \mathscr{P}_{m-1}} \left(⨍_{S_r(X)} |f_\alpha - \partial^\alpha P|^2 \, d\sigma \right)^{1/2} \right\} \quad \text{if } |\alpha| = m-1. \tag{3.389}$$

Since $\partial^\alpha P$ is a constant for each multi-index α of length $m-1$ and each $P \in \mathscr{P}_{m-1}$, estimate (3.389) readily implies that for each multi-index $\alpha \in \mathbb{N}_0^n$, $|\alpha| = m-1$, we have $f_\alpha \in \mathrm{bmo}(\partial\Omega)$ as desired.

We focus now on the right-to-left implication. To this end assume that the array $\dot{f} \in \dot{L}^2_{m-1,0}(\partial\Omega)$ is such that, for each $\alpha \in \mathbb{N}_0^n$, $|\alpha| = m-1$ we have $f_\alpha \in \mathrm{bmo}(\partial\Omega)$. Fix $X \in \partial\Omega$ and $r > 0$ and, for each $\alpha \in \mathbb{N}_0^n$ with $|\alpha| = m-1$, let us set

$$c_\alpha := ⨍_{S_r(X)} f_\alpha \, d\sigma \quad \text{and} \quad P_{m-1}(Y) := \sum_{|\alpha|=m-1} \frac{(m-1)! c_\alpha}{\alpha!} Y^\alpha, \quad \forall\, Y \in \mathbb{R}^n. \tag{3.390}$$

Next, for each $\alpha \in \mathbb{N}_0^n$ with $|\alpha| = m-2$ we set

$$c_\alpha := ⨍_{S_r(X)} \left\{ f_\alpha - \partial^\alpha P_{m-1} \right\} d\sigma \quad \text{and} \quad P_{m-2}(Y) := \sum_{|\alpha|=m-2} \frac{(m-2)! c_\alpha}{\alpha!} Y^\alpha, \tag{3.391}$$

for every $Y \in \mathbb{R}^n$, then repeat the process recursively. Specifically, assume that for some $0 \le k < m-1$ the constants c_α with $|\alpha| > k$ and the polynomials $P_{m-1}, \dots, P_{k+1}$ have been defined. Then, for each $\alpha \in \mathbb{N}_0^n$ with $|\alpha| = k$ we define

$$c_\alpha := \fint_{S_r(X)} \Big\{ f_\alpha(Y) - \partial^\alpha \Big(\sum_{j=1}^{m-k-1} P_{m-j}(Y) \Big) \Big\} \, d\sigma(Y) \tag{3.392}$$

$$P_{m-j}(Y) := \sum_{|\alpha|=m-j} \frac{(m-j)! c_\alpha}{\alpha!} Y^\alpha, \quad \forall\, Y \in \mathbb{R}^n, \tag{3.393}$$

$$P(Y) := \sum_{j=1}^{m} P_{m-j}(Y), \qquad \forall\, Y \in \mathbb{R}^n. \tag{3.394}$$

A useful observation is that, for a fixed $\gamma \in \mathbb{N}_0^n$, $|\gamma| \le m-1$, we have

$$\partial^\gamma P_k(Y) = \begin{cases} 0 & \text{if } \ |\gamma| > k, \\ c_\gamma & \text{if } \ |\gamma| = k, \end{cases} \qquad \text{and} \qquad \partial^\gamma P(Y) = \partial^\gamma \Big(\sum_{j=1}^{m-|\gamma|} P_{m-j}(Y) \Big), \tag{3.395}$$

for every $Y \in \mathbb{R}^n$.

For $Y \in \partial\Omega$, consider now the expression $|\dot{f}(Y) - \dot{P}(Y)|^2$, where P is as in (3.394). Then, using (3.395), we may write

$$\begin{aligned} |\dot{f}(Y) - \dot{P}(Y)|^2 &= \sum_{k=0}^{m-1} \sum_{|\alpha|=k} |f_\alpha(Y) - \partial^\alpha P(Y)|^2 \\ &= \sum_{k=0}^{m-1} \sum_{|\alpha|=k} \Big| f_\alpha(Y) - \partial^\alpha \Big(\sum_{j=1}^{m-|\alpha|} P_{m-j}(Y) \Big) \Big|^2, \end{aligned} \tag{3.396}$$

and make the important observation that, due to (3.392),

$$\int_{S_r(X)} \Big\{ f_\alpha(Y) - \partial^\alpha \Big(\sum_{j=|\alpha|}^{m-1} P_j(Y) \Big) \Big\} \, d\sigma(Y) = 0, \quad \forall\, \alpha \in \mathbb{N}_0^n \text{ with } |\alpha| \le m-1. \tag{3.397}$$

We now claim that iterated applications of Poincaré's inequality starting with

$$f_\emptyset - \Big(\sum_{j=1}^{m-1} P_j \Big) - c_\emptyset \tag{3.398}$$

(where the subindex $\emptyset$ denotes the multi-index $(0, 0, \dots, 0)$) give that there exists $C > 0$, which does not depend on $\dot{f}$, such that

$$\Big(⨍_{S_r(X)} |\dot{f}(Y) - \dot{P}(Y)|^2 \, d\sigma(Y)\Big)^{1/2} \tag{3.399}$$
$$\leq C \sum_{|\alpha|=m-1} \Big(⨍_{S_r(X)} \Big| f_\alpha(Y) - ⨍_{S_r(X)} f_\alpha(Z)\, d\sigma(Z)\Big|^2 \, d\sigma(Y)\Big)^{1/2}.$$

Indeed, let us illustrate how the Poincaré inequality works through the various levels. Fix $\alpha \in \mathbb{N}_0^n$ such that $|\alpha| \leq m-2$. Then, there exists $C > 0$ such that

$$\Big(⨍_{S_r(X)} \Big| f_\alpha(Y) - \partial^\alpha\Big(\sum_{j=|\alpha|+1}^{m-1} P_j(Y)\Big) - c_\alpha\Big|^2 \, d\sigma(Y)\Big)^{1/2} \tag{3.400}$$
$$\leq C\Big(⨍_{S_r(X)} \Big|\nabla_{tan}\Big(f_\alpha(Y) - \partial^\alpha\Big(\sum_{j=|\alpha|+1}^{m-1} P_j(Y)\Big) - c_\alpha\Big)\Big|^2 \, d\sigma(Y)\Big)^{1/2}.$$

Since $\nabla_{tan} = \{\nu_k \partial_{\tau_{kj}}\}_{1\leq j\leq n}$, employing the compatibility conditions (3.2), we have

$$\nabla_{tan}\Big(f_\alpha(Y) - \partial^\alpha\Big(\sum_{j=|\alpha|+1}^{m-1} P_j(Y)\Big) - c_\alpha\Big) \tag{3.401}$$
$$= \Big\{\nu_k(Y)\Big(\nu_k(Y) f_{\alpha+e_j}(Y) - \nu_j(Y) f_{\alpha+e_k}(Y)\Big)$$
$$- \nu_k(Y)\Big(\nu_k(Y)\partial^{\alpha+e_j}\Big(\sum_{j=|\alpha|+1}^{m-1} P_j(Y)\Big) - \nu_j(Y)\partial^{\alpha+e_k}\Big(\sum_{j=|\alpha|+1}^{m-1} P_j(Y)\Big)\Big)\Big\}_{1\leq j\leq n}.$$

Therefore, using that $\partial^{\alpha+e_\ell} P_{|\alpha|+1}(Y) = c_{\alpha+e_\ell}$ for $1 \leq \ell \leq n$ (which is seen from (3.395)), we obtain

$$\nabla_{tan}\Big(f_\alpha(Y) - \Big(\sum_{j=|\alpha|+1}^{m-1} P_j(Y)\Big) - c_\alpha\Big) \tag{3.402}$$
$$= \Big\{\nu_k(Y)^2\Big[f_{\alpha+e_j}(Y) - \partial^{\alpha+e_j}\Big(\sum_{j=|\alpha|+2}^{m-1} P_j(Y)\Big) - c_{\alpha+e_j}\Big]$$
$$- \nu_j(Y)\nu_k(Y)\Big[f_{\alpha+e_k}(Y) - \partial^{\alpha+e_k}\Big(\sum_{j=|\alpha|+2}^{m-1} P_j(Y)\Big) - c_{\alpha+e_k}\Big]\Big\}_{1\leq j\leq n}.$$

In conclusion, a combination of (3.400) and (3.402) gives

$$\Big(⨍_{S_r(X)}\Big|f_\alpha(Y)-\partial^\alpha\Big(\sum_{j=|\alpha|+1}^{m-1}P_j(Y)\Big)-c_\alpha\Big|^2\,d\sigma(Y)\Big)^{1/2} \tag{3.403}$$

$$\leq C\Big(⨍_{S_r(X)}\Big|f_{\alpha+e_j}(Y)-\partial^{\alpha+e_j}\Big(\sum_{j=|\alpha|+2}^{m-1}P_j(Y)\Big)-c_{\alpha+e_j}\Big|^2\,d\sigma(Y)\Big)^{1/2}$$

$$+C\Big(⨍_{S_r(X)}\Big|f_{\alpha+e_k}(Y)-\partial^{\alpha+e_k}\Big(\sum_{j=|\alpha|+2}^{m-1}P_j(Y)\Big)-c_{\alpha+e_k}\Big|^2\,d\sigma(Y)\Big)^{1/2}.$$

In other words, given $\alpha\in\mathbb{N}_0^n$, $|\alpha|\leq m-2$ one can control the left-hand side in (3.403) by a similar expression, however corresponding to a multi-index of length $|\alpha|+1$. Therefore, repeated applications of Poincaré's inequality will lead to

$$\Big(⨍_{S_r(X)}\Big|f_\alpha(Y)-\partial^\alpha\Big(\sum_{j=|\alpha|+1}^{m-1}P_j(Y)\Big)-c_\alpha\Big|^2\,d\sigma(Y)\Big)^{1/2} \tag{3.404}$$

$$\leq C\sum_{|\gamma|=m-1}\Big(⨍_{S_r(X)}|f_\gamma(Y)-c_\gamma|^2\,d\sigma(Y)\Big)^{1/2}.$$

Now (3.399) follows from this, (3.390) and (3.396). Finally, estimate (3.399) gives that $\dot{f}^\#(X)\leq C\sum_{|\gamma|=m-1}\|f_\gamma\|_{\mathrm{bmo}(\partial\Omega)}$, thus finishing the proof of the right-to-left implication in (3.388). This concludes the proof of (3.388). □

Next we record a useful, related version of Proposition 3.50.

Proposition 3.51. *Let Ω be a bounded Lipschitz domain in $\mathbb{R}^n$ and suppose that $m\in\mathbb{N}$. Then for every Whitney array $\dot{f}=(f_\alpha)_{|\alpha|\leq m-1}\in\dot{L}^2_{m-1,0}(\partial\Omega)$ the following equivalence holds:*

$$\dot{f}\in \dot{\mathrm{BMO}}_{m-1}(\partial\Omega)\Longleftrightarrow f_\alpha\in\mathrm{bmo}(\partial\Omega)\quad\forall\,\alpha\in\mathbb{N}_0^n,\ |\alpha|\leq m-1. \tag{3.405}$$

In fact,

$$\|\dot{f}\|_{\dot{\mathrm{BMO}}_{m-1}(\partial\Omega)}\approx\sum_{|\alpha|\leq m-1}\|f_\alpha\|_{\mathrm{bmo}(\partial\Omega)},\quad \textit{uniformly for}\ \ \dot{f}\in\dot{\mathrm{BMO}}_{m-1}(\partial\Omega). \tag{3.406}$$

Proof. Proposition 3.50 shows that the right-to-left implication holds. We are therefore left with proving the left-to-right implication in (3.405). To get started, fix a multi-index $\alpha\in\mathbb{N}_0^n$ with $|\alpha|=m-1$. Employing again Proposition 3.50, we obtain $f_\alpha \in \mathrm{bmo}(\partial\Omega)$. Also, $f_\alpha \in L^p_{loc}(\partial\Omega)$ for any $1 < p < \infty$, thanks to John–Nirenberg's inequality. Since $\partial\Omega$ is a compact set in $\mathbb{R}^n$ this further implies

$$f_\alpha \in \bigcap_{1<p<\infty} L^p(\partial\Omega). \tag{3.407}$$

Consider next a multi-index $\gamma \in \mathbb{N}_0^n$ of length $m-2$. Since $\dot{f} \in \dot{L}^2_{m-1,0}(\partial\Omega)$ we have $f_\gamma \in L^2_1(\partial\Omega) \hookrightarrow L^{p^*}(\partial\Omega)$ where $1/p^* := 1/2 - 1/(n-1)$. Using the compatibility conditions (3.2) and (3.407) we conclude that

$$\nabla_{tan} f_\gamma \in \bigcap_{1<p<\infty} L^p(\partial\Omega). \tag{3.408}$$

In particular,

$$f_\gamma \in L^{p^*}_1(\partial\Omega) \hookrightarrow L^{p^{**}}(\partial\Omega), \quad \text{where} \quad \frac{1}{p^{**}} := \frac{1}{p^*} - \frac{1}{n-1}. \tag{3.409}$$

By iterating (3.408)–(3.409) one finally arrives at

$$f_\gamma \in \bigcap_{1<p<\infty} L^p_1(\partial\Omega) \hookrightarrow \mathrm{bmo}(\partial\Omega), \qquad \forall\, \gamma \in \mathbb{N}_0^n,\ |\gamma| = m-2. \tag{3.410}$$

The embedding in (3.410) follows immediately from the general fact that, much as in the Euclidean setting, using Poincaré's inequality, one has

$$L^{n-1}_1(\partial\Omega) \hookrightarrow \mathrm{bmo}(\partial\Omega). \tag{3.411}$$

Reasoning step-by-step as before for multi-indices of length $\leq m-3$, the right-to-left implication in (3.405) follows. Finally, (3.406) is implicit in what we have done above. □

We next turn our attention to defining Whitney–VMO spaces. As a preliminary observation, we note that for each $s \in (0,1)$,

$$\dot{B}^{\infty,\infty}_{m-1,s}(\partial\Omega) \hookrightarrow \dot{L}^\infty_{m-1,0}(\partial\Omega) \hookrightarrow \mathrm{B\dot{M}O}_{m-1}(\partial\Omega), \tag{3.412}$$

linearly and continuously. Inspired by (2.448) we make the following definition.

Definition 3.52. Let Ω be a bounded Lipschitz domain in $\mathbb{R}^n$, and assume that the number $m \in \mathbb{N}$. Then $\mathrm{V\dot{M}O}_{m-1}(\partial\Omega)$ is the closed subspace of $\mathrm{B\dot{M}O}_{m-1}(\partial\Omega)$ given, for some $s \in (0,1)$, by

$$\mathrm{V\dot{M}O}_{m-1}(\partial\Omega) := \text{the closure of } \dot{B}^{\infty,\infty}_{m-1,s}(\partial\Omega) \text{ in } \mathrm{B\dot{M}O}_{m-1}(\partial\Omega). \tag{3.413}$$

The first order of business is to remark that the definition of $\dot{\mathrm{VMO}}_{m-1}(\partial\Omega)$ in (3.413) does not actually depend on $s \in (0,1)$. In turn, this is a direct consequence of the following result.

Proposition 3.53. *Assume that Ω is a bounded Lipschitz domain in $\mathbb{R}^n$, and fix some $m \in \mathbb{N}$. Then*

$$\dot{\mathrm{VMO}}_{m-1}(\partial\Omega) \text{ is the closure of } \mathrm{tr}_{m-1}\Big[C_c^\infty(\mathbb{R}^n)\Big] \text{ in } \dot{\mathrm{BMO}}_{m-1}(\partial\Omega). \tag{3.414}$$

Proof. The right-to-left inclusion in (3.414) follows immediately since for each number $s \in (0,1)$ we have $\mathrm{tr}_{m-1}\Big[C_c^\infty(\mathbb{R}^n)\Big] \hookrightarrow \dot{B}^{\infty,\infty}_{m-1,s}(\partial\Omega)$. Given that

$$\dot{L}^\infty_{m-1,0}(\partial\Omega) \hookrightarrow \dot{\mathrm{BMO}}_{m-1}(\partial\Omega), \tag{3.415}$$

the left-to-right inclusion also readily follows provided we establish that

$$\forall\, \dot{g} \in \dot{B}^{\infty,\infty}_{m-1,s}(\partial\Omega) \Longrightarrow \begin{cases} \exists\, \{G_j\}_{j\in\mathbb{N}} \subset C_c^\infty(\mathbb{R}^n) \text{ such that} \\ \lim\limits_{j\to\infty} \mathrm{tr}_{m-1} G_j = \dot{g} \text{ in } \dot{L}^\infty_{m-1,0}(\partial\Omega). \end{cases} \tag{3.416}$$

To see this, note that $\dot{g} \in \dot{B}^{\infty,\infty}_{m-1,s}(\partial\Omega) \hookrightarrow \dot{B}^{p,p}_{m-1,s}(\partial\Omega)$, for every $p \in (1,\infty)$. On the other hand, $\mathrm{tr}_{m-1}\Big[C_c^\infty(\mathbb{R}^n)\Big] \hookrightarrow \dot{B}^{p,p}_{m-1,s}(\partial\Omega)$ densely for every $1 < p < \infty$ and $s \in (0,1)$, by (3.85), whereas $\dot{B}^{p,p}_{m-1,s}(\partial\Omega) \hookrightarrow \dot{L}^\infty_{m-1,0}(\partial\Omega)$ continuously when p, s satisfy $1 < p < \infty$ and $sp > n-1$. Thus, given $s \in (0,1)$ and choosing p appropriately large, all inclusions are well-defined and, hence, (3.416) follows. This concludes the proof of the proposition. □

In closing, we wish to record the following.

Proposition 3.54. *Suppose that Ω is a bounded Lipschitz domain in $\mathbb{R}^n$, and fix $m \in \mathbb{N}$. Then*

$$\dot{g} = \{g_\alpha\}_{|\alpha|\le m-1} \in \dot{\mathrm{VMO}}_{m-1}(\partial\Omega) \Rightarrow g_\alpha \in \mathrm{vmo}\,(\partial\Omega),\ \ \forall\, \alpha : |\alpha| \le m-1. \tag{3.417}$$

Proof. This is a direct consequence of Definition 3.52 and (2.448). □

Chapter 4
The Double Multi-Layer Potential Operator

In this chapter we take on the task of introducing and studying what we call double multi-layer potential operators, associated with arbitrary elliptic, higher-order, homogeneous, constant (complex) matrix-valued coefficients. As a preamble, we first take a look at the nature of fundamental solutions associated with such operators.

4.1 Differential Operators and Fundamental Solutions

Fix an Euclidean space dimension $n \in \mathbb{N}$. For some fixed $M \in \mathbb{N}$ and $m \in \mathbb{N}$, consider a $M \times M$ system of homogeneous differential operators of order $2m$ in $\mathbb{R}^n$ with complex constant coefficients:

$$L = \sum_{|\alpha|=|\beta|=m} \partial^\alpha A_{\alpha\beta} \partial^\beta = \left(L_{jk}\right)_{1 \leq j,k \leq M} \tag{4.1}$$

where, for each multi-indices $\alpha, \beta \in \mathbb{N}_0^n$ of length m,

$$A_{\alpha\beta} := \left(a_{jk}^{\alpha\beta}\right)_{1 \leq j,k \leq M} \in \mathbb{C}^{M \times M} \quad \text{and} \quad L_{jk} := \sum_{|\alpha|=|\beta|=m} \partial^\alpha a_{jk}^{\alpha\beta} \partial^\beta, \tag{4.2}$$

for some coefficients $a_{jk}^{\alpha\beta} \in \mathbb{C}$, $j, k \in \{1, \ldots, M\}$. Throughout, we shall use the superscript t to indicate transposition. In particular,

$$L^t := \sum_{|\alpha|=|\beta|=m} \partial^\alpha A_{\beta\alpha}^t \partial^\beta, \tag{4.3}$$

I. Mitrea and M. Mitrea, *Multi-Layer Potentials and Boundary Problems*, Lecture Notes in Mathematics 2063, DOI 10.1007/978-3-642-32666-0_4,

and we shall call L `symmetric` if $L^t = L$. The complex conjugate of L is the operator $\overline{L}$ defined by the requirement that $\overline{L}u = \overline{L\overline{u}}$ for each smooth function u. Hence,

$$\overline{L} = \overline{L}^t = \sum_{|\alpha|=|\beta|=m} \partial^\alpha \overline{A_{\beta\alpha}} \partial^\beta. \tag{4.4}$$

The formal adjoint of L is then defined as

$$L^* := \overline{L}^t = \sum_{|\alpha|=|\beta|=m} \partial^\alpha \overline{A_{\beta\alpha}}^t \partial^\beta, \tag{4.5}$$

and we shall call the operator L `self-adjoint` if $L^* = L$. Given a tensor coefficient $A = \big(A_{\alpha\beta}\big)_{|\alpha|=|\beta|=m}$ with complex entries, define

$$A^t := \big(A^t_{\beta\alpha}\big)_{|\alpha|=|\beta|=m}, \quad \overline{A} := \big(\overline{A_{\alpha\beta}}\big)_{|\alpha|=|\beta|=m}, \quad A^* := \big(\overline{A_{\beta\alpha}}^t\big)_{|\alpha|=|\beta|=m}. \tag{4.6}$$

Also, call A `symmetric` if $A = A^t$, and `self-adjoint` if $A = A^*$. In particular,

$$A \text{ symmetric} \iff A_{\alpha\beta} = (A_{\beta\alpha})^t, \qquad \forall\, \alpha, \beta \in \mathbb{N}_0^n : |\alpha| = |\beta| = m, \tag{4.7}$$

$$A \text{ self-adjoint} \iff A_{\alpha\beta} = (\overline{A_{\beta\alpha}})^t, \qquad \forall\, \alpha, \beta \in \mathbb{N}_0^n : |\alpha| = |\beta| = m. \tag{4.8}$$

Let us temporarily use the notation L_A to indicate that the system L is associated with the tensor coefficient $A = \big(A_{\alpha\beta}\big)_{|\alpha|=|\beta|=m}$ as in (4.1). Then

$$(L_A)^t = L_{A^t}, \quad \overline{L_A} = L_{\overline{A}}, \quad (L_A)^* = L_{A^*}, \tag{4.9}$$

from which it is clear that

$$A \text{ symmetric} \Longrightarrow L_A \text{ symmetric}, \tag{4.10}$$

$$A \text{ self-adjoint} \Longrightarrow L_A \text{ self-adjoint}. \tag{4.11}$$

In general, the converse implications in (4.10) and (4.11) may fail, but

$$\left.\begin{array}{c} \text{if } L_A \text{ is symmetric and if} \\ \widetilde{A} := \Big(\tfrac{1}{2}\big(A_{\alpha\beta} + A^t_{\beta\alpha}\big)\Big)_{|\alpha|=|\beta|=m} \end{array}\right\} \Longrightarrow L_A = L_{\widetilde{A}} \text{ and } \widetilde{A} \text{ is symmetric.} \tag{4.12}$$

Likewise,

$$\left.\begin{array}{r}\text{if } L_A \text{ is self-adjoint and if} \\ \widetilde{A} := \left(\frac{1}{2}\big(A_{\alpha\beta} + \overline{A_{\beta\alpha}}^{\,t}\big)\right)_{|\alpha|=|\beta|=m}\end{array}\right\} \Longrightarrow L_A = L_{\widetilde{A}} \text{ and } \widetilde{A} \text{ is self-adjoint.} \tag{4.13}$$

We now proceed to define several concepts of ellipticity (listed below in increasing order of strength).

Definition 4.1. The operator L in (4.1) is said to be W-`elliptic` provided that

$$det\left[\Big(\sum_{|\alpha|=|\beta|=m} a_{jk}^{\alpha\beta}\xi^\alpha\xi^\beta\Big)_{1\leq j,k\leq M}\right] \neq 0, \qquad \forall\,\xi\in\mathbb{R}^n\setminus\{0\}. \tag{4.14}$$

Also, the differential operator L in (4.1) is said to be `elliptic` whenever the `Legendre--Hadamard condition` is satisfied. That is, there exists $C > 0$ such that

$$\mathrm{Re}\left(\sum_{|\alpha|=|\beta|=m}\sum_{j,k=1}^{M} a_{jk}^{\alpha\beta}\xi^\alpha\xi^\beta\,\overline{\eta_j}\,\eta_k\right) \geq C|\xi|^{2m}|\eta|^2, \quad \forall\,\xi\in\mathbb{R}^n,\ \forall\,\eta\in\mathbb{C}^M. \tag{4.15}$$

Finally, L is called S-`elliptic` provided there exists $C > 0$ such that

$$\mathrm{Re}\left(\sum_{|\alpha|=|\beta|=m}\sum_{j,k=1}^{M} a_{jk}^{\alpha\beta}\zeta_j^\alpha\overline{\zeta_k^\beta}\right) \geq C\sum_{|\alpha|=m}\sum_{j=1}^{M}\frac{\alpha!}{m!}|\zeta_j^\alpha|^2, \tag{4.16}$$

for all families of numbers $\zeta_j^\alpha\in\mathbb{C},\ |\alpha|=m,\ 1\leq j\leq M.$

Notice that the Legendre–Hadamard property (4.15) implies W-ellipticity, as introduced in (4.14). Also, (4.15) is equivalent to the strict positivity condition

$$\mathrm{Re}\left(\sum_{|\alpha|=|\beta|=m} A_{\alpha\beta}\xi^\alpha\xi^\beta\right) \geq C|\xi|^{2m} I_{M\times M}, \qquad \forall\,\xi\in\mathbb{R}^n, \tag{4.17}$$

in the sense of $M\times M$ matrices with complex entries (in which case we have $\mathrm{Re}\,A := (A + A^*)/2$). Let us also note that being S-elliptic is a stronger concept than mere ellipticity (in the sense of Legendre–Hadamard). Indeed, since in general the multinomial formula gives

$$\Big(\sum_{i=1}^{n} x_i\Big)^N = \sum_{|\alpha|=N}\frac{|\alpha|!}{\alpha!}X^\alpha, \qquad \forall\,X := (x_1,\ldots,x_n) \tag{4.18}$$

we conclude

$$\sum_{|\alpha|=m} \frac{m!}{\alpha!} \xi^{2\alpha} = \left(\sum_{k=1}^{n} \xi_k^2 \right)^m = |\xi|^{2m}, \qquad \forall \xi \in \mathbb{R}^n, \tag{4.19}$$

so (4.16) reduces to (4.15) when $\zeta_j^\alpha := \xi^\alpha \eta_j$ with $\xi \in \mathbb{R}^n$ and $\eta = (\eta_j)_{1 \le j \le M} \in \mathbb{C}^M$.

Finally, we remark that we will occasionally require only a much weaker condition than (4.16), namely the `semi-positivity condition` to the effect that

$$\mathrm{Re}\left(\sum_{|\alpha|=|\beta|=m} \sum_{j,k=1}^{M} a_{jk}^{\alpha\beta} \zeta_j^\alpha \overline{\zeta_k^\beta} \right) \ge 0, \quad \forall\, \zeta_j^\alpha \in \mathbb{C},\ |\alpha| = m,\ 1 \le j \le M. \tag{4.20}$$

We wish to note that given a differential operator $L = L_A$ as in (4.1), corresponding to some tensor coefficient $A = \left(A_{\alpha\beta}\right)_{|\alpha|=|\beta|=m}$, there exist infinitely many other tensor coefficients $B = \left(B_{\alpha\beta}\right)_{|\alpha|=|\beta|=m}$ such that $L = L_B$. In turn, the choice of these B's may affect some of the properties of the operator L discussed above. Concretely, while the quality of being W-elliptic (cf. (4.14)) or satisfying the Legendre–Hadamard condition (4.15) are unaffected by the choice of tensor coefficient B used in the writing of the given operator L, the property of being S-elliptic (cf. (4.16)) and the semi-positivity condition (4.20) may fail for L_B.

A few examples of homogeneous, constant coefficient, higher-order elliptic operators are as follows. First, the polyharmonic operator

$$\Delta^m = \sum_{|\gamma|=m} \frac{m!}{\gamma!} \partial^{2\gamma}, \tag{4.21}$$

has been extensively studied in the literature. In the three-dimensional setting, the fourth order operator

$$\partial_1^4 + \partial_2^4 + \partial_3^4 \tag{4.22}$$

has been considered by I. Fredholm who has computed an explicit fundamental solution for it in [48]. More generally, an explicit fundamental solution for the operator

$$\partial_1^4 + \partial_2^4 + \partial_3^4 + c\left(\partial_1^2\partial_2^2 + \partial_1^2\partial_3^2 + \partial_2^2\partial_3^2\right) \tag{4.23}$$

which is elliptic if $c \in (-\frac{1}{2}, \infty)$ has been found in [130] (where elliptic operators of the form $\sum_{j,k=1}^{3} a_{jk} \partial_j^2 \partial_k^2$ have also been considered).

Moving on, recall that the classical Malgrange–Ehrenpreis theorem asserts that any differential operator of the form $P(\partial) = \sum_{|\alpha|\leq m} a_\alpha \partial^\alpha$, with $a_\alpha \in \mathbb{C}$ not all zero, has a fundamental solution $E \in \mathscr{D}'(\mathbb{R}^n)$ in $\mathbb{R}^n$. In fact, as noted in [100], one may take

$$E(X) := \frac{1}{2\pi i\, \overline{P_m(\eta)}} \int_{z\in\mathbb{C},\,|z|=1} z^m e^{z\langle \eta, X\rangle} \mathscr{F}^{-1}_{\xi\to X}\left(\frac{\overline{P(i\xi+z\eta)}}{P(i\xi+z\eta)}\right)\frac{dz}{z}. \tag{4.24}$$

Above, $\mathscr{F}^{-1}_{\xi\to X}$ is the inverse of the Fourier transform, originally defined by the formula $\mathscr{F}_{X\to\xi}\phi(\xi) := \int_{\mathbb{R}^n} e^{-i\langle X,\xi\rangle}\phi(X)\,dX$ for $\phi \in C_c^\infty(\mathbb{R}^n)$ and $\xi \in \mathbb{R}^n$, then extended by continuity to tempered distributions. Also, $P(\zeta) = \sum_{|\alpha|\leq m} a_\alpha \zeta^\alpha$, $\zeta \in \mathbb{C}^n$, is the characteristic polynomial of $P(\partial)$, and P_m stands for the principal part of P. Finally, $\eta \in \mathbb{C}^n$ is a fixed vector with the property that $P_m(\eta) \neq 0$.

We shall need a result of a somewhat similar nature for systems of differential operators. The theorem below is essentially a collection of results proved in [60, pp. 72–76], [57, p. 169], [114], and [95, p. 104], in various degrees of generality. However, we feel that it is useful to have a unifying statement, accompanied by a fairly self-contained proof, presented here.

Theorem 4.2. *Let L be a homogeneous differential operator L in $\mathbb{R}^n$ of order $2m$, $m \in \mathbb{N}$, with complex matrix-valued constant coefficients as in* (4.1) *which satisfies* (4.14)*. Then, there exists a matrix of tempered distributions, $E = \big(E_{jk}\big)_{1\leq j,k\leq M}$, such that the following hold:*

(1) For each $1 \leq j,k \leq M$,

$$E_{jk} \in C^\infty(\mathbb{R}^n \setminus \{0\}) \quad and \quad E_{jk}(-X) = E_{jk}(X) \quad \forall\, X \in \mathbb{R}^n \setminus \{0\}. \tag{4.25}$$

(2) For each $1 \leq j,k \leq M$,

$$\sum_{r=1}^{M} L^X_{jr}\Big[E_{rk}(X-Y)\Big] = \begin{cases} 0 & if\ j \neq k, \\ \delta_Y(X) & if\ j = k, \end{cases} \tag{4.26}$$

where δ stands for the Dirac delta distribution at 0 and the superscript X denotes the fact that the operator L_{js} in (4.26) *is applied in the variable X.*

(3) If $1 \leq j,k \leq M$, then

$$E_{jk}(X) = \Phi_{jk}(X) + \Big(\log|X|\Big)P_{jk}(X), \qquad X \in \mathbb{R}^n \setminus \{0\}, \tag{4.27}$$

where $\Phi_{jk} \in C^\infty(\mathbb{R}^n \setminus \{0\})$ is a homogeneous function of degree $2m-n$, and P_{jk} is identically zero when either n is odd, or $n > 2m$, and is a homogeneous polynomial of degree $2m-n$ when $n \leq 2m$. In fact,

$$\left(P_{jk}(X)\right)_{1\leq j,k\leq M} = \frac{-1}{(2\pi\, i)^n(2m-n)!}\int\limits_{S^{n-1}} (X\cdot\xi)^{2m-n}\Big(\sum_{|\alpha|=|\beta|=m}\xi^{\alpha+\beta}A_{\alpha\beta}\Big)^{-1}\,d\sigma_\xi \tag{4.28}$$

for $X\in\mathbb{R}^n$.

(4) *For each* $\alpha\in\mathbb{N}_0^n$ *there exists* $C_\alpha>0$ *such that*

$$|\partial^\alpha E(X)|\leq\begin{cases}\dfrac{C_\alpha}{|X|^{n-2m+|\alpha|}} & \text{if either } n \text{ is odd, or } n>2m, \text{ or if } |\alpha|>2m-n,\\[2ex] \dfrac{C_\alpha(1+|\log|X||)}{|X|^{n-2m+|\alpha|}} & \text{if } 0\leq|\alpha|\leq 2m-n,\end{cases} \tag{4.29}$$

uniformly for $X\in\mathbb{R}^n\setminus\{0\}$.

(5) *One can assign to each differential operator L as in* (4.1)–(4.14) *a fundamental solution* E_L *which satisfies (1)-(4) above and, in addition,*

$$(E_L)^t = E_{L^t},\quad \overline{E_L}=E_{\overline{L}},\quad (E_L)^* = E_{L^*}. \tag{4.30}$$

(6) *The matrix-valued distribution E has entries which are tempered distributions in* $\mathbb{R}^n$. *When restricted to* $\mathbb{R}^n\setminus\{0\}$, *the matrix-valued distribution* $\widehat{E}$ *(with "hat" denoting the Fourier transform, defined as before) is a* C^∞ *function and, moreover,*

$$\widehat{E}(\xi) = (-1)^m\Big(\sum_{|\alpha|=|\beta|=m}\xi^{\alpha+\beta}A_{\alpha\beta}\Big)^{-1}\quad \text{for each}\quad \xi\in\mathbb{R}^n\setminus\{0\}. \tag{4.31}$$

(7) *For each multi-index* $\alpha\in\mathbb{N}_0^n$ *with* $|\alpha|=2m-1$, *the function* $\partial^\alpha E$ *is odd,* C^∞ *smooth, and homogeneous of degree* $n-1$ *in* $\mathbb{R}^n\setminus\{0\}$.

Proof. Let $\left(P^{jk}(\xi)\right)_{1\leq j,k\leq M}$ be the inverse of the characteristic matrix

$$L(\xi) := \sum_{|\alpha|=|\beta|=m} A_{\alpha\beta}\xi^{\alpha+\beta},\qquad \xi\in\mathbb{R}^n\setminus\{0\}. \tag{4.32}$$

Following [60, p. 76], if n is odd, define

$$E_{jk}(X) := \frac{1}{4(2\pi\, i)^{n-1}(2m-1)!}\Delta_X^{(n-1)/2}\int\limits_{S^{n-1}}\Big[(X\cdot\xi)^{2m-1}\cdot\mathrm{sign}\,(X\cdot\xi)\Big]P^{jk}(\xi)\,d\sigma_\xi, \tag{4.33}$$

for $X\in\mathbb{R}^n\setminus\{0\}$. Note that this expression is homogeneous of degree $2m-n$. On the other hand, if n is even, set

$$E_{jk}(X) := \frac{-1}{(2\pi i)^n(2m)!}\Delta_X^{n/2}\int_{S^{n-1}}\Big[(X\cdot\xi)^{2m}\cdot\log|X\cdot\xi|\Big]P^{jk}(\xi)\,d\sigma_\xi, \quad (4.34)$$

where $X \in \mathbb{R}^n \setminus \{0\}$. It is also clear from (4.33)–(4.34) that E is an even function.

Alternatively, transferring one Laplacian under the integral sign when n is odd, using the fact that $\log\left(\frac{X\cdot\xi}{i}\right) = \log|X\cdot\xi| - \frac{\pi i}{2}\,\mathrm{sign}\,(X\cdot\xi)$ and simple parity considerations, it can be checked that

$$E_{jk}(X) = \frac{-1}{(2\pi i)^n(2m+q)!}\Delta_X^{(n+q)/2}\int_{S^{n-1}}\Big[(X\cdot\xi)^{2m+q}\cdot\log\Big(\frac{X\cdot\xi}{i}\Big)\Big]P^{jk}(\xi)\,d\sigma_\xi, \quad (4.35)$$

for $X \in \mathbb{R}^n \setminus \{0\}$, where $q := 0$ if n is even and $q := 1$ if n is odd.

As a preamble to checking (4.26), consider $F(t) := t^{2m+q}\log(t/i)$, $t \in \mathbb{R}\setminus\{0\}$, and note that there exists a constant $C_{m,q}$ such that

$$\Big(\frac{d}{dt}\Big)^{2m}F(t) = \frac{(2m+q)!}{q!}t^q\log\frac{t}{i} + C_{m,q}\,t^q. \quad (4.36)$$

Consequently,

$$L_X F(X\cdot\xi) = \Big[\frac{(2m+q)!}{q!}(X\cdot\xi)^q\log\frac{X\cdot\xi}{i} + C_{m,q}\,(X\cdot\xi)^q\Big]L(\xi) \quad (4.37)$$

and, further,

$$L_X\Big(-\frac{1}{(2\pi i)^n(2m+q)!}\int_{S^{n-1}}\Big[(X\cdot\xi)^{2m+q}\cdot\log\Big(\frac{X\cdot\xi}{i}\Big)\Big][L(\xi)]^{-1}\,d\sigma_\xi\Big)$$
$$=\Big(B_{(n+q)/2}(X) + P_q(X)\Big)\cdot I_{M\times M}, \quad (4.38)$$

where

$$B_{(n+q)/2}(X) := -\frac{1}{(2\pi i)^n q!}\int_{S^{n-1}}(X\cdot\xi)^q\cdot\log\Big(\frac{X\cdot\xi}{i}\Big)\,d\sigma_\xi \quad (4.39)$$

is a fundamental solution for $\Delta^{(n+q)/2}$ (cf., e.g., Lemma 4.2 on p. 123 in [122]), $P_q(X)$ is a homogeneous polynomial of degree q, and $I_{M\times M}$ is the identity matrix. Applying $\Delta_X^{(n+q)/2}$ to both sides of (4.38) then gives

$$LE = \delta\cdot I_{M\times M} \quad (4.40)$$

i.e., (4.26) holds.

To show that each component E_{jk} of E belongs to $C^\infty(\mathbb{R}^n \setminus \{0\})$, for every number $\ell \in \{1,\dots,n\}$ and $X \in \mathscr{O}_\ell := \mathbb{R}^n \setminus \{\lambda\, e_\ell : \lambda \leq 0\}$, consider the rotation

$$R_{\ell,X}(\xi) := \xi + 2\xi_\ell \frac{X}{|X|} - \frac{\xi \cdot X + \xi_\ell |X|}{|X|(|X| + x_\ell)}(X + |X| e_\ell), \quad \xi \in \mathbb{R}^n. \tag{4.41}$$

Obviously, $R_{\ell,\lambda X} = R_{\ell,X}$ whenever $X \in \mathscr{O}_\ell$ and $\lambda > 0$, the application

$$\mathscr{O}_\ell \times \mathbb{R}^n \ni (X,\xi) \mapsto R_{\ell,X}(\xi) \in \mathbb{R}^n \tag{4.42}$$

is of class C^∞, and an elementary calculation shows that

$$X \cdot R_{\ell,X}(\xi) = |X|\xi_\ell, \qquad \forall\, X \in \mathscr{O}_\ell, \quad \forall\, \xi \in \mathbb{R}^n. \tag{4.43}$$

Using the rotation invariance of the operation of integrating over S^{n-1}, for each $X \in \mathscr{O}_\ell$ we may then write

$$\begin{aligned}
&\int_{S^{n-1}} \Big[(X\cdot\xi)^{2m+q} \cdot \log\Big(\frac{X\cdot\xi}{i}\Big)\Big] P^{jk}(\xi)\, d\sigma_\xi \\
&= \int_{S^{n-1}} \Big[(X\cdot R_{\ell,X}(\xi))^{2m+q} \cdot \log\Big(\frac{X\cdot R_{\ell,X}(\xi)}{i}\Big)\Big] P^{jk}(R_{\ell,X}(\xi))\, d\sigma_\xi \\
&= \int_{S^{n-1}} \Big[(|X|\xi_\ell)^{2m+q} \cdot \log\Big(\frac{|X|\xi_\ell}{i}\Big)\Big] P^{jk}(R_{\ell,X}(\xi))\, d\sigma_\xi \\
&= |X|^{2m+q} \int_{S^{n-1}} \xi_\ell^{2m+q} \Big\{\log|X| + \log\Big(\frac{\xi_\ell}{i}\Big)\Big\}\, P^{jk}(R_{\ell,X}(\xi))\, d\sigma_\xi.
\end{aligned} \tag{4.44}$$

From this representation and formula (4.35) it is clear that $E_{jk} \in C^\infty(\mathscr{O}_\ell)$. Since $\mathbb{R}^n \setminus \{0\} = \bigcup_{\ell=1}^n \mathscr{O}_\ell$, the first part of (4.25) follows.

Note that (4.44) can be re-written in the form

$$\int_{S^{n-1}} \Big[(X\cdot\xi)^{2m+q} \cdot \log\Big(\frac{X\cdot\xi}{i}\Big)\Big] P^{jk}(\xi)\, d\sigma_\xi = \Psi_{jk}(X) + \Big(\log|X|\Big) Q_{jk}(X) \tag{4.45}$$

where

$$Q_{jk}(X) := \int_{S^{n-1}} (X\cdot\xi)^{2m+q} P^{jk}(\xi)\, d\sigma_\xi, \qquad X \in \mathbb{R}^n, \tag{4.46}$$

is a polynomial of degree $2m+q$ which vanishes (by simple parity considerations) when n is odd, and we have set

$$\Psi_{jk}(X) := |X|^{2m+q} \int_{S^{n-1}} \xi_\ell^{2m+q} \log\Big(\frac{\xi_\ell}{i}\Big) P^{jk}(R_{\ell,X}(\xi))\, d\sigma_\xi, \quad X \in \mathscr{O}_\ell. \tag{4.47}$$

Let us observe that, thanks to (4.45)–(4.47), Ψ_{jk} is actually a smooth, unambiguously defined function in $\bigcup_{\ell=1}^{n} \mathscr{O}_\ell = \mathbb{R}^n \setminus \{0\}$. In particular, the decomposition (4.27) follows from the above discussion in concert with (4.45) and (4.35).

Concerning (4.28), we first observe that

$$P_{jk}(X) = \frac{-1}{(2\pi i)^n (2m+q)!} \Delta_X^{(n+q)/2} \int_{S^{n-1}} (X \cdot \xi)^{2m+q} P^{jk}(\xi)\, d\sigma_\xi, \tag{4.48}$$

as seen from (4.35), (4.27), (4.45)–(4.46), and the identity

$$\Delta_X^K [(X \cdot \xi)^N] = \frac{N!}{(N-2K)!} (X \cdot \xi)^{N-2K} |\xi|^{2K}, \tag{4.49}$$

valid for any $N, K \in \mathbb{N}$ with $N \geq 2K$.

The estimates in (4.29) are a direct consequence of the structure formula (4.27). Finally, (4.33)–(4.34) readily account for the transposition identity (4.30).

The proof of (6) relies on (4.27) which we shall abbreviate as $E = \Phi + \Psi$, where we have set $\Phi(X) := \big(\Phi_{jk}(X)\big)_{j,k}$ and $\Psi(X) := \log|X| \big(P_{jk}(X)\big)_{j,k}$. First, we note that $\Phi \in C^\infty(\mathbb{R}^n \setminus \{0\}) \cap L^1_{loc}(\mathbb{R}^n)$ is a homogeneous function and so, by Theorem 7.1.18 on p. 168 of [57], Φ is a (matrix-valued) tempered distribution in $\mathbb{R}^n$ whose Fourier transform $\widehat{\Phi}$ coincides with a C^∞ function on $\mathbb{R}^n \setminus \{0\}$. As for Ψ, pick a function $\theta \in C_c^\infty(B(0,2))$ such that $\theta \equiv 1$ on $B(0,1)$ and write $\Psi = \Psi_1 + \Psi_2$ with $\Psi_1 := (1-\theta)\Psi \in C^\infty(\mathbb{R}^n)$, a matrix-valued tempered distribution, and $\Psi_2 := \theta\Psi \in L^1_{comp}(\mathbb{R}^n)$. Hence, $\widehat{\Psi}_2 \in C^\infty(\mathbb{R}^n)$. Also, for every multi-index $\beta \in \mathbb{N}_0^n$, the function $X^\beta \partial^\alpha \Psi_1(X)$ belongs to $L^1(\mathbb{R}^n)$ if $\alpha \in \mathbb{N}_0^n$ is such that $|\alpha|$ is large enough. This readily implies that for any $k \in \mathbb{N}$ there exists $\alpha \in \mathbb{N}_0^n$ such that $\widehat{\partial^\alpha \Psi_1} \in C^k(\mathbb{R}^n)$. Thus, the function $\mathbb{R}^n \ni \xi \mapsto \xi^\alpha \widehat{\Psi}_1(\xi)$ belongs to $C^k(\mathbb{R}^n)$ if $|\alpha|$ is large enough, relative to k, forcing $\widehat{\Psi}_1 \in C^\infty(\mathbb{R}^n \setminus \{0\})$.

The above reasoning shows that E is a matrix-valued tempered distribution in $\mathbb{R}^n$ with $\widehat{E}$ a function of class C^∞, when restricted to $\mathbb{R}^n \setminus \{0\}$. Having established this, taking the Fourier transforms of both sides of (4.40) gives $(-1)^m L(\xi)\widehat{E}(\xi) = I_{M\times M}$ in the sense of tempered distributions in $\mathbb{R}^n$. Restricting this identity to $\mathbb{R}^n \setminus \{0\}$ gives (4.31).

Finally, the claim in part (7) is a direct consequence of (1) and (3), and this completes the proof of Theorem 4.2. □

4.2 The Definition of Double Multi-Layer and Non-tangential Maximal Estimates

Let L be a homogeneous differential operator of order $2m$, with constant (possibly matrix-valued) coefficients:

$$L := \sum_{|\alpha|=|\beta|=m} \partial^\alpha A_{\alpha\beta}\, \partial^\beta \tag{4.50}$$

and consider the quadratic form associated with (the representation of) L in (4.50)

$$\mathscr{B}(u,v) := \sum_{|\alpha|=|\beta|=m} \langle A_{\alpha\beta}\,\partial^\beta u, \partial^\alpha v\rangle. \tag{4.51}$$

Also, fix a bounded Lipschitz domain $\Omega \subset \mathbb{R}^n$ with outward unit normal ν and surface measure σ, as well as two arbitrary functions u, v. which are sufficiently well-behaved on Ω. Starting with $\int_\Omega \mathscr{B}(u,v)dX$ and successively integrating by parts (as to passing on to u all partial derivatives acting on v), we eventually arrive at

$$\sum_{|\alpha|=|\beta|=m} \int_\Omega \langle A_{\alpha\beta}\,\partial^\beta u, \partial^\alpha v\rangle\,dX = (-1)^m \int_\Omega \langle Lu, v\rangle\,dX + \text{boundary terms}. \tag{4.52}$$

It is precisely the boundary terms which will determine the actual form of the double multi-layer potential operator, which we shall define momentarily. For now, we record the following.

Proposition 4.3. *Assume that $\Omega \subset \mathbb{R}^n$ is a bounded Lipschitz domain with surface measure σ and outward unit normal $\nu = (\nu_1, \ldots, \nu_n)$, and suppose that L is as in (4.50). Then*

$$\sum_{\substack{|\alpha|=m\\|\beta|=m}} \int_\Omega \langle A_{\alpha\beta}\,\partial^\beta u, \partial^\alpha v\rangle\,dX = (-1)^m \sum_{|\alpha|=|\beta|=m} \int_\Omega \langle Lu, v\rangle\,dX \tag{4.53}$$

$$- \sum_{|\alpha|=|\beta|=m} \sum_{k=1}^m \sum_{\substack{\gamma+\delta+e_j=\alpha\\|\gamma|=k-1,\,|\delta|=m-k}} (-1)^k \frac{\alpha!(m-k)!(k-1)!}{m!\gamma!\delta!} \int_{\partial\Omega} \langle \nu_j A_{\alpha\beta}\,\partial^{\beta+\gamma} u, \partial^\delta v\rangle\,d\sigma,$$

for any $\mathbb{C}^M$-valued functions u, v which are reasonably behaved in Ω.

Proof. Let us focus on a generic term in the sum in the left hand-side of (4.53), say, corresponding to $\beta \in \mathbb{N}_0^N$ arbitrary and $\alpha = \sum_{k=1}^m e_{j_k}$ where $1 \le j_1, \ldots, j_m \le n$ (recall that e_j, $1 \le j \le n$, stands for the vector $(0,\ldots,1,\ldots,0) \in \mathbb{R}^n$ with the only non-zero component on the j-th position). Then

$$\int_\Omega \langle A_{\alpha\beta}\,\partial^\beta u, \partial^\alpha v\rangle\,dX = (-1)^m \int_\Omega \langle \partial^\alpha A_{\alpha\beta}\,\partial^\beta u, v\rangle\,dX \tag{4.54}$$

$$+ \sum_{k=1}^m (-1)^{k+1} \int_{\partial\Omega} \Big\langle \nu_{j_k} \partial^{(\sum_{r=1}^{k-1} e_{j_r})} A_{\alpha\beta}\,\partial^\beta u, \partial^{(\sum_{r=k+1}^m e_{j_r})} v\Big\rangle.$$

Based on a general identity, to the effect that for any function $G : \mathbb{N}_0^n \to \mathbb{R}$

$$\sum_{1\leq j_1,\dots,j_\ell\leq n} G\Big(\sum_{r=1}^{\ell} e_{j_r}\Big) = \sum_{\gamma\in\mathbb{N}_0^n:\,|\gamma|=\ell} \frac{\ell!}{\gamma!} G(\gamma), \tag{4.55}$$

and (4.54), we may then write

$$\sum_{\substack{|\alpha|=m\\|\beta|=m}} \int_\Omega \langle A_{\alpha\beta}\,\partial^\beta u, \partial^\alpha v\rangle\,dX$$

$$= \sum_{|\beta|=m} \int_\Omega \sum_{1\leq j_1,\dots,j_m\leq n} \frac{(\sum_{r=1}^m e_{j_r})!}{m!} \Big\langle A_{(\sum_{r=1}^m e_{j_r})\beta}\partial^\beta u, \partial^{(\sum_{r=1}^m e_{j_r})} v\Big\rangle\,dX$$

$$= (-1)^m \sum_{|\beta|=m} \sum_{1\leq j_1,\dots,j_m\leq n} \frac{(\sum_{r=1}^m e_{j_r})!}{m!} \int_\Omega \Big\langle \partial^{(\sum_{r=1}^m e_{j_r})} A_{(\sum_{r=1}^m e_{j_r})\beta}\partial^\beta u, v\Big\rangle\,dX$$

$$+ \sum_{|\beta|=m} \sum_{1\leq j_1,\dots,j_m\leq n} \frac{(\sum_{r=1}^m e_{j_r})!}{m!} \sum_{k=1}^{m} \Big\{(-1)^{k+1} \times$$

$$\times \int_{\partial\Omega} \Big\langle \nu_{j_k}\,\partial^{(\sum_{r=1}^{k-1} e_{j_r})} A_{(\sum_{r=1}^m e_{j_r})\beta}\partial^\beta u, \partial^{(\sum_{r=k+1}^m e_{j_r})} v\Big\rangle\,dX\Big\}$$

$$= (-1)^m \sum_{|\alpha|=|\beta|=m} \int_\Omega \langle \partial^\alpha A_{\alpha\beta}\,\partial^\beta u, v\rangle\,dX \tag{4.56}$$

$$+ \sum_{|\beta|=m} \sum_{k=1}^{m} \sum_{j=1}^{n} \sum_{|\gamma|=k-1} \sum_{|\delta|=m-k} \Big\{(-1)^{k+1} \frac{(\gamma+\delta+e_j)!(m-k)!(k-1)!}{m!\gamma!\delta!} \times$$

$$\times \int_{\partial\Omega} \langle \nu_j A_{(\gamma+\delta+e_j)\beta}\partial^{\beta+\gamma} u, \partial^\delta v\rangle\,d\sigma\Big\}.$$

Thus, (4.53) follows. □

The following definition is inspired by the format of the boundary integral in (4.53), in which formally, the function u is replaced by $E(X-\cdot)$, the family of derivatives $(\partial^\delta v)_{|\delta|\leq m-1}$ is replaced by the Whitney array $\dot{f} = (f_\delta)_{|\delta|\leq m-1}$ (in the process, it helps to keep in mind that $\partial^{\beta+\gamma}[E(X-\cdot)] = (-1)^{m+k-1}(\partial^{\beta+\gamma}E)(X-\cdot)$ for each $\beta\in\mathbb{N}_0$ with $|\beta|=1$ and each $\gamma\in\mathbb{N}_0^n$ with $|\gamma|=k-1$).

Definition 4.4. Let $\Omega\subset\mathbb{R}^n$ be a Lipschitz domain with outward unit normal vector $\nu=(\nu_j)_{1\leq j\leq n}$ and surface measure σ. Also, let L be a (complex) matrix-valued constant coefficient, homogeneous, W-elliptic differential operator of order $2m$ in $\mathbb{R}^n$ and denote by E a fundamental solution for L in $\mathbb{R}^n$. Then the `double multi-layer potential operator` acting on $\dot{f} = \{f_\delta\}_{|\delta|\leq m-1}$ is given by

$$\dot{\mathscr{D}}\dot{f}(X) := -\sum_{\substack{|\alpha|=m\\|\beta|=m}}\sum_{k=1}^{m}\sum_{\substack{|\delta|=m-k\\|\gamma|=k-1\\\gamma+\delta+e_j=\alpha}}\Big\{\frac{\alpha!(m-k)!(k-1)!}{m!\gamma!\delta!}\times \tag{4.57}$$
$$\times\int_{\partial\Omega}\nu_j(Y)(\partial^{\beta+\gamma}E)(X-Y)A_{\beta\alpha}f_\delta(Y)\,d\sigma(Y)\Big\}$$

for $X \in \mathbb{R}^n \setminus \partial\Omega$.

As a practical matter, $\dot{\mathscr{D}}\dot{f}(X)$ is obtained by formally replacing, in the boundary integral in (4.53), the function u by $E(X-\cdot)$, the family of derivatives $\{\partial^\delta v\}_{|\delta|\leq m-1}$ by the Whitney array $\dot{f} = \{f_\delta\}_{|\delta|\leq m-1}$, and then multiplying everything by $(-1)^{m-1}$. The reason for which this definition is natural is as follows.

Proposition 4.5. *Let $\Omega \subset \mathbb{R}^n$ be a Lipschitz domain and consider a (complex) matrix-valued constant coefficient, homogeneous, W-elliptic differential operator L of order $2m$ in $\mathbb{R}^n$. Assume that $\dot{\mathscr{D}}$ is associated with these as in* (4.57). *Then*

$$L(\dot{\mathscr{D}}\dot{f})(X) = 0 \quad \textit{for each} \quad X \in \mathbb{R}^n \setminus \partial\Omega. \tag{4.58}$$

Also, for each $F \in C_c^\infty(\mathbb{R}^n)$,

$$\sum_{|\alpha|=|\beta|=m}\int_\Omega (\partial^\beta E)(X-Y)A_{\beta\alpha}(\partial^\alpha F)(Y)\,dY$$
$$= \begin{cases} F(X) - \dot{\mathscr{D}}(\mathrm{tr}_{m-1}(F))(X) \ \textit{if}\ X \in \Omega, \\ -\dot{\mathscr{D}}(\mathrm{tr}_{m-1}(F))(X) \ \textit{if}\ X \in \mathbb{R}^n \setminus \overline{\Omega}. \end{cases} \tag{4.59}$$

Proof. The identity in (4.58) is clear from definitions. Next, writing (4.53) for L^t in place of L (recall that the superscript t indicates transposition), gives

$$\sum_{\substack{|\alpha|=m\\|\beta|=m}}\int_\Omega \langle (A_{\beta\alpha})^t\partial^\beta u, \partial^\alpha v\rangle\,dX = (-1)^m\sum_{|\alpha|=|\beta|=m}\int_\Omega \langle L^t u, v\rangle\,dX \tag{4.60}$$
$$-\sum_{|\alpha|=|\beta|=m}\sum_{k=1}^{m}\sum_{\substack{\gamma+\delta+e_j=\alpha\\|\gamma|=k-1,\,|\delta|=m-k}}(-1)^k\frac{\alpha!(m-k)!(k-1)!}{m!\gamma!\delta!}\int_{\partial\Omega}\langle \nu_j(A_{\beta\alpha})^t\partial^{\beta+\gamma}u, \partial^\delta v\rangle\,d\sigma.$$

Specializing this identity to the case when the function $v := F \in C_c^\infty(\mathbb{R}^n)$ and $u := E_{L^t}(X_o - \cdot)\eta$ where $X_o \in \Omega$ is a fixed point and $\eta \in \mathbb{C}^M$ is an arbitrary vector, yields, after some simple algebraic manipulations,

$$\Big\langle \eta, \sum_{\substack{|\alpha|=m \\ |\beta|=m}} \int_\Omega (\partial^\beta E)(X_o - \cdot) A_{\beta\alpha} (\partial^\alpha F)\, dX \Big\rangle \tag{4.61}$$

$$= \Big\langle \eta, F(X_o) \Big\rangle - \Big\langle \eta, \dot{\mathscr{D}}(\mathrm{tr}_{m-1} F)(X_o) \Big\rangle.$$

Above, property (5) in Theorem 4.2 is also used. Since $\eta \in \mathbb{C}^M$ was arbitrary, the version of (4.59) corresponding to $X_o \in \Omega$ is proved. The case when $X_o \in \mathbb{R}^n \setminus \overline{\Omega}$ is treated similarly, completing the proof of the proposition. □

Given $k \in \mathbb{N}_0$, recall that $\mathscr{P}_k$ stands for the space of polynomials of degree $\leq k$ in $\mathbb{R}^n$.

Proposition 4.6. *Suppose that $\Omega \subset \mathbb{R}^n$ is a bounded Lipschitz domain and consider a (complex) matrix-valued constant coefficient, homogeneous, W-elliptic differential operator L of order $2m$ in $\mathbb{R}^n$. Assume that $\dot{\mathscr{D}}$ is associated with these as in* (4.57). *Then for any $u \in \mathscr{P}_{m-1}$*

$$u = \dot{\mathscr{D}}(\mathrm{tr}_{m-1} u) \quad in \ \Omega. \tag{4.62}$$

Proof. Fix a function $\psi \in C_c^\infty(\mathbb{R}^n)$ with $\psi \equiv 1$ in a neighborhood of $\overline{\Omega}$. If $u \in \mathscr{P}_{m-1}$, set $F := \psi u \in C_c^\infty(\mathbb{R}^n)$. Since $F \equiv u$ near $\overline{\Omega}$, we have $\partial^\alpha F = 0$ in Ω whenever $|\alpha| = m$, by trivial degree considerations. Then (4.62) is a consequence of the first line in (4.59). □

We now proceed to establish non-tangential maximal function estimates for the double multi-layer operator $\dot{\mathscr{D}}$.

Theorem 4.7. *Given a Lipschitz domain $\Omega \subset \mathbb{R}^n$ and a (complex) matrix-valued constant coefficient, homogeneous, W-elliptic differential operator L of order $2m$ in $\mathbb{R}^n$, consider the double multi-layer potential operator defined in* (4.57). *Then for each $p \in (1, \infty)$ there exists a finite constant $C = C(\Omega, p, L) > 0$ with the property that*

$$\sum_{j=0}^{m-1} \|\mathcal{N}(\nabla^j \dot{\mathscr{D}} \dot{f})\|_{L^p(\partial\Omega)} \leq C \|\dot{f}\|_{\dot{L}^p_{m-1,0}(\partial\Omega)}, \qquad \forall\, \dot{f} \in \dot{L}^p_{m-1,0}(\partial\Omega), \tag{4.63}$$

$$\sum_{j=0}^{m} \|\mathcal{N}(\nabla^j \dot{\mathscr{D}} \dot{f})\|_{L^p(\partial\Omega)} \leq C \|\dot{f}\|_{\dot{L}^p_{m-1,1}(\partial\Omega)}, \qquad \forall\, \dot{f} \in \dot{L}^p_{m-1,1}(\partial\Omega). \tag{4.64}$$

Furthermore,

$$\partial^\gamma \dot{\mathscr{D}} \dot{f}\Big\lfloor_{\partial\Omega} \quad \textit{exists for every} \quad \dot{f} \in \dot{L}^p_{m-1,0}(\partial\Omega)$$
$$\textit{whenever} \quad \gamma \in \mathbb{N}_0^n \quad \textit{satisfies} \quad |\gamma| \leq m-1, \tag{4.65}$$

and

$$\partial^\gamma \dot{\mathscr{D}} \dot{f}\Big\lfloor_{\partial\Omega} \quad \textit{exists for every} \quad \dot{f} \in \dot{L}^p_{m-1,1}(\partial\Omega)$$
$$\textit{whenever} \quad \gamma \in \mathbb{N}_0^n \quad \textit{satisfies} \quad |\gamma| \leq m. \tag{4.66}$$

Proof. The starting point is the identity

$$\dot{\mathscr{D}}(\dot{f})(X) = \sum_{|\alpha|=|\beta|=m} \int_{\mathbb{R}^n\setminus\overline{\Omega}} (\partial^\beta E)(X-Y) A_{\beta\alpha} (\partial^\alpha F)(Y)\, dY, \quad \forall\, X \in \Omega, \tag{4.67}$$

valid whenever $F \in C_c^\infty(\mathbb{R}^n)$ and $\dot{f} := \mathrm{tr}_{m-1}(F)$. This is readily seen from (4.59) with the roles of Ω and $\mathbb{R}^n \setminus \overline{\Omega}$ reversed (here it is useful to keep in mind that the outward unit normal of the latter domain is $-\nu$).

Consider next $\gamma \in \mathbb{N}_0^n$, $|\gamma| \leq m-1$, $\gamma = \sum_{r=1}^{|\gamma|} e_{j_r}$, with $1 \leq j_1, \dots, j_n \leq n$, and using (4.55), for each $X \in \Omega$ write

$$\partial^\gamma \dot{\mathscr{D}}(\dot{f})(X) = \sum_{|\alpha|=|\beta|=m} \int_{\mathbb{R}^n\setminus\overline{\Omega}} (\partial^{\gamma+\beta} E)(X-Y) A_{\beta\alpha} (\partial^\alpha F)(Y)\, dY \tag{4.68}$$

$$= \sum_{\substack{|\beta|=m \\ 1\leq k_1,\dots,k_m\leq n}} \frac{(\sum_{s=1}^m e_{k_s})!}{m!} \int_{\mathbb{R}^n\setminus\overline{\Omega}} (\partial^{(\sum_{r=1}^{|\gamma|} e_{j_r})+\beta} E)(X-Y) A_{\beta(\sum_{s=1}^m e_{k_s})} \Big(\partial^{(\sum_{s=1}^m e_{k_s})} F\Big)(Y)\, dY.$$

Integrating in (4.68) by parts (to move $\partial^{e_{k_1}}$ from F on to E) and taking into account that the outward unit normal vector to $\mathbb{R}^n \setminus \overline{\Omega}$ is $-\nu$ further leads to

$$\partial^\gamma \dot{\mathscr{D}}(\dot{f})(X) = \sum_{\substack{|\beta|=m \\ 1\leq k_1,\dots,k_m\leq n}} \frac{(\sum_{s=1}^m e_{k_s})!}{m!} \times \tag{4.69}$$

$$\times \Bigg\{ \int_{\mathbb{R}^n\setminus\overline{\Omega}} (\partial^{(\sum_{r=1}^{|\gamma|} e_{j_r})+e_{k_1}+\beta} E)(X-Y) A_{\beta(\sum_{s=1}^m e_{k_s})} \times$$

$$\times \Big(\partial^{(\sum_{s=2}^m e_{k_s})} F\Big)(Y)\,dY - \int_{\partial\Omega} \nu_{k_1}(Y)(\partial^{(\sum_{r=1}^{|\gamma|} e_{j_r})+\beta} E)(X-Y) \times$$

$$\times A_{\beta(\sum_{s=1}^m e_{k_s})}\Big(\partial^{(\sum_{s=2}^m e_{k_s})} F\Big)(Y)\,d\sigma(Y)\Bigg\}.$$

We further integrate by parts in the first term in the curly brackets in the right hand-side of (4.69) and move $\partial^{e_{j_1}}$ from E onto F. This gives

$$\partial^\gamma \dot{\mathscr{D}}(\dot{f})(X) = \sum_{\substack{|\beta|=m \\ 1\le k_1,\dots,k_m\le n}} \frac{(\sum_{s=1}^m e_{k_s})!}{m!} \times \tag{4.70}$$

$$\times \Bigg\{ \int_{\mathbb{R}^n\setminus\overline{\Omega}} (\partial^{(\sum_{r=2}^{|\gamma|} e_{j_r})+e_{k_1}+\beta} E)(X-Y) \times$$

$$\times A_{\beta(\sum_{s=1}^m e_{k_s})}\Big(\partial^{(\sum_{s=2}^m e_{k_s})+e_{j_1}} F\Big)(Y)\,dY$$

$$- \int_{\partial\Omega} \nu_{k_1}(Y)(\partial^{(\sum_{r=1}^{|\gamma|} e_{j_r})+\beta} E)(X-Y) \times$$

$$\times A_{\beta(\sum_{s=1}^m e_{k_s})}\Big(\partial^{(\sum_{s=2}^m e_{k_s})} F\Big)(Y)\,d\sigma(Y)$$

$$+ \int_{\partial\Omega} \nu_{j_1}(Y)(\partial^{(\sum_{r=2}^{|\gamma|} e_{j_r})+e_{k_1}+\beta} E)(X-Y) \times$$

$$\times A_{\beta(\sum_{s=1}^m e_{k_s})}\Big(\partial^{(\sum_{s=2}^m e_{k_s})} F\Big)(Y)\,d\sigma(Y)\Bigg\}.$$

Next, recall that $\partial_{\tau_{k_1 j_1}} := \nu_{k_1}\partial^{e_{j_1}} - \nu_{j_1}\partial^{e_{k_1}}$ and notice that the sum of the last two terms in the curly brackets in (4.70) is

$$\int_{\partial\Omega} \partial_{\tau_{k_1 j_1}(Y)}\Big((\partial^{(\sum_{r=2}^{|\gamma|} e_{j_r})+\beta} E)(X-Y)\Big) A_{\beta(\sum_{s=1}^m e_{k_s})}\Big(\partial^{(\sum_{s=2}^m e_{k_s})} F\Big)(Y)\,d\sigma(Y). \tag{4.71}$$

Therefore,

$$\partial^\gamma \dot{\mathscr{D}}(\dot{f})(X) = \sum_{\substack{|\beta|=m \\ 1\leq k_1,\dots,k_m\leq n}} \frac{(\sum_{s=1}^m e_{k_s})!}{m!} \times \tag{4.72}$$

$$\times \Bigg\{ \int_{\mathbb{R}^n\setminus\overline{\Omega}} (\partial^{(\sum_{r=2}^{|\gamma|} e_{j_r})+e_{k_1}+\beta} E)(X-Y) \times$$

$$\times A_{\beta(\sum_{s=1}^m e_{k_s})}\Big(\partial^{(\sum_{s=2}^m e_{k_s})+e_{j_1}} F\Big)(Y)\,dY$$

$$+ \int_{\partial\Omega} \partial_{\tau_{k_1 j_1}(Y)}\Big((\partial^{(\sum_{r=2}^{|\gamma|} e_{j_r})+\beta} E)(X-Y)\Big) \times$$

$$\times A_{\beta(\sum_{s=1}^m e_{k_s})}\Big(\partial^{(\sum_{s=2}^m e_{k_s})} F\Big)(Y)\,d\sigma(Y)\Bigg\}.$$

Starting with the first integral inside the curly brackets in the right-hand side of (4.72), we now repeat the procedure described in (4.68)–(4.72) until we swap $\partial^{(\sum_{r=1}^{|\gamma|} e_{j_r})}$ from E with $\partial^{(\sum_{s=1}^{|\gamma|} e_{k_s})}$ from F. In this fashion, we obtain

$$\partial^\gamma \dot{\mathscr{D}}(\dot{f})(X) = \sum_{\substack{|\beta|=m \\ 1\leq k_1,\dots,k_m\leq n}} \frac{(\sum_{s=1}^m e_{k_s})!}{m!} \times \tag{4.73}$$

$$\times\Bigg\{ \int_{\mathbb{R}^n\setminus\overline{\Omega}} (\partial^{(\sum_{s=1}^{|\gamma|} e_{k_s})+\beta} E)(X-Y) \times$$

$$\times A_{\beta(\sum_{s=1}^m e_{k_s})}\Big(\partial^{(\sum_{r=1}^{|\gamma|} e_{j_r})+(\sum_{s=|\gamma|+1}^m e_{k_s})} F\Big)(Y)\,dY$$

$$+\sum_{\ell=1}^{|\gamma|} \int_{\partial\Omega} \partial_{\tau_{k_\ell j_\ell}(Y)}\Big((\partial^{(\sum_{s=1}^{\ell-1} e_{k_s})+(\sum_{r=\ell+1}^{|\gamma|} e_{j_r})+\beta} E)(X-Y)\Big) \times$$

$$\times A_{\beta(\sum_{s=1}^m e_{k_s})}\Big(\partial^{(\sum_{r=1}^{\ell-1} e_{j_r})+(\sum_{s=\ell+1}^m e_{k_s})} F\Big)(Y)\,d\sigma(Y)\Bigg\}.$$

At this point we use repeated integration by parts in the first integral inside the curly brackets in the right-hand side of (4.73) in order to move the remaining partial derivatives $\partial^{(\sum_{s=|\gamma|+1}^m e_{k_s})}$ from F onto E. Using that $(LE)(X-Y)=0$ for $X\neq Y$, this gives

$$\partial^\gamma \dot{\mathscr{D}}(\dot{f})(X) = \sum_{\substack{|\beta|=m \\ 1\le k_1,\dots,k_m\le n}} \frac{(\sum_{s=1}^m e_{k_s})!}{m!} \times \tag{4.74}$$

$$\times\Big\{\sum_{\ell=1}^{|\gamma|}\int_{\partial\Omega} \partial_{\tau_{k_\ell j_\ell}(Y)}\Big((\partial^{(\sum_{s=1}^{\ell-1} e_{k_s})+(\sum_{r=\ell+1}^{|\gamma|} e_{j_r})+\beta}E)(X-Y)\Big)\times$$

$$\times A_{\beta(\sum_{s=1}^m e_{k_s})}\Big(\partial^{(\sum_{r=1}^{\ell-1} e_{j_r})+(\sum_{s=\ell+1}^m e_{k_s})}F\Big)(Y)\,d\sigma(Y)$$

$$-\sum_{\ell=|\gamma|+1}^{m}\int_{\partial\Omega} \nu_{k_\ell}(Y)(\partial^{(\sum_{s=1}^{\ell-1} e_{k_s})+\beta}E)(X-Y)A_{\beta(\sum_{s=1}^m e_{k_s})}\times$$

$$\times\Big(\partial^{(\sum_{r=1}^{|\gamma|} e_{j_r})+(\sum_{s=\ell+1}^m e_{k_s})}F\Big)(Y)\,d\sigma(Y)\Big\}.$$

From the identity (4.74), Proposition 3.3 and a limiting argument, we can finally deduce that if $X\in\mathbb{R}^n\setminus\partial\Omega$ and if $\dot{f}=\{f_\delta\}_{|\delta|\le m-1}\in \dot{L}^p_{m-1,0}(\partial\Omega)$, $1<p<\infty$,

$$\partial^\gamma \dot{\mathscr{D}}(\dot{f})(X) = \sum_{\substack{|\beta|=m \\ 1\le k_1,\dots,k_m\le n}} \frac{(\sum_{s=1}^m e_{k_s})!}{m!} \times \tag{4.75}$$

$$\times\Big\{\sum_{\ell=1}^{|\gamma|}\int_{\partial\Omega} \partial_{\tau_{k_\ell j_\ell}(Y)}\Big((\partial^{(\sum_{s=1}^{\ell-1} e_{k_s})+(\sum_{r=\ell+1}^{|\gamma|} e_{j_r})+\beta}E)(X-Y)\Big)\times$$

$$\times A_{\beta(\sum_{s=1}^m e_{k_s})} f_{(\sum_{r=1}^{\ell-1} e_{j_r})+(\sum_{s=\ell+1}^m e_{k_s})}(Y)\,d\sigma(Y)$$

$$-\sum_{\ell=|\gamma|+1}^{m}\int_{\partial\Omega} \nu_{k_\ell}(Y)(\partial^{(\sum_{s=1}^{\ell-1} e_{k_s})+\beta}E)(X-Y)A_{\beta(\sum_{s=1}^m e_{k_s})}\times$$

$$\times f_{(\sum_{r=1}^{|\gamma|} e_{j_r})+(\sum_{s=\ell+1}^m e_{k_s})}(Y)\,d\sigma(Y)\Big\}.$$

Based on the identity (4.55) we may write

$$\partial^\gamma \dot{\mathscr{D}}(\dot{f})(X) = \sum_{\ell=1}^{|\gamma|}\sum_{\substack{|\beta|=m \\ |\delta|=\ell-1,|\eta|=m-\ell}}\sum_{k=1}^{n} \frac{(\delta+\eta+e_k)!(m-\ell)!(\ell-1)!}{m!\delta!\eta!}\times$$

$$\times\int_{\partial\Omega}\partial_{\tau_{kj_\ell}(Y)}\Big((\partial^{\delta+\omega_\ell+\beta}E)(X-Y)\Big)A_{\beta(\delta+\eta+e_k)} f_{\theta_\ell+\eta}(Y)\,d\sigma(Y)$$

$$-\sum_{\ell=|\gamma|+1}^{m}\sum_{\substack{|\beta|=m\\|\delta|=\ell-1,|\eta|=m-\ell}}\sum_{k=1}^{n}\frac{(\delta+\eta+e_k)!(m-\ell)!(\ell-1)!}{m!\delta!\eta!}\times$$

$$\times\int_{\partial\Omega}\nu_k(Y)(\partial^{\delta+\beta}E)(X-Y)A_{\beta(\delta+\eta+e_k)}f_{\gamma+\eta}(Y)\,d\sigma(Y). \quad (4.76)$$

Above, for each $\ell\in\mathbb{N}$, $1\leq\ell\leq|\gamma|$, we have set $\theta_\ell:=\sum_{r=1}^{\ell-1}e_{j_r}$, $\omega_\ell:=\sum_{r=\ell+1}^{|\gamma|}e_{j_r}$, and $\theta_\ell+\omega_\ell+e_{j_\ell}=\gamma$. Summing over all such representations of γ and taking into account multiplicities allows us to conclude the following. For every multi-index $\gamma\in\mathbb{N}_0^n$, $|\gamma|\leq m-1$, $\dot{f}\in\dot{L}^p_{m-1,0}(\partial\Omega)$, $X\in\Omega$, we have the identity

$$\partial^\gamma\dot{\mathscr{D}}(\dot{f})(X)=\sum_{\substack{|\alpha|=m\\|\beta|=m}}\sum_{\ell=1}^{|\gamma|}\sum_{\substack{\delta+\eta+e_k=\alpha\\|\delta|=\ell-1,|\eta|=m-\ell}}\Big\{\sum_{\substack{\theta+\omega+e_j=\gamma\\|\theta|=\ell-1,|\omega|=|\gamma|-\ell}}C_1(m,\ell,\alpha,\delta,\eta,\gamma,\theta,\omega)\times$$

$$\times\int_{\partial\Omega}\partial_{\tau_{kj}(Y)}\Big((\partial^{\delta+\omega+\beta}E)(X-Y)\Big)A_{\beta\alpha}f_{\theta+\eta}(Y)\,d\sigma(Y)\Big\}$$

$$-\sum_{\substack{|\alpha|=m\\|\beta|=m}}\sum_{\ell=|\gamma|+1}^{m}\sum_{\substack{\delta+\eta+e_k=\alpha\\|\delta|=\ell-1,|\eta|=m-\ell}}\Big\{C_2(m,\ell,\alpha,\delta,\eta)\times \quad (4.77)$$

$$\times\int_{\partial\Omega}\nu_k(Y)(\partial^{\delta+\beta}E)(X-Y)A_{\beta\alpha}f_{\gamma+\eta}(Y)\,d\sigma(Y)\Big\},$$

where

$$C_1(m,\ell,\alpha,\delta,\eta,\gamma,\theta,\omega):=\frac{\alpha!(m-\ell)!(\ell-1)!\gamma!(\ell-1)!(|\gamma|-\ell)!}{m!\,\delta!\,\eta!\,|\gamma|!\,\theta!\,\omega!}, \quad (4.78)$$

$$C_2(m,\ell,\alpha,\delta,\eta):=\frac{\alpha!(m-\ell)!(\ell-1)!}{m!\,\delta!\,\eta!}. \quad (4.79)$$

Now the estimate (4.63) is a simple consequence of (4.77) and (2.527) in Proposition 2.63. Furthermore, the claim in (4.65) also follows from (4.77) and formula (2.530) in Proposition 2.63.

As far as (4.64) is concerned, let us assume that $\dot{f}=\{f_\delta\}_{|\delta|\leq m-1}\in\dot{L}^p_{m-1,1}(\partial\Omega)$ for some $1<p<\infty$. In this case, thanks to the extra smoothness of $\dot{f}$, we can integrate by parts (cf. Corollary 2.12) one more time in the first integral in the right-hand side of (4.77) and write for each $\gamma \in \mathbb{N}_0^n$ of length $\leq m-1$

$$\partial^\gamma \dot{\mathscr{D}}(\dot{f})(X) = \sum_{\substack{|\alpha|=m\\|\beta|=m}} \sum_{\ell=1}^{|\gamma|} \sum_{\substack{\delta+\eta+e_k=\alpha\\|\delta|=\ell-1, |\eta|=m-\ell}} \sum_{\substack{\theta+\omega+e_j=\gamma\\|\theta|=\ell-1, |\omega|=|\gamma|-\ell}} \Big\{C_1(m,\ell,\alpha,\delta,\eta,\gamma,\theta,\omega) \times$$

$$\times \int_{\partial\Omega} (\partial^{\delta+\omega+\beta} E)(X-Y) A_{\beta\alpha} (\partial_{\tau_{jk}} f_{\theta+\eta})(Y)\, d\sigma(Y)\Big\}$$

$$- \sum_{\substack{|\alpha|=m\\|\beta|=m}} \sum_{\ell=|\gamma|+1}^{m} \sum_{\substack{\delta+\eta+e_k=\alpha\\|\delta|=\ell-1, |\eta|=m-\ell}} \Big\{C_2(m,\ell,\alpha,\delta,\eta) \times \tag{4.80}$$

$$\times \int_{\partial\Omega} \nu_k(Y) (\partial^{\delta+\beta} E)(X-Y) A_{\beta\alpha} f_{\gamma+\eta}(Y)\, d\sigma(Y)\Big\},$$

at each $X \in \mathbb{R}^n \setminus \partial\Omega$, where the coefficients C_1, C_2 are as in (4.78)–(4.79).

Note that the integrand in the first integral in (4.80) is only weakly singular (since it entails $m-1+|\gamma| \leq 2m-2$ derivatives on E). Hence, terms of this form can absorb yet another partial derivative, say ∂_{x_i}, and still yield integral operators with either weakly singular kernels, or kernels of Calderón–Zygmund type (cf. Proposition 2.63). In particular, their contribution when estimating $\|\mathscr{N}(\partial_i \partial^\gamma \dot{\mathscr{D}} \dot{f})\|_{L^p(\partial\Omega)}$ is bounded by a fixed multiple of $\|\dot{f}\|_{\dot{L}^p_{m-1,1}(\partial\Omega)}$.

As for the terms in the second part of (4.80), it suffices to consider only those which contain $2m-1$ derivatives on E (since the rest are treated as before). Note that this requires $\ell = m$ (hence, $\eta = 0$). After applying ∂_{x_i} to them, at each $X \in \mathbb{R}^n \setminus \partial\Omega$ we obtain

$$\sum_{\substack{|\alpha|=m\\|\beta|=m}} \sum_{\delta+e_k=\alpha} \frac{\alpha!}{m\,\delta!} \int_{\partial\Omega} \nu_k(Y) \partial_{y_i} [(\partial^{\delta+\beta} E)(X-Y)] A_{\beta\alpha} f_\gamma(Y)\, d\sigma(Y)$$

$$= - \sum_{\substack{|\alpha|=m\\|\beta|=m}} \sum_{\delta+e_k=\alpha} \frac{\alpha!}{m\,\delta!} \int_{\partial\Omega} \partial_{\tau_{ik}(Y)} [(\partial^{\delta+\beta} E)(X-Y)] A_{\beta\alpha} f_\gamma(Y)\, d\sigma(Y)$$

$$+ \sum_{\substack{|\alpha|=m\\|\beta|=m}} \sum_{\delta+e_k=\alpha} \frac{\alpha!}{m\,\delta!} \int_{\partial\Omega} \nu_i(Y) \partial_{y_k} [(\partial^{\delta+\beta} E)(X-Y)] A_{\beta\alpha} f_\gamma(Y)\, d\sigma(Y)$$

$$= -\sum_{\substack{|\alpha|=m\\|\beta|=m}} \sum_{\delta+e_k=\alpha} \frac{\alpha!}{m\,\delta!} \int_{\partial\Omega} (\partial^{\delta+\beta} E)(X-Y) A_{\beta\alpha} (\partial_{\tau_{ki}} f_\gamma)(Y)\, d\sigma(Y)$$

$$- \sum_{\substack{|\alpha|=m\\|\beta|=m}} \sum_{\delta+e_k=\alpha} \frac{\alpha!}{m\,\delta!} \int_{\partial\Omega} \nu_i(Y)(\partial^{\delta+\beta+e_k} E)(X-Y) A_{\beta\alpha} f_\gamma(Y)\, d\sigma(Y)$$

$$=: A + B. \tag{4.81}$$

Now, the contribution from the terms labeled A is handled as before since, under the assumptions on the indices involved, $(\partial^{\delta+\beta} E)(X-Y)$ is a Calderón–Zygmund kernel. As far as B is concerned, note that $\delta + \beta + e_k = \alpha + \beta$ and that

$$\sum_{\delta+e_k=\alpha} \frac{\alpha!}{m\,\delta!} = \sum_{k=1}^{n} \frac{\alpha_k}{m} = 1, \tag{4.82}$$

since $|\alpha| = m$. Hence,

$$B = -\sum_{\substack{|\alpha|=m\\|\beta|=m}} \int_{\partial\Omega} \nu_i(Y)(\partial^{\alpha+\beta} E)(X-Y) A_{\beta\alpha} f_\gamma(Y)\, d\sigma(Y)$$

$$= -\int_{\partial\Omega} \nu_i(Y) \Big[(L^t E_{L^t})(X-Y)\Big]^t f_\gamma(Y)\, d\sigma(Y)$$

$$= 0. \tag{4.83}$$

In summary, we have shown that for each multi-index $\gamma \in \mathbb{N}_0^n$ of length $\leq m-1$ and each $i \in \{1,\dots,n\}$, there holds

$$\partial_i \partial^\gamma \dot{\mathscr{D}}(\dot{f})(X) = \sum_{\substack{|\alpha|=m\\|\beta|=m}} \sum_{\ell=1}^{|\gamma|} \sum_{\substack{\delta+\eta+e_k=\alpha\\|\delta|=\ell-1,|\eta|=m-\ell}} \sum_{\substack{\theta+\omega+e_j=\gamma\\|\theta|=\ell-1,|\omega|=|\gamma|-\ell}} \Big\{C_1(m,\ell,\alpha,\delta,\eta,\gamma,\theta,\omega) \times$$

$$\times \int_{\partial\Omega} (\partial^{\delta+\omega+\beta+e_i} E)(X-Y) A_{\beta\alpha} (\partial_{\tau_{jk}} f_{\theta+\eta})(Y)\, d\sigma(Y)\Big\}$$

$$- \sum_{\substack{|\alpha|=m\\|\beta|=m}} \sum_{\ell=|\gamma|+1}^{m-1} \sum_{\substack{\delta+\eta+e_k=\alpha\\|\delta|=\ell-1,|\eta|=m-\ell}} \Big\{C_2(m,\ell,\alpha,\delta,\eta) \times \tag{4.84}$$

$$\times \int_{\partial\Omega} \nu_k(Y)(\partial^{\delta+\beta+e_i} E)(X-Y) A_{\beta\alpha} f_{\gamma+\eta}(Y)\, d\sigma(Y)\Big\}$$

$$- \sum_{\substack{|\alpha|=m \\ |\beta|=m}} \sum_{\delta+e_k=\alpha} \frac{\alpha!}{m\,\delta!} \int_{\partial\Omega} (\partial^{\delta+\beta} E)(X-Y) A_{\beta\alpha} (\partial_{\tau_{ki}} f_\gamma)(Y)\, d\sigma(Y)$$

at each $X \in \mathbb{R}^n \setminus \partial\Omega$, where the coefficients C_1, C_2 are as in (4.78)–(4.79). This justifies (4.64). To finish the proof of Theorem 4.7 there remains to observe that the claim made in (4.66) follows from (4.84) and formula (2.530) in Proposition 2.63. □

The formulas for the derivatives of the double multi-layer deduced during the course of the proof of Theorem 4.7 are also useful to establish the following smoothing properties of the double multi-layer, measured on the Besov scale.

Proposition 4.8. *Retain the same context as in Theorem 4.7. Then for each number $p \in (1,\infty)$ there exists a finite constant $C = C(\Omega, L, p) > 0$ such that*

$$\|\dot{\mathscr{D}} \dot{f}\|_{B^{p,p\vee 2}_{m-1+1/p}(\Omega)} \leq C \|\dot{f}\|_{\dot{L}^p_{m-1,0}(\partial\Omega)}, \qquad \forall\, \dot{f} \in \dot{L}^p_{m-1,0}(\partial\Omega), \tag{4.85}$$

$$\|\dot{\mathscr{D}} \dot{f}\|_{B^{p,p\vee 2}_{m+1/p}(\Omega)} \leq C \|\dot{f}\|_{\dot{L}^p_{m-1,1}(\partial\Omega)}, \qquad \forall\, \dot{f} \in \dot{L}^p_{m-1,1}(\partial\Omega). \tag{4.86}$$

Proof. Estimate (4.85) is a consequence of the (quantitative) lifting result from Proposition 2.24, Proposition 2.68, and formula (4.77). Likewise, estimate (4.86) is implied by Propositions 2.24, 2.68, and formula (4.80). □

Finally, we consider the action of the double multi-layer operator on the Whitney–Hardy spaces $\dot{h}^p_{m-1,1}(\partial\Omega)$.

Theorem 4.9. *Retain the same background assumptions as in Theorem 4.7. Then for any $p \in \left(\frac{n-1}{n}, 1\right]$ there holds*

$$\sum_{j=0}^{m} \|\mathscr{N}(\nabla^j \dot{\mathscr{D}} \dot{f})\|_{L^p(\partial\Omega)} \leq C \|\dot{f}\|_{\dot{h}^p_{m-1,1}(\partial\Omega)}, \tag{4.87}$$

for some finite constant $C > 0$ independent of $\dot{f} \in \dot{h}^p_{m-1,1}(\partial\Omega)$.

Proof. This is a consequence of the boundedness of the operator (2.482), the identities (4.80)–(4.83), and (2.528). □

4.3 Carleson Measure Estimates

Fix a bounded Lipschitz domain Ω in $\mathbb{R}^n$, along with a natural number $m \in \mathbb{N}$. As usual, let tr_{m-1} denote the multi-trace operator onto the boundary of Ω. Finally, recall (3.383), (3.384) and Definition 3.49. To proceed, for each $r > 0$ and $X_o \in \partial\Omega$, define

$$\dot{\mathscr{P}}_{m-1}(\partial\Omega) \ni \dot{\omega}_r(\dot{f}) := \text{ the best fit polynomial array}$$
$$\text{to } \dot{f} \in \dot{L}^2_{m-1,0}(\partial\Omega) \text{ on } S_r(X_o), \tag{4.88}$$

i.e., the orthogonal projection of $\dot{f}|_{S_r(X_o)}$ onto $\dot{\mathscr{P}}_{m-1}(S_r(X_o))$ with respect to the natural inner product in $L^2(S_r(X_o)) \oplus \cdots \oplus L^2(S_r(X_o))$, N times, where

$$N := \sum_{k=0}^{m-1} \binom{n+k-1}{n-1} \tag{4.89}$$

is the cardinality of $\{\alpha \in \mathbb{N}_0^n : |\alpha| \leq m-1\}$. In particular, for each $r > 0$,

$$\int_{S_r(X_o)} |\dot{f}(Q) - \dot{\omega}_r(\dot{f})(Q)|^2 \, d\sigma(Q) \tag{4.90}$$
$$= \inf_{\dot{P} \in \dot{\mathscr{P}}_{m-1}(\partial\Omega)} \int_{S_r(X_o)} |\dot{f}(Q) - \dot{P}(Q)|^2 \, d\sigma(Q),$$

and (recall (3.385)), we have

$$\dot{f}^{\#}(X_o) = \sup_{r>0} \left(⨍_{S_r(X_o)} |\dot{f}(Q) - \dot{\omega}_r(\dot{f})(Q)|^2 \, d\sigma(Q) \right)^{1/2}. \tag{4.91}$$

Next, recall that $\dot{\mathscr{P}}_{m-1}(S_r(X_o))$ stands for the space of restrictions of arrays in $\dot{\mathscr{P}}_{m-1}(\partial\Omega)$ to $S_r(X_o)$. Thus, any element $\dot{f}$ in $\dot{\mathscr{P}}_{m-1}(S_r(X_o))$ is of the form $(\mathrm{tr}_{m-1} P)|_{S_r(X_o)}$ for some $P \in \mathscr{P}_{m-1}$. It is important to note that such a polynomial P is unique. Indeed, by linearity, this is an immediate consequence of the fact that

$$P \in \mathscr{P}_{m-1} \text{ and } (\mathrm{tr}_{m-1} P)|_{S_r(X_o)} = 0 \Longrightarrow P \equiv 0 \text{ in } \mathbb{R}^n. \tag{4.92}$$

In turn, this is obvious from the fact that all derivatives of P vanish at, say, X_o. We shall refer to P as the polynomial extension to $\mathbb{R}^n$ of $\dot{f}$. In summary,

$$\mathrm{tr}_{m-1} : \mathscr{P}_{m-1} \longrightarrow \dot{\mathscr{P}}_{m-1}(S_r(X_o)) \text{ is an algebraic isomorphism.} \tag{4.93}$$

In the sequel, we shall frequently identify a generic element $\dot{f}$ from $\dot{\mathscr{P}}_{m-1}(S_r(X_o))$ with its canonical extension to the entire boundary $\partial\Omega$. By definition, the latter is taken to be $\mathrm{tr}_{m-1}P \in \dot{\mathscr{P}}_{m-1}(\partial\Omega)$, where $P \in \mathscr{P}_{m-1}$ is the polynomial extension to $\mathbb{R}^n$ of $\dot{f}$. We emphasize that, by design, the canonical extension of elements from $\dot{\mathscr{P}}_{m-1}(S_r(X_o))$ to elements in $\dot{\mathscr{P}}_{m-1}(\partial\Omega)$ is unique.

The space $\dot{\mathscr{P}}_{m-1}(S_r(X_o))$ has a Hilbert structure when equipped with the inner product naturally inherited from $L^2(S_r(X_o)) \oplus \cdots \oplus L^2(S_r(X_o))$, N times (with N as in (4.89)). Then, given (4.93), the restrictions to $S_r(X_o)$ of the Whitney arrays $\dot{p}_\alpha(X) := \mathrm{tr}_{m-1}[(X - X_o)^\alpha]$, $\alpha \in \mathbb{N}_0^n$ with $|\alpha| \leq m-1$, form a basis in $\dot{\mathscr{P}}_{m-1}(S_r(X_o))$. Applying the Gram–Schmidt process to $\{\dot{p}_\alpha\}_{|\alpha|\leq m-1}$ then yields an orthonormal (relative to $L^2(S_r(X_o)) \oplus \cdots \oplus L^2(S_r(X_o))$) basis for $\dot{\mathscr{P}}_{m-1}(S_r(X_o))$. We denote this basis by $\{\dot{e}_\alpha\}_{|\alpha|\leq m-1}$ and, for each multi-index $\alpha \in \mathbb{N}_0^n$ of length $\leq m-1$, we let $e_\alpha \in \mathscr{P}_{|\alpha|}$ be the polynomial extension to $\mathbb{R}^n$ of $\dot{e}_\alpha$. That is, e_α is a polynomial of degree $|\alpha|$ such that $\dot{e}_\alpha = [\mathrm{tr}_{m-1}e_\alpha]|_{S_r(X_o)}$. Then, for each $\beta \in \mathbb{N}_0^n$, $|\beta| \leq m-1$, it follows that at points in $S_r(X_o)$

$$(\dot{e}_\alpha)_\beta = (\mathrm{tr}_{m-1}e_\alpha)_\beta = \partial^\beta e_\alpha = 0 \quad \text{if} \quad |\beta| > |\alpha|, \tag{4.94}$$

by degree considerations. Also, by carefully keeping track of bounds for the various expressions appearing in the Gram–Schmidt orthonormalization process (as presented on, e.g., p. 120 of [121]) we obtain

$$|(\dot{e}_\alpha)_\beta(X)| \leq \frac{C}{r^{(n-1)/2}}\Big[|X - X_o| + r\Big]^{|\alpha|-|\beta|} \quad \text{if } |\beta| \leq |\alpha|, \quad \forall\, X \in \partial\Omega. \tag{4.95}$$

The point of the above considerations is to facilitate discussing a number of important properties of the best fit polynomial array (cf. (4.88)). To begin with, we agree that $\dot{\omega}_r(\dot{f})$ is always identified with its canonical extension to $\partial\Omega$. Thus,

$$\dot{\omega}_r(\dot{f}) \in \dot{\mathscr{P}}_{m-1}(\partial\Omega), \qquad \forall\, \dot{f} \in \dot{L}^2_{m-1,0}(\partial\Omega). \tag{4.96}$$

Let us also observe that

$$\dot{\omega}_r(\dot{f}) = \dot{f}, \qquad \forall\, \dot{f} \in \dot{\mathscr{P}}_{m-1}(\partial\Omega), \tag{4.97}$$

since, by definition, $\dot{\omega}_r(\dot{f})$ coincides with $\dot{f}|_{S_r(X_o)}$ and any $\dot{f} \in \dot{\mathscr{P}}_{m-1}(\partial\Omega)$ is the canonical extension of $\dot{f}|_{S_r(X_o)}$. As an immediate corollary of (4.96)–(4.97), we note that

$$\dot{\omega}_r(\dot{\omega}_{2r}(\dot{f})) = \dot{\omega}_{2r}(\dot{f}), \qquad \forall\, \dot{f} \in \dot{L}^2_{m-1,0}(\partial\Omega). \tag{4.98}$$

Finally, for each $\dot{f} = \{f_\beta\}_{|\beta|\leq m-1} \in \dot{L}^2_{m-1,0}(\partial\Omega)$, we have

$$\dot{\omega}_r(\dot{f}) = \sum_{|\alpha|,|\beta|\leq m-1} \langle f_\beta, (\dot{e}_\alpha)_\beta\rangle\, \dot{e}_\alpha \quad \text{on } \partial\Omega, \tag{4.99}$$

where $\{\dot{e}_\alpha\}_{|\alpha|\leq m-1}$ is as above and $\langle\cdot,\cdot\rangle$ stands here for the inner product in $L^2(S_r(X_o))$. At points in $S_r(X_o)$ this follows straight from definitions, so the desired result is implied by the uniqueness of the canonical extension to $\partial\Omega$.

Next, we augment the list of algebraic properties of the best fit polynomial array with some useful estimates, contained in the lemma below.

Lemma 4.10. *Consider a bounded Lipschitz domain $\Omega \subset \mathbb{R}^n$, and fix $X_o \in \partial\Omega$ and $r > 0$. Then, for each $\dot{f} \in \dot{\mathrm{BMO}}_{m-1}(\partial\Omega)$ and any multi-index $\alpha \in \mathbb{N}_0^n$ of length $\leq m-1$,*

$$\left(\fint_{S_r(X_o)}\left|\left(\dot{f}(Y) - \dot{\omega}_r(\dot{f})(Y)\right)_\alpha\right|^2 d\sigma(Y)\right)^{1/2} \leq Cr^{m-1-|\alpha|}\dot{f}^{\#}(X_o), \qquad (4.100)$$

and

$$|(\dot{\omega}_r(\dot{f}) - \dot{\omega}_{2r}(\dot{f}))_\alpha(X)| \leq C\dot{f}^{\#}(X_o)\Big[|X - X_o| + r\Big]^{m-1-|\alpha|}, \qquad \forall\, X \in \partial\Omega. \qquad (4.101)$$

Proof. Fix $r > 0$, $X_o \in \partial\Omega$ and consider the array $\dot{g} = \{g_\alpha\}_{|\alpha|\leq m-1}$ given by

$$\dot{g} := \dot{f} - \dot{\omega}_r(\dot{f}). \qquad (4.102)$$

Then, since the best fit polynomial array $\dot{\omega}_r(\dot{f})$ is the projection of $\dot{f}|_{S_r(X_o)}$ onto $\dot{\mathscr{P}}_{m-1}(S_r(X_o))$ with respect to the natural inner product in the space

$$L^2(S_r(X_o)) \oplus \cdots \oplus L^2(S_r(X_o)), \qquad (4.103)$$

for each $\alpha \in \mathbb{N}_0^n$ multi-index of length $\leq m-1$ we have

$$\int_{S_r(X_o)}\langle\dot{g}(Y), \dot{P}_\alpha(Y)\rangle\, d\sigma(Y) = \int_{S_r(X_o)}\sum_{|\gamma|\leq m-1} g_\gamma(Y)(\dot{P}_\alpha)_\gamma(Y)\, d\sigma(Y) = 0, \qquad (4.104)$$

where

$$P_\alpha(X) := (X - X_o)^\alpha \quad \text{and} \quad \dot{P}_\alpha := \mathrm{tr}_{m-1}P_\alpha \in \dot{\mathscr{P}}_{m-1}(\partial\Omega). \qquad (4.105)$$

Note that for each $X \in \mathbb{R}^n$ we have

$$(\dot{P}_\alpha)_\gamma(X) = \begin{cases} 0, & \text{if } \gamma > \alpha, \\ \frac{\alpha!}{(\alpha-\gamma)!}P_{\alpha-\gamma}(X), & \text{if } \gamma \leq \alpha. \end{cases} \qquad (4.106)$$

Therefore, from (4.104) and (4.106), we conclude that for each $|\alpha| \leq m-1$ there holds

$$0 = \int_{S_r(X_o)} \langle \dot{g}(Y), \dot{P}_\alpha(Y)\rangle \, d\sigma(Y) = \int_{S_r(X_o)} \sum_{\gamma\leq\alpha} g_\gamma(Y)(\dot{P}_\alpha)_\gamma(Y)\, d\sigma(Y)$$

$$= \int_{S_r(X_o)} \sum_{\gamma\leq\alpha} \frac{\alpha!}{(\alpha-\gamma)!} g_\gamma(Y) P_{\alpha-\gamma}(Y)\, d\sigma(Y). \tag{4.107}$$

In particular, making $\alpha = (0, 0, \cdots, 0)$ in (4.107) implies (recall that the subindex $\emptyset$ denotes the multi-index $(0, 0, \ldots, 0)$)

$$\int_{S_r(X_o)} g_\emptyset(Y)\, d\sigma(Y) = 0, \tag{4.108}$$

and recursively,

$$\alpha! \int_{S_r(X_o)} g_\alpha(Y)\, d\sigma(Y) = - \sum_{\substack{\gamma\leq\alpha \\ |\gamma|<|\alpha|}} \frac{\alpha!}{(\alpha-\gamma)!} \int_{S_r(X_o)} g_\gamma(Y) P_{\alpha-\gamma}(Y)\, d\sigma(Y). \tag{4.109}$$

Using (4.108) and applying Poincaré's inequality we obtain

$$\left(⨍_{S_r(X_o)} g_\emptyset(Y)\, d\sigma(Y) \right)^{1/2} \leq Cr \left(⨍_{S_r(X_o)} |\nabla_{tan} g_\emptyset(Y)|^2\, d\sigma(Y) \right)^{1/2}. \tag{4.110}$$

In turn, using the compatibility conditions (3.2),

$$\nabla_{tan} g_\alpha = \Big(\sum_{k=1}^n \nu_k \partial_{\tau_{kj}} g_\alpha \Big)_{1\leq j\leq n} \tag{4.111}$$

$$= \Big(\sum_{k=1}^n \nu_k (\nu_k g_{\alpha+e_j} - \nu_j g_{\alpha+e_k}) \Big)_{1\leq j\leq n}, \quad \text{if } |\alpha| \leq m-2.$$

Therefore,

$$\left(⨍_{S_r(X_o)} g_\emptyset(Y)\, d\sigma(Y) \right)^{1/2} \leq Cr \sum_{k=1}^n \left(⨍_{S_r(X_o)} |g_{e_k}(Y)|^2\, d\sigma(Y) \right)^{1/2}$$

$$\leq Cr \dot{f}^{\#}(X_o), \tag{4.112}$$

where the last inequality follows from (4.102) and the definition of the sharp maximal function (3.385). This is the conclusion of our first step. For the second step, fix $\alpha \in \mathbb{N}_0^n$, $|\alpha| = 1$. From (4.109),

$$\int_{S_r(X_o)} g_\alpha(Y)\, d\sigma(Y) = \int_{S_r(X_o)} g_\emptyset(Y) P_\alpha(Y)\, d\sigma(Y), \tag{4.113}$$

and since $|P_\alpha(Y)| \leq r^{|\alpha|} = r$ on $S_r(X_o)$, this implies

$$\begin{aligned} \Big| ⨍_{S_r(X_o)} g_\alpha(Y)\, d\sigma(Y)\Big| &\leq r ⨍_{S_r(X_o)} |g_\emptyset(Y)|\, d\sigma(Y) \\ &\leq r \Big(⨍_{S_r(X_o)} |g_\emptyset(Y)|^2\, d\sigma(Y)\Big)^{1/2} \\ &\leq C r^2 \dot{f}^{\#}(X_o), \end{aligned} \tag{4.114}$$

where the last inequality in (4.114) follows from (4.112). Applying the Poincaré inequality for g_α gives

$$\begin{aligned} \Big(⨍_{S_r(X_o)} |g_\alpha(Y)|^2\, d\sigma(Y)\Big)^{1/2} &\leq C r \Big(⨍_{S_r(X_o)} |\nabla_{tan} g_\alpha(Y)|^2\, d\sigma(Y)\Big)^{1/2} \\ &\quad + C \Big| ⨍_{S_r(X_o)} g_\alpha(Y)\, d\sigma(Y)\Big|. \end{aligned} \tag{4.115}$$

In turn, using (4.111) yields $|\nabla_{tan}\, g_\alpha|^2 \leq \sum_{|\gamma|=2} |g_\gamma|^2$ which, together with (4.114)–(4.115), implies

$$\begin{aligned} \Big(⨍_{S_r(X_o)} |g_\alpha(Y)|^2\, d\sigma(Y)\Big)^{1/2} &\leq C r \sum_{|\gamma|=2} \Big(⨍_{S_r(X_o)} |g_\gamma(Y)|^2\, d\sigma(Y)\Big)^{1/2} \\ &\quad + C \Big| ⨍_{S_r(X_o)} g_\alpha(Y)\, d\sigma(Y)\Big| \\ &\leq C r \dot{f}^{\#}(X_o) + C r^2 \dot{f}^{\#}(X_o) \\ &\leq C r \dot{f}^{\#}(X_o). \end{aligned} \tag{4.116}$$

Utilizing (4.116) instead of the definition of $\dot{f}^{\#}(X_o)$ in the second inequality in (4.112) improves (4.108) to

$$\Big(⨍_{S_r(X_o)} g_\emptyset(Y)\, d\sigma(Y)\Big)^{1/2} \leq C r^2 \dot{f}(X_o), \tag{4.117}$$

The conclusion of our second step is that

$$\left(\fint_{S_r(X_o)} g_\emptyset(Y)\,d\sigma(Y)\right)^{1/2} \le Cr^2 \dot{f}^\#(X_o) \quad \text{and}$$
$$\left(\fint_{S_r(X_o)} |g_\alpha(Y)|^2\,d\sigma(Y)\right)^{1/2} \le C r \dot{f}^\#(X_o), \quad \text{if } |\alpha| = 1. \tag{4.118}$$

In the third step we fix $\alpha \in \mathbb{N}_0^n$, $|\alpha| = 2$. Then, according to (4.109) we have

$$\int_{S_r(X_o)} g_\alpha(Y)\,d\sigma(Y) = -\int_{S_r(X_o)} g_\emptyset(Y) P_\alpha(Y)\,d\sigma(Y)$$
$$-\sum_{\substack{\gamma\le\alpha\\|\gamma|=1}} \frac{\alpha!}{(\alpha-\gamma)!}\int_{S_r(X_o)} g_\gamma(Y) P_{\alpha-\gamma}(Y)\,d\sigma(Y). \tag{4.119}$$

Since $|P_{\alpha-\gamma}(Y)| \le r$ and $|P_\alpha(Y)| \le r^2$ for $Y \in S_r(X_o)$, and using (4.118)

$$\left|\fint_{S_r(X_o)} g_\alpha(Y)\,d\sigma(Y)\right| \le r^2 \left(\fint_{S_r(X_o)} g_\emptyset(Y)\,d\sigma(Y)\right)^{1/2}$$
$$+r\sum_{\substack{\gamma\le\alpha\\|\gamma|=1}} \frac{\alpha!}{(\alpha-\gamma)!}\left(\fint_{S_r(X_o)} |g_\gamma(Y)|^2\,d\sigma(Y)\right)^{1/2}$$
$$\le Cr^4 \dot{f}^\#(X_o) + Cr^2 \dot{f}^\#(X_o)$$
$$\le Cr^2 \dot{f}^\#(X_o). \tag{4.120}$$

However, applying again Poincaré's inequality (4.115) and using (4.111) and (4.120), one has

$$\left(\fint_{S_r(X_o)} |g_\alpha(Y)|^2\,d\sigma(Y)\right)^{1/2} \le C r \sum_{|\gamma|=3}\left(\fint_{S_r(X_o)} |g_\gamma(Y)|^2\,d\sigma(Y)\right)^{1/2}$$
$$+C\left|\fint_{S_r(X_o)} g_\alpha(Y)\,d\sigma(Y)\right|$$
$$\le C r \dot{f}^\#(X_o) + Cr^2 \dot{f}^\#(X_o)$$
$$\le C r \dot{f}^\#(X_o). \tag{4.121}$$

Recall that $\alpha \in \mathbb{N}_0^n$ is an arbitrary multi-index of length $= 2$. Employing the inequality (4.121) in (4.116) to control

$$\sum_{|\gamma|=2} \Big(\fint_{S_r(X_o)} |g_\gamma(Y)|^2 \, d\sigma(Y) \Big)^{1/2}, \tag{4.122}$$

allows us to improve (4.116) to

$$\Big(\fint_{S_r(X_o)} |g_\delta(Y)|^2 \, d\sigma(Y) \Big)^{1/2} \leq Cr^2 \dot{f}^\#(X_o) \quad \text{if } |\delta| = 1. \tag{4.123}$$

In turn, using (4.123) in (4.112) gives

$$\Big(\fint_{S_r(X_o)} |g_\emptyset(Y)|^2 \, d\sigma(Y) \Big)^{1/2} \leq Cr^3 \dot{f}^\#(X_o). \tag{4.124}$$

This concludes the third step of our analysis at the end of which we have

$$\Big(\fint_{S_r(X_o)} g_\emptyset(Y) \, d\sigma(Y) \Big)^{1/2} \leq Cr^3 \dot{f}^\#(X_o), \tag{4.125}$$

$$\begin{aligned} \Big(\fint_{S_r(X_o)} |g_\gamma(Y)|^2 \, d\sigma(Y) \Big)^{1/2} &\leq Cr^2 \dot{f}^\#(X_o), \quad \text{if } |\gamma| = 1, \\ \Big(\fint_{S_r(X_o)} |g_\alpha(Y)|^2 \, d\sigma(Y) \Big)^{1/2} &\leq C r \dot{f}^\#(X_o), \quad \text{if } |\alpha| = 2. \end{aligned} \tag{4.126}$$

Thus, inductively, we obtain (4.100).

We are left with showing (4.101). Let $r > 0$ and $X_o \in \partial\Omega$ be fixed and notice that, by the linearity of $\dot{\omega}_r$, (4.98) and (4.99),

$$\begin{aligned} \dot{\omega}_r(\dot{f}) - \dot{\omega}_{2r}(\dot{f}) &= \dot{\omega}_r(\dot{f} - \dot{\omega}_{2r}(\dot{f})) \\ &= \sum_{|\beta|,|\gamma| \leq m-1} \Big\langle (\dot{f} - \dot{\omega}_{2r}(\dot{f}))_\beta, (\dot{e}_\gamma)_\beta \Big\rangle \dot{e}_\gamma, \end{aligned} \tag{4.127}$$

where the collection $\{\dot{e}_\alpha\}_{|\alpha| \leq m-1}$ is the orthonormal basis in $\dot{\mathscr{P}}_{m-1}(S_r(X_o))$ relative to the Hilbert space $L^2(S_r(X_o)) \oplus \cdots \oplus L^2(S_r(X_o))$, N times, discussed in the preamble of this lemma, and $\langle \cdot, \cdot \rangle$ stands here for the inner product in $L^2(S_r(X_o))$. Then, for every $X \in \partial\Omega$,

$$
\begin{aligned}
&\Big|\Big(\dot{\omega}_r(\dot{f}) - \dot{\omega}_{2r}(\dot{f})\Big)_\alpha(X)\Big| \\
&\quad \le \sum_{|\beta|,|\gamma|\le m-1} \Big|\Big\langle(\dot{f} - \dot{\omega}_{2r}(\dot{f}))_\beta, (\dot{e}_\gamma)_\beta\Big\rangle(\dot{e}_\gamma)_\alpha(X)\Big| \\
&\quad \le Cr^{\frac{n-1}{2}} \sum_{|\beta|\le|\gamma|\le m-1} \Big(\fint_{S_r(X_o)} \Big|\Big(\dot{f} - \dot{\omega}_{2r}(\dot{f})\Big)_\beta(Y)\Big|^2 d\sigma(Y)\Big)^{1/2} \|(\dot{e}_\gamma)_\alpha\|_{L^\infty(S_r(X_o))} \\
&\quad \le C \sum_{|\beta|\le|\gamma|\le m-1} \Big(\fint_{S_{2r}(X_o)} \Big|\Big(\dot{f} - \dot{\omega}_{2r}(\dot{f})\Big)_\beta(Y)\Big|^2 d\sigma(Y)\Big)^{1/2} \Big[|X - X_o| + r\Big]^{|\gamma|-|\alpha|} \\
&\quad \le C \sum_{|\beta|\le|\gamma|\le m-1} r^{m-1-|\beta|} \dot{f}^{\#}(X_o)\Big[|X - X_o| + r\Big]^{|\gamma|-|\alpha|},
\end{aligned}
\tag{4.128}
$$

where the first inequality in (4.128) follows from (4.127), the second is a consequence of Hölder's inequality, the normality of $\{\dot{e}_\gamma\}_{|\gamma|\le m-1}$ and (4.94), the third is implied by (4.95) and $|S_r(X_o)| \approx |S_{2r}(X_o)|$, while the last follows from (4.100). Then (4.101) is an immediate consequence of (4.128) and the fact that $r \le |X - X_o| + r$ and that $\partial\Omega$ is bounded. □

Next, we present the main result of this section. Before stating it, recall (2.43).

Theorem 4.11. *Consider a W-elliptic homogeneous differential operator L of order 2m with (complex) matrix-valued constant coefficients. Let Ω be a bounded Lipschitz domain in $\mathbb{R}^n$ and for each $X \in \Omega$ denote by $\rho(X)$ its distance to $\partial\Omega$. Let $\dot{\mathscr{D}}$ be a double multi-layer potential associated with L (as in § 4.2). Then there exists a finite constant $C = C(\Omega, L) > 0$ with the property that*

$$
\begin{aligned}
&|\nabla^m \dot{\mathscr{D}} \dot{f}(X)|^2 \rho(X)\, dX \quad \textit{is a Carleson measure on } \Omega, \textit{ with} \\
&\textit{Carleson constant} \le C\|\dot{f}\|^2_{\dot{\mathrm{BMO}}_{m-1}(\partial\Omega)} \textit{ for each } \dot{f} \in \dot{\mathrm{BMO}}_{m-1}(\partial\Omega).
\end{aligned}
\tag{4.129}
$$

Proof. Let $\dot{f} \in \dot{\mathrm{BMO}}_{m-1}(\partial\Omega)$ and fix $X_o \in \partial\Omega$ and $r > 0$. Consider next a function η in $\mathbb{R}^n$ with the following properties

$$
\eta \in C_c^\infty(\mathbb{R}^n), \quad 0 \le \eta \le 1, \quad \eta \equiv 1 \text{ on } B_{2r}(X_o), \ \operatorname{supp}\eta \subset B_{4r}(X_o), \tag{4.130}
$$

$$
\text{and} \quad |\partial^\alpha \eta| \le \frac{C_\alpha}{r^{|\alpha|}}, \qquad \text{for each multi-index } \alpha \in \mathbb{N}_0^n. \tag{4.131}
$$

Also, denote by $\Delta_r(X_o) := \Omega \cap B(X_o, r)$ the Carleson box associated with the surface ball $S_r(X_o)$, and denote by $|\Delta_r(X_o)|$ its n-dimensional Lebesgue measure. Let us agree that $|S_r(X_o)|$ denotes the surface measure of $S_r(X_o)$.

Throughout the proof, we shall abbreviate $\dot{\omega}_r(\dot{f})$, $\dot{\omega}_{2r}(\dot{f})$, etc., as $\dot{\omega}_r$, $\dot{\omega}_{2r}$, etc. Recall (3.9). Writing now $\dot{f} = \eta(\dot{f} - \dot{\omega}_{2r}) + (1-\eta)(\dot{f} - \dot{\omega}_{2r}) + \dot{\omega}_{2r}$ and using the reproducing property (4.62) of $\mathscr{D}$ we obtain

$$\nabla^m \dot{\mathscr{D}} \dot{f} = \nabla^m \dot{\mathscr{D}}[\eta(\dot{f} - \dot{\omega}_{2r})] + \nabla^m \dot{\mathscr{D}}[(1-\eta)(\dot{f} - \dot{\omega}_{2r})], \tag{4.132}$$

as $\nabla^m \dot{\mathscr{D}} \dot{\omega}_{2r} = \nabla^m \omega_{2r} = 0$. Consequently, using that $(a+b)^2 \le 2a^2 + 2b^2$, we have,

$$\begin{aligned}
&\frac{1}{|S_r(X_o)|}\int_{\Delta_r(X_o)} \rho(X)|\nabla^m \dot{\mathscr{D}} \dot{f}(X)|^2\,dX \\
&\quad \le \frac{2}{|S_r(X_o)|}\int_{\Delta_r(X_o)} \rho(X)|\nabla^m \dot{\mathscr{D}}[\eta(\dot{f} - \dot{\omega}_{2r})](X)|^2\,dX \\
&\quad + \frac{2}{|S_r(X_o)|}\int_{\Delta_r(X_o)} \rho(X)|\nabla^m \dot{\mathscr{D}}[(1-\eta)(\dot{f} - \dot{\omega}_{2r})](X)|^2\,dX \\
&\quad =: I + II.
\end{aligned} \tag{4.133}$$

We claim next that there exists a finite constant $C = C(\Omega, L) > 0$ such that the following are satisfied

$$\frac{1}{|S_r(X_o)|}\int_{\Delta_r(X_o)} \rho(X)|\nabla^m \dot{\mathscr{D}}[\eta(\dot{f} - \dot{\omega}_{2r})](X)|^2\,dX \le C(\dot{f}^{\#}(X_o))^2, \tag{4.134}$$

$$|\nabla^m \dot{\mathscr{D}}[(1-\eta)(\dot{f} - \dot{\omega}_{2r})](X)| \le \frac{C}{r}\|\dot{f}\|_{\dot{\mathrm{BMO}}_{m-1}(\partial\Omega)} \quad \text{if } X \in \Delta_r(X_o). \tag{4.135}$$

Let us assume for the moment that (4.134) and (4.135) hold and continue with the proof of (4.129). Using (3.387), then (4.134) readily implies

$$I \le C\|\dot{f}\|^2_{\dot{\mathrm{BMO}}_{m-1}(\partial\Omega)}. \tag{4.136}$$

Also, based on (4.135), $|S_r(X_o)| \approx r^{n-1}$, $|\Delta_r(X_o)| \approx r^n$, and $|\rho(X)| \le r$ whenever $X \in \Delta_r(X_o)$, we have

$$II \le \frac{C}{r^2}\|\dot{f}\|^2_{\dot{\mathrm{BMO}}_{m-1}(\partial\Omega)} \frac{1}{|S_r(X_o)|}\int_{\Delta_r(X_o)} |\rho(X)|\,dX \le C\|\dot{f}\|^2_{\dot{\mathrm{BMO}}_{m-1}(\partial\Omega)}. \tag{4.137}$$

Then (4.129) follows from (4.133) and (4.136)–(4.137).

We are therefore left with proving (4.134) and (4.135), a task to which we turn now. Starting with (4.134) let us notice that since $|S_r(X_o)| \approx r^{n-1}$,

$$I \le \frac{C}{r^{n-1}} \int_{\Omega} \rho(X)|\nabla^m \dot{\mathscr{D}}[\eta(\dot{f} - \dot{\omega}_{2r})(X)]|^2 dX. \tag{4.138}$$

Next, from Proposition 2.67 and the integral identity in (4.77)–(4.79) we may estimate

$$\int_{\Omega} \rho(X)|\nabla^m \dot{\mathscr{D}}[\eta(\dot{f} - \dot{\omega}_{2r})(X)]|^2\, dX \le C \int_{\partial\Omega} |\eta(\dot{f} - \dot{\omega}_{2r})(Q)|^2\, d\sigma(Q). \tag{4.139}$$

Consequently,

$$\begin{aligned} I &\le \frac{C}{r^{n-1}} \int_{\partial\Omega} |\eta(\dot{f} - \dot{\omega}_{2r})(Q)|^2\, d\sigma(Q) \\ &= \frac{C}{r^{n-1}} \int_{S_{4r}(X_o)} |(\dot{f} - \dot{\omega}_{2r})(Q)|^2\, d\sigma(Q), \end{aligned} \tag{4.140}$$

where the equality above follows from the fact that η is supported in $B_{4r}(X_o)$. Hence,

$$I \le \frac{C}{r^{n-1}} \int_{S_{4r}(X_o)} |(\dot{f} - \dot{\omega}_{2r})(Q)|^2\, d\sigma(Q) \le C(\dot{f}^{\#}(X_o))^2, \tag{4.141}$$

where the last inequality is a consequence of the property (4.90) and the definition of the sharp maximal function $\dot{f}^{\#}(X_o)$. The proof of (4.136) is therefore completed.

Now we focus on proving (4.137) and to this end, let $X \in \Delta_r(X_o)$. Then, since

$$|\nabla^m \dot{\mathscr{D}}[(1-\eta)(\dot{f} - \dot{\omega}_{2r})](X)| = |\nabla(\nabla^{m-1}\dot{\mathscr{D}}[(1-\eta)(\dot{f} - \dot{\omega}_{2r})])(X)|, \tag{4.142}$$

employing (4.77) we obtain

$$\begin{aligned} &|\nabla^m \dot{\mathscr{D}}[(1-\eta)(\dot{f} - \dot{\omega}_{2r})](X)| \\ &\le C \sum_{\substack{|\gamma|=m-1,|\alpha|=m \\ |\beta|=m,|\delta|=m-1}} \Big|\nabla_X \int_{\partial\Omega} \nu_j(Y)(\partial^{\beta+\gamma}E)(X-Y)A_{\beta\alpha} \times \\ &\qquad\qquad \times((1-\eta)(\dot{f} - \dot{\omega}_{2r}))_\delta(Y)\, d\sigma(Y)\Big|, \end{aligned} \tag{4.143}$$

where $C > 0$ is a universal constant. Using the support properties of η from (4.130), we can replace $\partial\Omega$ by $\partial\Omega \setminus S_{2r}(X_o)$ as the domain of integration in the right hand side of (4.142) since $1 - \eta \equiv 0$ on $S_{2r}(X_o)$. Also, using (4.29),

$$|\nabla^{2m} E(X-Y)| \le \frac{C}{|X-Y|^n}. \tag{4.144}$$

Therefore, since

$$|(1-\eta)(\dot{f}-\dot{\omega}_{2r})(Y)| \le C \sum_{|\mu|\le m-1} |(\dot{f}-\dot{\omega}_{2r})_\mu(Y)| \cdot |\nabla^{m-1-|\mu|}(1-\eta)(Y)|, \tag{4.145}$$

we obtain

$$\begin{aligned} &|\nabla^m \dot{\mathscr{D}}[(1-\eta)(\dot{f}-\dot{\omega}_{2r})](X)| \\ &\le C \sum_{|\mu|\le m-1} \int_{\partial\Omega\setminus S_{2r}(X_o)} \frac{|(\dot{f}-\dot{\omega}_{2r})_\mu(Y)|}{|X-Y|^n} |\nabla^{m-1-|\mu|}(1-\eta)(Y)|\, d\sigma(Y). \end{aligned} \tag{4.146}$$

A simple observation is that there exist constants $C_j = C_j(\Omega)$, $j = 1, 2$, such that for $X \in \Delta_r(X_o)$ and $Y \in \partial\Omega \setminus S_{2r}(X_o)$ we have

$$C_1|X-Y| \le |X_o - Y| \le C_2|X-Y|. \tag{4.147}$$

Employing (4.147) in (4.146), it follows that

$$\begin{aligned} &|\nabla^m \dot{\mathscr{D}}[(1-\eta)(\dot{f}-\dot{\omega}_{2r})](X)| \\ &\le C \sum_{|\mu|=m-1} \int_{\partial\Omega\setminus S_{2r}(X_o)} \frac{1}{|X_o-Y|^n} |(\dot{f}-\dot{\omega}_{2r})_\mu(Y)|\, |(1-\eta)(Y)| d\sigma(Y) \\ &\quad + C \sum_{|\mu|<m-1} \int_{\partial\Omega\setminus S_{2r}(X_o)} \frac{1}{|X_o-Y|^n} |(\dot{f}-\dot{\omega}_{2r})_\mu(Y)|\, |\nabla^{m-1-|\mu|}(1-\eta)(Y)|\, d\sigma(Y) \\ &=: III + IV. \end{aligned} \tag{4.148}$$

For each multi-index $\mu \in \mathbb{N}_0^n$, $|\mu| < m-1$, the function $\nabla^{m-1-|\mu|}(1-\eta)$ is supported in $S_{4r}(X_o) \setminus S_{2r}(X_o)$. This, together with (4.131) imply

$$III \le C \sum_{|\mu|=m-1} \int_{\partial\Omega\setminus S_{2r}(X_o)} \frac{1}{|X_o-Y|^n} |(\dot{f}-\dot{\omega}_{2r})_\mu(Y)|\, d\sigma(Y), \tag{4.149}$$

$$IV \le C \sum_{|\mu|<m-1} \frac{1}{r^{m-1-|\mu|}} \int_{S_{4r}(X_o)\setminus S_{2r}(X_o)} \frac{1}{|X_o-Y|^n} |(\dot{f}-\dot{\omega}_{2r})_\mu(Y)|\, d\sigma(Y). \tag{4.150}$$

Using that $|S_{4r}(X_o)| \approx r^{n-1}$ and for each point $Y \in S_{4r}(X_o) \setminus S_{2r}(X_o)$ we have $2r \leq |X_o - Y| \leq 4r$, (4.150) gives

$$IV \leq C \sum_{|\mu|<m-1} \frac{1}{r^{m-|\mu|}} \cdot \frac{1}{|S_{4r}(X_o)|} \int_{S_{4r}(X_o)} |(\dot{f} - \dot{\omega}_{2r})_\mu(Y)| \, d\sigma(Y). \tag{4.151}$$

$$\leq C \sum_{|\mu|<m-1} \frac{1}{r^{m-|\mu|}} \left(\frac{1}{|S_{4r}(X_o)|} \int_{S_{4r}(X_o)} |(\dot{f} - \dot{\omega}_{2r})_\mu(Y)| \, d\sigma(Y) \right)^{1/2}.$$

Now (4.100) allows us to conclude,

$$IV \leq C \sum_{|\mu|\leq m-1} \frac{1}{r^{m-|\mu|}} \cdot r^{m-1-|\mu|} \dot{f}^{\#}(X_o)$$

$$\leq \frac{C}{r} \dot{f}^{\#}(X_o) \leq \frac{C}{r} \|\dot{f}\|_{\dot{\mathrm{BMO}}_{m-1}(\partial\Omega)}. \tag{4.152}$$

As for III, by writing the domain of integration in the right-hand side (4.149) as the union of disjoint annuli $S_{2^{j+1}r}(X_o) \setminus S_{2^j r}(X_o)$, $j \geq 1$, we have

$$III \leq C \sum_{|\mu|=m-1} \sum_{j=1}^{\infty} \int_{S_{2^{j+1}r}(X_o) \setminus S_{2^j r}(X_o)} \frac{1}{(2^j r)^n} |(\dot{f} - \dot{\omega}_{2r})_\mu(Y)| \, d\sigma(Y), \tag{4.153}$$

since, for each $j \geq 1$ and $Y \in S_{2^{j+1}r} \setminus S_{2^j r}$, one has $2^j r \leq |X_o - Y|$. We then write

$$III \leq C \sum_{|\mu|=m-1} \sum_{j=1}^{\infty} \frac{1}{2^j r} \fint_{S_{2^{j+1}r}(X_o)} \Big[|(\dot{f} - \dot{\omega}_{2^{j+1}r})_\mu(Y)|$$

$$+ \sum_{i=1}^{j} |(\dot{\omega}_{2^{i+1}r} - \dot{\omega}_{2^i r})_\mu(Y)| \Big] \, d\sigma(Y)$$

$$\leq C \sum_{|\mu|=m-1} \sum_{j=1}^{\infty} \Big[\frac{1}{2^j r} \Big(\fint_{S_{2^{j+1}r}(X_o)} |(\dot{f} - \dot{\omega}_{2^{j+1}r})_\mu(Y)|^2 \, d\sigma(Y) \Big)^{\frac{1}{2}}$$

$$+ \sum_{i=1}^{j} \frac{1}{2^j r} \fint_{S_{2^{j+1}r}(X_o)} |(\dot{\omega}_{2^{i+1}r} - \dot{\omega}_{2^i r})_\mu(Y)| \, d\sigma(Y) \Big]$$

$$\leq C \sum_{|\mu|=m-1} \sum_{j=1}^{\infty} \Big(\frac{1}{2^j r} \dot{f}^{\#}(X_o) + \frac{j}{2^j r} \dot{f}^{\#}(X_o) \Big)$$

$$\leq \frac{C}{r} \dot{f}^{\#}(X_o) \leq \frac{C}{r} \|\dot{f}\|_{\dot{\mathrm{BMO}}_{m-1}(\partial\Omega)}. \tag{4.154}$$

Above, the first inequality in (4.154) follows from (4.153) by writing $\dot{f} - \dot{\omega}_{2r}$ as a telescoping sum and applying the triangle inequality, while the second one is obtained from Hölder's inequality. The third inequality follows from Lemma 4.10, whereas the last inequality is a consequence of (3.387). Now, (4.154), (4.152) and (4.148) give (4.135) and complete the proof of Theorem 4.11. □

We conclude by establishing the counterpart of Theorem 4.11 in the VMO context. Specifically, we have:

Theorem 4.12. *Retain the setting of Theorem 4.11. Then*

$$\begin{aligned} &|\nabla^m \dot{\mathscr{D}} \dot{f}(X)|^2 \rho(X)\, dX \ \text{ is a vanishing Carleson} \\ &\text{measure on } \Omega \text{ for each } \dot{f} \in \dot{\mathrm{VMO}}_{m-1}(\partial\Omega). \end{aligned} \tag{4.155}$$

Proof. Fix an arbitrary $\dot{f} \in \dot{\mathrm{VMO}}_{m-1}(\partial\Omega)$ along with a threshold $\varepsilon > 0$. Pick a number $s \in (0,1)$ such that $s > \frac{n-1}{2n}$, and choose $\dot{g} \in \dot{B}^{\infty,\infty}_{m-1,s}(\partial\Omega)$ such that $\|\dot{f} - \dot{g}\|_{\dot{\mathrm{BMO}}_{m-1}(\partial\Omega)} < \varepsilon$. That this is possible is ensured by Definition 3.52 and Proposition 3.53. Then, for every $X_o \in \partial\Omega$ and $0 < r < \mathrm{diam}\,(\Omega)$, we have

$$\begin{aligned} r^{1-n} \int_{\Delta(X_o,r)} |\nabla^m \dot{\mathscr{D}} \dot{f}|^2 \rho\, dX &\le 2r^{1-n} \int_{\Delta(X_o,r)} |\nabla^m \dot{\mathscr{D}} (\dot{f} - \dot{g})|^2 \rho\, dX \\ &\quad + 2r^{1-n} \int_{\Delta(X_o,r)} |\nabla^m \dot{\mathscr{D}} \dot{g}|^2 \rho\, dX =: I + II. \end{aligned} \tag{4.156}$$

Now, Theorem 4.11 gives that

$$I \le 2\|\, |\nabla^m \dot{\mathscr{D}} (\dot{f} - \dot{g})|^2 \rho\, dX\,\|_{Car} \le C\|\dot{f} - \dot{g}\|^2_{\dot{\mathrm{BMO}}_{m-1}(\partial\Omega)} \le C\varepsilon^2. \tag{4.157}$$

As for term II, anticipating an estimate which will be proved later, we remark that the function $u := |\nabla^m \dot{\mathscr{D}} \dot{g}|^2 \rho$ satisfies $|u| \le C\rho^{2s-1}$ in Ω, thanks to (4.243), where $C := \|\dot{g}\|^2_{\dot{B}^{\infty,\infty}_{m-1,s}(\partial\Omega)}$ is a finite constant. Consequently, by (2.58), $\mu := u\,dX$ is a vanishing Carleson measure, since $2s - 1 > -1/n$ by our choice of s. In turn, this entails that there exists $R > 0$ such that

$$II \le \varepsilon \ \text{ if } \ 0 < r < R, \tag{4.158}$$

uniformly in $X_o \in \partial\Omega$. Altogether, (4.156)–(4.158) show that

$$\lim_{R\to 0+} \left(\sup_{0<r<R,\, X_o\in\partial\Omega} r^{1-n} \int_{\Delta(X_o,r)} |\nabla^m \dot{\mathscr{D}} \dot{f}|^2 \rho\, dX \right) = 0. \tag{4.159}$$

This proves (4.155). □

4.4 Jump Relations

Throughout this section we define and study the boundary principal-value version of the double multi-layer potential operator $\dot{\mathscr{D}}$ introduced in (4.57).

Definition 4.13. Assume that $\Omega \subset \mathbb{R}^n$ is a bounded Lipschitz domain, and that L is a (complex) matrix-valued constant coefficient, homogeneous, W-elliptic differential operator L of order $2m$ in $\mathbb{R}^n$. For each $\dot{f} \in \dot{L}^p_{m-1,0}(\partial\Omega)$, $1 < p < \infty$, define

$$\dot{K}\dot{f} := \Big\{\big(\dot{K}\dot{f}\big)_\gamma\Big\}_{|\gamma|\leq m-1} \tag{4.160}$$

where, for each $\gamma \in \mathbb{N}_0^n$ of length $\leq m-1$, we have set

$$\begin{aligned}
\big(\dot{K}\dot{f}\big)_\gamma(X) := &\sum_{\substack{|\alpha|=m\\|\beta|=m}} \sum_{\ell=1}^{|\gamma|} \sum_{\substack{\delta+\eta+e_k=\alpha\\|\delta|=\ell-1,|\eta|=m-\ell}} \sum_{\substack{\theta+\omega+e_j=\gamma\\|\theta|=\ell-1,|\omega|=|\gamma|-\ell}} \Big\{C_1(m,\ell,\alpha,\delta,\eta,\gamma,\theta,\omega)\times\\
&\times \lim_{\varepsilon\to 0+}\int_{\substack{Y\in\partial\Omega\\|X-Y|>\varepsilon}} \partial_{\tau_{kj}(Y)}\Big(\big(\partial^{\delta+\omega+\beta}E\big)(X-Y)\Big)A_{\beta\alpha}f_{\theta+\eta}(Y)\,d\sigma(Y)\Big\}\\
&-\sum_{\substack{|\alpha|=m\\|\beta|=m}} \sum_{\ell=|\gamma|+1}^{m} \sum_{\substack{\delta+\eta+e_k=\alpha\\|\delta|=\ell-1,|\eta|=m-\ell}} \Big\{C_2(m,\ell,\alpha,\delta,\eta)\times\\
&\times \lim_{\varepsilon\to 0+}\int_{\substack{Y\in\partial\Omega\\|X-Y|>\varepsilon}} \nu_k(Y)\big(\partial^{\delta+\beta}E\big)(X-Y)A_{\beta\alpha}f_{\gamma+\eta}(Y)\,d\sigma(Y)\Big\},
\end{aligned} \tag{4.161}$$

for $X \in \partial\Omega$. Above C_1, C_2 are as in (4.78)–(4.79).

To state our next result, recall the non-tangential boundary multi-trace from (3.272). Also, recall the double multi-layer potential operator defined in (4.57), the convention (2.11), and set

$$\dot{\mathscr{D}}^{\pm}\dot{f} := \Big(\dot{\mathscr{D}}\dot{f}\Big)\Big|_{\Omega_\pm}. \tag{4.162}$$

In the sequel we will sometimes abbreviate $\dot{\mathscr{D}}^{+}$ by $\dot{\mathscr{D}}$. Also, I will denote the identity operator.

Theorem 4.14. *Let $\Omega \subset \mathbb{R}^n$ be a bounded Lipschitz domain and consider a (complex) matrix-valued constant coefficient, homogeneous, W-elliptic differential* *operator L of order $2m$ in $\mathbb{R}^n$. Assume that $\dot{K}$ is associated with these as in (4.160)–(4.161).*

Then for each $\dot{f} \in \dot{L}^p_{m-1,0}(\partial\Omega)$, $1 < p < \infty$, the expression $\dot{K}\dot{f}(X)$ is meaningful at almost every point $X \in \partial\Omega$. Moreover, the operator

$$\dot{K} : \dot{L}^p_{m-1,0}(\partial\Omega) \longrightarrow \dot{L}^p_{m-1,0}(\partial\Omega) \tag{4.163}$$

is well-defined, linear and bounded for each $p \in (1,\infty)$, and the following jump-relation holds

$$\dot{\mathscr{D}}^{\pm}\dot{f}\Big|^{m-1}_{\partial\Omega} = (\pm\tfrac{1}{2}I + \dot{K})\dot{f}, \tag{4.164}$$

for each $\dot{f} \in \dot{L}^p_{m-1,0}(\partial\Omega)$, $1 < p < \infty$.

Furthermore, for each $p \in (1,\infty)$, the operator

$$\dot{K} : \dot{L}^p_{m-1,1}(\partial\Omega) \longrightarrow \dot{L}^p_{m-1,1}(\partial\Omega) \tag{4.165}$$

is also well-defined, linear and bounded. As a corollary,

$$\begin{aligned} \dot{K}^* &: \left(\dot{L}^p_{m-1,0}(\partial\Omega)\right)^* \longrightarrow \left(\dot{L}^p_{m-1,0}(\partial\Omega)\right)^* \\ \dot{K}^* &: \left(\dot{L}^p_{m-1,1}(\partial\Omega)\right)^* \longrightarrow \left(\dot{L}^p_{m-1,1}(\partial\Omega)\right)^* \end{aligned} \tag{4.166}$$

are also well-defined, linear, bounded operators for each $p \in (1,\infty)$.

Proof. The fact that the limits in (4.161) exist at a.e. $X \in \partial\Omega$ is a consequence of Proposition 2.63. This proposition also gives that

$$\sum_{|\gamma|\leq m-1} \|(\dot{K}\dot{f})_\gamma\|_{L^p(\partial\Omega)} \leq C\|\dot{f}\|_{\dot{L}^p_{m-1,0}(\partial\Omega)} \tag{4.167}$$

for some finite constant $C = C(\Omega, L, p) > 0$.

We shall now prove that (4.164) holds which requires that

$$(\dot{\mathscr{D}}^{\pm}\dot{f})_\gamma\Big\lfloor_{\partial\Omega} = \Big[\big(\pm\tfrac{1}{2}I + \dot{K}\big)\dot{f}\Big]_\gamma, \qquad \forall\,\gamma \in \mathbb{N}_0^n,\ |\gamma| \leq m-1. \tag{4.168}$$

We consider first the case of superscript $+$ in (4.168) and fix for the moment $\gamma \in \mathbb{N}_0^n$ with length $\leq m-1$. Notice that in (4.77), a typical kernel for the first group of integral operators is of the type

$$k(X,Y) = \partial_{\tau_{kj}(Y)}\Big(k_o(X-Y)\Big) \quad \text{where } k_o(Z) := (\partial^{\delta+\omega+\beta}E)(Z)A_{\beta\alpha}, \tag{4.169}$$

where $\alpha, \beta, \delta, \omega \in \mathbb{N}_0^n$, $|\alpha| = |\beta| = m$, $|\delta| = \ell - 1$ and $|\omega| = |\gamma| - \ell$, for some number $\ell \in \{1, \dots, |\gamma|\}$. In particular $|\delta| + |\omega| + |\beta| = m + |\gamma| - 1 \leq 2m - 2$ and, consequently, the kernel $k(X, Y)$ is either weakly singular, or amenable to the Calderón–Zygmund theory reviewed in § 2.8. The integral operators with weakly singular kernels do not jump thus, when restricted to the boundary, they contribute to the left-hand side of (4.168) the boundary operators in the first part of (4.161). In the case when the multi-index $\gamma \in \mathbb{N}_0^n$ satisfies $|\gamma| = m - 1$, we employ the jump relations (2.530) to write

$$\Big(\int_{\partial\Omega} \partial_{\tau_{jk}(Y)}[k_o(\cdot - Y)]g(Y)\,d\sigma(Y)\Big)\Big|_{\partial\Omega}(X)$$
$$= \frac{1}{2\sqrt{-1}}[\nu_j \nu_k \widehat{k_o}(\nu) - \nu_k \nu_j \widehat{k_o}(\nu)](X)g(X)$$
$$+ \lim_{\varepsilon \to 0+} \int_{\substack{Y \in \partial\Omega \\ |X-Y|>\varepsilon}} \partial_{\tau_{jk}(Y)}[k_o(X - Y)]g(Y)\,d\sigma(Y)$$
$$= \lim_{\varepsilon \to 0+} \int_{\substack{Y \in \partial\Omega \\ |X-Y|>\varepsilon}} \partial_{\tau_{jk}(Y)}[k_o(X - Y)]g(Y)\,d\sigma(Y). \tag{4.170}$$

for each $g \in L^p(\partial\Omega)$ and almost every $X \in \partial\Omega$. This shows that the restriction to the boundary of the first part in (4.77) gives precisely the first part of (4.161).

Let us now focus on the restriction to the boundary of the second part of (4.77). Notice that whenever $\ell < m$ the kernels in the second part of (4.77) are weakly singular and hence the integral operators with these kernels do not jump. In turn, the restriction to the boundary of these operators give the corresponding terms (i.e., with $\ell < m$) in the second part of (4.161).

Finally, we are left with considering the boundary behavior of the following sum of integral operators (corresponding to the choice $\ell = m$ in the second part of (4.77)):

$$-\sum_{\substack{|\alpha|=m \\ |\beta|=m}} \sum_{\delta+e_k=\alpha} \frac{\alpha!}{m\,\delta!} \int_{\partial\Omega} \nu_k(Y)(\partial^{\delta+\beta}E)(X - Y)A_{\beta\alpha} f_\gamma(Y)\,d\sigma(Y). \tag{4.171}$$

Since $|\beta| + |\delta| = 2m - 1$, the kernels in (4.171) are Calderón–Zygmund and employing (2.530) the non-tangential limit on $\partial\Omega$ of (4.171) is

$$
-\frac{1}{2\sqrt{-1}}\sum_{\substack{|\alpha|=m\\|\beta|=m}}\sum_{\delta+e_k=\alpha}\frac{\alpha!}{m\,\delta!}\nu_k(X)(\widehat{\partial^{\delta+\beta}E})(\nu(X))A_{\beta\alpha}f_\gamma(X) \tag{4.172}
$$

$$
-\lim_{\varepsilon\to 0+}\sum_{\substack{|\alpha|=m\\|\beta|=m}}\sum_{\delta+e_k=\alpha}\frac{\alpha!}{m\,\delta!}\int_{\partial\Omega}\nu_k(Y)(\partial^{\delta+\beta}E)(X-Y)A_{\beta\alpha}f_\gamma(Y)\,d\sigma(Y),
$$

where as before "hat" denotes the Fourier transform. Based on (4.77) and (4.161), in the light of our previous analysis, in order to prove (4.168) it suffices to show

$$
-\frac{1}{2\sqrt{-1}}\sum_{\substack{|\alpha|=m\\|\beta|=m}}\sum_{\delta+e_k=\alpha}\frac{\alpha!}{m\,\delta!}\nu_k(X)(\widehat{\partial^{\delta+\beta}E})(\nu(X))A_{\beta\alpha}f_\gamma(X)=\tfrac{1}{2}f_\gamma(X). \tag{4.173}
$$

Using that

$$
\begin{aligned}
(\widehat{\partial^{\delta+\beta}E})(\nu(X)) &= \sqrt{-1}^{|\delta|+|\beta|}\nu(X)^{\delta+\beta}\widehat{E}(\nu(X))\\
&= (\sqrt{-1})^{2m-1}\nu(X)^{\delta+\beta}\widehat{E}(\nu(X)),
\end{aligned} \tag{4.174}
$$

we obtain

$$
\begin{aligned}
&-\tfrac{1}{2\sqrt{-1}}\sum_{\substack{|\alpha|=m\\|\beta|=m}}\sum_{\delta+e_k=\alpha}\frac{\alpha!}{m\,\delta!}\nu_k(X)(\widehat{\partial^{\delta+\beta}E})(\nu(X))A_{\beta\alpha}f_\gamma(X)\\
&\quad=\tfrac{1}{2}(-1)^m\sum_{\substack{|\alpha|=m\\|\beta|=m}}\sum_{\delta+e_k=\alpha}\frac{\alpha!}{m\,\delta!}\nu(X)^{\delta+\beta+e_k}\widehat{E}(\nu(X))A_{\beta\alpha}f_\gamma(X)\\
&\quad=\tfrac{1}{2}(-1)^m\sum_{\substack{|\alpha|=m\\|\beta|=m}}\Bigg(\sum_{\delta+e_k=\alpha}\frac{\alpha!}{m\,\delta!}\Bigg)\nu(X)^{\alpha+\beta}\widehat{E}(\nu(X))A_{\beta\alpha}f_\gamma(X)\\
&\quad=\tfrac{1}{2}(-1)^m\sum_{\substack{|\alpha|=m\\|\beta|=m}}\nu(X)^{\alpha+\beta}\widehat{E}(\nu(X))A_{\beta\alpha}f_\gamma(X)\\
&\quad=\tfrac{1}{2}(-1)^m\widehat{E}(\nu(X))\Big[\sum_{\substack{|\alpha|=m\\|\beta|=m}}\nu(X)^{\alpha+\beta}A_{\alpha\beta}\Big]f_\gamma(X),
\end{aligned} \tag{4.175}
$$

where the next-to-last identity follows from (4.82). Finally, (4.31) gives

$$\widehat{E}(\nu(X)) = (-1)^m \Big[\sum_{|\alpha|=|\beta|=m} \nu(X)^{\alpha+\beta} A_{\alpha\beta} \Big]^{-1} \tag{4.176}$$

so the identity (4.173) readily follows. This in turn gives (4.168) and finishes the proof of (4.164).

In order to complete the proof of (4.163) it remains to show that for each Whitney array $\dot{f} \in \dot{L}^p_{m-1,0}(\partial\Omega)$ for some $p \in (1,\infty)$ we have that $\dot{K}\dot{f} \in CC$. This is however a direct consequence of (4.164), the non-tangential maximal function estimate (4.63) and (3.275) in Proposition 3.29.

We turn now to proving that the operator (4.165) is also well-defined and bounded. To this end, fix $p \in (1,\infty)$. Then, for each $\dot{f} \in \dot{L}^p_{m-1,1}(\partial\Omega)$ the jump formula (4.164) holds. Then, the desired properties of the operator (4.165) are a consequence of the non-tangential maximal function estimates (4.64), (3.279) from Proposition 3.30 and (3.265). □

Theorem 4.15. *Retain the same background assumptions as in Theorem 4.14. Then the operator*

$$\dot{K} : \dot{h}^p_{m-1,1}(\partial\Omega) \longrightarrow \dot{h}^p_{m-1,1}(\partial\Omega) \tag{4.177}$$

is well-defined, linear and bounded for each $p \in (\frac{n-1}{n},\infty)$.

Proof. When $1 < p < \infty$, it follows that $\dot{h}^p_{m-1,1}(\partial\Omega) = \dot{L}^p_{m-1,1}(\partial\Omega)$ and (4.177) follows from (4.165). On the other hand, the embedding $\dot{h}^p_{m-1,1}(\partial\Omega) \hookrightarrow \dot{L}^{p^*}_{m-1,0}(\partial\Omega)$ holds if $\frac{n-1}{n} < p \leq 1$, where p^* is as in (2.433). Hence, in either case, (4.164) holds for $\dot{f} \in \dot{h}^p_{m-1,1}(\partial\Omega)$, so the boundedness of (4.163) is a consequence of this, (4.87), and (3.279) from Proposition 3.30. □

We continue by presenting a structure theorem for the principal-value double multi-layer in the case when the underlying differential operator factors through the Laplacian.

Theorem 4.16. *Retain the same background assumptions as in Theorem 4.14 and, in addition, assume that L is a W-elliptic differential operator which factors as*

$$L = \Delta\tilde{L}, \tag{4.178}$$

where $\tilde{L}$ *is a constant coefficient differential operator of order* $2m-2$. *Then, for each* $\gamma \in \mathbb{N}_0^n$, *with* $|\gamma| \leq m-1$, *and* $\dot{f} = \{f_\gamma\}_{|\gamma|\leq m-1} \in \dot{L}^p_{m-1,0}(\partial\Omega)$, $1 < p < \infty$, *one has*

$$(\dot{K}\dot{f})_\gamma(X) = \sum_{|\theta|\leq m-1}\sum_{j,k=1}^{n}\lim_{\varepsilon\to 0+}\int_{\substack{Y\in\partial\Omega\\|X-Y|>\varepsilon}} \partial_{\tau_{jk}(Y)}\Big([p_{\gamma,\theta,j,k}(\partial)E](X-Y)\Big)f_\theta(Y)\,d\sigma(Y)$$

$$+ K_\Delta f_\gamma(X) + \textit{ lower order terms}, \tag{4.179}$$

where each $p_{\gamma,\theta,j,k}(\partial)$ *is a differential operator of order* $2m-2$.

Proof. Fix $\gamma \in \mathbb{N}_0^n$ such that $|\gamma| \leq m-1$. A simple analysis of (4.161) reveals that for (4.179) it suffices to show that

$$-\sum_{\substack{|\alpha|=m\\|\beta|=m}}\sum_{\ell=|\gamma|+1}^{m}\sum_{\substack{\delta+\eta+e_k=\alpha\\|\delta|=\ell-1,|\eta|=m-\ell}} C_2(m,\ell,\alpha,\delta,\eta)\times \tag{4.180}$$

$$\times\lim_{\varepsilon\to 0+}\int_{\substack{Y\in\partial\Omega\\|X-Y|>\varepsilon}}\nu_k(Y)(\partial^{\delta+\beta}E)(X-Y)A_{\beta\alpha}f_{\gamma+\eta}(Y)\,d\sigma(Y),$$

can be written as in the right-hand side of (4.179) as the remaining terms in (4.161) are already in this form. As for (4.180), we notice that whenever $\ell \neq m$ in (4.180) one obtains a lower order term. Therefore matters are reduced to showing that if

$$I := -\sum_{\substack{|\alpha|=m\\|\beta|=m}}\sum_{\delta+e_k=\alpha}\frac{\alpha_k}{m}\lim_{\varepsilon\to 0+}\int_{\substack{Y\in\partial\Omega\\|X-Y|>\varepsilon}}\nu_k(Y)(\partial^{\delta+\beta}E)(X-Y)A_{\beta\alpha}f_\gamma(Y)\,d\sigma(Y) \tag{4.181}$$

then

$$I = \sum_{|\theta|\leq m-1}\sum_{j,k=1}^{n}\lim_{\varepsilon\to 0+}\int_{\substack{Y\in\partial\Omega\\|X-Y|>\varepsilon}}\partial_{\tau_{jk}(Y)}\Big([p_{\gamma,\theta,j,k}(\partial)E](X-Y)\Big)f_\theta(Y)\,d\sigma(Y)$$

$$+K_\Delta f_\gamma(X) + \text{ lower order terms }, \tag{4.182}$$

with $p_{\gamma,\theta,j,k}(\partial)$ as in the hypothesis. Above we have used the fact that

$$C_2(m,m,\alpha,\delta,\emptyset) = \frac{\alpha_k}{m} \quad\text{whenever } \delta+e_k=\alpha. \tag{4.183}$$

However, based on the definition of I,

$$I = -\sum_{\substack{|\alpha|=m\\|\beta|=m}}\sum_{k=1}^{n}\frac{\alpha_k}{m}\lim_{\varepsilon\to 0+}\int_{\substack{Y\in\partial\Omega\\|X-Y|>\varepsilon}}\nu_k(Y)(\partial^{\alpha+\beta-e_k}E)(X-Y)A_{\beta\alpha}f_\gamma(Y)\,d\sigma(Y), \tag{4.184}$$

where we make the convention that $\partial^{\alpha+\beta-e_k} E = 0$ if any of the components of the multi-index $\alpha + \beta - e_k$ is negative.

Consider next a generic homogeneous differential operator $\tilde{L}$ of order $2m - 2$,

$$\tilde{L} := \sum_{|\alpha'|=|\beta'|=m-1} \partial^{\alpha'} B_{\alpha'\beta'} \partial^{\beta'}. \tag{4.185}$$

Then, due to (4.178), we have

$$\sum_{|\alpha|=|\beta|=m} A_{\alpha\beta}\, \partial^{\alpha+\beta} = \sum_{|\alpha'|=|\beta'|=m-1} \sum_{j=1}^{n} B_{\alpha'\beta'} \partial^{\alpha'+\beta'+2e_j}. \tag{4.186}$$

A moment's reflection shows then that, for each $\alpha, \beta \in \mathbb{N}_0^n$, with $|\alpha| = |\beta| = m$, we have

$$A_{\alpha\beta} = \sum_{j=1}^{n} B_{(\alpha-e_j)(\beta-e_j)}, \tag{4.187}$$

where $B_{(\alpha-e_j)(\beta-e_j)} = 0$ whenever either $\alpha - e_i$ or $\beta - e_i$ fails to be a multi-index in $\mathbb{N}_0^n$, otherwise they are as in (4.185). Employing (4.187) in (4.184) we obtain

$$I = -\sum_{\substack{|\alpha|=m\\|\beta|=m}} \sum_{k=1}^{n} \frac{\alpha_k}{m} \lim_{\varepsilon\to 0+} \int_{\substack{Y\in\partial\Omega\\|X-Y|>\varepsilon}} \nu_k(Y) \sum_{j=1}^{n} (\partial^{(\alpha-e_j)+(\beta-e_j)+2e_j-e_k} E)(X-Y) \times$$
$$\times B_{(\beta-e_j)(\alpha-e_i)} f_\gamma(Y)\, d\sigma(Y). \tag{4.188}$$

Writing $\nu_k \partial^{e_j} = \partial_{\tau_{kj}} + \nu_j \partial^{e_k}$ we get

$$I = II + III, \tag{4.189}$$

where

$$II := \sum_{\substack{|\alpha|=m\\|\beta|=m}} \sum_{k=1}^{n} \frac{\alpha_k}{m} \lim_{\varepsilon\to 0+} \int_{\substack{Y\in\partial\Omega\\|X-Y|>\varepsilon}} \partial_{\tau_{kj}(Y)} \sum_{j=1}^{n} (\partial^{(\alpha-e_j)+(\beta-e_j)+e_j-e_k} E)(X-Y) \times$$
$$\times B_{(\beta-e_j)(\alpha-e_j)} f_\gamma(Y)\, d\sigma(Y), \tag{4.190}$$

and

$$III := \sum_{\substack{|\alpha|=m\\|\beta|=m}} \sum_{k=1}^{n} \frac{\alpha_k}{m} \lim_{\varepsilon\to0+} \int_{\substack{Y\in\partial\Omega\\|X-Y|>\varepsilon}} \sum_{j=1}^{n} \nu_j(Y)\partial_Y^{e_j}(\partial^{(\alpha-e_j)+(\beta-e_j)}E)(X-Y)\times$$

$$\times B_{(\beta-e_j)(\alpha-e_j)} f_\gamma(Y)\,d\sigma(Y). \tag{4.191}$$

Observe next that II is of the form

$$\sum_{|\theta|\le m-1} \sum_{j,k=1}^{n} \lim_{\varepsilon\to0+} \int_{\substack{Y\in\partial\Omega\\|X-Y|>\varepsilon}} \partial_{\tau_{jk}(Y)}\Big([p_{\gamma,\theta,j,k}(\partial)E](X-Y)\Big) f_\theta(Y)\,d\sigma(Y). \tag{4.192}$$

As far as III is concerned, notice first that for a fixed $\alpha\in\mathbb{N}_0^n$, $|\alpha|=m$ we have that $\sum_{k=1}^{m}\frac{\alpha_k}{m}=1$. Also, based on (4.185), for each fixed $j\in\{1,\dots,n\}$ we have

$$\sum_{|\alpha|=|\beta|=m} B_{(\alpha-e_j)(\beta-e_j)}\partial^{\alpha+\beta-2e_j} = \tilde{L}. \tag{4.193}$$

Finally, (4.191) and the preceding observations imply

$$III = \sum_{j=1}^{n} \lim_{\varepsilon\to0+} \int_{\substack{Y\in\partial\Omega\\|X-Y|>\varepsilon}} \nu_j(Y)\partial_Y^{e_j}\Big[(\tilde{L}^t E_{L^t})(X-Y)\Big]^t f_\gamma(Y)\,d\sigma(Y), \tag{4.194}$$

and since $\tilde{L}^t E_{L^t} = \Gamma$ (as $L^t = \Delta\tilde{L}^t$), where Γ is the fundamental solution of the Laplacian, (4.194) readily gives that

$$III = K_\Delta f_\gamma(X). \tag{4.195}$$

This finishes the proof of (4.182) and completes the proof of Theorem 4.16. □

Our next theorem addresses the issue of the boundedness of the principal-value double multi-layer on Whitney–BMO spaces. As a preamble, we shall prove the following useful result.

Proposition 4.17. *Let $\Omega\subset\mathbb{R}^n$ be a bounded Lipschitz domain and consider a (complex) matrix-valued constant coefficient, homogeneous, W-elliptic differential operator L of order $2m$ in $\mathbb{R}^n$. In this context, define $\dot{K}$ as in* (4.160)–(4.161)*. Then*

$$\dot{K}(\mathrm{tr}_{m-1}u) = \tfrac{1}{2}\,\mathrm{tr}_{m-1}u \quad on \quad \partial\Omega, \quad for\ any \quad u\in\mathscr{P}_{m-1}. \tag{4.196}$$

Proof. This follows by taking the non-tangential limit in (4.62) and using the jump-formula (4.164). □

The action of the principal-value double multi-layer on Whitney–BMO spaces is considered next. The theorem below extends work done in the case when $n = 2$ and $L = \Delta^2$ in [24] and answers the question posed by J. Cohen at the top of page 111 in [24].

Theorem 4.18. *Assume that $\Omega \subset \mathbb{R}^n$ is a bounded Lipschitz domain, L is a (complex) matrix-valued constant coefficient, homogeneous, W-elliptic differential operator L of order $2m$ in $\mathbb{R}^n$, and consider $\dot{K}$ as in* (4.160)–(4.161). *Then*

$$\dot{K} : \dot{\mathrm{BMO}}_{m-1}(\partial\Omega) \longrightarrow \dot{\mathrm{BMO}}_{m-1}(\partial\Omega), \tag{4.197}$$

is well-defined, linear and bounded.

Proof. The first observation is that, given an arbitrary $\dot{f} \in \dot{\mathrm{BMO}}_{m-1}(\partial\Omega)$, due to (4.163),

$$\dot{K}\dot{f} \in \dot{L}^2_{m-1,0}(\partial\Omega). \tag{4.198}$$

Therefore matters reduce to showing

$$(\dot{K}\dot{f})^{\#} \in L^\infty(\partial\Omega). \tag{4.199}$$

To this end, fix $X_o \in \partial\Omega$, $r > 0$ and $\dot{f} \in \dot{\mathrm{BMO}}_{m-1}(\partial\Omega)$. For each $R > 0$, recall the best fit polynomial array $\dot{\omega}_R(\dot{f})$ introduced as in (4.88). We consider next a function η in $\mathbb{R}^n$ as in (4.130). In particular the following hold

$$\eta \in C_c^\infty(\mathbb{R}^n), \quad 0 \le \eta \le 1, \quad \eta \equiv 1 \ \text{ on } \ S_{2r}(X_o), \quad \mathrm{supp}\,\eta \subset S_{4r}(X_o), \tag{4.200}$$

$$|\partial^\alpha \eta| \le \frac{C_\alpha}{r^{|\alpha|}}, \qquad \text{for each } \alpha \in \mathbb{N}_0^n. \tag{4.201}$$

We write

$$\dot{f} = \eta(\dot{f} - \dot{\omega}_{2r}) + (1-\eta)(\dot{f} - \dot{\omega}_{2r}) + \dot{\omega}_{2r}, \tag{4.202}$$

where for an array $\dot{g}$, the multiplication $\eta\dot{g}$ is in the sense of (3.309). Using the linearity of the operator $\dot{K}$ on $L^2_{m-1,0}(\partial\Omega)$, formulas (4.202) and (4.196) give

$$\begin{aligned}(\tfrac{1}{2}I + \dot{K})\dot{f} &= (\tfrac{1}{2}I + \dot{K})\Big(\eta(\dot{f} - \dot{\omega}_{2r})\Big) \\ &\quad + (\tfrac{1}{2}I + \dot{K})\Big((1-\eta)(\dot{f} - \dot{\omega}_{2r})\Big) + \tfrac{3}{2}\dot{\omega}_{2r}.\end{aligned} \tag{4.203}$$

Therefore

$$(\tfrac{1}{2}I+\dot{K})\dot{f}-\tfrac{3}{2}\dot{\omega}_{2r}=(\tfrac{1}{2}I+\dot{K})\Big(\eta(\dot{f}-\dot{\omega}_{2r})\Big)+(\tfrac{1}{2}I+\dot{K})\Big((1-\eta)(\dot{f}-\dot{\omega}_{2r})\Big)$$
$$=:I+II. \tag{4.204}$$

We claim that there exists $C>0$, independent of X_o, $\dot{f}$ and $r>0$ such that

$$\Big(⨍_{S_r(X_o)}|I(Y)|^2\,d\sigma(Y)\Big)^{1/2}\leq C\,\dot{f}^{\#}(X_o), \tag{4.205}$$

and that

$$\begin{gathered}\text{there exists } \dot{P}\in\dot{\mathscr{P}}_{m-1}(\partial\Omega)\text{ with the property that}\\ \Big(⨍_{S_r(X_o)}|II(Y)-\dot{P}(Y)|^2\,d\sigma(Y)\Big)^{1/2}\leq C\,\dot{f}^{\#}(X_o).\end{gathered} \tag{4.206}$$

We start with showing (4.205) and, in this regard, first notice that

$$\begin{aligned}\int_{S_r(X_o)}|I(Y)|^2\,d\sigma(Y)&=\int_{S_r(X_o)}\Big|(\tfrac{1}{2}I+\dot{K})\Big(\eta(Y)(\dot{f}(Y)-\dot{\omega}_{2r}(Y))\Big)\Big|^2\,d\sigma(Y)\\&\leq\int_{\partial\Omega}\Big|(\tfrac{1}{2}I+\dot{K})\Big(\eta(Y)(\dot{f}(Y)-\dot{\omega}_{2r}(Y))\Big)\Big|^2\,d\sigma(Y).\end{aligned} \tag{4.207}$$

Based on Theorem 4.14, the operator $\frac{1}{2}I+\dot{K}$ is bounded on $L^2_{m-1,0}(\partial\Omega)$, and together with (4.207), this implies that there exists a finite constant $C>0$ such that

$$\begin{aligned}&\int_{S_r(X_o)}|I(Y)|^2\,d\sigma(Y)\\&\quad\leq C\int_{\partial\Omega}|\eta(Y)(\dot{f}(Y)-\dot{\omega}_{2r}(Y))|^2\,d\sigma(Y)\\&\quad\leq C\sum_{|\alpha|\leq m-1}\sum_{\beta+\gamma=\alpha}\frac{\alpha!}{\beta!\gamma!}\int_{S_{2r}(X_o)}|\partial^\beta\eta(Y)|^2|\big(\dot{f}(Y)-\dot{\omega}_{2r}(Y)\big)_\gamma|^2\,d\sigma(Y).\end{aligned} \tag{4.208}$$

The last identity above follows from (3.309) and the Leibniz rule of differentiation. Employing next (4.201), we have $|\partial^\beta\eta(Y)|\leq Cr^{-|\beta|}$ and hence

$$\int_{S_r(X_o)} |I(Y)|^2\,d\sigma(Y) \le C \sum_{|\alpha|\le m-1}\sum_{\beta+\gamma=\alpha} \frac{1}{r^{2|\beta|}} \int_{S_{2r}(X_o)} |(\dot f(Y) - \dot\omega_{2r}(Y))_\gamma|^2\,d\sigma(Y). \tag{4.209}$$

Due to (4.100) from Lemma 4.10 the above further gives

$$\begin{aligned}\int_{S_r(X_o)} |I(Y)|^2\,d\sigma(Y) &\le C \sum_{|\alpha|\le m-1}\sum_{\beta+\gamma=\alpha} \frac{r^{2(m-1-|\gamma|)}}{r^{2|\beta|}} \Big(\dot f^{\#}(X_o)\Big)^2 \cdot |S_{2r}(X_o)| \\ &\le C \sum_{|\alpha|\le m-1} r^{2(m-1-|\alpha|)} \Big(\dot f^{\#}(X_o)\Big)^2 \cdot |S_r(X_o)|, \end{aligned}\tag{4.210}$$

where in the last inequality in (4.210) we have used that $|\beta| + |\gamma| = |\alpha|$ as well as $|S_r(X_o)| \approx |S_{2r}(X_o)|$. Finally (4.205) now readily follows from (4.210) and the fact that $r^{2(m-1-|\alpha|)} \le C = C(\partial\Omega)$, since $|\alpha| \le m-1$.

We shall focus next on proving (4.206). For each array $\dot g \in L^2_{m-1,0}(\partial\Omega)$, $\dot g = \{g_\gamma\}_{|\gamma|\le m-1}$ and $\alpha \in \mathbb{N}_0^n$, multi-index of length $\le m-1$, we introduce

$$P_\alpha(\dot g; X, X_o) := \sum_{|\beta|\le m-1-|\alpha|} \frac{1}{\beta!} g_{\alpha+\beta}(X_o)(X - X_o)^\beta, \quad X \in \partial\Omega, \tag{4.211}$$

and

$$R_\alpha(\dot g; X, X_o) := g_\alpha(X) - P_\alpha(\dot g; X, X_o), \quad X \in \partial\Omega. \tag{4.212}$$

Going further, we set

$$\begin{aligned}\dot P(\dot g; X, X_o) &:= \Big(P_\alpha(\dot g; X, X_o)\Big)_{|\alpha|\le m-1} \quad \text{and} \\ \dot R(\dot g; X, X_o) &:= \Big(R_\alpha(\dot g; X, X_o)\Big)_{|\alpha|\le m-1},\end{aligned}\tag{4.213}$$

and straightforward algebraic manipulations show that $\dot P(\dot g; X, X_o)$ satisfies the compatibility conditions (3.2). This and degree considerations guarantee that the array $\dot P(\dot g; X, X_o) \in \dot{\mathscr{P}}_{m-1}(\partial\Omega)$, where as before the latter stands for the set of polynomial arrays of degree less than or equal to $m-1$. In turn, based on (4.212), $\dot R(\dot g; X, X_o)$ also satisfies the compatibility conditions (3.2). In the notation we just introduced,

$$\begin{aligned}(\tfrac12 I + \dot K)\Big((1-\eta)(\dot f - \dot\omega_{2r})\Big) &= \dot P\Big((\tfrac12 I + \dot K)((1-\eta)(\dot f - \dot\omega_{2r})); X, X_o\Big) \\ &\quad + \dot R\Big((\tfrac12 I + \dot K)((1-\eta)(\dot f - \dot\omega_{2r})); X, X_o\Big),\end{aligned}\tag{4.214}$$

and we claim that there exists $C > 0$ (independent of $\dot{f}$, X_o, $r > 0$) such that for each multi-index $\alpha \in \mathbb{N}_0^n$, $|\alpha| \leq m-1$, we have

$$\left| R_\alpha \left((\tfrac{1}{2}I + \dot{K})((1-\eta)(\dot{f} - \dot{\omega}_{2r})); X, X_o \right) \right| \leq C \dot{f}^{\#}(X_o). \tag{4.215}$$

The starting point for proving (4.215) is the observation that, based on (4.212) and the jump relations (4.164), for a given $\dot{g} \in L^2_{m-1,0}(\partial\Omega)$ and for almost every point $X \in \partial\Omega$, we have

$$\begin{aligned} & R_\alpha \left((\tfrac{1}{2}I + \dot{K})\dot{g}; X, X_o \right) \\ & = \mathrm{Tr}\Big[\text{Taylor remainder of order } m-1-|\alpha| \text{ of } \partial^\alpha \dot{\mathscr{D}}\dot{g} \text{ at } X_o \Big](X). \end{aligned} \tag{4.216}$$

Next, if $\dot{g}$ vanishes near X_o, the Taylor formula implies

$$\begin{aligned} & \left| R_\alpha \left((\tfrac{1}{2}I + \dot{K})\dot{g}; X, X_o \right) \right| \\ & \leq C \sum_{|\gamma|=m-|\alpha|} \int_0^1 |X - X_o|^{m-|\alpha|} \left| \left(\partial^{\gamma+\alpha} \dot{\mathscr{D}} \right) \dot{g}(X_o + t(X - X_o)) \right| dt. \end{aligned} \tag{4.217}$$

It is important to point out that Taylor's formula require that the function $\dot{\mathscr{D}}\dot{g}$ be sufficiently smooth in a neighborhood of X_o. In general, $\dot{\mathscr{D}}\dot{g}$ is not even continuous near X_o, however, it is so if $\dot{g}$ vanishes near X_o. Now, based on (4.57), (4.144) and the support condition on η,

$$\begin{aligned} & \left| \left(\partial^{\gamma+\alpha} \dot{\mathscr{D}} \right) \dot{g}(X_o + t(X - X_o)) \right| \\ & \leq C \int_{\partial\Omega \setminus S_{2r}(X_o)} \frac{1}{|Y - (X_o + t(X - X_o))|^n} \sum_{|\delta| \leq m-1} |\dot{g}_\delta(Y)| \, d\sigma(Y). \end{aligned} \tag{4.218}$$

Furthermore, a simple inspection reveals that there exist $C_j > 0$, $j = 1, 2$, independent of r, X_o and $X \in S_r(X_o)$, such that for each $X \in S_r(X_o)$ the following holds

$$C_1 |Y - (X_o + t(X - X_o))| \leq |X_o - Y| \leq C_2 |Y - (X_o + t(X - X_o))|. \tag{4.219}$$

Collectively, (4.217)–(4.219) further imply

$$\left| R_\alpha \left((\tfrac{1}{2}I + \dot{K})\dot{g}; X, X_o \right) \right| \leq C \int_{\partial\Omega \setminus S_{2r}(X_o)} \frac{r^{m-|\alpha|}}{|Y - X_o|^n} \sum_{|\delta| \leq m-1} |g_\delta(Y)| \, d\sigma(Y), \tag{4.220}$$

whenever $X \in S_r(X_o)$. However, proceeding as in the proof of Theorem 4.11 when having to estimate the right hand side of (4.137), we have

$$\int_{\partial\Omega\setminus S_{2r}(X_o)} \frac{1}{|Y - X_o|^n} \sum_{|\delta|\leq m-1} |(1-\eta(Y))(\dot{f}(Y) - \dot{\omega}_{2r}(Y))_\delta| \, d\sigma(Y) \leq \frac{C}{r} \dot{f}^{\#}(X_o). \tag{4.221}$$

Above we considered $\dot{g} = (1-\eta)(\dot{f} - \dot{\omega}_{2r})$ which vanishes in a neighborhood of X_o due to the fact that $\eta \equiv 1$ on $S_{2r}(X_o)$. Then, from (4.220) and (4.221) we deduce

$$\begin{aligned} \left| R_\alpha\left((\tfrac{1}{2}I + \dot{K})((1-\eta)(\dot{f} - \dot{\omega}_{2r})); X, X_o\right)\right| &\leq C r^{m-1-|\alpha|} \dot{f}^{\#}(X_o) \\ &\leq C \dot{f}^{\#}(X_o). \end{aligned} \tag{4.222}$$

That (4.206) holds can now be justified by invoking (4.214) and (4.222). Finally, (4.204)–(4.206) give that there exists $\dot{P} \in \dot{\mathscr{P}}_{m-1}(\partial\Omega)$ such that

$$\left(⨍_{S_r(X_o)} \left| (\tfrac{1}{2}I + \dot{K})\dot{f}(Y) - \dot{P}(Y) \right|^2 d\sigma(Y) \right)^{1/2} \leq C \dot{f}^{\#}(X_o), \tag{4.223}$$

from which (4.199) readily follows. This finishes the proof of Theorem 4.18. □

4.5 Estimates on Besov, Triebel–Lizorkin, and Weighted Sobolev Spaces

The aim of this section is to study the action of the multi-layer operators $\dot{\mathscr{D}}$ and $\dot{K}$, introduced in Definition 4.4 and Definition 4.13, respectively, on Besov and Triebel–Lizorkin scales in Lipschitz domains.

Theorem 4.19. *Suppose that Ω is a bounded Lipschitz domain in $\mathbb{R}^n$, and assume that L is a W-elliptic homogeneous differential operator of order $2m$ with (complex) matrix-valued constant coefficients. Then if the numbers p, q, s satisfy the inequalities $\frac{n-1}{n} < p \leq \infty$, $0 < q \leq \infty$, $(n-1)(\frac{1}{p} - 1)_+ < s < 1$, the operator*

$$\dot{\mathscr{D}} : \dot{B}^{p,q}_{m-1,s}(\partial\Omega) \longrightarrow B^{p,q}_{m-1+s+1/p}(\Omega) \tag{4.224}$$

is well-defined, linear and bounded. Moreover,

$$\dot{\mathscr{D}} : \dot{B}^{p,p}_{m-1,s}(\partial\Omega) \longrightarrow F^{p,q}_{m-1+s+1/p}(\Omega) \tag{4.225}$$

is also well-defined, linear and bounded (with the additional convention that $q = \infty$ if $p = \infty$).

Finally, similar properties hold for $\psi\dot{\mathscr{D}}^{-}$ *(cf. the convention* (4.162)*), for any cutoff function* $\psi \in C_c^\infty(\mathbb{R}^n)$.

Proof. Consider the task of proving that the operator (4.224) is well-defined and bounded when $p, q \neq \infty$. In this case, fix $\dot{f} = \{f_\delta\}_{|\delta| \leq m-1} \in \dot{B}^{p,q}_{m-1,s}(\partial\Omega)$ and note that, by (3.224), it suffices to show that

$$\partial^\alpha \dot{\mathscr{D}} \dot{f} \in B^{p,q}_{-1+s+1/p}(\Omega), \qquad \forall\, \alpha \in \mathbb{N}_0^n,\ |\alpha| \leq m, \tag{4.226}$$

with appropriate control of the norm. Let us indicate how this is done in the case when $|\alpha| = m$. Pick $i \in \{1, \ldots, n\}$ and write $\alpha = e_i + \gamma$ where γ has length $m-1$. In particular, the discussion in the next couple of paragraphs following (4.80) shows that $\partial^\alpha(\dot{\mathscr{D}} \dot{f})(X)$ can be written in the form

$$\sum_{|\beta|=2m-1} \sum_{|\delta| \leq m-1} \sum_{j,k=1}^{n} \int_{\partial\Omega} (\partial^\beta E)(X-Y) C_{\beta,\delta,j,k} (\partial_{\tau_{jk}} f_\delta)(Y)\, d\sigma(Y), \tag{4.227}$$

where $C_{\beta,\delta,j,k}$ are suitable matrices. Since Proposition 2.58 gives $\partial_{\tau_{jk}} f_\delta \in B^{p,q}_{s-1}(\partial\Omega)$, for every δ, j, k, it suffices to show that the assignment

$$B^{p,q}_{s-1}(\partial\Omega) \ni f \mapsto \nabla R_{\beta'} f \in B^{p,q}_{-1+s+1/p}(\Omega) \tag{4.228}$$

is bounded, where for each $\beta' \in \mathbb{N}_0^n$ of length $2m-2$ we have set

$$R_{\beta'} f(X) := \int_{\partial\Omega} k(X-Y) f(Y)\, d\sigma(Y), \quad X \in \Omega. \tag{4.229}$$

where $k(X)$ is a generic entry in $(\partial^{\beta'} E)(X-Y)$. We shall show that, whenever $|\beta'| = 2m-2$,

$$R_{\beta'} : B^{p,q}_{s-1}(\partial\Omega) \longrightarrow B^{p,q}_{s+1/p}(\Omega) \quad \text{boundedly}, \tag{4.230}$$

from which (4.228) follows.

In turn, via real interpolation, (4.230) will be a consequence of the fact that

$$R_{\beta'} : B^{p,p}_{s-1}(\partial\Omega) \longrightarrow B^{p,p}_{s+1/p}(\Omega) \quad \text{boundedly}, \tag{4.231}$$

for each multi-index $\beta' \in \mathbb{N}_0^N$ with $|\beta'| = 2m-2$, if $\frac{n-1}{n} < p \leq \infty$, as well as $(n-1)(\frac{1}{p}-1)_+ < s < 1$. Finally, given that $L(R_{\beta'} f) = 0$ in Ω, the claim (4.231) follows from Proposition 2.64 and Theorem 2.41. This finishes the treatment of the case when $|\alpha| = m$. The case when $|\alpha| \leq m-1$ is similar and simpler, completing the proof of (4.224) in the case when $p, q \neq \infty$.

The case when $p \neq \infty$ and $0 < q \leq \infty$ can then be covered from what we have proved so far and real interpolation; cf. (3.314) and (3.317). By the same token, matters can be reduced to considering the case when $p = q = \infty$. To this end, let us fix $\dot{f} = \{f_\eta\}_{|\eta|\leq m-1} \in \dot{B}^{\infty,\infty}_{m-1,s}(\partial\Omega)$ of norm one, and abbreviate (4.4) as

$$\dot{\mathscr{D}}\dot{f}(X) = \sum_{\substack{|\alpha|=|\beta|=m \\ \gamma+\eta+e_j=\alpha}} C^{\alpha,\beta}_{\gamma,\eta,j} \int_{\partial\Omega} \nu_j(Y)(\partial^{\beta+\gamma}E)(X-Y)A_{\beta\alpha} f_\eta(Y)\,d\sigma(Y), \quad X \in \Omega, \tag{4.232}$$

where the $C^{\alpha,\beta}_{\gamma,\eta,j}$'s are real constants. For each multi-index $\gamma \in \mathbb{N}_0^n$ of length $\leq m-1$, consider next, in analogy with (3.78) and (3.73),

$$P_\gamma(X,Y) := \sum_{|\delta|\leq m-1-|\gamma|} \frac{1}{\delta!} f_{\delta+\gamma}(Y)(X-Y)^\delta, \quad X \in \mathbb{R}^n,\ Y \in \partial\Omega, \tag{4.233}$$

$$R_\gamma(X,Y) := f_\gamma(X) - P_\gamma(X,Y), \qquad X, Y \in \partial\Omega. \tag{4.234}$$

For ease of reference, for each $Y \in \partial\Omega$ let us also set

$$\dot{P}(\cdot,Y) := \{P_\gamma(\cdot,Y)\}_{|\gamma|\leq m-1}, \qquad \dot{R}(\cdot,Y) := \{R_\gamma(\cdot,Y)\}_{|\gamma|\leq m-1}. \tag{4.235}$$

A direct calculation shows that $\dot{P}(\cdot,Y) \in CC$ for every $Y \in \partial\Omega$ (cf. (3.81), or the discussion on p. 177 in [119]). Since $f_\gamma(X) = P_\gamma(X,Z) + R_\gamma(X,Z)$ if $|\gamma| \leq m-1$ and $X, Z \in \partial\Omega$, we have

$$\dot{f} = \dot{P}(\cdot,Z) + \dot{R}(\cdot,Z), \quad \forall\, Z \in \partial\Omega. \tag{4.236}$$

In particular, $\dot{R}(\cdot,Y) \in CC$ for every $Y \in \partial\Omega$. Furthermore, since $\dot{P}(\cdot,Z) =$ has polynomial components of degree $\leq m-1$, it follows from (4.236) and the reproducing property (4.62) that

$$\begin{aligned} \dot{\mathscr{D}}\dot{f}(X) &= \dot{\mathscr{D}}(\dot{P}(\cdot,Z))(X) + \dot{\mathscr{D}}(\dot{R}(\cdot,Z))(X) \\ &= P_{(0,\dots,0)}(X,Z) + \dot{\mathscr{D}}(\dot{R}(\cdot,Z))(X) \end{aligned} \tag{4.237}$$

for every $X \in \Omega$ and $Z \in \partial\Omega$. Consequently, for each multi-index $\theta \in \mathbb{N}_0^n$ with $|\theta| = m$, we arrive at the identity

$$\begin{aligned} \partial^\theta \dot{\mathscr{D}}\dot{f}(X) &= \partial^\theta \dot{\mathscr{D}}(\dot{R}(\cdot,Z))(X) \\ &= \sum_{\substack{|\alpha|=|\beta|=m \\ \gamma+\eta+e_j=\alpha}} C^{\alpha,\beta}_{\gamma,\eta,j} \int_{\partial\Omega} \nu_j(Y)(\partial^{\beta+\gamma+\theta}E)(X-Y)A_{\beta\alpha} R_\eta(Y,Z)\,d\sigma(Y), \end{aligned} \tag{4.238}$$

valid for each $X \in \Omega$ and $Z \in \partial\Omega$. Next, given a point $X \in \Omega$, we specialize (4.238) by choosing $Z := \pi(X)$, where $\pi : \Omega \to \partial\Omega$ is a mapping chosen with the property that $\operatorname{dist}(X, \partial\Omega) \approx |X - \pi(X)|$, uniformly for $X \in \Omega$. In turn, for each $\theta \in \mathbb{N}_0^n$ with $|\theta| = m$, this and (4.29) entail

$$|\partial^\theta \dot{\mathscr{D}} \dot{f}(X)| \leq C \sum_{|\eta| \leq m-1} \int_{\partial\Omega} \frac{1}{|X - Y|^{n-1+m-|\eta|}} |R_\eta(Y, \pi(X))| \, d\sigma(Y), \tag{4.239}$$

for every $X \in \Omega$. We may therefore estimate

$$\begin{aligned} \rho^{1-s}(X)|\partial^\theta \dot{\mathscr{D}} \dot{f}(X)| &\leq C \sum_{|\eta| \leq m-1} \int_{\partial\Omega} \frac{\rho(X)^{1-s}}{|X - Y|^{n-1+m-|\eta|}} |R_\eta(Y, \pi(X))| \, d\sigma(Y) \\ &\leq C \sum_{|\eta| \leq m-1} \int_{\partial\Omega} \frac{\rho(X)^{1-s}}{|X - Y|^{n-1+m-|\eta|} |Y - \pi(X)|^{m-1-|\eta|+s}} \, d\sigma(Y) \\ &= C \sum_{|\eta| \leq m-1} \int_{|Y - \pi(X)| > c\rho(X)} \cdots + C \sum_{|\eta| \leq m-1} \int_{|Y - \pi(X)| < c\rho(X)} \cdots \\ &=: I_1 + I_2, \end{aligned} \tag{4.240}$$

where the second inequality above utilizes (3.75). Localizing and passing from graph to local (Euclidean) coordinates, we see that

$$\begin{aligned} I_1 &\leq C \sum_{|\eta| \leq m-1} \int_{\substack{|y-x| > ct \\ y \in \mathbb{R}^{n-1}}} \frac{t^{1-s}}{[t + |x - y|]^{n-1+m-|\eta|}} \cdot |x - y|^{m-1-|\eta|+s} \, dy \\ &\leq C \sum_{|\eta| \leq m-1} \frac{t^{1-s}}{t^{n-1+m-|\eta|}} \cdot t^{m-1-|\eta|+s} \cdot t^{n-1} \cdot \int_{\substack{|z| \geq c \\ z \in \mathbb{R}^{n-1}}} \frac{|z|^{m-1-|\eta|+s}}{[1 + |z|]^{n-1+m-|\eta|}} \, dz \leq C, \end{aligned} \tag{4.241}$$

where in the second step we have changed variables twice (first from $y - x$ to y, then from y to tz). In a similar fashion,

$$\begin{aligned} I_2 &\leq C \sum_{|\eta| \leq m-1} \int_{|y-x| < ct} \frac{t^{-s} \cdot |y - x|}{|x - y|^{n-1+m-|\eta|}} |x - y|^{m-1-|\eta|+s} \, dy \\ &\leq C \sum_{|\eta| \leq m-1} \frac{t^{-s}}{t^{n-1-s}} \cdot t^{n-1} \cdot \int_{\substack{|z| \leq c \\ z \in \mathbb{R}^{n-1}}} \frac{dz}{|z|^{n-1-s}} \leq C. \end{aligned} \tag{4.242}$$

Altogether, (4.240)–(4.242) show that there exists $C = C(\Omega, L, s)$ such that

$$\sup_{X\in\Omega}\Big[\rho(X)^{1-s}|\nabla^m \dot{\mathscr{D}}\dot{f}(X)|\Big] \leq C\|\dot{f}\|_{\dot{B}^{\infty,\infty}_{m-1,s}(\partial\Omega)}. \tag{4.243}$$

Next, we remark that for each $0 < s < 1$,

$$\dot{B}^{\infty,\infty}_{m-1,s}(\partial\Omega) \hookrightarrow \dot{B}^{p,p}_{m-1,r}(\partial\Omega) \xrightarrow{\nabla^j \dot{\mathscr{D}}} B^{p,p}_{m-1-j+r+1/p}(\Omega) \hookrightarrow L^\infty(\Omega), \tag{4.244}$$

whenever $j \in \{1,\dots,m-1\}$, $0 < r < s$ and $\frac{n-1}{r} < p < \infty$ Indeed, the first inclusion in (4.244) is a consequence of Proposition 2.51, the boundedness of the second arrow is based (4.224), and the last inclusion is a standard embedding result. Consequently, there exists $C > 0$ such that for each $\dot{f} \in \dot{B}^{\infty,\infty}_{m-1,s}(\partial\Omega)$,

$$\sup_{X\in\Omega}\Big[\sum_{j=0}^{m-1}|\nabla^j \dot{\mathscr{D}}\dot{f}(X)|\Big] \leq C\|\dot{f}\|_{\dot{B}^{\infty,\infty}_{m-1,s}(\partial\Omega)}. \tag{4.245}$$

Together, estimates (4.243), (4.245) and (2.227) give that $\nabla^{m-1}\dot{\mathscr{D}}\dot{f} \in C^s(\Omega)$. Hence $\dot{\mathscr{D}}\dot{f} \in C^{m-1+s}(\Omega)$ which, by (2.185), gives $\dot{\mathscr{D}}\dot{f} \in B^{\infty,\infty}_{m-1+s}(\Omega)$, as desired.

Moving on, (4.225) is a consequence of (4.224) with $p = q$, the fact that the diagonal of the Besov scale coincides with the diagonal of the Triebel–Lizorkin scale, and (2.265) from Theorem 2.41.

Consider now the last claim in the statement of Theorem 4.19. Given a function $\psi \in C^\infty_c(\mathbb{R}^n)$, pick $R > 0$ large enough so that $B_R(0)$ contains $\overline{\Omega} \cup \operatorname{supp}\psi$. Set now $D_R := B_R(0) \setminus \overline{\Omega}$ and denote by $\dot{\mathscr{D}}_R$ the double multi-layer associated with the bounded Lipschitz domain $D_R \subset \mathbb{R}^n$. Finally, let

$$\iota_R : \dot{B}^{p,q}_{m-1,s}(\partial\Omega) \to \dot{B}^{p,q}_{m-1,s}(\partial\Omega) \oplus \dot{B}^{p,q}_{m-1,s}(\partial B_R(0)) \equiv \dot{B}^{p,q}_{m-1,s}(\partial D_R),$$

$$\pi_R : \dot{B}^{p,q}_{m-1,s}(\partial D_R) \equiv \dot{B}^{p,q}_{m-1,s}(\partial\Omega) \oplus \dot{B}^{p,q}_{m-1,s}(\partial B_R(0)) \to \dot{B}^{p,q}_{m-1,s}(\partial\Omega), \tag{4.246}$$

act according to $\iota_R(\dot{f}) := (\dot{f}, \dot{0})$ and $\pi_R(\dot{f}, \dot{g}) := \dot{f}$, for every $\dot{f}$.

Based on definitions, it is straightforward to check that, when acting on Whitney–Besov spaces,

$$\psi\dot{\mathscr{D}}^- = -(\psi\dot{\mathscr{D}}_R) \circ \iota_R \quad \text{in } D_R. \tag{4.247}$$

With this in hand, the desired conclusions about $\psi\dot{\mathscr{D}}^-$ follow from what we have established in the first part of the proof. □

In the following result we establish mapping properties for the double multi-layer acting on Whitney–Besov spaces and taking values in the weighted Sobolev spaces introduced in § 2.5.

Theorem 4.20. *Retain the same background hypotheses as in Theorem 4.19. In addition, assume that* $1 < p < \infty$, $0 < s < 1$, *and set* $a := 1 - s - 1/p \in (-1/p, 1 - 1/p)$. *Then the operator*

$$\dot{\mathscr{D}} : \dot{B}^{p,q}_{m-1,s}(\partial\Omega) \longrightarrow W^{m,p}_a(\Omega) \tag{4.248}$$

is well-defined, linear and bounded. Moreover, a similar boundedness result holds for $\psi\dot{\mathscr{D}}^-$ *(cf. the convention* (4.162)*), for any cutoff function* $\psi \in C^\infty_c(\mathbb{R}^n)$.

Proof. This is a consequence of Proposition 2.64 (used with $k = 1$) and the fact that for each $\dot{f} = \{f_\delta\}_{|\delta| \le m-1} \in \dot{B}^{p,q}_{m-1,s}(\partial\Omega)$ and each $\alpha \in \mathbb{N}^n_0$ with $|\alpha| \le m$ we may represent $\partial^\alpha \dot{\mathscr{D}} \dot{f}$ as a finite linear combination of terms of the form (4.227). □

Our next result deals with the mapping properties of the boundary multiple layer $\dot{K}$ and its relation with $\dot{\mathscr{D}}$, when considered on Whitney–Besov spaces. Recall (4.162).

Theorem 4.21. *Let* Ω *be a bounded Lipschitz domain in* $\mathbb{R}^n$, *and assume that* L *is a W-elliptic homogeneous differential operator of order* $2m$ *with (complex) matrix-valued constant coefficients.*

Then the boundary layer potential operator $\dot{K}$, *originally introduced in Definition 4.13, extends to a bounded mapping*

$$\dot{K} : \dot{B}^{p,q}_{m-1,s}(\partial\Omega) \longrightarrow \dot{B}^{p,q}_{m-1,s}(\partial\Omega) \tag{4.249}$$

whenever $\frac{n-1}{n} < p \le \infty$, $(n-1)\big(1/p - 1\big)_+ < s < 1$, *and* $0 < q \le \infty$. *In addition, with the same assumptions on* p, q *and* s,

$$\operatorname{tr}_{m-1} \circ \dot{\mathscr{D}}^{\pm} = \pm\tfrac{1}{2}I + \dot{K} \quad \text{in } \dot{B}^{p,q}_{m-1,s}(\partial\Omega). \tag{4.250}$$

Proof. Consider first the case when $1 < p = q < \infty$, and $0 < s < 1$. The fact that the operator (4.249) is bounded in this situation follows by interpolating by the real method (cf. (3.319) in Theorem 3.39 between (4.163) and (4.165) in Theorem 4.14). Furthermore, the jump formula (4.250) holds since the two operators involved are bounded on $\dot{B}^{p,p}_{m-1,s}(\partial\Omega)$ and coincide on $\dot{L}^p_{m-1,1}(\partial\Omega)$ which, under the current assumptions on the indices, is densely embedded into $\dot{B}^{p,p}_{m-1,s}(\partial\Omega)$, thanks to (3.118). Once (4.249)–(4.250) have been dealt with in the range of indices $1 < p = q < \infty$, $0 < s < 1$, their validity can be further extended to the case when $\frac{n-1}{n} < p = q < \infty$ and $(n-1)(1/p - 1)_+ < s < 1$ by arguing as follows. First, for each such indices p, s, it is possible to find indices $p^* \in (1, \infty)$ and $s^* \in (0, 1)$ with the property that $\dot{B}^{p,p}_{m-1,s}(\partial\Omega) \hookrightarrow \dot{B}^{p^*,p^*}_{m-1,s^*}(\partial\Omega)$ in a bounded fashion. As a result, for each $\dot{f} \in \dot{B}^{p,p}_{m-1,s}(\partial\Omega)$ we may write

$$\|(\pm\tfrac{1}{2}I + \dot{K})\dot{f}\|_{\dot{B}^{p,p}_{m-1,s}(\partial\Omega)} = \|\mathrm{tr}_{m-1}\dot{\mathscr{D}}^{\pm}\dot{f}\|_{\dot{B}^{p,p}_{m-1,s}(\partial\Omega)} \leq C\|\dot{f}\|_{\dot{B}^{p,p}_{m-1,s}(\partial\Omega)}, \quad (4.251)$$

where $C > 0$ is a constant that does not depend on $\dot{f}$. The identity in (4.251) follows from the fact that $\dot{f} \in \dot{B}^{p^*,q^*}_{m-1,s^*}(\partial\Omega)$ while the inequality follows from the fact that $\dot{f}$ belongs to $\dot{B}^{p,q}_{m-1,s}(\partial\Omega)$, Theorems 4.19 and 3.9.

Going further, the range $\frac{n-1}{n} < p = q < \infty$ and $(n-1)(\frac{1}{p}-1)_+ < s < 1$, where (4.249)–(4.250) are valid, can be extended to $\frac{n-1}{n} < p < \infty, 0 < q \leq \infty$ and $(n-1)(\frac{1}{p}-1)_+ < s < 1$, by invoking Theorem 3.39. In fact, by employing the same interpolation result, there only remains to deal with the case when $p = q = \infty$ and $0 < s < 1$.

However, if $\dot{f} \in \dot{B}^{\infty,\infty}_{m-1,s}(\partial\Omega)$ then $\dot{f} \in \dot{L}^p_{m-1,0}(\partial\Omega)$ for any $p \in (1, \infty)$. Consequently,

$$(\pm\tfrac{1}{2}I + \dot{K})\dot{f} = \dot{\mathscr{D}}^{\pm}\dot{f}\Big|^{m-1}_{\partial\Omega} = \mathrm{tr}_{m-1}\dot{\mathscr{D}}^{\pm}\dot{f}, \quad (4.252)$$

by (4.164), the fact that $\dot{\mathscr{D}} : \dot{B}^{\infty,\infty}_{m-1,s}(\partial\Omega) \to C^{m-1+s}(\bar{\Omega})$ boundedly (plus a similar statement for Ω_-), and (3.102) with $p = q = \infty$. This concludes the proof of the theorem. □

For further reference, let us also record here the following useful result.

Corollary 4.22. *In the same setting as in Theorem 4.21, the adjoint of the boundary layer potential operator $\dot{K}$ extends to a bounded mapping*

$$\dot{K}^* : \left(\dot{B}^{p,q}_{m-1,s}(\partial\Omega)\right)^* \longrightarrow \left(\dot{B}^{p,q}_{m-1,s}(\partial\Omega)\right)^*, \quad (4.253)$$

whenever $1 \leq p, q \leq \infty$ *and* $0 < s < 1$.

Proof. This is an immediate consequence of Theorem 4.21 and duality. □

For later use, let us point out that, by taking boundary traces in (4.247) and duality, we obtain

$$\iota_R \circ \dot{K} = -\dot{K}_R \circ \iota_R \quad \text{and} \quad \dot{K}^* \circ \iota_R^* = -\iota_R^* \circ \dot{K}_R^*, \quad (4.254)$$

where $\dot{K}_R$ is the boundary-to-boundary version of the double multi-layer integral operator for the domain $D_R = B_R(0) \setminus \overline{\Omega}$.

Our next result deals with the mapping properties of the operator $\dot{K}$ on the space $\dot{\mathrm{VMO}}_{m-1}(\partial\Omega)$. Specifically we have

Theorem 4.23. *Assume that Ω is a bounded Lipschitz domain in $\mathbb{R}^n$, and that L is a W-elliptic homogeneous differential operator of order $2m$ with (complex) matrix-valued constant coefficients. Then the boundary layer potential operator $\dot{K}$, originally introduced in Definition 4.13, induces a mapping*

$$\dot{K} : \mathrm{V\dot{M}O}_{m-1}(\partial\Omega) \longrightarrow \mathrm{V\dot{M}O}_{m-1}(\partial\Omega) \tag{4.255}$$

which is well-defined, linear and bounded.

Proof. This is an immediate consequence of Theorems 4.21 and 4.18 and the definition of the space $\mathrm{V\dot{M}O}_{m-1}(\partial\Omega)$ given in (3.413). □

The last result in this section deals with the limiting cases $s = 0$ and $s = 1$ in (4.224), (4.225). To state it, recall that $p \vee 2 = \max\{p, 2\}$.

Theorem 4.24. *Let Ω be a bounded Lipschitz domain in $\mathbb{R}^n$, and suppose that L is a W-elliptic homogeneous differential operator of order $2m$ with (complex) matrix-valued constant coefficients. Then the operators*

$$\dot{\mathscr{D}} : \dot{L}^p_{m-1,s}(\partial\Omega) \longrightarrow B^{p,p\vee 2}_{m-1+s+1/p}(\Omega), \qquad 1 < p < \infty, \tag{4.256}$$

$$\dot{\mathscr{D}} : \dot{L}^p_{m-1,s}(\partial\Omega) \longrightarrow F^{p,q}_{m-1+s+1/p}(\Omega), \quad 2 \leq p < \infty, \ 0 < q \leq \infty, \tag{4.257}$$

are well-defined, linear and bounded if either $s = 0$, or $s = 1$.

Proof. Consider first (4.256) in which case, by Proposition 2.24, it suffices to show that $\nabla^{m-1+s}\dot{\mathscr{D}}\dot{f} \in B^{p,p\vee 2}_{1/p}(\Omega)$ for each $\dot{f} \in \dot{L}^p_{m-1,s}(\partial\Omega)$ (assuming that $s \in \{0, 1\}$). This, however, is a consequence of (4.80)–(4.83) and Proposition 2.68.

Next, the claim about (4.257) is a consequence of what we have just proved, the fact that $B^{p,p\vee 2}_{m-1+s+1/p}(\Omega) \hookrightarrow F^{p,p}_{m-1+s+1/p}(\Omega)$ if $2 \leq p < \infty$, and the last part in Theorem 2.41. □

Chapter 5
The Single Multi-Layer Potential Operator

The general goal in this chapter is to define and study the main properties of the single multi-layer potential operator associated with arbitrary elliptic, higher-order, homogeneous, constant (complex) matrix-valued coefficients differential operators.

5.1 The Definition of Single Multi-Layer and Non-tangential Maximal Estimates

We begin by making the following basic definition.

Definition 5.1. Fix $m, M \in \mathbb{N}$ and consider a W-elliptic, homogeneous differential operator L of order $2m$, with $\mathbb{C}^{M\times M}$-valued coefficients, and denote by E the fundamental solution of L constructed in Theorem 4.2. In addition, assume that Ω is a bounded Lipschitz domain in $\mathbb{R}^n$, and fix $p \in (1,\infty)$. The action of the single multi-layer potential operator $\dot{\mathscr{S}}$ on an arbitrary functional $\Lambda \in \left(\dot{L}^p_{m-1,1}(\partial\Omega)\right)^*$ is given by

$$(\dot{\mathscr{S}}\Lambda)(X) := \Big\langle \mathrm{tr}_{m-1}[E(X-\cdot)]\,,\,\Lambda\Big\rangle \tag{5.1}$$

$$= \left(\left\langle\left\{(-1)^{|\alpha|}(\partial^\alpha E_{j\bullet})(X-\cdot)\Big|_{\partial\Omega}\right\}_{|\alpha|\le m-1}\,,\,\Lambda\right\rangle\right)_{1\le j\le M},$$

where $X \in \mathbb{R}^n \setminus \partial\Omega$ and $E_{j\bullet}$ is the j-th row in E.

In (5.1), $\mathrm{tr}_{m-1} E$ is defined as a $M \times M$ matrix with entries arrays, whose rows are obtained by applying tr_{m-1} to each row of E. Upon recalling the convention (2.11), we set

$$\dot{\mathscr{S}}^{\pm}\Lambda := \left(\dot{\mathscr{S}}\Lambda\right)\Big|_{\Omega_\pm}. \tag{5.2}$$

I. Mitrea and M. Mitrea, *Multi-Layer Potentials and Boundary Problems*, Lecture Notes in Mathematics 2063, DOI 10.1007/978-3-642-32666-0_5,

We shall occasionally abbreviate $\dot{\mathscr{S}}^{+}$ by $\dot{\mathscr{S}}$.

Next, it can then be easily checked that, if $1/p + 1/p' = 1$,

$$\dot{\mathscr{S}}\Lambda(X) = \sum_{|\alpha|\leq m-1} (-1)^{|\alpha|} \int_{\partial\Omega} (\partial^{\alpha} E)(X-Y) f_{\alpha}(Y)\, d\sigma(Y), \quad \forall\, X \in \mathbb{R}^n \setminus \partial\Omega, \tag{5.3}$$

whenever $\Lambda = \{f_{\alpha}\}_{|\alpha|\leq m-1} \in L^p(\partial\Omega) \oplus \cdots \oplus L^p(\partial\Omega) \hookrightarrow \left(\dot{L}^{p'}_{m-1,1}(\partial\Omega)\right)^{*}$.

For future reference, let us point out here that, as a simple consequence of the Hahn–Banach theorem, we have that

$$L^p(\partial\Omega) \oplus \cdots \oplus L^p(\partial\Omega) \hookrightarrow \left(\dot{L}^{p'}_{m-1,1}(\partial\Omega)\right)^{*} \text{ densely.} \tag{5.4}$$

Next, clearly,

$$L(\dot{\mathscr{S}}\Lambda) = 0 \quad \text{in } \mathbb{R}^n \setminus \partial\Omega, \tag{5.5}$$

for every $\Lambda \in \left(\dot{L}^{p}_{m-1,1}(\partial\Omega)\right)^{*}$. Since $\left(\dot{L}^{p}_{m-1,0}(\partial\Omega)\right)^{*} \hookrightarrow \left(\dot{L}^{p}_{m-1,1}(\partial\Omega)\right)^{*}$, it follows that $\dot{\mathscr{S}}$ has also a well-defined pointwise action on $\left(\dot{L}^{p}_{m-1,0}(\partial\Omega)\right)^{*}$, and (5.5) holds in that context as well.

Proposition 5.2. *Let Ω be a bounded Lipschitz domain in $\mathbb{R}^n$ and assume that $1 < p, p' < \infty$ satisfy $\frac{1}{p} + \frac{1}{p'} = 1$. In addition, consider a W-elliptic, homogeneous differential operator L of order $2m$, with (complex) matrix-valued coefficients, and recall the single multi-layer potential operator $\dot{\mathscr{S}}$ introduced in this context as in Definition 5.1. Then there exists a finite constant $C = C(\Omega, L, p) > 0$ such that*

$$\sum_{j=0}^{m} \|\mathscr{N}(\nabla^j \dot{\mathscr{S}}\Lambda)\|_{L^p(\partial\Omega)} \leq C\|\Lambda\|_{(\dot{L}^{p'}_{m-1,0}(\partial\Omega))^*}, \quad \forall\, \Lambda \in \left(\dot{L}^{p'}_{m-1,0}(\partial\Omega)\right)^{*}, \tag{5.6}$$

and

$$\sum_{j=0}^{m-1} \|\mathscr{N}(\nabla^j \dot{\mathscr{S}}\Lambda)\|_{L^p(\partial\Omega)} \leq C\|\Lambda\|_{(\dot{L}^{p'}_{m-1,1}(\partial\Omega))^*}, \quad \forall\, \Lambda \in \left(\dot{L}^{p'}_{m-1,1}(\partial\Omega)\right)^{*}. \tag{5.7}$$

Proof. Denote by E the fundamental solution associated with the operator L as in Theorem 4.2. We being by noticing that $\dot{L}^{p'}_{m-1,0}(\partial\Omega)$ is a closed subspace of $\oplus_1^N L^{p'}(\partial\Omega)$, where the number N is as in (4.89) (i.e., N is the cardinality of $\{\alpha \in \mathbb{N}_0^n : |\alpha| \leq m-1\}$). If a functional $\Lambda \in (\dot{L}^{p'}_{m-1,0}(\partial\Omega))^*$ is given, consider the linear, bounded extension $\tilde{\Lambda} : \oplus_1^N L^{p'}(\partial\Omega) \to \mathbb{C}$ given by the Hahn–Banach

Theorem. By the Riesz Representation Theorem, there exists a unique N-tuple, $\boldsymbol{g} := (g_\gamma)_{|\gamma|\le m-1} \in \oplus_1^N L^p(\partial\Omega)$, such that

$$\begin{aligned}\tilde{\Lambda}\Big((f_\alpha)_{|\alpha|\le m-1}\Big) = \sum_{|\alpha|\le m-1} \int_{\partial\Omega} f_\alpha(Y) g_\alpha(Y)\, d\sigma(Y),\\ \forall\, (f_\alpha)_{|\alpha|\le m-1} \in \oplus_1^N L^{p'}(\partial\Omega),\end{aligned} \tag{5.8}$$

and

$$\|\boldsymbol{g}\|_{\oplus_1^N L^p(\partial\Omega)} = \|\tilde{\Lambda}\|_{(\oplus_1^N L^{p'}(\partial\Omega))^*} \le \|\Lambda\|_{(\dot{L}^{p'}_{m-1,0}(\partial\Omega))^*}. \tag{5.9}$$

Then, for each $X \in \mathbb{R}^n \setminus \partial\Omega$,

$$\begin{aligned}\dot{\mathscr{S}}\Lambda(X) &= \langle \mathrm{tr}_{m-1}[E(X-\cdot)]\,,\,\Lambda\rangle = \langle \mathrm{tr}_{m-1}[E(X-\cdot)]\,,\,\tilde{\Lambda}\rangle\\ &= \sum_{|\alpha|\le m-1} \int_{\partial\Omega} \big(\mathrm{tr}_{m-1}[E(X-\cdot)]\big)_\alpha\, g_\alpha\, d\sigma\\ &= \sum_{|\alpha|\le m-1} (-1)^{|\alpha|} \int_{\partial\Omega} (\partial^\alpha E)(X-Y) g_\alpha(Y)\, d\sigma(Y)\end{aligned} \tag{5.10}$$

so that further, by Calderón–Zygmund theory and (5.9),

$$\sum_{j=0}^{m} \big\|\mathcal{N}(\nabla^j \dot{\mathscr{S}}\Lambda)\big\|_{L^p(\partial\Omega)} \le C\|\boldsymbol{g}\|_{\oplus_1^N L^p(\partial\Omega)} \le C\|\Lambda\|_{(\dot{L}^{p'}_{m-1,0}(\partial\Omega))^*}, \tag{5.11}$$

establishing (5.6).

Consider next $\Lambda \in (\dot{L}^{p'}_{m-1,1}(\partial\Omega))^*$. By viewing $\dot{L}^{p'}_{m-1,1}(\partial\Omega)$ as a closed subspace of $\oplus_1^N L_1^{p'}(\partial\Omega)$ and using that $\Big(\oplus_1^N L_1^{p'}(\partial\Omega)\Big)^* = \oplus_1^N L^p_{-1}(\partial\Omega)$, the same type of reasoning based on Hahn–Banach and Riesz Theorems, leads to the following conclusion. There exists an N-tuple $\boldsymbol{g} = \{g_\alpha\}_{|\alpha|\le m-1} \in \oplus_1^N L^p_{-1}(\partial\Omega)$ with the property that

$$\Big\langle \Lambda, \dot{f}\Big\rangle = \sum_{|\alpha|\le m-1} \langle f_\alpha, g_\alpha\rangle \qquad \forall\, \dot{f} \in \dot{L}^{p'}_{m-1,1}(\partial\Omega), \tag{5.12}$$

where $\langle\cdot,\cdot\rangle$ in the right hand-side of (5.12) is the natural duality pairing between $L_1^{p'}(\partial\Omega)$ and $L^p_{-1}(\partial\Omega)$, and for which

$$\|\boldsymbol{g}\|_{\oplus_1^N L^p_{-1}(\partial\Omega)} \le \|\Lambda\|_{(\dot{L}^{p'}_{m-1,1}(\partial\Omega))^*}. \tag{5.13}$$

From Corollary 2.14 we know that, for each $\alpha \in \mathbb{N}_0^n$ with $|\alpha| \leq m-1$, there exist $g_*^\alpha, g_{ij}^\alpha \in L^p(\partial\Omega)$ such that

$$g_\alpha = \sum_{i,j=1}^n \partial_{\tau_{ij}} g_{ij}^\alpha + g_*^\alpha, \quad \text{and}$$
$$\sum_{i,j=1}^n \|g_{ij}^\alpha\|_{L^p(\partial\Omega)} + \|g_*^\alpha\|_{L^p(\partial\Omega)} \leq C\|g_\alpha\|_{L^p_{-1}(\partial\Omega)}. \tag{5.14}$$

Proceeding as before, for each $X \in \mathbb{R}^n \setminus \partial\Omega$ we obtain

$$\begin{aligned}
\dot{\mathscr{S}}\Lambda(X) &= \sum_{|\alpha|\leq m-1} \sum_{i,j=1}^n (-1)^{|\alpha|} \Big\langle (\partial^\alpha E)(X-\cdot)\,,\, (\partial_{\tau_{ij}} g_{ij}^\alpha) \Big\rangle \\
&\quad + \sum_{|\alpha|\leq m-1} (-1)^{|\alpha|} \int_{\partial\Omega} (\partial^\alpha E)(X-Y) g_*^\alpha(Y)\, d\sigma(Y) \\
&= \sum_{|\alpha|\leq m-1} \sum_{i,j=1}^n (-1)^{|\alpha|} \int_{\partial\Omega} \partial_{\tau_{ji}(Y)}[(\partial^\alpha E)(X-Y)] g_{ij}^\alpha(Y)\, d\sigma(Y) \\
&\quad + \sum_{|\alpha|\leq m-1} (-1)^{|\alpha|} \int_{\partial\Omega} (\partial^\alpha E)(X-Y) g_*^\alpha(Y)\, d\sigma(Y),
\end{aligned} \tag{5.15}$$

after integrating by parts on the boundary; cf. Corollary 2.13 and its proof. Hence,

$$\begin{aligned}
\sum_{j=0}^{m-1} \|\mathscr{N}(\nabla^j \dot{\mathscr{S}}\Lambda)\|_{L^p(\partial\Omega)} &\leq C \sum_{|\alpha|\leq m-1} \Big(\sum_{i,j=1}^n \|g_{ij}^\alpha\|_{L^p(\partial\Omega)} + \|g_*^\alpha\|_{L^p(\partial\Omega)} \Big) \\
&\leq C\|\boldsymbol{g}\|_{\oplus_1^N L^p_{-1}(\partial\Omega)} \leq C\|\Lambda\|_{(\dot{L}^{p'}_{m-1,1}(\partial\Omega))^*},
\end{aligned}$$

by Calderón–Zygmund theory, (5.14) and (5.13). □

Having proved Proposition 5.2, it is useful to also record the following smoothing properties of the single multi-layer, measured on the Besov scale.

Proposition 5.3. *Retain the same context as in Proposition 5.2. Then for each number $p \in (1,\infty)$ there exists a finite constant $C = C(\Omega, L, p) > 0$ such that*

$$\|\dot{\mathscr{S}}\Lambda\|_{B^{p,p\vee 2}_{m+1/p}(\Omega)} \leq C\|\Lambda\|_{\left(\dot{L}^{p'}_{m-1,0}(\partial\Omega)\right)^*}, \qquad \forall\, \Lambda \in \left(\dot{L}^{p'}_{m-1,0}(\partial\Omega)\right)^*, \tag{5.16}$$

$$\|\dot{\mathscr{S}}\Lambda\|_{B^{p,p\vee 2}_{m-1+1/p}(\Omega)} \leq C\|\Lambda\|_{\left(\dot{L}^{p'}_{m-1,1}(\partial\Omega)\right)^*}, \qquad \forall\,\Lambda \in \left(\dot{L}^{p'}_{m-1,1}(\partial\Omega)\right)^*, \quad (5.17)$$

where, as usual, $p' \in (1,\infty)$ is such that $\frac{1}{p}+\frac{1}{p'}=1$.

Proof. We shall use notation introduced in the course of the proof of Proposition 5.2. In particular, from the integral representation formula (5.10) we deduce that

$$\|\dot{\mathscr{S}}\Lambda\|_{B^{p,p\vee 2}_{m+1/p}(\Omega)} \leq C\sum_{j=0}^{m}\|\nabla^j\dot{\mathscr{S}}\Lambda\|_{B^{p,p\vee 2}_{1/p}(\Omega)} \leq C\|g\|_{\oplus_1^N L^p(\partial\Omega)}$$

$$\leq C\|\Lambda\|_{(L^{p'}_{m-1,0}(\partial\Omega))^*}, \quad (5.18)$$

by the lifting result from Propositions 2.24, 2.68, and (5.9). This establishes (5.16). The estimate (5.17) is proved in a similar manner, this time starting with the integral representation formula (5.15) (and taking at most $m-1$ derivatives on the single multi-layer). □

We now proceed to define the boundary version of the single multi-layer, and study its mapping properties.

Theorem 5.4. *Let Ω be a bounded Lipschitz domain in $\mathbb{R}^n$ and fix $p\in(1,\infty)$. In addition, consider a W-elliptic, homogeneous differential operator L of order $2m$, with (complex) matrix-valued coefficients, and denote by E the fundamental solution of L constructed in Theorem 4.2. Finally, recall the single multi-layer potential operator $\dot{\mathscr{S}}$ introduced in this context as in Definition 5.1. Then the pointwise non-tangential traces $\dot{\mathscr{S}}\Lambda\Big|^{m-1}_{\partial\Omega_\pm}$ are well-defined for each $\Lambda\in\left(\dot{L}^p_{m-1,1}(\partial\Omega)\right)^*$, and*

$$\dot{\mathscr{S}}\Lambda\Big|^{m-1}_{\partial\Omega_+} = \dot{\mathscr{S}}\Lambda\Big|^{m-1}_{\partial\Omega_-} =: \dot{S}\Lambda, \qquad \forall\,\Lambda\in\left(\dot{L}^p_{m-1,1}(\partial\Omega)\right)^*. \quad (5.19)$$

Moreover, if $1/p+1/p'=1$, then

$$\dot{S}:\left(\dot{L}^{p'}_{m-1,1}(\partial\Omega)\right)^* \longrightarrow \dot{L}^p_{m-1,0}(\partial\Omega) \text{ boundedly, and} \quad (5.20)$$

$$\dot{S}:\left(\dot{L}^{p}_{m-1,0}(\partial\Omega)\right)^* \longrightarrow \dot{L}^{p'}_{m-1,1}(\partial\Omega) \text{ boundedly.} \quad (5.21)$$

Also, if $\dot{S}_L$ indicates that the single multi-layer is associated with L, etc., then

$$\left(\dot{S}_L\right)^t = \dot{S}_{L^t} \quad (5.22)$$

where $\dot{S}_L$, $\dot{S}_{L^t}$ are taken in the context of (5.20) *and* (5.21), *respectively (or vice-versa). In particular, when $A = A^t$ (cf.* (4.7)), *then $\dot{S}$ is invariant under transposition (in the sense that the transposed of* (5.20) *is* (5.21)).

Proof. Fix $\Lambda \in \left(\dot{L}^{p'}_{m-1,1}(\partial\Omega)\right)^*$ and let $g = \{g_\alpha\}_{|\alpha|\le m-1} \in \oplus_1^N L^p_{-1}(\partial\Omega)$ be as in (5.12)–(5.14). Then, using (5.15), for each multi-index $\gamma \in \mathbb{N}_0^n$ of length $\le m-1$ we may write

$$\partial^\gamma \dot{\mathscr{S}}\Lambda(X) = \sum_{|\alpha|\le m-1}\sum_{i,j=1}^{n}(-1)^{|\alpha|}\int_{\partial\Omega}\partial_{\tau_{ji}(Y)}[(\partial^{\alpha+\gamma}E)(X-Y)]g^\alpha_{ij}(Y)\,d\sigma(Y) \tag{5.23}$$

$$+\sum_{|\alpha|\le m-1}(-1)^{|\alpha|}\int_{\partial\Omega}(\partial^{\alpha+\gamma}E)(X-Y)g^\alpha_*(Y)\,d\sigma(Y), \quad X\in\mathbb{R}^n\setminus\partial\Omega,$$

where E is the fundamental solution associated with the operator L as in Theorem 4.2. Let $\nu = (\nu_j)_{1\le j\le n}$ denote the outward unit normal to Ω. Then the integral operators on the right-hand-side of (5.23) have either kernels of the type

$$k^1_{\alpha,\gamma}(X,Y) := -\nu_j(Y)(\partial^{\alpha+\gamma+e_i}E)(X-Y) + \nu_i(Y)(\partial^{\alpha+\gamma+e_j}E)(X-Y), \tag{5.24}$$

which are amenable to the Calderón–Zygmund theory reviewed in § 2.8 whenever the multi-index $\gamma \in \mathbb{N}_0^n$ satisfies $|\gamma| = m-1$, or have weakly singular kernels of the type

$$k^2_{\alpha,\gamma}(X,Y) := (\partial^{\alpha+\gamma}E)(X-Y) \tag{5.25}$$

whenever $\gamma \in \mathbb{N}_0^n$ has length $\le m-2$. Now, with the "hat" denoting the Fourier transform,

$$-\nu_j(X)\widehat{(\partial^{\alpha+\gamma+e_i}E)}(\nu(X)) + \nu_i(X)\widehat{(\partial^{\alpha+\gamma+e_j}E)}(\nu(X)) = 0, \tag{5.26}$$

so thanks to the jump-formula (2.530) we obtain

$$\left(\partial^\gamma\dot{\mathscr{S}}\Lambda\right)\Big|_{\partial\Omega_+} = \left(\partial^\gamma\dot{\mathscr{S}}\Lambda\right)\Big|_{\partial\Omega_-}, \tag{5.27}$$

from which the well-definiteness of (5.19) follows. Next, based on formula (5.23) written for a multi-index $\gamma \in \mathbb{N}_0^n$ with $|\gamma| = m-1$ we can conclude that for each $\Lambda \in \left(\dot{L}^{p'}_{m-1,1}(\partial\Omega)\right)^*$ we have $\mathscr{N}(\nabla^{m-1}\dot{\mathscr{S}}\Lambda) \in L^p(\partial\Omega)$. Then, using (5.5) and Proposition 3.29 we obtain that $\dot{\mathscr{S}}\Lambda\Big|^{m-1}_{\partial\Omega} \in \dot{L}^p_{m-1,0}(\partial\Omega)$. From this and the accompanying estimate, (5.20) follows.

As regards the claim made in (5.21), we start by fixing $\Lambda \in \left(\dot{L}^p_{m-1,0}(\partial\Omega)\right)^*$. Next, let $\boldsymbol{g} = \{g_\alpha\}_{|\alpha|\leq m-1} \in \oplus_1^N L^{p'}(\partial\Omega)$ be as in (5.8)–(5.9) but with the roles of indices p and p' reversed. Then, based on (5.10), for each multi-index $\gamma \in \mathbb{N}_0^n$ with $|\gamma| = m$ and each $X \in \mathbb{R}^n \setminus \partial\Omega$, we may write

$$\partial^\gamma(\dot{\mathscr{S}}\Lambda)(X) = \sum_{|\alpha|\leq m-1} (-1)^{|\alpha|} \int_{\partial\Omega} (\partial^{\alpha+\gamma}E)(X-Y)g_\alpha(Y)\,d\sigma(Y). \tag{5.28}$$

Since the integral operators in the right-hand-side are either weakly singular, or of Calderón–Zygmund type, it follows that $\mathscr{N}(\nabla^m\dot{\mathscr{S}}\Lambda) \in L^{p'}(\partial\Omega)$. This, (5.5) and Proposition 3.30 then give $\dot{\mathscr{S}}\Lambda\Big|_{\partial\Omega}^{m-1} \in \dot{L}^{p'}_{m-1,1}(\partial\Omega)$, plus a natural estimate. Now, (5.21) follows from this.

Let us finally consider the last claim in the statement of the theorem. Fix a functional $\Lambda \in \left(\dot{L}^{p'}_{m-1,1}(\partial\Omega)\right)^*$ and assume that g_α, g^α_{ij}, g^α_* are as in (5.12)–(5.14). Also, pick an arbitrary functional $\Lambda' \in \left(\dot{L}^p_{m-1,0}(\partial\Omega)\right)^*$ and assume that $(g'_\gamma)_{|\gamma|\leq m-1} \in \oplus_1^N L^{p'}(\partial\Omega)$ are such that

$$\left\langle \Lambda', \dot{f}\right\rangle = \sum_{|\gamma|\leq m-1} \int_{\partial\Omega} f_\gamma g'_\gamma\,d\sigma, \quad \forall\, \dot{f} = (f_\gamma)_{|\gamma|\leq m-1} \in \dot{L}^p_{m-1,0}(\partial\Omega). \tag{5.29}$$

In the next calculation, let us stress that certain single multi-layers are associated with the differential operator L by temporarily denoting them by $\dot{\mathscr{S}}_L$, $\dot{S}_L$, etc. Using (5.19), (5.29), (5.23) and natural manipulations, we may write

$$\begin{aligned}
\left\langle \dot{S}_L\Lambda, \Lambda'\right\rangle &= \left\langle \left\{\partial^\gamma(\dot{\mathscr{S}}_L\Lambda)\Big|_{\partial\Omega}\right\}_{|\gamma|\leq m-1}, \Lambda'\right\rangle = \sum_{|\gamma|\leq m-1} \int_{\partial\Omega} \left((\partial^\gamma\dot{\mathscr{S}}_L\Lambda)\Big|_{\partial\Omega}\right) g'_\gamma\,d\sigma \\
&= \sum_{\substack{|\alpha|\leq m-1\\|\gamma|\leq m-1}} \sum_{i,j=1}^n (-1)^{|\alpha|} \int_{\partial\Omega}\int_{\partial\Omega} \left\langle A^{\alpha\gamma}_{ij}(X,Y), g'_\gamma(X)\right\rangle d\sigma(X)\,d\sigma(Y) \\
&\quad + \sum_{\substack{|\alpha|\leq m-1\\|\gamma|\leq m-1}} (-1)^{|\alpha|} \int_{\partial\Omega}\int_{\partial\Omega} \left\langle B^{\alpha\gamma}(X,Y), g'_\gamma(X)\right\rangle d\sigma(X)\,d\sigma(Y) \\
&= \sum_{\substack{|\alpha|\leq m-1\\|\gamma|\leq m-1}} \sum_{i,j=1}^n (-1)^{|\gamma|} \int_{\partial\Omega}\int_{\partial\Omega} \left\langle g^\alpha_{ij}(X), C^{\alpha\gamma}_{ij}(X,Y)\right\rangle d\sigma(Y)\,d\sigma(X) \\
&\quad + \sum_{\substack{|\alpha|\leq m-1\\|\gamma|\leq m-1}} (-1)^{|\gamma|} \int_{\partial\Omega}\int_{\partial\Omega} \left\langle g^\alpha_*(X), (\partial^{\alpha+\gamma}E^t)(X-Y)g'_\gamma(Y)\right\rangle d\sigma(Y)\,d\sigma(X)
\end{aligned}$$

$$= \Big\langle \Lambda, \Big\{(-1)^{|\gamma|}\int_{\partial\Omega}(\partial^\gamma E^t)(\cdot - Y)g'_\gamma(Y)\,d\sigma(Y)\Big\}_{|\gamma|\le m-1}\Big\rangle$$

$$= \big\langle \Lambda, \dot S_{L^t}\Lambda'\big\rangle, \tag{5.30}$$

where for each $i, j \in \{1,\dots,n\}$ and each $\alpha, \gamma \in \mathbb{N}_0^n$ with $|\alpha|, |\gamma| \le m-1$ we have set

$$\begin{aligned} A_{ij}^{\alpha\gamma}(X,Y) &:= \partial_{\tau_{ji}(Y)}[(\partial^{\alpha+\gamma}E)(X-Y)]g_{ij}^{\alpha}(Y), \\ B^{\alpha\gamma}(X,Y) &:= (\partial^{\alpha+\gamma}E)(X-Y)g_*^{\alpha}(Y), \\ C_{ij}^{\alpha\gamma}(X,Y) &:= \partial_{\tau_{ji}(X)}[(\partial^{\alpha+\gamma}E^t)(X-Y)]g'_\gamma(Y), \end{aligned} \tag{5.31}$$

and the last step in (5.30) utilizes (5) in Theorem 4.2. This justifies (5.22) and concludes the proof of the theorem. □

5.2 Carleson Measure Estimates

This section is devoted to proving the following theorem, which may be regarded as the analogue of Theorem 4.11 (i.e., the BMO-to-Carleson mapping property of the double multi-layer) for the single multi-layer operator introduced in Definition 5.1.

Theorem 5.5. *Let L be a W-elliptic homogeneous differential operator of order $2m$ with complex matrix-valued constant coefficients. Assume that Ω is a bounded Lipschitz domain in $\mathbb{R}^n$ and, for each $X \in \Omega$, denote by $\rho(X)$ its distance to $\partial\Omega$. Then there exists a finite constant $C = C(\Omega, L) > 0$ with the property that*

$$\begin{aligned} &|\nabla^{m-1}\dot{\mathscr{S}}\Lambda(X)|^2\rho(X)\,dX \quad \textit{is a Carleson measure in } \Omega \textit{ with} \\ &\textit{Carleson constant} \le C\,\|\Lambda\|^2_{(\dot h^1_{m-1,1}(\partial\Omega))^*}, \textit{ for each } \Lambda \in \big(\dot h^1_{m-1,1}(\partial\Omega)\big)^*. \end{aligned} \tag{5.32}$$

Proof. We start by considering the map

$$J : \dot h^1_{m-1,1}(\partial\Omega) \longrightarrow \oplus_1^N\Big[L^{\frac{n-1}{n-2}}(\partial\Omega) \oplus \big(\oplus_1^{\frac{n(n-1)}{2}} h^1_{at}(\partial\Omega)\big)\Big] \tag{5.33}$$

given by

$$J(\dot f) := \big(f_\alpha, (\partial_{\tau_{jk}} f_\alpha)_{1<\le j,k\le n}\big)_{|\alpha|\le m-1}, \quad \forall\, \dot f = \{f_\alpha\}_{|\alpha|\le m-1} \in \dot h^1_{m-1,1}(\partial\Omega). \tag{5.34}$$

It is immediate that J is one-to-one and has closed range. Thus the image of J is a closed subspace of $\oplus_1^N\Big[L^{\frac{n-1}{n-2}}(\partial\Omega)\oplus\big(\oplus_1^{\frac{n(n-1)}{2}}h^1_{at}(\partial\Omega)\big)\Big]$ and J is an isomorphism when acting from $\dot{h}^1_{m-1,1}(\partial\Omega)$ onto its image. Denote by J^{-1} the inverse map and, for each given $\Lambda\in\big(\dot{h}^1_{m-1,1}(\partial\Omega)\big)^*$ let

$$
\widetilde{\Lambda}\in\Big(\oplus_1^N\Big[L^{\frac{n-1}{n-2}}(\partial\Omega)\oplus\big(\oplus_1^{\frac{n(n-1)}{2}}h^1_{at}(\partial\Omega)\big)\Big]\Big)^*
$$

$$
=\oplus_1^N\Big[L^{n-1}(\partial\Omega)\oplus\big(\oplus_1^{\frac{n(n-1)}{2}}\mathrm{bmo}(\partial\Omega)\big)\Big] \tag{5.35}
$$

be the Hahn–Banach extension of $\Lambda\circ J^{-1}$. Using the Riesz Representation Theorem we therefore conclude that for each $\Lambda\in\big(\dot{h}^1_{m-1,1}(\partial\Omega)\big)^*$

$$
\begin{aligned}
&\exists\,\big(g^\alpha_*,(g^\alpha_{jk})_{1\le j,k\le n}\big)_{|\alpha|\le m-1}\in\oplus_1^N\Big[L^{n-1}(\partial\Omega)\oplus\big(\oplus_1^{\frac{n(n-1)}{2}}\mathrm{bmo}(\partial\Omega)\big)\Big]\\
&\text{such that}\quad \Lambda(\dot{f})=\sum_{|\alpha|\le m-1}\langle f_\alpha,g^\alpha_*\rangle+\sum_{\substack{|\alpha|\le m-1\\ j,k\in\{1,\dots,n\}}}\langle\partial_{\tau_{jk}}f_\alpha,g^\alpha_{jk}\rangle,
\end{aligned}\tag{5.36}
$$

for every $\dot{f}=\{f_\alpha\}_{|\alpha|\le m-1}\in\dot{h}^1_{m-1,1}(\partial\Omega)$. In addition, there exists $C\in(0,\infty)$, independent of Λ, such that

$$
\sum_{|\alpha|\le m-1}\|g^\alpha_*\|_{L^{n-1}(\partial\Omega)}+\sum_{\substack{|\alpha|\le m-1\\ j,k\in\{1,\dots,n\}}}\|g^\alpha_{jk}\|_{\mathrm{bmo}(\partial\Omega)}\le C\|\Lambda\|_{\left(\dot{h}^1_{m-1,1}(\partial\Omega)\right)^*}. \tag{5.37}
$$

Based on this analysis, and by reasoning much as in (5.3), we arrive at the representation formula

$$
\begin{aligned}
\dot{\mathscr{S}}\Lambda(X)=&\sum_{|\alpha|\le m-1}(-1)^{|\alpha|}\int_{\partial\Omega}(\partial^\alpha E)(X-Y)g^\alpha_*(Y)\,d\sigma(Y)\\
&+\sum_{\substack{|\alpha|\le m-1\\ j,k\in\{1,\dots,n\}}}(-1)^{|\alpha|}\int_{\partial\Omega}\partial_{\tau_{jk}(Y)}\big[(\partial^\alpha E)(X-Y)\big]g^\alpha_{jk}(Y)\,d\sigma(Y),
\end{aligned}\tag{5.38}
$$

for every $X\in\mathbb{R}^n\setminus\partial\Omega$.

Consider next $\nabla^{m-1}\dot{\mathscr{S}}\Lambda$ when $\dot{\mathscr{S}}\Lambda$ is represented as in (5.38). This gives rise to two types of terms. On the one hand, the terms containing top singularities (and involving the functions $g^\alpha_{jk}\in\mathrm{bmo}(\partial\Omega)$), are precisely of the form discussed in

Theorem 2.66. As such, the contribution of these terms in $|\nabla^{m-1}\dot{\mathscr{S}}\Lambda(X)|^2\rho(X)\,dX$ is a Carleson measure in Ω with Carleson constant dominated by

$$C\sum_{\substack{|\alpha|\leq m-1\\ j,k\in\{1,\dots,n\}}}\|g^{\alpha}_{jk}\|^2_{\mathrm{bmo}(\partial\Omega)}\leq C\,\|\Lambda\|^2_{\left(\dot{h}^1_{m-1,1}(\partial\Omega)\right)^*},\tag{5.39}$$

where the last inequality uses (5.37). This, of course, suits our purpose. On the other hand, the remaining terms in $\nabla^{m-1}\dot{\mathscr{S}}\Lambda$ involve operators whose integral kernels are only weakly singular. More specifically, the kernels in question are $\mathscr{O}(|X-Y|^{2-n})$, uniformly for $X,Y\in\partial\Omega$. The fact that these terms contribute Carleson measures (with appropriate control of constants) in the context of $|\nabla^{m-1}\dot{\mathscr{S}}\Lambda(X)|^2\rho(X)\,dX$, may be easily checked using (2.44) and the fractional integration result from [92, Lemma 7.3, p. 34]. This finishes the proof of the theorem. □

5.3 Estimates on Besov, Triebel–Lizorkin, and Weighted Sobolev Spaces

In this section we study the mapping properties of the single multi-layer operators introduced in (5.1) and (5.19) on Besov and Triebel–Lizorkin scales on Lipschitz domains.

Proposition 5.6. *Let Ω be a bounded Lipschitz domain in $\mathbb{R}^n$ and assume that L is a W-elliptic homogeneous differential operator of order $2m$ with complex matrix-valued constant coefficients. Also, fix $1<p,q<\infty$ and $s\in(0,1)$. Then*

$$\dot{S}:\left(\dot{B}^{p,q}_{m-1,s}(\partial\Omega)\right)^*\longrightarrow\dot{B}^{p',q'}_{m-1,1-s}(\partial\Omega),\tag{5.40}$$

where $1/p+1/p'=1/q+1/q'=1$. Furthermore, $\dot{S}$ is a self-adjoint operator in this context, in the sense that the dual of the operator (5.40) *is*

$$\dot{S}:\left(\dot{B}^{p',q'}_{m-1,1-s}(\partial\Omega)\right)^*\to\dot{B}^{p,q}_{m-1,s}(\partial\Omega).\tag{5.41}$$

Proof. The claim in first part of the statement is a consequence of (5.20), (5.21), Theorem 3.39 and the Duality Theorem for the real method of interpolation. The second part follows from this, the last claim in Theorem 5.4, and a density argument. With regard to the latter, we note that Propositions 3.3 and 3.7 yield

$$\dot{L}^p_{m-1,1}(\partial\Omega)\hookrightarrow\dot{B}^{p,q}_{m-1,s}(\partial\Omega)\hookrightarrow\dot{L}^p_{m-1,0}(\partial\Omega)\quad\text{densely},\tag{5.42}$$

whenever $1<p,q<\infty$ and $s\in(0,1)$. □

Theorem 5.7. *Let Ω be a bounded Lipschitz domain in $\mathbb{R}^n$ and consider a W-elliptic homogeneous differential operator L of order $2m$ with complex matrix-valued constant coefficients. Also, fix $1 \leq p, q < \infty$ and $s \in (0,1)$. Then*

$$\dot{\mathscr{S}} : \left(\dot{B}^{p,q}_{m-1,1-s}(\partial\Omega)\right)^* \longrightarrow B^{p',q'}_{m-1+s+1/p'}(\Omega), \tag{5.43}$$

linearly and boundedly, where $1/p + 1/p' = 1/q + 1/q' = 1$, and

$$\dot{S} = \mathrm{tr}_{m-1} \circ \dot{\mathscr{S}} \quad on \quad \left(\dot{B}^{p,q}_{m-1,s}(\partial\Omega)\right)^*. \tag{5.44}$$

Moreover, for each $1 \leq p < \infty$, $0 < q < \infty$ and $s \in (0,1)$, the operator

$$\dot{\mathscr{S}} : \left(\dot{B}^{p,p}_{m-1,1-s}(\partial\Omega)\right)^* \longrightarrow F^{p',q}_{m-1+s+1/p'}(\Omega) \tag{5.45}$$

is also well-defined, linear and bounded.

Finally, similar properties hold for $\psi\dot{\mathscr{S}}^-$ (cf. the convention (5.2)), for any cutoff function $\psi \in C_c^\infty(\mathbb{R}^n)$, and

$$\mathrm{tr}_{m-1} \circ \dot{\mathscr{S}}^+ = \mathrm{tr}_{m-1} \circ \dot{\mathscr{S}}^- \quad on \quad \left(\dot{B}^{p,q}_{m-1,s}(\partial\Omega)\right)^*. \tag{5.46}$$

Proof. Via real interpolation, cf. Theorem 3.39, it suffices to establish (5.43) in the case when $1 \leq p = q < \infty$ and $s \in (0,1)$. Thus, fix $1 \leq p < \infty$, $s \in (0,1)$ and let $\Lambda \in \left(\dot{B}^{p,p}_{m-1,1-s}(\partial\Omega)\right)^*$. Notice that the space $\dot{B}^{p,p}_{m-1,1-s}(\partial\Omega)$ is a closed subspace of $\oplus_1^N B^{p,p}_{1-s}(\partial\Omega)$, where N is as in (4.89) and let $\tilde{\Lambda} : \oplus_1^N B^{p,p}_{1-s}(\partial\Omega) \to \mathbb{C}$ be the linear bounded extension of Λ given by the Hahn–Banach Theorem. Proposition 2.52 then guarantees the existence of an (unique) N-tuple $\{g_\alpha\}_{|\alpha|\leq m-1} \in B^{p',p'}_{s-1}(\partial\Omega)$, such that

$$\langle\tilde{\Lambda}, \{f_\alpha\}_{|\alpha|\leq m-1}\rangle = \sum_{|\alpha|\leq m-1} \langle f_\alpha, g_\alpha\rangle, \quad \forall\{f_\alpha\}_{|\alpha|\leq m-1} \in \oplus_1^N B^{p,p}_{1-s}(\partial\Omega), \tag{5.47}$$

and

$$\|\{g_\alpha\}_{|\alpha|\leq m-1}\|_{\oplus_1^N B^{p',p'}_{s-1}(\partial\Omega)} \leq \|\tilde{\Lambda}\|_{(\oplus_1^N B^{p,p}_{1-s}(\partial\Omega))^*} \leq \|\Lambda\|_{(\dot{B}^{p,p}_{m-1,1-s}(\partial\Omega))^*}. \tag{5.48}$$

In particular, for each $\dot{f} \in \dot{B}^{p,p}_{m-1,1-s}(\partial\Omega)$ we have $\langle\Lambda, \dot{f}\rangle = \sum_{|\alpha|\leq m-1}\langle f_\alpha, g_\alpha\rangle$. Then, using (5.1), for each multi-index $\gamma \in \mathbb{N}_0$ of length $\leq m-1$, we may write

$$\begin{aligned}\partial^\gamma \dot{\mathscr{S}}\Lambda(X) &= \sum_{|\alpha|\leq m-1} (-1)^{|\alpha|}\Big\langle(\partial^{\alpha+\gamma}E)(X-\cdot), g_\alpha\Big\rangle \\ &=: (R_{\alpha\gamma}g_\alpha)(X), \qquad X \in \Omega.\end{aligned} \tag{5.49}$$

Using (4.29), it is easy to check that, for each α and γ multi-indices in $\mathbb{N}_0^n$ with $|\alpha|, |\gamma| \leq m-1$, the operators $R_{\alpha\gamma}$ introduced above satisfy the hypothesis of Theorem 2.64. Next, choosing $k=1$ in (2.533) and using (5.48), we obtain

$$\|\rho^{1-\frac{1}{p'}-s}|\nabla^m \dot{\mathscr{S}}\Lambda|\|_{L^{p'}(\Omega)} + \|\nabla^{m-1}\dot{\mathscr{S}}\Lambda\|_{L^{p'}(\Omega)} \leq C \sum_{|\alpha|\leq m-1} \|g_\alpha\|_{B_{s-1}^{p',p'}(\partial\Omega)} \leq C\|\Lambda\|_{\left(\dot{B}_{m-1,1-s}^{p,p}(\partial\Omega)\right)^*}. \tag{5.50}$$

Fix $q \in (1,\infty)$ with $q \geq p$. Since $\Lambda \in \left(\dot{B}_{m-1,1-s}^{p,p}(\partial\Omega)\right)^* \hookrightarrow \left(\dot{L}_{m-1,1}^{q}(\partial\Omega)\right)^*$, using (5.7) we may write

$$\sum_{j=0}^{m-1} \|\mathscr{N}(\nabla^j \dot{\mathscr{S}}\Lambda)\|_{L^{q'}(\partial\Omega)} \leq C\|\Lambda\|_{\left(\dot{B}_{m-1,1-s}^{p,p}(\partial\Omega)\right)^*}. \tag{5.51}$$

By Proposition 2.3, we then have $\dot{\mathscr{S}}\Lambda \in W^{m-1,\frac{nq'}{n-1}}(\Omega) \hookrightarrow L^{p'}(\Omega)$, plus a natural estimate, granted that $q > p$ is chosen sufficiently close to p. In particular,

$$\|\dot{\mathscr{S}}\Lambda\|_{L^{p'}(\Omega)} \leq C\|\Lambda\|_{\left(\dot{B}_{m-1,1-s}^{p,p}(\partial\Omega)\right)^*}. \tag{5.52}$$

Recall the weighted Sobolev space $W_a^{k,p}(\Omega)$ introduced in (2.248). From (5.50), (5.52) and (2.249) we deduce that

$$\dot{\mathscr{S}}\Lambda \in W_{1-1/p'-s}^{m,p'}(\Omega) \text{ and } \|\dot{\mathscr{S}}\Lambda\|_{W_{1-1/p'-s}^{m,p'}(\Omega)} \leq C\|\Lambda\|_{\left(\dot{B}_{m-1,1-s}^{p,p}(\partial\Omega)\right)^*}. \tag{5.53}$$

All together, (5.53), (5.5), (2.263), (2.264) and (2.305) show that

$$\dot{\mathscr{S}}\Lambda \in B_{m-1+1/p'+s}^{p',p'}(\Omega) \text{ and } \|\dot{\mathscr{S}}\Lambda\|_{B_{m-1+1/p'+s}^{p',p'}(\Omega)} \leq C\|\Lambda\|_{\left(\dot{B}_{m-1,1-s}^{p,p}(\partial\Omega)\right)^*}, \tag{5.54}$$

which finishes the proof of the claim about (5.43). As regards (5.45), this is a direct consequence of the boundedness of (5.45) when $p=q$, (2.264) and (5.5).

Going further, (5.44) is a consequence of (5.19), the inclusion

$$\left(\dot{B}_{m-1,s}^{p,q}(\partial\Omega)\right)^* \hookrightarrow \left(\dot{L}_{m-1,1}^{r}(\partial\Omega)\right)^*, \tag{5.55}$$

valid whenever $r \in (1,\infty)$ is sufficiently large, the fact that (5.43) is bounded, Theorem 3.9 and the remark following Theorem 2.53.

That $\psi\dot{\mathscr{S}}^-$ enjoys properties which are similar to those of $\dot{\mathscr{S}}$ is then seen much as we have argued in the last part of the proof of Theorem 4.19. Finally, (5.46) is a consequence of (5.19), density and the boundedness of the operators involved. □

The argument in the above proof breaks down when $p = \infty$, since this end-point is not covered by Proposition 2.52. Nonetheless, the corresponding claim about (5.43) continues to hold. More specifically, we have the following.

Theorem 5.8. *Consider a bounded Lipschitz domain $\Omega \subset \mathbb{R}^n$ and assume that L is a W-elliptic homogeneous differential operator of order $2m$ with complex matrix-valued constant coefficients. Then for each $s \in (0,1)$ the operator*

$$\dot{\mathscr{S}} : \Big(\dot{B}^{\infty,\infty}_{m-1,1-s}(\partial\Omega)\Big)^* \longrightarrow B^{1,1}_{m+s}(\Omega) \tag{5.56}$$

is well-defined, linear and bounded. Furthermore, a similar mapping property holds for $\psi\dot{\mathscr{S}}^-$ (cf. the convention (5.2)), for any cutoff function $\psi \in C^\infty_c(\mathbb{R}^n)$, and the identity from (5.46) continues to hold in this case as well.

Proof. By (2.263) and (2.264), it suffices to show that there exists a finite constant $C > 0$ such that the estimate $\|\dot{\mathscr{S}}\Lambda\|_{W^{m,1}_{s-1}(\Omega)} \leq C\|\Lambda\|_{(\dot{B}^{\infty,\infty}_{m-1,s}(\partial\Omega))^*}$ holds for each functional $\Lambda \in (\dot{B}^{\infty,\infty}_{m-1,s}(\partial\Omega))^*$. The lower order terms in the definition of the norm in $W^{m,1}_{s-1}(\Omega)$ can be controlled as before, so we will focus on establishing that

$$\sum_{|\gamma|=m} \int_\Omega \rho(X)^{s-1}|\partial^\gamma \dot{\mathscr{S}}\Lambda(X)|\,dX \leq C\|\Lambda\|_{(\dot{B}^{\infty,\infty}_{m-1,s}(\partial\Omega))^*}, \tag{5.57}$$

or, equivalently,

$$\Big|\int_\Omega \rho(X)^{s-1}g(X)\Big\langle \Lambda\,,\,\mathrm{tr}_{m-1}[(\partial^\gamma E)(X-\cdot)]\Big\rangle\,dX\Big| \leq C\|\Lambda\|_{(\dot{B}^{\infty,\infty}_{m-1,s}(\partial\Omega))^*} \tag{5.58}$$

for every $\gamma \in \mathbb{N}^n_0$ with $|\gamma| = m$, and $g \in L^\infty_c(\Omega)$ with $\|g\|_{L^\infty(\Omega)} \leq 1$. By interchanging integration over Ω with the action of Λ, we see that it is enough to establish the following estimate:

$$\Big\|\mathrm{tr}_{m-1}\Big[\int_\Omega \rho(X)^{s-1}g(X)(\partial^\gamma E)(X-\cdot)\,dX\Big]\Big\|_{\dot{B}^{\infty,\infty}_{m-1,s}(\partial\Omega)} \leq C = C(\Omega,s) < \infty, \tag{5.59}$$

for each fixed $\gamma \in \mathbb{N}^n_0$ with $|\gamma| = m$, and $g \in L^\infty_{comp}(\Omega)$ with $\|g\|_{L^\infty(\Omega)} \leq 1$. Fix γ, g as above and consider

$$F(Y) := \int_\Omega \rho(X)^{s-1}g(X)(\partial^\gamma E)(X-Y)\,dX, \quad Y \in \Omega, \tag{5.60}$$

and, for each $\alpha \in \mathbb{N}^n_0$ with $|\alpha| \leq m-1$, denote by $R_\alpha(Y,Z)$ the reminder of order $m-1-|\alpha|$ in the Taylor expansion of F at Z (and evaluated at Y). Also, we shall set $f_\alpha := \mathrm{Tr}\,[\partial^\alpha F]$, i.e.,

$$f_\alpha(Y) := \int_\Omega \rho(X)^{s-1}g(X)(\partial^{\alpha+\gamma}E)(X-Y)\,dX, \quad Y \in \partial\Omega,\ |\alpha| \leq m-1. \tag{5.61}$$

We wish to establish that

$$\sum_{|\alpha|\leq m-1}\Big(\|f_\alpha\|_{L^\infty(\partial\Omega)}+\sup_{Y\neq Z\in\partial\Omega}\frac{|R_\alpha(Y,Z)|}{|Y-Z|^{m-1+s-|\alpha|}}\Big)\leq C=C(\Omega,s)<\infty. \tag{5.62}$$

To set the stage, we express the Taylor reminder R_α as

$$R_\alpha(Y,Z)=\int_\Omega \rho(X)^{s-1}g(X)\Big[(\partial^{\alpha+\gamma}E)(X-Y)$$

$$-\sum_{|\beta|\leq m-1-|\alpha|}\tfrac{1}{\beta!}(\partial^{\alpha+\beta+\gamma}E)(X-Z)(Y-Z)^\beta\Big]\,dX. \tag{5.63}$$

Assuming that $Y,Z\in\partial\Omega$ are arbitrary and fixed, we shall estimate the expression in the square brackets above in two ways. In the first scenario, we combine the first term with the ensuing sum performed only over $|\beta|\leq m-2-|\alpha|$ (while retaining the terms corresponding to $|\beta|=m-1-|\alpha|$ as they are) and use the Taylor remainder formula in order to produce a bound of the form

$$C|Z-Y|^{m-1-|\alpha|}\int_0^1|\nabla^{m-1-|\alpha|}(\partial^{\alpha+\gamma}E)(X-Q_\tau)|\,d\tau$$

$$+C|Z-Y|^{m-1-|\alpha|}|\nabla^{m-1-|\alpha|}(\partial^{\alpha+\gamma}E)(X-Z)|$$

$$\leq C|Z-Y|^{m-1-|\alpha|}\Big[\frac{1}{|X-Z|^{n-1}}+\int_0^1\frac{1}{|X-Q_\tau|^{n-1}}\,d\tau\Big], \tag{5.64}$$

where we have used the first line in (4.29), and have set

$$Q_\tau:=\tau Y+(1-\tau)Z\in[Y,Z],\qquad 0\leq\tau\leq 1. \tag{5.65}$$

In the second scenario, we simply use Taylor's remainder formula for the entire expression in the brackets in order to estimate it from above by

$$C|Z-Y|^{m-|\alpha|}\Big[\sup_{0\leq\tau\leq 1}|\nabla^m(\partial^\gamma E)(X-Q_\tau)|\Big] \tag{5.66}$$

$$\leq C|Z-Y|^{m-|\alpha|}\Big[\sup_{0\leq\tau\leq 1}|X-Q_\tau|^{-n}\Big],$$

by once again appealing to the first line in (4.29). In the next step, we decompose the domain of integration in (5.63) into two pieces, namely $\{X\in\Omega:\ |X-Y|\leq 2|Y-Z|\}$ and $\{X\in\Omega:\ |X-Y|>2|Y-Z|\}$, then use the bound (5.64) for the first resulting term, and the bound (5.66) for the second one. The estimate we arrive at in this fashion then reads:

$$
\begin{aligned}
|R_\alpha(Y,Z)| \;&\le\; C|Z-Y|^{m-1-|\alpha|} \int\limits_{\substack{X\in\Omega\\ |X-Y|\le 2|Y-Z|}} \frac{\rho(X)^{s-1}}{|X-Z|^{n-1}}\,dX \\
&\quad + C|Z-Y|^{m-1-|\alpha|}\int_0^1\Big(\int\limits_{\substack{X\in\Omega\\ |X-Y|\le 2|Y-Z|}} \frac{\rho(X)^{s-1}}{|X-Q_\tau|^{n-1}}\,dX\Big)d\tau \\
&\quad + C|Z-Y|^{m-|\alpha|}\int\limits_{\substack{X\in\Omega\\ |X-Y|>2|Y-Z|}} \rho(X)^{s-1}\Big[\sup_{0\le\tau\le 1}|X-Q_\tau|^{-n}\Big]\,dX \\
&=: I+II+III.
\end{aligned}
\tag{5.67}
$$

Next, observe that

$$
|X-Y|\le 2|Y-Z| \Longrightarrow |X-Z|\le 3|Y-Z|. \tag{5.68}
$$

Thus, after localizing and changing variables (from $\partial\Omega$ to $\mathbb{R}^{n-1}$), we obtain

$$
\begin{aligned}
|I| &\le C|Z-Y|^{m-1-|\alpha|}\int\limits_{\substack{X\in\Omega\\ |X-Z|\le 3|Y-Z|}} \frac{\rho(X)^{s-1}}{|X-Z|^{n-1}}\,dX \\
&\le C|Z-Y|^{m-1-|\alpha|}\int\limits_{\substack{(x',t)\in\mathbb{R}^n_+\\ t+|x'|<C|Y-Z|}} \frac{t^{s-1}}{[t+|x'|]^{n-1}}\,dx'dt \\
&\le C|Z-Y|^{m-1-|\alpha|+s}\int\limits_{\substack{(x',t)\in\mathbb{R}^n_+\\ t+|x'|<C}} \frac{t^{s-1}}{[t+|x'|]^{n-1}}\,dx'dt,
\end{aligned}
\tag{5.69}
$$

after rescaling $t\mapsto t|Z-Y|$, $x'\mapsto x'|Z-Y|$, and since

$$
\int\limits_{\substack{(x',t)\in\mathbb{R}^n_+\\ t+|x'|<C}} \frac{t^{s-1}}{[t+|x'|]^{n-1}}\,dx'dt \le \int\limits_{\substack{x'\in\mathbb{R}^{n-1}\\ |x'|<C}}\int_0^\infty \frac{t^{s-1}}{(1+t)^{n-1}}\cdot\frac{1}{|x'|^{n-1-s}}\,dt\,dx' < \infty,
\tag{5.70}
$$

we conclude that

$$
|I|\le C|Z-Y|^{m-1-|\alpha|+s}. \tag{5.71}
$$

For each fixed $\tau \in (0,1)$,

$$|X-Y| \leq 2|Y-Z| \Longrightarrow |X-Q_\tau| \leq |X-Y| + |Y-Q_\tau|$$
$$\leq 2|Y-Z| + |Y-Z| = 3|Y-Z|, \quad (5.72)$$

so the inner integral in II may be estimated in a similar fashion by $C|Z-Y|^s$. Thus,

$$|II| \leq C|Z-Y|^{m-1-|\alpha|+s}, \tag{5.73}$$

which is of the right order.

As for III in (5.67) we first observe that, on the domain of integration,

$$|X-Y| \leq |X-Q_\tau| + |Y-Z| \leq |X-Q_\tau| + 2^{-1}|X-Y| \tag{5.74}$$

and, therefore, $|X-Y| \leq 2|X-Q_\tau|$. Consequently, the integrand in III is dominated by $C\rho(X)^{s-1}|X-Y|^{-n}$. Thus, after localizing and flattening the boundary (via a bi-Lipschitz mapping), we obtain

$$|III| \leq C|Z-Y|^{m-|\alpha|} \int_{\substack{x' \in \mathbb{R}^{n-1}, t>0 \\ t+|x'| \geq c|Y-Z|}} \frac{t^{s-1}}{[t+|x'|]^n}\, dx'dt \leq C|Y-Z|^{m-|\alpha|+s-1}. \tag{5.75}$$

All in all,

$$|R_\alpha(Y,Z)| \leq C|Y-Z|^{m-|\alpha|+s-1}, \tag{5.76}$$

by (5.71), (5.73) and (5.75).

Finally, given that Ω is bounded, elementary considerations based on (4.29) and (5.61) yield

$$\sum_{|\alpha| \leq m-1} \|f_\alpha\|_{L^\infty(\partial\Omega)} \leq C = C(\Omega, s) < \infty. \tag{5.77}$$

Hence, (5.62) follows from (5.76) and (5.77). This finishes the proof of the fact that the operator (5.56) is well-defined and bounded. □

We can now further augment the results in Proposition 5.6 and Theorem 5.7 with the following.

Corollary 5.9. *In the context of Theorem 5.7, the operator* (5.45) *is in fact well-defined and bounded for* $1 \leq p \leq \infty$, $0 < q < \infty$, $0 < s < 1$. *Furthermore, the operator* (5.40) *is in fact well-defined and bounded whenever* $1 \leq p, q \leq \infty$, $0 < s < 1$, *and* (5.44) *holds in this range.*

Proof. This can be deduced from Theorem 5.8 by reasoning as before. □

Let us also record the following analogue of Theorem 4.20 for the single multi-layer operator.

Corollary 5.10. *Assume that Ω is a bounded Lipschitz domain in $\mathbb{R}^n$ and suppose that L is a W-elliptic homogeneous differential operator of order $2m$ with complex matrix-valued constant coefficients. Recall the weighted Sobolev spaces from* (2.248)*. Then for each $p \in (1, \infty)$ and $s \in (0, 1)$, the operator*

$$\dot{\mathscr{S}} : \left(\dot{B}^{p,p}_{m-1,1-s}(\partial\Omega)\right)^* \longrightarrow W^{m,p'}_{1-1/p'-s}(\Omega) \tag{5.78}$$

is well-defined, linear and bounded. Moreover, a similar boundedness result holds for $\psi\dot{\mathscr{S}}^-$ (cf. the convention (5.2)*), for any cutoff function $\psi \in C^\infty_c(\mathbb{R}^n)$.*

Proof. This is implicit in (5.53). □

We conclude with a useful global regularity result for the single multi-layer potential operator.

Proposition 5.11. *Let L be a W-elliptic homogeneous differential operator of order $2m$ with complex matrix-valued constant coefficients. Also, assume that $\Omega \subset \mathbb{R}^n$ is a bounded Lipschitz domain and, as usual, set $\Omega_+ := \Omega$ and $\Omega_- := \mathbb{R}^n \setminus \overline{\Omega}$. Finally, suppose that $1 \leq p, q \leq \infty$, $0 < s < 1$, and for some arbitrary $\Lambda \in \left(\dot{B}^{p,q}_{m-1,1-s}(\partial\Omega)\right)^*$ set $u^\pm := \dot{S}\Lambda$ in $\Omega_\pm$. Then, for every $\psi \in C^\infty_c(\mathbb{R}^n)$, it follows that*

$$u := \begin{cases} u^+ \text{ in } \Omega_+, \\ u^- \text{ in } \Omega_-, \end{cases} \Longrightarrow \psi\, u \in B^{p',q'}_{m-1+s+1/p'}(\mathbb{R}^n), \tag{5.79}$$

plus a natural estimate, where $p', q' \in [1, \infty]$ satisfy $1/p + 1/p' = 1/q + 1/q' = 1$. Furthermore, $\psi\, u \in F^{p',q}_{m-1+s+1/p'}(\mathbb{R}^n)$ in the case when $\Lambda \in \left(\dot{B}^{p,p}_{m-1,1-s}(\partial\Omega)\right)^*$.

Proof. This is a direct consequence of Theorem 5.7 and Proposition 3.26. □

5.4 The Conormal Derivative

To motivate the subsequent considerations, we would like to elaborate on the following phenomenon. Given a Lipschitz domain $\Omega \subseteq \mathbb{R}^n$, for any bilinear form

$$\mathscr{B}(u, v) := (-1)^m \sum_{|\alpha|=|\beta|=m} \int_\Omega \langle A_{\alpha\beta}\, \partial^\beta u\, ,\, \partial^\alpha v\rangle\, dX, \tag{5.80}$$

where the $A_{\alpha\beta}$'s are $M \times M$ matrices with complex entries, there exists a unique differential operator L of order $2m$ – in fact, given by (1.1) – such that

$$\mathscr{B}(u,v) = \int_\Omega \langle Lu, v\rangle\, dX, \qquad \forall\, u, v \in C_c^\infty(\Omega). \tag{5.81}$$

In the converse direction, it is clear that given a differential operator L as in (1.1) there exist infinitely many a bilinear forms $\mathscr{B}$ such that (5.81) holds. A concrete example, in the case when $L = \Delta^2$, is discussed later, in (5.86) and in (6.351), (6.353).

As has been already noted in § 4.2, a given bilinear form $\mathscr{B}$ as in (5.80) and a choice of the fundamental solution E for the differential operator L associated with $\mathscr{B}$ as in (5.81), determine a unique double multi-layer. On the other hand, starting with a differential operator L as in (1.1), there are infinitely many bilinear forms satisfying (5.81) and, therefore, infinitely many double multi-layers associated with L. We summarize this discussion in the following diagram:

a differential operator → many bilinear forms → many double multi-layers;

$$\text{a bilinear form} \longrightarrow \begin{cases} \text{a unique differential operator} \\ \qquad\qquad \text{and} \\ \text{a unique double multi-layer.} \end{cases} \tag{5.82}$$

For instance, a given differential operator

$$Lu = \sum_{|\alpha|=|\beta|=m} \partial^\alpha A_{\alpha\beta}\, \partial^\beta u, \tag{5.83}$$

is associated with *any* bilinear form of the type

$$\mathscr{B}(u,v) = (-1)^m \sum_{|\alpha|=|\beta|=m} \int_\Omega \langle (A_{\alpha\beta} + A'_{\alpha\beta})\, \partial^\beta u\,,\, \partial^\alpha v\rangle\, dX, \tag{5.84}$$

where the tensor coefficient $A' = (A'_{\alpha\beta})_{|\alpha|=|\beta|=m}$ satisfies

$$\sum_{\substack{\alpha+\beta=\gamma \\ |\alpha|=|\beta|=m}} A'_{\alpha\beta} = 0, \qquad \forall\, \gamma \in \mathbb{N}_0^n \;\text{ with }\; |\gamma| = 2m. \tag{5.85}$$

A concrete example in $\mathbb{R}^2$ is offered by the infinite family of bilinear forms, indexed by $\eta \in \mathbb{R}$,

$$\mathscr{B}_\eta(u,v) = \int_\Omega \Big\{\Delta u \Delta v + \eta\Big(\frac{\partial^2 u}{\partial^2 x}\frac{\partial^2 v}{\partial^2 y} + \frac{\partial^2 u}{\partial^2 y}\frac{\partial^2 v}{\partial^2 x} - 2\frac{\partial^2 u}{\partial x \partial y}\frac{\partial^2 v}{\partial x \partial y}\Big)\Big\}\,dxdy, \tag{5.86}$$

all of which are associated with the bi-Laplacian Δ^2. The significance of this type of example is going to become more apparent in § 6.5.

In this section we wish to further elaborate on this matter by underscoring the role played by the *conormal derivative* associated with each of the bilinear forms mentioned above. Turning to specifics, we make the following definition (the reader is reminded that all functions involved are vector-valued).

Definition 5.12. Assume that Ω is a bounded Lipschitz domain in $\mathbb{R}^n$, with outward unit normal $\nu = (\nu_j)_{1\le j\le n}$. Suppose that a tensor coefficient $A = (A_{\alpha\beta})_{|\alpha|=|\beta|=m}$ (of $M \times M$ matrices with complex entries) has been given. For a sufficiently nice function u in Ω then define

$$\partial_\nu^A u := \Big\{(\partial_\nu^A u)_\delta\Big\}_{|\delta|\le m-1} \quad \text{with the } \delta\text{-component given by the formula}$$

$$(\partial_\nu^A u)_\delta := \sum_{|\alpha|=|\beta|=m}\sum_{j=1}^{n}(-1)^{|\delta|}\frac{\alpha!|\delta|!(m-|\delta|-1)!}{m!\delta!(\alpha-\delta-e_j)!}\nu_j A_{\alpha\beta}\Big(\partial^{\alpha+\beta-\delta-e_j}u\Big)\Big|_{\partial\Omega}, \tag{5.87}$$

with the convention that the sum in α and j is only performed over those α's and j's such that $\alpha - \delta - e_j$ does not have any negative components.

Equivalently, if u and v are sufficiently well-behaved functions in Ω then, with σ denoting the surface measure on $\partial\Omega$,

$$\int_{\partial\Omega}\Big\langle \partial_\nu^A u, \mathrm{tr}_{m-1} v\Big\rangle\,d\sigma = \sum_{|\alpha|=|\beta|=m}\sum_{k=1}^{m}\sum_{\substack{|\gamma|=k-1,|\delta|=m-k\\ \gamma+\delta+e_j=\alpha}} C(k,m,\alpha,\delta,\gamma)\times \tag{5.88}$$

$$\times\int_{\partial\Omega}\Big\langle \nu_j A_{\alpha\beta}\Big(\partial^{\beta+\gamma}u\Big)\Big|_{\partial\Omega}, \Big(\partial^\delta v\Big)\Big|_{\partial\Omega}\Big\rangle\,d\sigma,$$

where

$$C(k,m,\alpha,\delta,\gamma) := \frac{(-1)^{m-k}\alpha!(m-k)!(k-1)!}{m!\delta!\gamma!}. \tag{5.89}$$

In the spirit of diagram (5.82) we note that, as is apparent from Definition 5.12,

$$\begin{array}{c} \text{a given differential} \\ \text{operator } L \end{array} \Longrightarrow \begin{array}{c} \text{infinitely many coefficient tensors} \\ A \text{ yielding the same operator } L \end{array} \tag{5.90}$$

$$\Longrightarrow \begin{array}{c} \text{infinitely many} \\ \text{conormals } \partial_\nu^A. \end{array}$$

The relationship between conormals and bilinear forms is highlighted in the result below.

Proposition 5.13. *Assume that Ω is a bounded Lipschitz domain in $\mathbb{R}^n$, with outward unit normal $\nu = (\nu_j)_{1\leq j\leq n}$, and consider a tensor coefficient $A = (A_{\alpha\beta})_{|\alpha|=|\beta|=m}$.*

If u and v are two reasonably behaved functions in Ω, then the following Green formula holds:

$$\sum_{|\alpha|=|\beta|=m} \int_\Omega \Big\langle A_{\alpha\beta}\, \partial^\beta u, \partial^\alpha v\Big\rangle\, dX \tag{5.91}$$

$$= (-1)^{m+1} \int_{\partial\Omega} \Big\langle \partial_\nu^A u, \operatorname{tr}_{m-1} v\Big\rangle\, d\sigma + (-1)^m \int_\Omega \langle Lu, v\rangle\, dX,$$

where L is the differential operator associated with A according to

$$Lu := \sum_{|\alpha|=|\beta|=m} \partial^\alpha A_{\alpha\beta}\, \partial^\beta u. \tag{5.92}$$

Proof. This is a direct consequence of (5.88)–(5.89) and Proposition 4.3. □

Recall the double multi-layer from (4.57) and the number N from (4.89), which stands for the cardinality of the set of multi-indices with n components and of length $\leq m-1$.

Definition 5.14. Let Ω be a bounded Lipschitz domain in $\mathbb{R}^n$ with outward unit normal $\nu = (\nu_j)_{1\leq j\leq n}$. Assume that $A = (A_{\alpha\beta})_{|\alpha|=|\beta|=m}$ is a tensor coefficient of $M \times M$ matrices with complex entries with the property that the differential operator L associated with A as in (5.92) is W-elliptic. Finally, let E be the fundamental solution of L described in Theorem 4.2 (hence, E is an $M \times M$ matrix-valued function).

In this context, define the conormal $\partial_\nu^A E$ as the $(NM) \times M$-matrix (with N as above) whose i-th column is ∂_ν^A acting (according to (5.87)) on the i-th column in E.

In the context of the above definition it is then elementary to check that

$$(\partial_\nu^A E)\eta = \partial_\nu^A (E\eta), \qquad \forall\, \eta \in \mathbb{C}^M. \tag{5.93}$$

In the sequel, the notation $\partial_\nu^{A^t}$ is chosen to indicate that the conormal is taken with respect to the bilinear form associated with A^t.

Our next result shows that the integral kernel of the double multi-layer is the conormal derivative of the fundamental solution. Recall that L^t is the transposed of L.

Proposition 5.15. *Assume that Ω is a bounded Lipschitz domain in $\mathbb{R}^n$ and suppose that $A = (A_{\alpha\beta})_{|\alpha|=|\beta|=m}$ is a tensor coefficient with the property that the differential operator L associated with A as in (5.92) is W-elliptic. Let E be the fundamental solution of L described in Theorem 4.2 and let $\dot{\mathscr{D}}$ be the double multi-layer defined in this context as in Definition 4.4.*

Then for any Whitney array $\dot f$ on $\partial\Omega$,

$$\dot{\mathscr{D}}\dot f(X) = \int_{\partial\Omega} \Big[\partial_{\nu(Y)}^{A^t}[E_{L^t}(X-Y)]\Big]^t \dot f(Y)\, d\sigma(Y), \qquad X \in \mathbb{R}^n \setminus \partial\Omega. \tag{5.94}$$

Proof. This follows by inspecting (5.88) and (4.57). □

Recall the conventions and results from (4.6)–(4.10).

Proposition 5.16. *Retain the same setting as in Proposition 5.15. Then*

$$\int_{\partial\Omega} \Big\langle \partial_\nu^A u, \mathrm{tr}_{m-1} v\Big\rangle\, d\sigma - \int_{\partial\Omega} \Big\langle \mathrm{tr}_{m-1} u, \partial_\nu^{A^t} v\Big\rangle\, d\sigma$$
$$= \int_\Omega \Big\langle Lu, v\Big\rangle\, dX - \int_\Omega \Big\langle u, L^t v\Big\rangle\, dX, \tag{5.95}$$

for any two reasonably behaved functions u and v in Ω.

Proof. Write formula (5.91) twice: once in its original form, then with u, v, A, L replaced by v, u, A^t, L^t, respectively. The identity (5.95) is then obtained by subtracting the latter from the former and canceling the two bilinear forms. □

Let us also associate with L a Newtonian-like potential, by setting

$$\big(\Pi_\Omega\, u\big)(X) := \int_\Omega E(X-Y)u(Y)\, dY, \qquad X \in \mathbb{R}^n, \tag{5.96}$$

for any reasonable function u in Ω.

Corollary 5.17. *In the same setting as above, for any sufficiently nice function u in Ω, the following integral representation formula holds:*

$$u = \dot{\mathscr{D}}(\mathrm{tr}_{m-1} u) - \dot{\mathscr{S}}(\partial_\nu^A u) + \Pi_\Omega(Lu) \quad \text{in } \Omega. \tag{5.97}$$

In particular, if u is also a null-solution of L in Ω, then

$$u = \dot{\mathscr{D}}(\mathrm{tr}_{m-1}u) - \dot{\mathscr{S}}(\partial_\nu^A u) \quad \text{in } \Omega. \tag{5.98}$$

Proof. Fix $X_o \in \Omega$ and $\eta \in \mathbb{C}^M$. Taking $v := E_{L^t}(X_o - \cdot)\eta$ in (5.95) then leads to (5.97), by making use of (5.93), (5.94) and

$$\Big\langle \eta\,,\, \dot{\mathscr{S}}\dot{f}(X_o)\Big\rangle = \Big\langle \dot{f}\,,\, \mathrm{tr}_{m-1}\left[E_{L^t}(X_o - \cdot)\eta\right]\Big\rangle \tag{5.99}$$

(used with $\dot{f} = \partial_\nu^A u$) which, in turn, can be checked with the help of (5.3). □

Proposition 5.18. *Let Ω be a Lipschitz domain in $\mathbb{R}^n$ and assume that the functions $u, v \in C_c^\infty(\mathbb{R}^n)$. Also, let Λ be a reasonable vector-valued function defined on $\partial\Omega$ (regarded as a linear functional on Whitney arrays). Consider a constant coefficient W-elliptic differential operator L of order 2m, $m \in \mathbb{N}$, and denote by ∂_ν^A the conormal associated with a particular choice of the tensor coefficient $A = (A_{\alpha\beta})_{|\alpha|=|\beta|=m}$ in the writing of L as in (5.92). If $\dot{\mathscr{D}}$ is as in (4.57) and $\dot{\mathscr{S}}$ is as in (5.1), then*

$$\int_\Omega \big\langle \dot{\mathscr{S}}\Lambda, v\big\rangle\, dX = \int_{\partial\Omega} \big\langle \Lambda, \mathrm{tr}_{m-1}(\Pi_\Omega v)\big\rangle\, d\sigma. \tag{5.100}$$

If, in addition, L is also symmetric, then

$$\int_\Omega \big\langle \Pi_\Omega u, v\big\rangle\, dX = \int_\Omega \big\langle u, \Pi_\Omega v\big\rangle\, dX, \tag{5.101}$$

$$\int_{\partial\Omega} \big\langle \partial_\nu^A \Pi_\Omega u, \mathrm{tr}_{m-1} v\big\rangle\, d\sigma = \int_\Omega \big\langle u, \dot{\mathscr{D}}(\mathrm{tr}_{m-1} v)\big\rangle\, dX, \tag{5.102}$$

and

$$\int_{\partial\Omega} \big\langle \partial_\nu^A(\dot{\mathscr{D}}(\mathrm{tr}_{m-1}u))\,,\, \mathrm{tr}_{m-1}v\big\rangle\, d\sigma = \int_{\partial\Omega} \big\langle \mathrm{tr}_{m-1}u, \partial_\nu^A(\dot{\mathscr{D}}(\mathrm{tr}_{m-1}v))\big\rangle\, d\sigma. \tag{5.103}$$

Proof. To begin with, (5.100) is a direct consequence of (5.3), (3.100) and (5.96). Next, employing (5.95) and the symmetry of L we may write

$$\begin{aligned}\int_{\partial\Omega} \big\langle \partial_\nu^A \Pi_\Omega u, \mathrm{tr}_{m-1} v\big\rangle\, d\sigma &= \int_{\partial\Omega} \big\langle \mathrm{tr}_{m-1}(\Pi_\Omega u), \partial_\nu^A v\big\rangle\, d\sigma \\ &\quad + \int_\Omega \big\langle L(\Pi_\Omega u), v\big\rangle\, dX - \int_\Omega \big\langle \Pi_\Omega u, Lv\big\rangle\, dX.\end{aligned} \tag{5.104}$$

Since $L(\Pi_\Omega u) = u$ in Ω, from (5.104) and (5.100)–(5.101) we conclude

$$\int_{\partial\Omega} \langle \partial_\nu^A \Pi_\Omega u, \mathrm{tr}_{m-1} v\rangle \, d\sigma = \int_\Omega \langle u, v + \dot{\mathscr{S}}(\partial_\nu^A v) - \Pi_\Omega(Lv)\rangle \, dX. \tag{5.105}$$

Invoking (5.97), this further implies (5.102).

As for (5.103) we appeal again to (5.97) to write

$$\dot{\mathscr{D}}(\mathrm{tr}_{m-1} u) = u + \dot{\mathscr{S}}(\partial_\nu^A u) - \Pi_\Omega(Lu). \tag{5.106}$$

Since a direct calculation based on (5.165) shows that

$$\partial_\nu^A\big(u + \dot{\mathscr{S}}(\partial_\nu^A u)\big) = \big(\tfrac{1}{2}I + \dot{K}^*\big)(\partial_\nu^A u), \tag{5.107}$$

we have

$$\int_{\partial\Omega} \langle \partial_\nu^A(\dot{\mathscr{D}}(\mathrm{tr}_{m-1} u)), \mathrm{tr}_{m-1} v\rangle \, d\sigma = I - II, \tag{5.108}$$

where

$$\begin{aligned} I &:= \int_{\partial\Omega} \langle (\tfrac{1}{2}I + \dot{K}^*)(\partial_\nu^A u), \mathrm{tr}_{m-1} v\rangle \, d\sigma, \\ II &:= \int_{\partial\Omega} \langle \partial_\nu^A(\Pi_\Omega(Lu)), \mathrm{tr}_{m-1} v\rangle \, d\sigma. \end{aligned} \tag{5.109}$$

However, since by duality and (4.164),

$$\begin{aligned} I &= \int_{\partial\Omega} \langle \partial_\nu^A u, (\tfrac{1}{2}I + \dot{K})(\mathrm{tr}_{m-1} v)\rangle \, d\sigma \\ &= \int_{\partial\Omega} \big\langle \partial_\nu^A u, \mathrm{tr}_{m-1}\big(\dot{\mathscr{D}}(\mathrm{tr}_{m-1} v)\big)\big\rangle \, d\sigma, \end{aligned} \tag{5.110}$$

whereas, using (5.102),

$$II = \int_\Omega \langle Lu, \dot{\mathscr{D}}(\mathrm{tr}_{m-1} v)\rangle \, dX. \tag{5.111}$$

Using (5.110), (5.111), Green's identity (5.95), and the fact that $L(\dot{\mathscr{D}}(\mathrm{tr}_{m-1} v)) = 0$ in Ω, we obtain that

$$I - II = \int_{\partial\Omega} \langle \mathrm{tr}_{m-1} u, \partial_\nu^A(\dot{\mathscr{D}}(\mathrm{tr}_{m-1} v))\rangle \, d\sigma. \tag{5.112}$$

Finally, (5.103) is a direct consequence of (5.108) and (5.112). This completes the proof of the proposition. □

Let Ω be a Lipschitz domain in $\mathbb{R}^n$ and set $\Omega_+ := \Omega$ and $\Omega_- := \mathbb{R}^n \setminus \overline{\Omega}$. In the sequel we denote by $\partial^A_{\nu_\pm}$ the conormal derivative (5.87) where the restriction to $\partial\Omega$ is taken from Ω_+ and Ω_-, respectively.

Proposition 5.19. *Assume that Ω is a Lipschitz domain in $\mathbb{R}^n$ and recall* (2.11). *Also, consider a constant coefficient W-elliptic differential operator L of order $2m$, $m \in \mathbb{N}$, which is symmetric. Then, with the double multi-layer defined as in* (4.57) *and the Newtonian potential operator defined as in* (5.96)*, one has*

$$\partial^A_{\nu_+} \Pi_{\Omega_+} u = \partial^A_{\nu_-} \Pi_{\Omega_+} u, \qquad \forall\, u \in C^\infty_c(\mathbb{R}^n), \tag{5.113}$$

and for each $u, v \in C^\infty_c(\mathbb{R}^n)$,

$$\int_{\partial\Omega} \Big\langle \partial^A_{\nu_+}(\dot{\mathscr{D}}(\mathrm{tr}_{m-1} u))\,,\,\mathrm{tr}_{m-1} v \Big\rangle\, d\sigma = \int_{\partial\Omega} \Big\langle \partial^A_{\nu_-}(\dot{\mathscr{D}}(\mathrm{tr}_{m-1} u))\,,\,\mathrm{tr}_{m-1} v \Big\rangle\, d\sigma. \tag{5.114}$$

Proof. Consider (5.113). To this end, we debut with the observation that, due to the weak singularity in the kernel of $\partial^\alpha \Pi_{\Omega_+} u$, for α multi-index in $\mathbb{N}^n_0$ of length $\leq m-1$, there holds

$$\mathrm{tr}^+_{m-1} \Pi_{\Omega_+} u = \mathrm{tr}^-_{m-1} \Pi_{\Omega_+} u \quad \text{on } \partial\Omega, \tag{5.115}$$

where $\mathrm{tr}^\pm_{m-1}$ denote the multi-trace operators in $\Omega_\pm$. Next, fix $v \in C^\infty_c(\mathbb{R}^n)$ and write Green's identity (5.95) in the domain Ω_-. Using the fact that

$$L(\Pi_{\Omega_+} u) = u\chi_{\Omega_+} \quad \text{in} \quad \mathbb{R}^n \tag{5.116}$$

we thus obtain (keeping in mind that L is symmetric)

$$\begin{aligned}\int_{\partial\Omega} \Big\langle \partial^A_{\nu_-}(\Pi_{\Omega_+} u), \mathrm{tr}_{m-1} v \Big\rangle\, d\sigma &= \int_{\partial\Omega} \Big\langle \mathrm{tr}^-_{m-1}(\Pi_{\Omega_+} u), \partial^A_\nu v \Big\rangle\, d\sigma \\ &\quad + \int_{\Omega_-} \langle \Pi_{\Omega_+} u, Lv \rangle\, dX.\end{aligned} \tag{5.117}$$

Going further, Green's identity (5.95) in the domain Ω_+ yields

$$\begin{aligned}\int_{\partial\Omega} \Big\langle \partial^A_{\nu_+}(\Pi_{\Omega_+} u), \mathrm{tr}_{m-1} v \Big\rangle\, d\sigma &= \int_{\partial\Omega} \Big\langle \mathrm{tr}^+_{m-1}(\Pi_{\Omega_+} u), \partial^A_\nu v \Big\rangle\, d\sigma \\ &\quad + \int_{\Omega_+} \langle u, v \rangle\, dX - \int_{\Omega_+} \langle \Pi_{\Omega_+} u, Lv \rangle\, dX.\end{aligned} \tag{5.118}$$

Here we made use again of (5.116). However,

$$\begin{aligned}\int_{\Omega_-}\langle \Pi_{\Omega_+}u, Lv\rangle\,dX + \int_{\Omega_+}\langle \Pi_{\Omega_+}u, Lv\rangle\,dX &= \int_{\mathbb{R}^n}\langle \Pi_{\Omega_+}u, Lv\rangle\,dX \\ &= \int_{\Omega_+}\langle L(\Pi_{\Omega_+}u), v\rangle\,dX \\ &= \int_{\Omega_+}\langle u, v\rangle\,dX. \qquad (5.119)\end{aligned}$$

where the first identity is trivial, the second follows by integrating by parts, and the third identity follows by appealing again to (5.116). Combining (5.117)–(5.119), we conclude that

$$\int_{\partial\Omega}\Big\langle \partial^A_{\nu+}(\Pi_{\Omega_+}u), \mathrm{tr}_{m-1}v\Big\rangle\,d\sigma = \int_{\partial\Omega}\Big\langle \partial^A_{\nu-}(\Pi_{\Omega_+}u), \mathrm{tr}_{m-1}v\Big\rangle\,d\sigma, \qquad (5.120)$$

and since $v \in C^\infty_c(\Omega)$ was arbitrary, the identity in (5.113) now follows.

Turning attention to the claim made in (5.114) we fix $u, v \in C^\infty_c(\mathbb{R}^n)$ and note that, since $\Pi_{\mathbb{R}^n}(Lv) = v$, there holds

$$\Pi_{\Omega_-}(Lv) = \Pi_{\mathbb{R}^n}(Lv) - \Pi_{\Omega_+}(Lv) = v - \Pi_{\Omega_+}(Lv). \qquad (5.121)$$

Hence, taking the conormal derivative $\partial^A_{\nu-}$ in (5.121) and employing (5.113), we obtain

$$\partial^A_{\nu-}\Pi_{\Omega_-}(Lv) = \partial^A_\nu v - \partial^A_{\nu-}\Pi_{\Omega_+}(Lv) = \partial^A_\nu v - \partial^A_{\nu+}\Pi_{\Omega_+}(Lv). \qquad (5.122)$$

Next, applying Green's formula (5.95) in the domain Ω_+ along with (4.250) and (4.58), we have

$$\int_{\partial\Omega}\Big\langle \partial^A_{\nu+}(\dot{\mathscr{D}}(\mathrm{tr}_{m-1}u)), \mathrm{tr}_{m-1}v\Big\rangle\,d\sigma = I - II, \qquad (5.123)$$

where

$$\begin{aligned} I &:= \int_{\partial\Omega}\Big\langle (\tfrac{1}{2}I + \dot{K})\mathrm{tr}_{m-1}u, \partial^A_\nu v\Big\rangle\,d\sigma, \\ II &:= \int_{\Omega}\Big\langle \dot{\mathscr{D}}(\mathrm{tr}_{m-1}u), Lv\Big\rangle\,dX. \end{aligned} \qquad (5.124)$$

At this stage we invoke (5.102), (5.122) and again (5.102) (the latter applied in the domain Ω_-), to write

$$\begin{aligned} II &= \int_{\partial\Omega} \Big\langle \mathrm{tr}_{m-1} u, \partial^A_{\nu+}(\Pi_{\Omega_+}(Lv))\Big\rangle \, d\sigma \\ &= \int_{\partial\Omega} \Big\langle \mathrm{tr}_{m-1} u, \partial^A_{\nu} v\Big\rangle \, d\sigma - \int_{\partial\Omega} \Big\langle \mathrm{tr}_{m-1} u, \partial^A_{\nu-}(\Pi_{\Omega_-}(Lv))\Big\rangle \, d\sigma \\ &= \int_{\partial\Omega} \Big\langle \mathrm{tr}_{m-1} u, \partial^A_{\nu} v\Big\rangle \, d\sigma - \int_{\Omega_-} \Big\langle \dot{\mathscr{D}}(\mathrm{tr}_{m-1}), Lv\Big\rangle \, dX. \end{aligned} \tag{5.125}$$

Using Green's formula (5.95) along with (4.250) and (4.58) in the domain Ω_-, we can further expand the last term in the third line of (5.125) as

$$\begin{aligned} \int_{\Omega_-} \Big\langle \dot{\mathscr{D}}(\mathrm{tr}_{m-1} u), Lv\Big\rangle \, dX &= -\int_{\partial\Omega} \Big\langle (-\tfrac{1}{2} I + \dot{K})(\mathrm{tr}_{m-1} u), \partial^A_{\nu} v\Big\rangle \, d\sigma \\ &\quad + \int_{\partial\Omega} \Big\langle \partial^A_{\nu-}(\dot{\mathscr{D}}(\mathrm{tr}_{m-1} u)), \mathrm{tr}_{m-1} v\Big\rangle \, d\sigma. \end{aligned} \tag{5.126}$$

Finally, (5.114) follows by combining (5.123)–(5.126). □

Moving on, we define the action of the conormal in the context of Besov spaces. As a preamble, recall that given a nonempty open subset Ω of $\mathbb{R}^n$, we denote by $\mathscr{D}'(\Omega)$ the space of distributions in Ω, and by $\langle \cdot, \cdot \rangle_{\mathscr{D}'(\Omega)-\mathscr{D}(\Omega)}$ the pairing between distributions and test functions in Ω. Also, if $w \in \Big(A^{p,q}_s(\Omega)\Big)^*$, where $A \in \{B, F\}$, $0 < p, q \leq \infty$ and $s \in \mathbb{R}$, we shall denote by $w\big\lfloor_{\Omega}$ the distribution in $\mathscr{D}'(\Omega)$ defined by

$$\langle w\lfloor_{\Omega}, \varphi\rangle_{\mathscr{D}'(\Omega)-\mathscr{D}(\Omega)} := \langle w, \varphi\rangle, \qquad \forall\, \varphi \in C^\infty_c(\Omega), \tag{5.127}$$

where the bracket $\langle \cdot, \cdot \rangle$ in the right-hand side of (5.127) denotes the duality pairing between functionals from the space $\Big(A^{p,q}_s(\Omega)\Big)^*$ and functions from $A^{p,q}_s(\Omega)$ (with $\varphi \in C^\infty_c(\Omega)$ viewed, this time, as a function in the latter space).

The definition below is modeled upon Green's formula deduced in Proposition 5.13 for sufficiently regular functions.

Definition 5.20. Let Ω be a bounded Lipschitz domain in $\mathbb{R}^n$ and consider L a constant coefficient W-elliptic differential operator of order $2m$. Fix a particular choice of the tensor coefficient $A = (A_{\alpha\beta})_{|\alpha|=|\beta|=m}$ such that $L = \sum_{|\alpha|=|\beta|=m} \partial^\alpha A_{\alpha\beta}\, \partial^\beta$. Finally, suppose that $1 < p, q < \infty$, $0 < s < 1$ and let $p', q' \in (1, \infty)$ be such that $1/p + 1/p' = 1/q + 1/q' = 1$.

In this context, define the conormal derivative operator ∂_ν^A as the mapping acting on the space

$$\Big\{(u,w)\in B^{p,q}_{m-s+1/p}(\Omega)\oplus\Big(B^{p',q'}_{m+s-1+1/p'}(\Omega)\Big)^*:\ Lu=w\big\lfloor_{\Omega}\ \text{ in }\ \mathscr{D}'(\Omega)\Big\}\tag{5.128}$$

(where the convention introduced in (5.127) has been used), and taking values in $\Big(\dot{B}^{p',q'}_{m-1,s}(\partial\Omega)\Big)^*$, by setting, for each $\dot{f}\in\dot{B}^{p',q'}_{m-1,s}(\partial\Omega)$,

$$\Big\langle\partial_\nu^A(u,w),\dot{f}\Big\rangle:=(-1)^{m+1}\sum_{|\alpha|=|\beta|=m}\Big\langle A_{\alpha\beta}\partial^\beta u,\partial^\alpha F\Big\rangle_{\Omega}+\langle w,F\rangle,\tag{5.129}$$

where $F\in B^{p',q'}_{m-1+s+1/p'}(\Omega)$ is such that $\mathrm{tr}_{m-1}F=\dot{f}$. In (5.129), the bracket $\langle\cdot,\cdot\rangle_\Omega$ denotes the duality pairing between elements of the space $B^{p,q}_{-s+1/p}(\Omega)$ and elements in its dual, $B^{p',q'}_{s-1+1/p'}(\Omega)$, and $\langle\cdot,\cdot\rangle$ denotes the duality pairing between elements of the space $B^{p',q'}_{m-1+s+1/p'}(\Omega)$ and its dual, $\Big(B^{p',q'}_{m+s-1+1/p'}(\Omega)\Big)^*$.

In the same framework, one may also define the conormal derivative operator ∂_ν^A as the mapping acting on the space (defined using the convention introduced in (5.127))

$$\Big\{(u,w)\in F^{p,q}_{m-s+1/p}(\Omega)\oplus\Big(F^{p',q'}_{m+s-1+1/p'}(\Omega)\Big)^*:\ Lu=w\big\lfloor_{\Omega}\ \text{ in }\ \mathscr{D}'(\Omega)\Big\},\tag{5.130}$$

and taking values in the space $\Big(\dot{F}^{p',q'}_{m-1,s}(\partial\Omega)\Big)^*$, by formally employing the same formula (5.129) where, given $\dot{f}\in\dot{B}^{p',p'}_{m-1,s}(\partial\Omega)$, the function $F\in F^{p',q'}_{m-1+s+1/p'}(\Omega)$ is such that $\mathrm{tr}_{m-1}F=\dot{f}$, and where the duality pairings in the right-hand side of (5.129) are suitably interpreted on the Triebel–Lizorkin scale.

Finally, a conormal derivative for the exterior domain $\Omega_-:=\mathbb{R}^n\setminus\overline{\Omega}$ (in place of Ω) may be introduced similarly (in which case (5.129) is altered by changing the sign of the entire left-hand side). When necessary to distinguish the conormals thus defined in Ω_+ and Ω_-, we shall denote the former by $\partial^A_{\nu_+}$ and denote the latter by $\partial^A_{\nu_-}$.

A couple of comments are in order here.

Remark 5.21. (1) It is important to point out that definition (5.129) is independent of the choice of the function $F\in B^{p',q'}_{m-1+s+1/p'}(\Omega)$ such that $\mathrm{tr}_{m-1}F=\dot{f}$. Indeed, in order to see this, by linearity, it suffices to show that the right-hand side in (5.129) equals zero whenever $F\in B^{p',q'}_{m-1+s+1/p'}(\Omega)$ is such that $\mathrm{tr}_{m-1}F=0$. This latter fact easily follows from integration by parts whenever $F\in C_c^\infty(\Omega)$ (here we use

the fact that $Lu = w$ as distributions in Ω). Since, due to (3.129), $C_c^\infty(\Omega)$ is dense in the subspace of $B^{p',q'}_{m-1+s+1/p'}(\Omega)$ consisting of functions with vanishing multi-trace, a routine limiting argument shows that the right-hand side in (5.129) vanishes whenever $F \in B^{p',q'}_{m-1+s+1/p'}(\Omega)$ satisfies $\mathrm{tr}_{m-1} F = 0$, as desired. A similar comment applies to the conormal defined on Triebel–Lizorkin spaces.
(2) The two cornormals defined, respectively, on (5.128) and (5.130), act in a compatible manner with one another (i.e., they agree on the intersection of their domains; this justifies using the same notation). This may be proved using a limiting argument using the density result from (2.208). ■

Below we consider an important special case of Definition 5.20.

Proposition 5.22. *Let Ω be a bounded Lipschitz domain in $\mathbb{R}^n$ and consider a constant coefficient W-elliptic differential operator L of order $2m$. For each triplet of numbers p, q, s satisfying $1 < p, q < \infty$ and $0 < s < 1$, the conormal derivative operator ∂^A_ν from Definition 5.20 induces a linear, bounded operator*

$$\partial^A_\nu : \left\{u \in B^{p,q}_{m-s+1/p}(\Omega) : Lu = 0 \text{ in } \mathscr{D}'(\Omega)\right\} \longrightarrow \left(\dot{B}^{p',q'}_{m-1,s}(\partial\Omega)\right)^*, \tag{5.131}$$

where $1/p + 1/p' = 1/q + 1/q' = 1$, according to

$$\left\langle \partial^A_\nu u, \dot{f}\right\rangle = (-1)^{m+1} \sum_{|\alpha|=|\beta|=m} \left\langle A_{\alpha\beta}\, \partial^\beta u, \partial^\alpha F\right\rangle_\Omega, \tag{5.132}$$

for each $\dot{f} \in \dot{B}^{p',q'}_{m-1,s}(\partial\Omega)$ and $F \in B^{p',q'}_{m-1+s+1/p'}(\Omega)$ such that $\mathrm{tr}_{m-1} F = \dot{f}$ (as before, $\langle\cdot,\cdot\rangle_\Omega$ denote the duality pairing between elements of the space $B^{p,q}_{-s+1/p}(\Omega)$ and elements in its dual, $B^{p',q'}_{s-1+1/p'}(\Omega)$). Moreover, a similar result is valid for

$$\partial^A_\nu : \left\{u \in F^{p,q}_{m-s+1/p}(\Omega) : Lu = 0 \text{ in } \mathscr{D}'(\Omega)\right\} \longrightarrow \left(\dot{B}^{p',p'}_{m-1,s}(\partial\Omega)\right)^*, \tag{5.133}$$

formally by the same formula (5.132), suitably interpreted. Finally, analogous results hold for the conormal derivative acting on null-solutions of L in $\Omega_- := \mathbb{R}^n \setminus \overline{\Omega}$, in place of $\Omega_+ := \Omega$ (in which scenario, (5.132) is altered by changing the sign of the left-hand side).

As before, definition (5.132) is independent on the choice of the extension of $\dot{f}$, i.e., on the choice of function $F \in B^{p',q'}_{m-1+s+1/p'}(\Omega)$ satisfying $\mathrm{tr}_{m-1} F = \dot{f}$. When necessary to distinguish the conormals considered in Proposition 5.22 for the domains Ω_+, Ω_-, as before, we shall denote them by $\partial^A_{\nu+}$ and $\partial^A_{\nu-}$, respectively.

Proof of Proposition 5.22. This is a direct consequence of (5.128)–(5.129) (cf. also the comments made immediately thereafter) and the fact that the mapping $u \mapsto (u, 0)$ is an embedding of the domain of ∂^A_ν in (5.131) into (5.128). □

The following observation is going to be useful later on.

Remark 5.23. In the context of Proposition 5.22, the conormal derivative operator from (5.131)–(5.132) actually takes values in the space

$$\Big\{\Lambda \in \Big(\dot{B}^{p',q'}_{m-1,s}(\partial\Omega)\Big)^* : \langle \Lambda, \dot{P}\rangle = 0 \ \text{ for every } \dot{P} \in \dot{\mathscr{P}}_1(\partial\Omega)\Big\}. \tag{5.134}$$

This is clear from (5.132) and simple degree considerations.

Our next result further elaborates on the nature of the conormal operator acting on null-solutions of L belonging to "diagonal" Besov spaces.

Proposition 5.24. *Assume that Ω is a bounded Lipschitz domain in $\mathbb{R}^n$ and let L be a constant coefficient W-elliptic differential operator of order $2m$. Also, suppose that the indices $p, p' \in (1,\infty)$ satisfy $1/p + 1/p' = 1$, and that $s \in (0,1)$. Then the conormal derivative operator*

$$\partial^A_\nu : \Big\{u \in B^{p,p}_{m-s+1/p}(\Omega) : Lu = 0 \ \text{ in } \ \mathscr{D}'(\Omega)\Big\} \longrightarrow \Big(\dot{B}^{p',p'}_{m-1,s}(\partial\Omega)\Big)^*, \tag{5.135}$$

defined as in Proposition 5.22 (with $q := p$), has the property that

$$\big\langle \partial^A_\nu u, \dot{f}\big\rangle = (-1)^{m+1} \sum_{|\alpha|=|\beta|=m} \int_\Omega \big\langle A_{\alpha\beta}\partial^\beta u, \partial^\alpha F\big\rangle\, dX, \tag{5.136}$$

with an absolutely convergent integral, whenever $u \in B^{p,p}_{m-s+1/p}(\Omega)$ satisfies $Lu = 0$ in Ω, while $\dot{f} \in \dot{B}^{p',p'}_{m-1,s}(\partial\Omega)$ and $F \in W^{m,p'}_{1-s-1/p'}(\Omega)$ are such that $\mathrm{tr}_{m-1} F = \dot{f}$.

Furthermore, any function $u \in B^{p,p}_{m-s+1/p}(\Omega)$ satisfying $Lu = 0$ in Ω actually belongs to the weighted Sobolev space $W^{m,p}_{s-1/p}(\Omega)$, and there exists a finite constant $C = C(\Omega, p, s) > 0$ with the property that for any such function

$$\big\|\partial^A_\nu u\big\|_{\big(\dot{B}^{p',p'}_{m-1,s}(\partial\Omega)\big)^*} \le C\,\|u\|_{W^{m,p}_{s-1/p}(\Omega)}. \tag{5.137}$$

Proof. The first observation we make is that, in the case when $p = q \in (1,\infty)$, the domain of the conormal ∂^A_ν in (5.131) becomes

$$\begin{aligned}\Big\{u \in B^{p,p}_{m-s+1/p}(\Omega) : Lu = 0 \ \text{ in } \ \mathscr{D}'(\Omega)\Big\} &= B^{p,p}_{m-s+1/p}(\Omega) \cap \mathrm{Ker}\, L \\ &= W^{m,p}_{s-1/p}(\Omega) \cap \mathrm{Ker}\, L,\end{aligned} \tag{5.138}$$

thanks to (2.306). For each $u \in W^{m,p}_{s-1/p}(\Omega) \cap \mathrm{Ker}\, L$, let us temporarily denote by $\widetilde{\partial^A_\nu} u$ the functional

$$\dot{B}^{p',p'}_{m-1,s}(\partial\Omega) \ni \dot{f} \longmapsto (-1)^{m+1} \sum_{|\alpha|=|\beta|=m} \int_\Omega \big\langle A_{\alpha\beta}\partial^\beta u, \partial^\alpha F\big\rangle\, dX, \tag{5.139}$$

where, for each $\dot{f}$, the function $F \in W^{m,p'}_{1-s-1/p'}(\Omega)$ is assumed to satisfy $\mathrm{tr}_{m-1} F = \dot{f}$. That such a function F exists is guaranteed by (3.123) in Theorem 3.11. Moreover, the fact that the particular choice of the function $F \in W^{m,p'}_{1-s-1/p'}(\Omega)$ with the property that $\mathrm{tr}_{m-1} F = \dot{f}$ is irrelevant follows with the help of the density result proved in Theorem 3.17 by arguing as in part (1) of Remark 5.21.

Going further, given $u \in W^{m,p}_{s-1/p}(\Omega) \cap \mathrm{Ker}\, L$ and two functions $\dot{f} \in \dot{B}^{p',p'}_{m-1,s}(\partial\Omega)$ and $F \in W^{m,p'}_{1-s-1/p'}(\Omega)$ such that $\mathrm{tr}_{m-1} F = \dot{f}$, we estimate

$$
\begin{aligned}
&\sum_{|\alpha|=|\beta|=m} \int_\Omega \left|\langle A_{\alpha\beta}\partial^\beta u, \partial^\alpha F\rangle\right| dX \\
&\quad\le \sum_{|\alpha|=|\beta|=m} \int_\Omega |A_{\alpha\beta}| \left(|\partial^\beta u|\rho^{s-1/p}\right)\left(|\partial^\alpha F|\rho^{1/p-s}\right) dX \\
&\quad\le \sum_{|\alpha|=|\beta|=m} |A_{\alpha\beta}| \Big[\int_\Omega \left(|\partial^\beta u|\rho^{s-1/p}\right)^p dX\Big]^{1/p} \\
&\qquad\times \Big[\int_\Omega \left(|\partial^\alpha F|\rho^{1-1/p'-s}\right) dX\Big]^{1/p'} \\
&\quad\le C(A)\|u\|_{W^{m,p}_{s-1/p}(\Omega)} \|F\|_{W^{m,p'}_{1-s-1/p'}(\Omega)} < +\infty.
\end{aligned}
\tag{5.140}
$$

This shows that the integral in the right-hand side of (5.136) is absolutely convergent. Finally, the consistency of $\partial^A_\nu u$ with $\partial^A_\nu u$ from Proposition 5.22 is a consequence of the density of $C^\infty(\overline{\Omega})$ both in weighted Sobolev spaces and Besov spaces in Ω, established earlier. □

We are now well positioned to discuss a version of the Green representation formula (5.98) in the context of Besov spaces.

Theorem 5.25. *Let Ω be a bounded Lipschitz domain in $\mathbb{R}^n$ and consider L a constant coefficient W-elliptic differential operator of order 2m. Fix a particular choice of the tensor coefficient $A = (A_{\alpha\beta})_{|\alpha|=|\beta|=m}$, consisting of $M \times M$ matrices with complex entries, such that $L = \sum_{|\alpha|=|\beta|=m} \partial^\alpha A_{\alpha\beta}\, \partial^\beta$, In this context, associate the multi-layer potential operators $\dot{\mathscr{D}}$ and $\dot{\mathscr{S}}$ as in* (4.57) *and* (5.1), *respectively. Finally, suppose that $1 < p, q < \infty$, $0 < s < 1$, and that the function $u \in B^{p,q}_{m-s+1/p}(\Omega)$ satisfies $Lu = 0$ in $\mathscr{D}'(\Omega)$.*

Then with ∂^A_ν defined as in (5.131)–(5.132), *the following Green-type representation formula holds:*

$$
u = \dot{\mathscr{D}}(\mathrm{tr}_{m-1} u) - \dot{\mathscr{S}}(\partial^A_\nu u) \quad \text{in } \Omega. \tag{5.141}
$$

Proof. Via embeddings of Besov spaces, there is no loss of generality in assuming that $p = q$. Assume that this is the case and fix an arbitrary point $X_o \in \Omega$ along with an arbitrary vector $\eta \in \mathbb{C}^M$. Then, based on (5.3), (5.136), and (4.61), we may write

$$\begin{aligned}\langle \eta, \dot{\mathscr{S}}(\partial_\nu^A u)(X_o)\rangle &= \left\langle \partial_\nu^A u, \operatorname{tr}_{m-1}[E_{L^t}(X_o-\cdot)\eta]\right\rangle \\ &= -\sum_{|\alpha|=|\beta|=m}\int_\Omega \left\langle A_{\alpha\beta}\partial^\beta u, (\partial^\alpha E_{L^t})(X_o-\cdot)\eta\right\rangle dX \\ &= -\left\langle \eta, \sum_{\substack{|\alpha|=m\\|\beta|=m}}\int_\Omega (\partial^\beta E)(X_o-\cdot)A_{\beta\alpha}(\partial^\alpha u)\,dX\right\rangle \\ &= \left\langle \eta, \dot{\mathscr{D}}(\operatorname{tr}_{m-1}u)(X_o)\right\rangle - \left\langle \eta, u(X_o)\right\rangle. \end{aligned} \tag{5.142}$$

This proves that $u(X_o) = \dot{\mathscr{D}}(\operatorname{tr}_{m-1}u)(X_o) - \dot{\mathscr{S}}(\partial_\nu^A u)(X_o)$ and since $X_o \in \Omega$ was arbitrary, (5.141) follows. □

For further use, it is also convenient to formulate the action of the conormal derivative operator acting from weighted Sobolev spaces, in the manner described below.

Proposition 5.26. *Let Ω be a bounded Lipschitz domain in $\mathbb{R}^n$ and consider L a constant coefficient W-elliptic differential operator of order $2m$. Fix a particular choice of the tensor coefficient $A = (A_{\alpha\beta})_{|\alpha|=|\beta|=m}$ such that there holds $L = \sum_{|\alpha|=|\beta|=m}\partial^\alpha A_{\alpha\beta}\,\partial^\beta$. Finally, suppose that $1 < p < \infty$, $0 < s < 1$ and let $p' \in (1,\infty)$ be such that $1/p + 1/p' = 1$.*

Then the conormal derivative operator ∂_ν^A, originally introduced in Definition 5.20, induces a linear and bounded mapping from the space (defined using a similar convention to (5.127))

$$\left\{(u,w) \in W^{m,p}_{s-1/p}(\Omega) \oplus \left(W^{m,p'}_{s-1/p'}(\Omega)\right)^* : Lu = w\lfloor_\Omega \text{ in } \mathscr{D}'(\Omega)\right\} \tag{5.143}$$

and taking values in the space $\left(\dot{B}^{p',p'}_{m-1,1-s}(\partial\Omega)\right)^*$ which, for every Whitney array $\dot{f} \in \dot{B}^{p',p'}_{m-1,1-s}(\partial\Omega)$, acts according to the formula

$$\begin{aligned}\left\langle \partial_\nu^A(u,w), \dot{f}\right\rangle := (-1)^{m+1}&\sum_{|\alpha|=|\beta|=m}\int_\Omega \left\langle A_{\alpha\beta}\partial^\beta u(X), \partial^\alpha F(X)\right\rangle dX \\ &+\langle w, F\rangle,\end{aligned} \tag{5.144}$$

where $F \in W^{m,p'}_{s-1/p'}(\Omega)$ *is such that* $\mathrm{tr}_{m-1} F = \dot{f}$. *Above, the last bracket* $\langle \cdot, \cdot \rangle$ *denotes the duality pairing between elements of the space* $W^{m,p'}_{s-1/p'}(\Omega)$ *and its dual,* $\left(W^{m,p'}_{s-1/p'}(\Omega)\right)^*$.

Finally, a conormal derivative for the exterior domain $\Omega_- := \mathbb{R}^n \setminus \overline{\Omega}$ *(in place of* Ω*) may be introduced similarly (in which case (5.144) is altered by changing the sign of the entire left-hand side). When necessary to distinguish the conormals thus defined in* Ω_+ *and* Ω_-, *we shall denote the former by* $\partial^A_{\nu_+}$ *and denote the latter by* $\partial^A_{\nu_-}$.

Proof. This is largely established in the same manner in which Proposition 5.24 has been proved to be consistent with Proposition 5.22. In the process, the multi-trace result from Theorem 3.11 plays a basic role. □

5.5 Jump Relations for the Conormal Derivative

The first order of business is to actually define the conormal derivative of the single multi-layer potential operator.

Proposition 5.27. *Assume that* Ω *is a bounded Lipschitz domain in* $\mathbb{R}^n$. *Also, consider a constant coefficient W-elliptic differential operator* L *of order* $2m$, $m \in \mathbb{N}$, *and denote by* ∂^A_ν *the conormal associated with a particular choice of the tensor coefficient* $A = (A_{\alpha\beta})_{|\alpha|=|\beta|=m}$ *in the writing of* L *as in (5.92).*

Then for any $p, q \in (1, \infty)$ *and* $s \in (0, 1)$ *one can define the conormal derivative of the single multi-layer potential (associated with* Ω*) in such a way that*

$$\partial^A_\nu \dot{\mathscr{S}} : \left(\dot{B}^{p,q}_{m-1,s}(\partial\Omega)\right)^* \longrightarrow \left(\dot{B}^{p,q}_{m-1,s}(\partial\Omega)\right)^* \tag{5.145}$$

becomes a linear, bounded operator.

Similar considerations apply to the conormal derivative of the single multi-layer associated with $\Omega_- := \mathbb{R}^n \setminus \overline{\Omega}$. *When necessary to distinguish this from (5.145), we shall denote this by* $\partial^A_{\nu_-} \dot{\mathscr{S}}$, *and denote the former by* $\partial^A_{\nu_+} \dot{\mathscr{S}}$.

Proof. Note that, by (5.43) and (5.5),

$$\dot{\mathscr{S}} : \left(\dot{B}^{p,q}_{m-1,s}(\partial\Omega)\right)^* \longrightarrow \left\{u \in B^{p',q'}_{m-s+1/p'}(\Omega) : Lu = 0 \text{ in } \Omega\right\} \tag{5.146}$$

is a well-defined, linear and bounded operator. We define $\partial^A_\nu \dot{\mathscr{S}}$ as the composition of ∂^A_ν (from Proposition 5.22) with this operator. Thanks to (5.131), we may conclude that the operator (5.145) is indeed well-defined, linear and bounded.

When Ω_- is considered in place of Ω, we define

$$\begin{aligned}
&\partial^A_{\nu_-}\dot{\mathscr{S}} : \left(\dot{B}^{p,q}_{m-1,s}(\partial\Omega)\right)^* \longrightarrow \left(\dot{B}^{p,q}_{m-1,s}(\partial\Omega)\right)^*, \\
&\partial^A_{\nu_-}\dot{\mathscr{S}}\Lambda := \partial^A_{\nu_-}\left(\psi\dot{\mathscr{S}}^-\Lambda\,,\,L(\psi\dot{\mathscr{S}}^-\Lambda)\right), \\
&\forall\,\Lambda \in \left(\dot{B}^{p,q}_{m-1,s}(\partial\Omega)\right)^*
\end{aligned} \tag{5.147}$$

in the sense of Definition 5.20, where $\psi \in C^\infty_c(\mathbb{R}^n)$ is an arbitrary function with the property that $\psi \equiv 1$ near $\partial\Omega$. In particular, this ensures that $L(\psi\dot{\mathscr{S}}^-\Lambda) \in C^\infty_c(\Omega_-)$ for every functional $\Lambda \in \left(\dot{B}^{p,q}_{m-1,s}(\partial\Omega)\right)^*$. More specifically, (5.147) says that

$$\begin{aligned}
\left\langle\partial^A_{\nu_-}\dot{\mathscr{S}}\Lambda\,,\,\dot{f}\right\rangle &:= (-1)^m \sum_{|\alpha|=|\beta|=m}\left\langle A_{\alpha\beta}\,\partial^\beta(\psi\dot{\mathscr{S}}^-\Lambda), \partial^\alpha F\right\rangle_{\Omega_-} \\
&\quad - \int_{\Omega_-}\left\langle L(\psi\dot{\mathscr{S}}^-\Lambda), F\right\rangle dX,
\end{aligned} \tag{5.148}$$

whenever $\dot{f} \in \dot{B}^{p,q}_{m-1,s}(\partial\Omega)$ and $F \in B^{p,q}_{m-1+s+1/p}(\Omega_-)$ are such that $\mathrm{tr}_{m-1} F = \dot{f}$. Note that the right-hand side of (5.148) vanishes if $\psi \equiv 0$ near $\partial\Omega$ since, in this case, $\psi\dot{\mathscr{S}}^-\Lambda \in C^\infty_c(\Omega_-)$. By linearity, this shows that the definition of $\partial^A_{\nu_-}\dot{\mathscr{S}}$ in (5.147) is independent of the particular function $\psi \in C^\infty_c(\mathbb{R}^n)$ with $\psi \equiv 1$ near $\partial\Omega$.

Finally, by the last part in Theorem 5.7, the assignment

$$\Lambda \mapsto \left(\psi\dot{\mathscr{S}}^-\Lambda,\,L(\psi\dot{\mathscr{S}}^-\Lambda)\right) \tag{5.149}$$

maps $\left(\dot{B}^{p,q}_{m-1,s}(\partial\Omega)\right)^*$ linearly and boundedly into the analogue of the space in (5.128), written for Ω_- in place of Ω. This, in concert with earlier arguments and Definition 5.20, shows that (5.147) is indeed a well-defined, bounded operator. Hence, the proof of the proposition is complete. Parenthetically, we wish to point out that (5.148) reduces to

$$\left\langle\partial^A_{\nu_-}\dot{\mathscr{S}}\Lambda\,,\,\dot{f}\right\rangle = (-1)^m \sum_{|\alpha|=|\beta|=m}\left\langle A_{\alpha\beta}\,\partial^\beta(\dot{\mathscr{S}}^-\Lambda), \partial^\alpha F\right\rangle_{\Omega_-}, \tag{5.150}$$

whenever $\dot{f} \in \dot{B}^{p,q}_{m-1,s}(\partial\Omega)$ and $F \in B^{p,q}_{m-1+s+1/p}(\Omega_-)$ are such that $\mathrm{tr}_{m-1} F = \dot{f}$ and F has bounded support. Indeed, in this situation, we simply take $\psi \in C^\infty_c(\mathbb{R}^n)$ such that $\psi \equiv 1$ near $\partial\Omega$, as well as near $\mathrm{supp}\,F$. □

We next discuss the boundary behavior of the conormal derivative of the single multi-layer.

Proposition 5.28. *Retain the hypotheses of Proposition 5.27 and fix* $1 < p,q < \infty$ *and* $0 < s < 1$. *Then for each* $\Lambda \in \left(\dot{B}^{p',q'}_{m-1,s}(\partial\Omega)\right)^*$ *one has*

$$\left\langle \partial^A_{\nu_\pm} \dot{\mathscr{S}}\Lambda, \dot{g}\right\rangle = \left\langle \Lambda, \left(\mp\tfrac{1}{2}I + \dot{K}\right)\dot{g}\right\rangle, \qquad \forall\, \dot{g} \in \dot{B}^{p',q'}_{m-1,s}(\partial\Omega), \tag{5.151}$$

where $1/p + 1/p' = 1/q + 1/q' = 1$. *In particular,*

$$\partial^A_{\nu_\pm} \dot{\mathscr{S}} = \mp\tfrac{1}{2}I + \dot{K}^* \quad \text{as operators on} \quad \left(\dot{B}^{p',q'}_{m-1,s}(\partial\Omega)\right)^*, \tag{5.152}$$

and, hence,

$$\partial^A_{\nu_+} \dot{\mathscr{S}} - \partial^A_{\nu_-} \dot{\mathscr{S}} = I \quad \text{on} \quad \left(\dot{B}^{p',p'}_{m-1,s}(\partial\Omega)\right)^*. \tag{5.153}$$

Proof. Notice that, since all operators involved are continuous, it suffices to prove (5.151) only for Λ belonging to a dense subclass in $\left(\dot{B}^{p',q'}_{m-1,s}(\partial\Omega)\right)^*$. Thus, thanks to (5.4), there is no loss in generality in assuming that

$$\Lambda = \{f_\gamma\}_{|\gamma|\leq m-1} \in L^p(\partial\Omega) \oplus \cdots \oplus L^p(\partial\Omega), \tag{5.154}$$

(cf. (5.4)). Also, thanks to (3.118), we may assume that $\dot{g} = \mathrm{tr}_{m-1} G$ for some function $G \in C^\infty_c(\mathbb{R}^n)$. Using (5.132) we may write

$$(-1)^{m+1}\left\langle \partial^A_{\nu_+} \dot{\mathscr{S}}\Lambda, \dot{g}\right\rangle = \sum_{|\alpha|=|\beta|=m} \left\langle A_{\alpha\beta}\, \partial^\beta \dot{\mathscr{S}}\Lambda, \partial^\alpha G\right\rangle_\Omega. \tag{5.155}$$

Let next $\{\Omega_j\}_{j\in\mathbb{N}}$ be a nested sequence of subdomains of Ω exhausting Ω from within, in the manner described in the paragraph containing (2.114). A useful observation is that for $u \in B^{p,q}_\theta(\Omega)$ and $w \in B^{p',q'}_{-\theta}(\Omega)$, where $-1 + \frac{1}{p} < \theta < \frac{1}{p}$, we have (with R_{Ω_j} denoting the restriction of distributions in Ω to Ω_j)

$$\langle u, v\rangle_\Omega = \lim_{j\to\infty} \left\langle R_{\Omega_j} u, R_{\Omega_j} v\right\rangle_{\Omega_j}, \tag{5.156}$$

where $\langle\cdot,\cdot\rangle_\Omega$ denotes the duality pairing between elements of the space $B^{p,q}_\theta(\Omega)$ and elements of its dual, and $\langle\cdot,\cdot\rangle_{\Omega_j}$ denotes the duality pairing between elements of the space $B^{p,q}_\theta(\Omega_j)$ and elements of its dual, respectively. Indeed, (5.156) is easily seen to hold for $u, v \in C^\infty_c(\Omega)$ and (5.156) follows then by continuity. Therefore,

$$(-1)^{m+1}\left\langle \partial^A_{\nu_+} \dot{\mathscr{S}}\Lambda, \dot{g}\right\rangle = \lim_{j\to\infty} \sum_{|\alpha|=|\beta|=m} \left\langle A_{\alpha\beta}\, \partial^\beta \dot{\mathscr{S}}\Lambda, \partial^\alpha G\right\rangle_{\Omega_j}. \tag{5.157}$$

Employing (5.3) we can further expand (with $\langle\cdot,\cdot\rangle_{\partial\Omega}$ standing for the integral pairing on $\partial\Omega$)

$$(-1)^{m+1}\Big\langle \partial^A_{\nu+}\dot{\mathscr{S}}\Lambda, \dot{g}\Big\rangle \tag{5.158}$$

$$= \lim_{j\to\infty} \sum_{\substack{|\alpha|=|\beta|=m \\ |\gamma|\leq m-1}} \Big\langle\Big\langle A_{\alpha\beta}\langle\partial^\gamma_\cdot \partial^\beta_X[E(X-\cdot)], f_\gamma(\cdot)\Big\rangle_{\partial\Omega}, (\partial^\alpha G)(X)\Big\rangle_{\Omega_j}.$$

On a separate note, using (4.59) for each Ω_j yields

$$\Big\langle \sum_{|\alpha|=|\beta|=m} A_{\alpha\beta}\,\partial^\beta_X[E(X-\cdot)], (\partial^\alpha G)(X)\Big\rangle_{\Omega_j} = (-1)^{m+1}\dot{\mathscr{D}}^-_j(\mathrm{tr}_{m-1,j}G)(\cdot), \tag{5.159}$$

where $\mathrm{tr}_{m-1,j}$ stands for the generalized trace operator to $\partial\Omega_j$, while $\dot{\mathscr{D}}^-_j$ stands for the double multi-layer potential associated to the domain $\mathbb{R}^n\setminus\overline{\Omega}_j$.

Returning to (5.158) observe that it is possible to change the order of integration in the right-hand side since, for each fixed j, the variables X and "dot" are geometrically separated and, as such, there are no singularities to deal with. Thus, interchanging the pairings $\langle\cdot,\cdot\rangle_{\Omega_j}$ and $\langle\cdot,\cdot\rangle_{\partial\Omega}$ and using (5.159), allows us to transform formula (5.158) into

$$\Big\langle \partial^A_{\nu+}\dot{\mathscr{S}}\Lambda, \dot{g}\Big\rangle = \lim_{j\to\infty}\Big\langle \Lambda, \mathrm{tr}_{m-1}\Big(\dot{\mathscr{D}}^-_j(\mathrm{tr}_{m-1,j}G)\Big)\Big\rangle_{\partial\Omega}. \tag{5.160}$$

Thus, in order to prove (5.151) it suffices to show that

$$\mathrm{tr}_{m-1}\Big(\dot{\mathscr{D}}^-_j(\mathrm{tr}_{m-1,j}G)\Big) \to \mathrm{tr}_{m-1}\Big(\dot{\mathscr{D}}^-(\mathrm{tr}_{m-1}G)\Big) \quad \text{in } \dot{B}^{p',q'}_{m-1,s}(\partial\Omega) \text{ as } j\to\infty. \tag{5.161}$$

In fact we claim that the stronger convergence result,

$$\mathrm{tr}_{m-1}\Big(\dot{\mathscr{D}}^-_j(\mathrm{tr}_{m-1,j}G)\Big) \to \mathrm{tr}_{m-1}\Big(\dot{\mathscr{D}}^-(\mathrm{tr}_{m-1}G)\Big) \quad \text{in } \dot{L}^{p'}_{m-1,1}(\partial\Omega) \text{ as } j\to\infty, \tag{5.162}$$

holds, from each (5.161) readily follows. The convergence in (5.162) is, however, a direct consequence of Proposition 2.65 and formula (4.84).

This finishes the proof of the versions of (5.151), (5.152) corresponding to choosing the "top" sign in both sides of each identity. There remains to treat the case when the "bottom" sign is considered. To this end, recall from (5.150) that if the array $\dot{g}\in\dot{B}^{p,q}_{m-1,s}(\partial\Omega)$ and the function $G\in B^{p,q}_{m-1+s+1/p}(\Omega_-)$ are such that

$\mathrm{tr}_{m-1} G = \dot{g}$ and G has bounded support, say, $\operatorname{supp} G \subset B_R(0)$ (with R large), then for every $\Lambda \in \left(\dot{B}^{p',q'}_{m-1,s}(\partial\Omega)\right)^*$ we have

$$\begin{aligned}\left\langle \partial^A_{\nu_-} \dot{\mathscr{S}}\Lambda\,,\, \dot{g}\right\rangle &= (-1)^m \sum_{|\alpha|=|\beta|=m} \left\langle A_{\alpha\beta}\, \partial^\beta (\dot{\mathscr{S}}^- \Lambda), \partial^\alpha G\right\rangle_{\Omega_-}, \\ &= (-1)^m \sum_{|\alpha|=|\beta|=m} \left\langle A_{\alpha\beta}\, \partial^\beta (\dot{\mathscr{S}}_R(\Lambda \circ \pi_R)), \partial^\alpha G\right\rangle_{D_R}, \end{aligned} \tag{5.163}$$

where $\dot{\mathscr{S}}_R$ is the single multi-layer associated with the domain $D_R := B_R(0) \setminus \overline{\Omega}$, and π_R is as in (4.246). Based on Proposition 5.27 and on the version of (5.152) for bounded domains, the last expression in (5.163) can be further transformed (with $(\partial^A_\nu)_R$ and I_R playing, respectively, the role of the conormal derivative and identity operator on ∂D_R) into

$$\begin{aligned}(-1)^m &\sum_{|\alpha|=|\beta|=m} \left\langle A_{\alpha\beta}\, \partial^\beta (\dot{\mathscr{S}}_R(\Lambda \circ \pi_R)), \partial^\alpha G\right\rangle_{D_R} \\ &= -\left\langle (\partial^A_\nu)_R \dot{\mathscr{S}}_R(\Lambda \circ \pi_R)\,,\, \iota_R(\dot{g})\right\rangle \\ &= -\left\langle \left(-\tfrac{1}{2} I_R + \dot{K}^*_R\right)(\Lambda \circ \pi_R)\,,\, \iota_R(\dot{g})\right\rangle \\ &= -\left\langle \iota^*_R \circ \left(-\tfrac{1}{2} I_R + \dot{K}^*_R\right)(\Lambda \circ \pi_R)\,,\, \dot{g}\right\rangle \\ &= \left\langle \left(\tfrac{1}{2} I + \dot{K}^*\right)(\iota^*_R \circ \Lambda \circ \pi_R)\,,\, \dot{g}\right\rangle \\ &= \left\langle \left(\tfrac{1}{2} I + \dot{K}^*\right)\Lambda, \dot{g}\right\rangle. \end{aligned} \tag{5.164}$$

Together with (5.163), this shows that $\left\langle \partial^A_{\nu_-} \dot{\mathscr{S}}\Lambda, \dot{g}\right\rangle = \left\langle \left(\tfrac{1}{2} I + \dot{K}^*\right)\Lambda, \dot{g}\right\rangle$, thus finishing the proof of (5.151)–(5.152). Finally, (5.153) is immediate from (5.152). □

We next study the conormal of the single multi-layer acting on duals of Whitney–Lebesgue and Whitney–Sobolev spaces.

Corollary 5.29. *Retain the hypotheses of Proposition 5.27. Then for each* $1 < p < \infty$ *the conormal derivative of the single multi-layer, initially considered in the sense of* (5.145)*, extends to a bounded operator from* $\left(\dot{L}^p_{m-1,0}(\partial\Omega)\right)^*$ *into itself, and from* $\left(\dot{L}^p_{m-1,1}(\partial\Omega)\right)^*$ *into itself. Hence,*

$$\partial^A_{\nu_\pm}\mathscr{S} = \mp\tfrac{1}{2}I + \dot{K}^* \tag{5.165}$$

considered as either operators on $\left(\dot{L}^p_{m-1,0}(\partial\Omega)\right)^*$, *or on* $\left(\dot{L}^p_{m-1,1}(\partial\Omega)\right)^*$.

Proof. This is an immediate consequence of Proposition 5.28 and Theorem 4.14. □

The jump-relations proved in Proposition 5.28 make it possible to prove the intertwining identity formulated below.

Proposition 5.30. *Retain the hypotheses of Proposition 5.27 and assume that* p, p' *satisfy* $1 < p, p' < \infty$ *and* $1/p + 1/p' = 1$. *Then, for each* $s \in (0,1)$,

$$\dot{S}\dot{K}^* = \dot{K}\dot{S} \tag{5.166}$$

as (linear, bounded) operators from $\left(\dot{B}^{p,q}_{m-1,s}(\partial\Omega)\right)^*$ *into* $\dot{B}^{p',q'}_{m-1,1-s}(\partial\Omega)$. *As a consequence,*

$$\begin{aligned} &\text{the adjoint of } \dot{S}\dot{K}^* : \left(\dot{B}^{p,q}_{m-1,s}(\partial\Omega)\right)^* \longrightarrow \dot{B}^{p',q'}_{m-1,1-s}(\partial\Omega) \\ &\text{is the operator } \dot{S}\dot{K}^* : \left(\dot{B}^{p',q'}_{m-1,1-s}(\partial\Omega)\right)^* \longrightarrow \dot{B}^{p,q}_{m-1,s}(\partial\Omega) \end{aligned} \tag{5.167}$$

and, hence,

$$\dot{S}\dot{K}^* : \left(\dot{B}^{2,2}_{m-1,1/2}(\partial\Omega)\right)^* \longrightarrow \dot{B}^{2,2}_{m-1,1/2}(\partial\Omega) \text{ is self-adjoint.} \tag{5.168}$$

Finally, the intertwining formula (5.166) *is also valid when both sides are viewed as operators from* $\left(\dot{L}^p_{m-1,0}(\partial\Omega)\right)^*$ *into* $\dot{L}^{p'}_{m-1,1}(\partial\Omega)$, *or as operators from* $\left(\dot{L}^p_{m-1,1}(\partial\Omega)\right)^*$ *into* $\dot{L}^{p'}_{m-1,0}(\partial\Omega)$.

Proof. The starting point is to specialize (5.141) to the case when $u := \dot{\mathscr{S}}\Lambda$, for some arbitrary, fixed $\Lambda \in \left(\dot{B}^{p,q}_{m-1,s}(\partial\Omega)\right)^*$. This yields

$$\dot{\mathscr{S}}\Lambda = \dot{\mathscr{D}}(\dot{S}\Lambda) - \dot{\mathscr{S}}\Big[\Big(-\tfrac{1}{2}I + \dot{K}^*\Big)\Lambda\Big] \quad \text{in } B^{p',q'}_{m-s+1/p'}(\Omega), \tag{5.169}$$

by (5.44), (5.40), (4.224) (5.152) and (4.253). Applying tr_{m-1} to both sides now readily yields (5.166), as bounded operators from the space $\left(\dot{B}^{p,q}_{m-1,s}(\partial\Omega)\right)^*$ into the space $\dot{B}^{p',q'}_{m-1,1-s}(\partial\Omega)$, on account of (5.44) and (4.250). The final claim in the statement of the proposition then follows from what we have proved up to this point, Theorems 5.4, 4.14, and a density argument (cf. (5.42) for the latter). □

We now proceed to define the conormal derivative of the double multi-layer potential operator.

Proposition 5.31. *Suppose that Ω is a bounded Lipschitz domain in $\mathbb{R}^n$, and consider a constant coefficient W-elliptic differential operator L of order $2m$, $m \in \mathbb{N}$, and denote by ∂_ν^A the conormal associated with an arbitrary choice of the tensor coefficient $A = (A_{\alpha\beta})_{|\alpha|=|\beta|=m}$ in the writing of L as in* (5.92).

Finally, fix indices $p, q, p'q' \in (1, \infty)$ with $1/p + 1/p' = 1/q + 1/q' = 1$, along with $0 < s < 1$. Then it is possible to define the conormal derivative of the double multi-layer (associated with Ω and A) in such a way that

$$\partial_\nu^A \dot{\mathscr{D}} : \dot{B}^{p,q}_{m-1,s}(\partial\Omega) \longrightarrow \left(\dot{B}^{p',q'}_{m-1,1-s}(\partial\Omega)\right)^* \tag{5.170}$$

becomes a linear, bounded operator. Analogously, one can define the conormal derivative of the double multi-layer associated with $\Omega_- := \mathbb{R}^n \setminus \overline{\Omega}$. When necessary to distinguish this from (5.170)*, we shall denote this by $\partial_{\nu_-}^A \dot{\mathscr{D}}$, and denote the former by $\partial_{\nu_+}^A \dot{\mathscr{D}}$.*

Proof. Thanks to Theorem 4.19 and (4.58),

$$\dot{\mathscr{D}} : \dot{B}^{p,q}_{m-1,s}(\partial\Omega) \longrightarrow \left\{u \in B^{p,q}_{m-1+s+1/p}(\Omega) : Lu = 0 \text{ in } \Omega\right\} \tag{5.171}$$

is a well-defined, linear and bounded operator. We define $\partial_\nu^A \dot{\mathscr{D}}$ as the composition of ∂_ν^A with this operator. Thanks to (5.131), we may conclude that the operator (5.170) is indeed well-defined, linear and bounded.

The reasoning when Ω is replaced by Ω_- is similar. More specifically, we consider

$$\begin{aligned} &\partial_{\nu_-}^A \dot{\mathscr{D}} : \dot{B}^{p,q}_{m-1,s}(\partial\Omega) \longrightarrow \left(\dot{B}^{p',q'}_{m-1,1-s}(\partial\Omega)\right)^*, \\ &\partial_{\nu_-}^A \dot{\mathscr{D}} \dot{f} := \partial_{\nu_-}^A \left(\psi \dot{\mathscr{D}}^- \dot{f}\,,\, L(\psi \dot{\mathscr{D}}^- \dot{f})\right), \end{aligned} \tag{5.172}$$

in the sense of Definition 5.20, where $\psi \in C_c^\infty(\mathbb{R}^n)$ is an arbitrary function with the property that $\psi \equiv 1$ near $\partial\Omega$. In particular, this ensures that $L\left(\psi \dot{\mathscr{D}}^- \dot{f}\right) \in C_c^\infty(\Omega_-)$ for every Whitney array $\dot{f} \in \dot{B}^{p,q}_{m-1,s}(\partial\Omega)$. Note that (5.172) can be rephrased as

$$\begin{aligned} \left\langle \partial_{\nu_-}^A \dot{\mathscr{D}} \dot{f}, \dot{g}\right\rangle := (-1)^m \sum_{|\alpha|=|\beta|=m} \left\langle A_{\alpha\beta}\, \partial^\beta \left(\psi \dot{\mathscr{D}}^- \dot{f}\right), \partial^\alpha G\right\rangle_{\Omega_-} \\ - \int_{\Omega_-} \left\langle L\left(\psi \dot{\mathscr{D}}^- \dot{f}\right), G\right\rangle dX, \end{aligned} \tag{5.173}$$

whenever $\dot{g} \in \dot{B}^{p',q'}_{m-1,1-s}(\partial\Omega)$ and $G \in B^{p',q'}_{m-s+1/p'}(\Omega_-)$ are such that $\mathrm{tr}_{m-1}G = \dot{g}$. That this definition is meaningful and unambiguous can be shown as before (cf. the last part of the proof of Proposition 5.27). Once again, (5.173) reduces to

$$\left\langle \partial^A_{\nu_-}\dot{\mathscr{D}}\dot{f}, \dot{g}\right\rangle = (-1)^m \sum_{|\alpha|=|\beta|=m} \left\langle A_{\alpha\beta}\,\partial^\beta\left(\dot{\mathscr{D}}^-\dot{f}\right), \partial^\alpha G\right\rangle_{\Omega_-}, \tag{5.174}$$

in the case when the array $\dot{g} \in \dot{B}^{p',q'}_{m-1,1-s}(\partial\Omega)$ and the function $G \in B^{p',q'}_{m-s+1/p'}(\Omega_-)$ are such that $\mathrm{tr}_{m-1}G = \dot{g}$ and $\mathrm{supp}\,G$ is bounded. □

We now describe some of the basic properties of the conormal derivative of the double multi-layer potential introduced above.

Proposition 5.32. *Retain the same basic background hypotheses as in Proposition 5.31 and, in addition, assume that L is self-adjoint. Also, fix $s \in (0,1)$, and suppose that $1 < p, p', q, q' < \infty$ satisfy $1/p + 1/p' = 1/q + 1/q' = 1$. Then*

$$\partial^A_{\nu_+}\dot{\mathscr{D}}\dot{f} = \partial^A_{\nu_-}\dot{\mathscr{D}}\dot{f} \ \text{ in } \left(\dot{B}^{p',q'}_{m-1,1-s}(\partial\Omega)\right)^*, \quad \forall\,\dot{f} \in \dot{B}^{p,q}_{m-1,s}(\partial\Omega). \tag{5.175}$$

Also, the conormal derivative of the double multi-layer potential is self-adjoint in the sense that the adjoint of (5.170) *is* $\partial^A_\nu\dot{\mathscr{D}} : \dot{B}^{p',q'}_{m-1,1-s}(\partial\Omega) \longrightarrow \left(\dot{B}^{p,q}_{m-1,s}(\partial\Omega)\right)^*$.

Finally,

$$\partial^A_\nu\dot{\mathscr{D}} \circ \dot{S} = \left(\tfrac{1}{2}I + \dot{K}^*\right) \circ \left(-\tfrac{1}{2}I + \dot{K}^*\right) \ \text{ on } \left(\dot{B}^{p,q}_{m-1,s}(\partial\Omega)\right)^*, \tag{5.176}$$

and

$$\dot{S} \circ \partial^A_\nu\dot{\mathscr{D}} = \left(\tfrac{1}{2}I + \dot{K}\right) \circ \left(-\tfrac{1}{2}I + \dot{K}\right) \ \text{ on } \dot{B}^{p,q}_{m-1,s}(\partial\Omega). \tag{5.177}$$

Proof. The claims in the first part of the statement are direct consequences of (5.114), (5.103) and Proposition 5.31. As for the second part, it suffices to only prove (5.177) since (5.176) will follow from this, duality, and the formal self-adjointness of the conormal derivative of the double multi-layer and single multi-layer, respectively. With (5.177) in mind, we start by specializing (5.141) to the situation when $u := \dot{\mathscr{D}}\dot{f}$, for some arbitrary, fixed $\dot{f} \in \dot{B}^{p,q}_{m-1,s}(\partial\Omega)$. This gives

$$\dot{\mathscr{D}}\dot{f} = \dot{\mathscr{D}}\left[\left(\tfrac{1}{2}I + \dot{K}\right)\dot{f}\right] - \dot{\mathscr{S}}\left(\partial^A_\nu\dot{\mathscr{D}}\dot{f}\right) \quad \text{in } B^{p,q}_{m-1+s+1/p}(\Omega), \tag{5.178}$$

by (4.250), (4.224), (4.249) and (5.43). Applying tr_{m-1} to both sides then gives (5.177) after some simple algebra, on account of (4.250) and (5.44). □

Chapter 6
Functional Analytic Properties of Multi-Layer Potentials and Boundary Value Problems

This chapter has a twofold goal. In a first stage, we shall study the Fredholm properties of multi-layer potentials introduced in earlier chapters, while in a second stage we shall proceed to use these results as a tool for establishing the well-posedness of boundary value problems associated with higher-order operators.

6.1 Fredholm Properties of Boundary Multi-Layer Potentials

We begin by proving that, in an appropriate context, the single multi-layer is bounded from below, modulo compact operators.

Lemma 6.1. *Let Ω be a bounded Lipschitz domain in $\mathbb{R}^n$ and assume that the differential operator L is as in* (4.1)–(4.2)*, and satisfies the Legendre–Hadamard ellipticity condition* (4.15)*. Then there exists a finite constant $C > 0$ and a linear compact operator* Comp *mapping $\big(\dot{B}^{2,2}_{m-1,1/2}(\partial\Omega)\big)^*$ into a Banach space, such that*

$$(-1)^m \operatorname{Re}\Big\langle \Lambda, \dot{\overline{S\Lambda}} \Big\rangle + \|\mathrm{Comp}(\Lambda)\|^2 \geq C\,\|\Lambda\|^2_{\left(\dot{B}^{2,2}_{m-1,1/2}(\partial\Omega)\right)^*} \tag{6.1}$$

for every $\Lambda \in \Big(\dot{B}^{2,2}_{m-1,1/2}(\partial\Omega)\Big)^*$, where $\langle\cdot,\cdot\rangle$ above is the duality pairing between $\big(\dot{B}^{2,2}_{m-1,1/2}(\partial\Omega)\big)^*$ and $\dot{B}^{2,2}_{m-1,1/2}(\partial\Omega)$.

Proof. Fix $\Lambda \in \big(\dot{B}^{2,2}_{m-1,1/2}(\partial\Omega)\big)^*$ and, as usual, set $\Omega_+ := \Omega$ and $\Omega_- := \mathbb{R}^n \setminus \overline{\Omega}$. Consider next $u := u^\pm$ in $\Omega_\pm$, where $u^\pm := \dot{\mathscr{S}}\Lambda$ in $\Omega_\pm$. Thus, $u^\pm = u|_{\Omega_\pm}$ and, by Proposition 5.11, for every $\psi \in C^\infty_c(\mathbb{R}^n)$,

$$\psi u \in B^{2,2}_m(\mathbb{R}^n) = F^{2,2}_m(\mathbb{R}^n) = W^{m,2}(\mathbb{R}^n). \tag{6.2}$$

I. Mitrea and M. Mitrea, *Multi-Layer Potentials and Boundary Problems*, Lecture Notes in Mathematics 2063, DOI 10.1007/978-3-642-32666-0_6,

For the remainder of this proof, we fix some $\psi \in C_c^\infty(\mathbb{R}^n)$, $\psi \equiv 1$ near $\overline{\Omega}$. Then Proposition 5.27 (and its proof; cf. (5.148)) gives

$$\|\partial_\nu^A u^+\|_{\left(\dot{B}^{2,2}_{m-1,1/2}(\partial\Omega)\right)^*} \leq C\|u\|_{W^{m,2}(\Omega_+)}, \quad \text{and} \tag{6.3}$$

$$\|\partial_\nu^A u^-\|_{\left(\dot{B}^{2,2}_{m-1,1/2}(\partial\Omega)\right)^*} \leq C\|\psi u\|_{W^{m,2}(\Omega_-)} + C\|L(\psi u)\|_{\left(W^{m,2}(\Omega_-)\right)^*}. \tag{6.4}$$

Based on these, the jump relations (5.153) and the triangle inequality, we may thus estimate

$$\begin{aligned}\|\Lambda\|_{\left(\dot{B}^{2,2}_{m-1,1/2}(\partial\Omega)\right)^*} &\leq \|\partial_\nu^A u^+\|_{\left(\dot{B}^{2,2}_{m-1,1/2}(\partial\Omega)\right)^*} + \|\partial_\nu^A u^-\|_{\left(\dot{B}^{2,2}_{m-1,1/2}(\partial\Omega)\right)^*} \\ &\leq C\Big(\|\psi u\|_{W^{m,2}(\mathbb{R}^n)} + \|L(\psi u)\|_{\left(W^{m,2}(\Omega_-)\right)^*}\Big). \end{aligned} \tag{6.5}$$

It is relevant to note that, since $\psi \equiv 1$ near $\partial\Omega$, the function $L(\psi u)$ is supported in a compact subset of Ω_-. In turn, this readily leads to the conclusion that the assignment

$$\left(\dot{B}^{2,2}_{m-1,1/2}(\partial\Omega)\right)^* \ni \Lambda \mapsto L(\psi u) \in \left(W^{m,2}(\Omega_-)\right)^* \tag{6.6}$$

is compact.

Next, we employ again the jump relations (5.152), (5.44), (5.46), the definition of the conormal derivative in Ω from (5.132) and formula (5.150) to write

$$\begin{aligned}(-1)^m \operatorname{Re}\left\langle \Lambda, \overline{\dot{S}\Lambda}\right\rangle &= (-1)^m \left\{\operatorname{Re}\left\langle (\tfrac{1}{2}I + \dot{K}^*)\Lambda, \overline{\dot{S}\Lambda}\right\rangle - \operatorname{Re}\left\langle (-\tfrac{1}{2}I + \dot{K}^*)\Lambda, \overline{\dot{S}\Lambda}\right\rangle\right\} \\ &= (-1)^m \operatorname{Re}\left\langle \partial_\nu^A u^-, \overline{\operatorname{tr}_{m-1}u^-}\right\rangle - (-1)^m \left\langle \partial_\nu^A u^+, \overline{\operatorname{tr}_{m-1}u^+}\right\rangle \\ &= \operatorname{Re}\int_{\Omega_-} \sum_{|\alpha|=|\beta|=m} \langle A_{\alpha\beta}\,(\partial^\alpha u^-)(X), \overline{\partial^\beta(\psi^2 u^-)(X)}\rangle\, dX \\ &\quad +\operatorname{Re}\int_{\Omega_+} \sum_{|\alpha|=|\beta|=m} \langle A_{\alpha\beta}\,(\partial^\alpha u^+)(X), \overline{\partial^\beta u^+(X)}\rangle\, dX \\ &= \operatorname{Re}\int_{\mathbb{R}^n} \sum_{|\alpha|=|\beta|=m} \langle A_{\alpha\beta}\,\partial^\alpha(\psi u)(X), \overline{\partial^\beta(\psi u)(X)}\rangle\, dX + R(u), \end{aligned} \tag{6.7}$$

where, with M_ψ denoting the operator of multiplication by ψ, the residual term is given by

$$R(u) := \mathrm{Re} \int_{\Omega_-} \sum_{|\alpha|=|\beta|=m} \big\langle A_{\alpha\beta}\, [\partial^\beta, M_\psi](\psi u)\, ,\, \overline{\partial^\alpha u} \big\rangle\, dX$$

$$-\mathrm{Re} \int_{\Omega_-} \sum_{|\alpha|=|\beta|=m} \big\langle A_{\alpha\beta}\, \partial^\beta(\psi u)\, ,\, \overline{[\partial^\alpha, M_\psi] u} \big\rangle\, dX. \tag{6.8}$$

Here, $[A, B] := AB - BA$ is the usual commutator bracket. Given that $\psi \equiv 1$ near $\partial\Omega$ and $Lu = 0$ in Ω_-, it follows that there exists a compact subset $\mathscr{O}$ of Ω_- with the property that

$$|R(u)| \leq C \int_{\mathscr{O}} \sum_{|\gamma|\leq m} |(\partial^\gamma u)(X)|^2\, dX. \tag{6.9}$$

As a consequence, the assignment

$$\Big(\dot{B}^{2,2}_{m-1,1/2}(\partial\Omega)\Big)^* \ni \Lambda \mapsto \Big((\partial^\gamma \dot{S}\Lambda)|_{\mathscr{O}}\Big)_{|\gamma|\leq m} \in L^2(\mathscr{O}) \oplus \cdots \oplus L^2(\mathscr{O}) \tag{6.10}$$

is compact. On the other hand, for every $v \in C^\infty_c(\mathbb{R}^n)$ we may write

$$\begin{aligned}
\mathrm{Re} &\int_{\mathbb{R}^n} \sum_{|\alpha|=|\beta|=m} \langle A_{\alpha\beta}\, \partial^\alpha v(X)\, ,\, \overline{\partial^\beta v(X)} \rangle\, dX \\
&= \mathrm{Re} \int_{\mathbb{R}^n} \sum_{|\alpha|=|\beta|=m} \langle A_{\alpha\beta}\, \widehat{\partial^\alpha v}(\xi)\, ,\, \overline{\widehat{\partial^\beta v}(\xi)} \rangle\, d\xi \\
&= \int_{\mathbb{R}^n} \mathrm{Re} \Big\langle \Big(\sum_{|\alpha|=|\beta|=m} A_{\alpha\beta}\, \xi^\alpha \xi^\beta \Big) \widehat{v}(\xi)\, ,\, \overline{\widehat{v}(\xi)} \Big\rangle\, d\xi \\
&\geq C \int_{\mathbb{R}^n} |\xi|^{2m} |\widehat{v}(\xi)|^2\, d\xi \\
&= C \int_{\mathbb{R}^n} \sum_{|\gamma|=m} \frac{m!}{\gamma!} \xi^{2\gamma} |\widehat{v}(\xi)|^2\, d\xi = C \int_{\mathbb{R}^n} \sum_{|\gamma|=m} \frac{m!}{\gamma!} |\widehat{\partial^\gamma v}(\xi)|^2\, d\xi \\
&\geq C \int_{\mathbb{R}^n} \sum_{|\gamma|=m} |\partial^\gamma v(X)|^2\, dX,
\end{aligned} \tag{6.11}$$

by Plancherel's formula, the ellipticity condition (4.17) and (4.19). Consequently, using (6.11) and the fact that $C_c^\infty(\mathbb{R}^n)$ is dense in $W^{m,2}(\mathbb{R}^n)$, we arrive at the conclusion that

$$\mathrm{Re}\int_{\mathbb{R}^n}\sum_{\substack{|\alpha|=m\\|\beta|=m}}\langle A_{\alpha\beta}\,\partial^\alpha v\,,\,\overline{\partial^\beta v}\rangle\,dX \geq C\int_{\mathbb{R}^n}\sum_{|\gamma|=m}|\partial^\gamma v|^2\,dX,\qquad \forall\, v\in W^{m,2}(\mathbb{R}^n). \tag{6.12}$$

In particular, there exists $C = C(\Omega, L) > 0$ such that

$$\begin{aligned} v\in W^{m,2}(\mathbb{R}^n) \Longrightarrow (-1)^m\,\mathrm{Re}\int_{\mathbb{R}^n}\sum_{|\alpha|=|\beta|=m}&\langle A_{\alpha\beta}\,\partial^\alpha v(X)\,,\,\overline{\partial^\beta v(X)}\rangle\,dX\\ &+\int_{\mathbb{R}^n}\sum_{|\gamma|\leq m-1}|\partial^\gamma v(X)|^2\,dX \geq C\,\|v\|^2_{W^{m,2}(\mathbb{R}^n)}. \end{aligned}\tag{6.13}$$

We shall use (6.13) for $v := \psi u$. In this regard, it is also useful to observe that the assignment $\big(\dot{B}^{2,2}_{m-1,1/2}(\partial\Omega)\big)^* \ni \Lambda \mapsto \Big(\partial^\gamma(\psi\dot{S}\Lambda)\Big)_{|\gamma|\leq m-1} \in L^2(\mathbb{R}^n)\oplus\cdots\oplus L^2(\mathbb{R}^n)$ is compact (cf. Theorem 5.7). Thus, based on this, (6.7), (6.5) and the various comments made about compact operators, (6.1) follows. □

It is useful to further isolate the estimate contained in the lemma below.

Lemma 6.2. *Let Ω be a bounded Lipschitz domain in $\mathbb{R}^n$ and assume that the differential operator L is as in* (4.1)–(4.2)*, and satisfies the Legendre–Hadamard ellipticity condition* (4.15)*. Then there exists a finite constant $C > 0$ and a linear compact operator* Comp *mapping $\big(\dot{B}^{2,2}_{m-1,1/2}(\partial\Omega)\big)^*$ into a Banach space, such that for every $\Lambda \in \Big(\dot{B}^{2,2}_{m-1,1/2}(\partial\Omega)\Big)^*$,*

$$\|\dot{S}\Lambda\|_{\dot{B}^{2,2}_{m-1,1/2}(\partial\Omega)} + \|\mathrm{Comp}(\Lambda)\| \geq C\,\|\Lambda\|_{\left(\dot{B}^{2,2}_{m-1,1/2}(\partial\Omega)\right)^*}. \tag{6.14}$$

As a corollary,

$$\dot{S} : \Big(\dot{B}^{2,2}_{m-1,1/2}(\partial\Omega)\Big)^* \longrightarrow \dot{B}^{2,2}_{m-1,1/2}(\partial\Omega) \tag{6.15}$$

has closed range and finite dimensional kernel.

Proof. This is an immediate consequence of estimate (6.1) and the fact that, given any $\varepsilon > 0$, there exists $C_\varepsilon \in (0,\infty)$ such that

$$\begin{aligned} \left|\left\langle \Lambda, \overline{\dot{S}\Lambda} \right\rangle\right| &\leq \|\Lambda\|_{\left(\dot{B}^{2,2}_{m-1,1/2}(\partial\Omega)\right)^*} \|\dot{S}\Lambda\|_{\dot{B}^{2,2}_{m-1,1/2}(\partial\Omega)} \\ &\leq \varepsilon \|\Lambda\|^2_{\left(\dot{B}^{2,2}_{m-1,1/2}(\partial\Omega)\right)^*} + C_\varepsilon \|\dot{S}\Lambda\|^2_{\dot{B}^{2,2}_{m-1,1/2}(\partial\Omega)}, \end{aligned} \tag{6.16}$$

for every $\Lambda \in \left(\dot{B}^{2,2}_{m-1,1/2}(\partial\Omega)\right)^*$. □

The main result pertaining to the Fredholm properties of the single multi-layer is contained in the next theorem.

Theorem 6.3. *Let Ω be a bounded Lipschitz domain in $\mathbb{R}^n$ and assume that the differential operator L is as in* (4.1)–(4.2)*, and satisfies the Legendre–Hadamard ellipticity condition* (4.15)*. Then there exists $\varepsilon > 0$ with the property that*

$$\begin{aligned} &\dot{S} : \left(\dot{B}^{p,q}_{m-1,s}(\partial\Omega)\right)^* \longrightarrow \dot{B}^{p',q'}_{m-1,1-s}(\partial\Omega) \ \text{ is Fredholm if} \\ &|p-2| + |q-2| + \left|s - \tfrac{1}{2}\right| < \varepsilon \ \text{ and } \ \tfrac{1}{p} + \tfrac{1}{p'} = \tfrac{1}{q} + \tfrac{1}{q'} = 1. \end{aligned} \tag{6.17}$$

Moreover,

$$L = L^* \Longrightarrow \dot{S} \ \text{ in (6.17) has index zero.} \tag{6.18}$$

Proof. It suffices to treat only the case when $p = q = 2$, $s = 1/2$, since the more general case in (6.17)–(6.18) follows from this and general perturbation results (cf. [63]) on complex interpolation scales (cf. Theorem 3.38). Thus, assume that $p = q = 2$, $s = 1/2$ and, in order to stress the dependence of $\dot{S}$ on L let us agree to temporarily use the notation $\dot{S}_L$. Note that if L is as in the statement of the theorem, then so is $L^* := \bar{L}^t$. Lemma 6.2 used for L^* in place of L then shows that the operator $\dot{S}_{L^*} : \left(\dot{B}^{2,2}_{m-1,1/2}(\partial\Omega)\right)^* \to \dot{B}^{2,2}_{m-1,1/2}(\partial\Omega)$ has closed range and a finite dimensional kernel. This, duality, and (5.22) then give (cf. Theorem 5.13 on p. 234 in [67]) that $\dot{S}_L : \left(\dot{B}^{2,2}_{m-1,1/2}(\partial\Omega)\right)^* \to \dot{B}^{2,2}_{m-1,1/2}(\partial\Omega)$ has closed range, of finite codimension. In concert with (6.15), this shows that the single multi-layer operator in (6.17) is Fredholm. The implication (6.18) is then a consequence of (6.17), the last part in the statement of Theorem 5.4 (which ensures that, in our context, $\dot{S}$ is self-adjoint), and Corollary 5.14 on p. 234 in [67]. □

Basic spectral theory results for the principal-value double multi-layer are contained in the theorem below.

Theorem 6.4. *Let Ω be a bounded Lipschitz domain in $\mathbb{R}^n$ and let L be a homogeneous differential operator of order $2m$, $m \in \mathbb{N}$, as in* (4.1)–(4.2) *and such that* (4.14) *holds. For $\lambda \in \mathbb{C}$ and* Comp *a linear, compact operator mapping $\left(\dot{B}^{2,2}_{m-1,1/2}(\partial\Omega)\right)^*$ into a Banach space, consider the estimate*

$$\|(\lambda I - \dot{K}^*)\Lambda\|_{\left(\dot{B}^{2,2}_{m-1,1/2}(\partial\Omega)\right)^*} + \|\mathrm{Comp}(\Lambda)\| \geq C\|\Lambda\|_{\left(\dot{B}^{2,2}_{m-1,1/2}(\partial\Omega)\right)^*},$$
$$\textit{for all functionals} \quad \Lambda \in \left(\dot{B}^{2,2}_{m-1,1/2}(\partial\Omega)\right)^*. \tag{6.19}$$

Then the following claims are true:

(i) If L is self-adjoint then there exists Comp *as above and finite constant $C > 0$ such that* (6.19) *holds for every $\lambda \in \mathbb{C}$ with* $\mathrm{Im}\,\lambda \neq 0$.
(ii) If L is self-adjoint and its tensor coefficient satisfies the semi-positivity condition (4.20)*, then there exists* Comp *as above and finite constant $C > 0$ such that* (6.19) *holds for every $\lambda \in \mathbb{C} \setminus [-1/2, 1/2]$.*
(iii) If L is self-adjoint and S-elliptic (cf. (4.16)*) then there exists* Comp *as above and finite constant $C > 0$ such that* (6.19) *holds for every $\lambda \in \mathbb{C}\setminus(-1/2, 1/2)$.*

Furthermore, in each of the situations (i)–(iii) described above, there exists some small $\varepsilon > 0$ with the property that

$$\lambda I - \dot{K} : \dot{B}^{p,q}_{m-1,s}(\partial\Omega) \to \dot{B}^{p,q}_{m-1,s}(\partial\Omega) \ \textit{and}$$
$$\lambda I - \dot{K}^* : \left(\dot{B}^{p,q}_{m-1,s}(\partial\Omega)\right)^* \to \left(\dot{B}^{p,q}_{m-1,s}(\partial\Omega)\right)^* \tag{6.20}$$
$$\textit{are Fredholm operators with index zero}$$
$$\textit{whenever } |p-2| + |q-2| + |s-1/2| < \varepsilon.$$

Proof. Fix $\lambda \in \mathbb{C}$ with $\mathrm{Im}\,\lambda \neq 0$ and pick an arbitrary $\Lambda \in \left(\dot{B}^{2,2}_{m-1,1/2}(\partial\Omega)\right)^*$. Thanks to Proposition 5.30 and (5.22) in Theorem 5.4 the operators $\dot{S}\dot{K}^*$ and $\dot{S}$ are formally self-adjoint. In particular,

$$\left\langle \dot{K}^*\Lambda, \overline{\dot{S}\Lambda}\right\rangle \in \mathbb{R} \quad \text{and} \quad \left\langle \dot{f}, \overline{\dot{S}\Lambda}\right\rangle \in \mathbb{R}, \tag{6.21}$$

hence,

$$\mathrm{Im}\left\langle(\lambda I - \dot{K}^*)\Lambda, \overline{\dot{S}\Lambda}\right\rangle = (\mathrm{Im}\,\lambda)\left\langle\Lambda, \overline{\dot{S}\Lambda}\right\rangle. \tag{6.22}$$

On the other hand,

$$\left|\mathrm{Im}\left\langle(\lambda I - \dot{K}^*)\Lambda, \overline{\dot{S}\Lambda}\right\rangle\right| \leq \|(\lambda I - \dot{K}^*)\Lambda\|_{\left(\dot{B}^{2,2}_{m-1,1/2}(\partial\Omega)\right)^*}\|\dot{S}\Lambda\|_{\dot{B}^{2,2}_{m-1,1/2}(\partial\Omega)}. \tag{6.23}$$

Thus, for each $\varepsilon > 0$ there exists a finite constant $C_\varepsilon > 0$ such that

$$\begin{aligned}
&C_\varepsilon \|(\lambda I - \dot{K}^*)\Lambda\|^2_{(\dot{B}^{2,2}_{m-1,1/2}(\partial\Omega))^*} + \varepsilon\, C\, \|\Lambda\|^2_{(\dot{B}^{2,2}_{m-1,1/2}(\partial\Omega))^*} \\
&\quad \geq C_\varepsilon \|(\lambda I - \dot{K}^*)\Lambda\|^2_{(\dot{B}^{2,2}_{m-1,1/2}(\partial\Omega))^*} + \varepsilon \|\dot{S}\Lambda\|^2_{\dot{B}^{2,2}_{m-1,1/2}(\partial\Omega)} \\
&\quad \geq \Big|\mathrm{Im}\Big\langle(\lambda I - \dot{K}^*)\Lambda, \overline{\dot{S}\Lambda}\Big\rangle\Big| \\
&\quad \geq |\mathrm{Im}\,\lambda| \Big\langle \Lambda, \overline{\dot{S}\Lambda}\Big\rangle \\
&\quad \geq C\,\|\Lambda\|^2_{(\dot{B}^{2,2}_{m-1,1/2}(\partial\Omega))^*} - \|\mathrm{Comp}(\Lambda)\|^2, \qquad (6.24)
\end{aligned}$$

by Proposition 5.6, (6.23), (6.22) and (6.1). Then (6.19) follows from (6.24) (for some $C = C(\lambda) > 0$, finite constant depending only on λ) by choosing ε small enough. This finishes the treatment of case (*i*).

Consider next the scenario described in (*ii*). Given what we have proved already, it suffices to treat the case when $\lambda \in \mathbb{R} \setminus [-1/2, 1/2]$. In this situation, write

$$\begin{aligned}
\pm(-1)^m\Big\langle(\lambda I - \dot{K}^*)\Lambda, \overline{\dot{S}\Lambda}\Big\rangle &= (-1)^m\Big\langle(\tfrac{1}{2}I \mp \dot{K}^*)\Lambda, \overline{\dot{S}\Lambda}\Big\rangle \\
&\quad + (-1)^m(-\tfrac{1}{2} \pm \lambda)\Big\langle \Lambda, \overline{\dot{S}\Lambda}\Big\rangle. \qquad (6.25)
\end{aligned}$$

At this point we claim that, in the current scenario, there exists a linear compact operator, Comp, mapping $\big(\dot{B}^{2,2}_{m-1,1/2}(\partial\Omega)\big)^*$ into some Banach space, such that for every choice of the sign,

$$\|\mathrm{Comp}\,(\Lambda)\|^2 + (-1)^m\Big\langle(\tfrac{1}{2}I \pm \dot{K}^*)\Lambda, \overline{\dot{S}\Lambda}\Big\rangle \geq 0. \qquad (6.26)$$

To justify this, set $u := \dot{\mathscr{S}}\Lambda$ in Ω_-, and pick some $\psi \in C^\infty_c(\mathbb{R}^n)$ with $\psi \equiv 1$ near $\partial\Omega$. Based on the jump relations (5.152) and the definition of the conormal derivative from (5.148), we may then write

$$\begin{aligned}
&(-1)^m\Big\langle(\tfrac{1}{2}I + \dot{K}^*)\Lambda, \overline{\dot{S}\Lambda}\Big\rangle \\
&\quad = (-1)^m\,\mathrm{Re}\Big\langle(\tfrac{1}{2}I + \dot{K}^*)\Lambda, \overline{\dot{S}\Lambda}\Big\rangle = (-1)^m\,\mathrm{Re}\Big\langle \partial^A_{\nu_-}u, \overline{\mathrm{tr}_{m-1}u}\Big\rangle \\
&\quad = \mathrm{Re}\int_{\Omega_-}\sum_{|\alpha|=|\beta|=m}\big\langle A_{\alpha\beta}\,\partial^\alpha u\,,\, \overline{\partial^\beta(\psi^2 u)}\big\rangle\, dX - \int_{\Omega_-}\big\langle L(\psi^2 u), \bar{u}\big\rangle\, dX
\end{aligned}$$

$$= \mathrm{Re} \int_{\Omega_-} \sum_{|\alpha|=|\beta|=m} \langle A_{\alpha\beta}\, \partial^\alpha(\psi u)(X)\, ,\, \overline{\partial^\beta(\psi u)(X)}\rangle\, dX$$

$$-\mathrm{Re} \int_{\Omega_-} \langle L(\psi^2 u), \bar{u}\rangle\, dX + R(u), \tag{6.27}$$

where $R(u)$ as in (6.8). Since, as before, the last line in (6.27) contributes terms of the form $\|\mathrm{Comp}\,(\Lambda)\|^2$, estimate (6.26) corresponding to the "top" choice of signs readily follows from the above identity and (4.20). Its other version, corresponding to the "bottom" choice of signs, is proved analogously, by working in Ω_+ rather than Ω_-. Thus, from (6.25) and (6.26) we obtain

$$\pm(-1)^m \Big\langle (\lambda I - \dot{K}^*)\Lambda, \overline{\dot{S}\Lambda} \Big\rangle \geq (-1)^m\big(-\tfrac{1}{2} \pm \lambda\big)\Big\langle \Lambda, \overline{\dot{S}\Lambda} \Big\rangle. \tag{6.28}$$

By observing that the membership of λ to $\mathbb{R} \setminus [-1/2, 1/2]$ is equivalent with the statement that there exists a choice of the sign for which $-\frac{1}{2} \pm \lambda > 0$, it follows from (6.28), (6.1) and Proposition 5.6 that, for every $\varepsilon > 0$,

$$\begin{aligned}
C_\varepsilon \|(\lambda I - \dot{K}^*)\Lambda\|^2_{(\dot{B}^{2,2}_{m-1,1/2}(\partial\Omega))^*} &+ \varepsilon C\, \|\Lambda\|^2_{(\dot{B}^{2,2}_{m-1,1/2}(\partial\Omega))^*} \\
&\geq \|(\lambda I - \dot{K}^*)\Lambda\|_{(\dot{B}^{2,2}_{m-1,1/2}(\partial\Omega))^*} \|\dot{S}\Lambda\|_{\dot{B}^{2,2}_{m-1,1/2}(\partial\Omega)} \\
&\geq \Big|\Big\langle (\lambda I - \dot{K}^*)\Lambda, \overline{\dot{S}\Lambda} \Big\rangle\Big| \\
&\geq (-1)^m\big(-\tfrac{1}{2} \pm \lambda\big)\Big\langle \Lambda, \overline{\dot{S}\Lambda} \Big\rangle \\
&\geq \|\Lambda\|^2_{(\dot{B}^{2,2}_{m-1,1/2}(\partial\Omega))^*} - \|\mathrm{Comp}(\Lambda)\|^2,
\end{aligned} \tag{6.29}$$

for some finite $C_\varepsilon > 0$, and $C > 0$ independent of ε. Choosing $\varepsilon > 0$ sufficiently small, we may conclude that (6.19) holds in the situation described in (ii).

In the final part of the proof, we consider the situation described in case (iii) in the statement of the theorem. Granted what we have proved so far, it suffices to only treat the values $\lambda = \pm 1/2$. In this scenario, (6.14) is useful. To be specific, fix $\Lambda \in \big(\dot{B}^{2,2}_{m-1,1/2}(\partial\Omega)\big)^*$ and set $u := \dot{\mathscr{S}}\Lambda$ in the domain Ω_-. Then, based on (6.27) and the comments right after, for every $\psi \in C^\infty_c(\mathbb{R}^n)$ with $\psi \equiv 1$ near $\partial\Omega$ we may write

$$
\begin{aligned}
(-1)^m\Big\langle(\tfrac{1}{2}I+\dot{K}^*)\Lambda,\overline{\dot{S}\Lambda}\Big\rangle &= \mathrm{Re}\int\limits_{\Omega_-}\sum_{|\alpha|=|\beta|=m}\big\langle A_{\alpha\beta}\,\partial^\alpha(\psi u)(X)\,,\,\overline{\partial^\beta(\psi u)(X)}\big\rangle\,dX \\
&\quad -\mathrm{Re}\int\limits_{\Omega_-}\big\langle L(\psi^2 u),\bar{u}\big\rangle\,dX + R(u) \\
&\ge C\,\|\psi u\|^2_{W^{m,2}(\Omega_-)} - \|\mathrm{Comp}\,(\Lambda)\|^2 \\
&\ge C\,\|\mathrm{tr}_{m-1}u\|^2_{\dot{B}^{2,2}_{m-1,1/2}(\partial\Omega)} - \|\mathrm{Comp}\,(\Lambda)\|^2 \\
&= C\,\|\dot{S}\Lambda\|^2_{\dot{B}^{2,2}_{m-1,1/2}(\partial\Omega)} - \|\mathrm{Comp}\,(\Lambda)\|^2 \\
&\ge C\,\|\Lambda\|^2_{\left(\dot{B}^{2,2}_{m-1,1/2}(\partial\Omega)\right)^*} - \|\mathrm{Comp}(\Lambda)\|^2.
\end{aligned}
\tag{6.30}
$$

Above, we have also used the strict positive definiteness property of the tensor of coefficients $(A_{\alpha\beta})_{|\alpha|=|\beta|=m}$ from (4.16), the boundedness of the trace operator (cf. Theorem 3.9), formulas (5.46), (5.44), as well as (6.14). Proceeding as before and estimating the first term in (6.30) from above using the Cauchy–Schwarz inequality, then using the Cauchy inequality with ε, and finally employing the boundedness of the operator $\dot{S}$ we conclude that (6.19) holds for $\lambda = 1/2$. The case when $\lambda = -1/2$ is analogous (and simpler), finishing the proof of case (iii). All in all, the above reasoning shows that (6.19) holds in each of the cases (i)–(iii).

As for (6.20), once again it suffices to consider only the case $p = q = 2$, $s = 1/2$ since, as before, the more general case discussed in (6.20) follows from this, abstract perturbation results (cf. [63]), and our interpolation results from Theorem 3.38. In this scenario, (6.19) yields that $\lambda I - \dot{K}^*$ is semi-Fredholm when acting from $\big(\dot{B}^{2,2}_{m-1,1/2}(\partial\Omega)\big)^*$ into itself for each λ belonging to an unbounded, connected subset of $\mathbb{C}$ (described concretely in each of the cases (i)–(iii)). Since, for $|\lambda|$ large, $\lambda I - \dot{K}^*$ can be inverted using a Neumann series, it follows that the index of $\lambda I - \dot{K}^*$ is zero whenever $|\lambda|$ is sufficiently large. Given that the index is homotopic invariant and that the assignment $\lambda \mapsto \lambda I - \dot{K}^*$ is obviously continuous, we may finally conclude that $\lambda I - \dot{K}^* : \big(\dot{B}^{2,2}_{m-1,1/2}(\partial\Omega)\big)^* \to \big(\dot{B}^{2,2}_{m-1,1/2}(\partial\Omega)\big)^*$ is Fredholm with index zero for each λ belonging to each of the subsets of $\mathbb{C}$ described in the cases (i)–(iii). Finally, the corresponding claim for $\lambda I - \dot{K}$ follows from this and duality. □

Recall next that, given a Banach space $\mathscr{X}$ along with $T \in \mathscr{L}\big(\mathscr{X} \to \mathscr{X}\big)$, the `Fredholm radius` of the operator T on $\mathscr{X}$ is defined as

$$
\inf\Big\{r > 0 : \lambda I - T \text{ is Fredholm on } \mathscr{X} \text{ for every } \lambda \in \mathbb{C} \text{ with } |\lambda| > r\Big\}. \tag{6.31}
$$

In this regard, we have the following result.

Corollary 6.5. *Assume that Ω is a bounded Lipschitz domain in $\mathbb{R}^n$ and that L is a homogeneous differential operator of order $2m$, $m \in \mathbb{N}$ as in* (4.1)–(4.2)*, which is S-elliptic (cf.* (4.16)*) and self-adjoint. Then there exists $\varepsilon > 0$ with the property that*

$$\begin{aligned} &\textit{the Fredholm radius of } \dot{K} \text{ on } \dot{B}^{p,q}_{m-1,s}(\partial\Omega) \text{ is } < \tfrac{1}{2} \\ &\textit{whenever } |p-2| + |q-2| + |s-1/2| < \varepsilon. \end{aligned} \tag{6.32}$$

Proof. This is a direct consequence of Theorem 6.4. □

Our next task is to prove the injectivity of the operator $\frac{1}{2}I + \dot{K}^*$ on the dual of Whitney–Besov space $\dot{B}^{2,2}_{m-1,1/2}(\partial\Omega)$. This requires a number of preliminaries which we now begin to address. Recall (3.383)–(3.384).

Lemma 6.6. *Suppose that Ω is a bounded Lipschitz domain in $\mathbb{R}^n$, and assume that L is a W-elliptic homogeneous differential operator of order $2m$ with (complex) matrix-valued constant coefficients. Consider $\Lambda \in \big(\dot{B}^{2,2}_{m-1,1/2}(\partial\Omega)\big)^*$ with the property that $\big(\lambda I - \dot{K}^*\big)\Lambda = 0$ for some $\lambda \in \mathbb{C} \setminus \{\frac{1}{2}\}$. Then*

$$\langle \Lambda, \dot{f}\rangle = 0 \quad \textit{for every} \quad \dot{f} \in \dot{\mathscr{P}}_{m-1}(\partial\Omega). \tag{6.33}$$

Proof. Indeed, by Proposition 4.17, for every $\dot{f} \in \dot{\mathscr{P}}_{m-1}(\partial\Omega)$ we may write

$$\langle \Lambda, \dot{f}\rangle = \big\langle \Lambda, (\tfrac{1}{2}I + \dot{K})\dot{f}\big\rangle = \big\langle (\tfrac{1}{2}I + \dot{K}^*)\Lambda, \dot{f}\big\rangle = (\tfrac{1}{2} + \lambda)\langle \Lambda, \dot{f}\,\rangle. \tag{6.34}$$

Hence, $\big(-\frac{1}{2} + \lambda\big)\langle \Lambda, \dot{f}\,\rangle = 0$ and, ultimately, $\langle \Lambda, \dot{f}\,\rangle = 0$ since $\lambda \neq \frac{1}{2}$. □

Going further, we shall also need decay properties for the single multi-layer, an issue addressed in our next lemma.

Lemma 6.7. *Suppose that Ω is a bounded Lipschitz domain in $\mathbb{R}^n$, and assume that L is a W-elliptic homogeneous differential operator of order $2m$ with (complex) matrix-valued constant coefficients. Let $\Lambda \in \big(\dot{B}^{2,2}_{m-1,1/2}(\partial\Omega)\big)^*$ satisfy $\langle \Lambda, \dot{f}\rangle = 0$ for every $\dot{f} \in \dot{\mathscr{P}}_{m-1}(\partial\Omega)$. Then*

$$\begin{aligned} &|\partial^\alpha \dot{\mathscr{S}}^- \Lambda(X)| \le C|X|^{m-n-|\alpha|}\log|X|, \quad \textit{if} \quad 0 \le |\alpha| \le m-n, \ \textit{and} \\ &|\partial^\alpha \dot{\mathscr{S}}^- \Lambda(X)| \le C|X|^{m-n-|\alpha|}, \quad \textit{if } |\alpha| > m-n, \text{ or } n \text{ odd, or } n > 2m, \end{aligned} \tag{6.35}$$

uniformly for $|X|$ large.

Proof. Recall that, by definition, $\dot{\mathscr{S}}^-\Lambda(X) = \langle \mathrm{tr}_{m-1}E(X-\cdot), \Lambda\rangle$ for $X \in \mathbb{R}^n \setminus \overline{\Omega}$ and, using (6.33), for every $\alpha \in \mathbb{N}_0^n$ we may write

$$\begin{aligned} \partial^\alpha \dot{\mathscr{S}}^- \Lambda(X) &= \langle \mathrm{tr}_{m-1}(\partial^\alpha E)(X-\cdot), \Lambda\rangle \\ &= \Big\langle \mathrm{tr}_{m-1}\Big(\partial^\alpha E(X-\cdot) - P_X^{m-1,\alpha}(\cdot)\Big), \Lambda\Big\rangle \end{aligned} \tag{6.36}$$

for each $X \in \mathbb{R}^n \setminus \overline{\Omega}$, where $P_X^{m-1,\alpha}(\cdot)$ is the Taylor polynomial of degree $m-1$ for the function $(\partial^\alpha E)(X-\cdot)$ about 0. It is straightforward to check that for each $X \in \mathbb{R}^n \setminus \overline{\Omega}$ we have

$$\partial_Y^\beta [P_X^{m-1,\alpha}(Y)] = (-1)^{|\beta|} P_X^{m-1-|\beta|,\alpha+\beta}(Y), \quad \text{for every } \beta \in \mathbb{N}_0^n. \tag{6.37}$$

Therefore, using (6.37) and Taylor's theorem, we infer that, for each $X \in \mathbb{R}^n \setminus \overline{\Omega}$,

$$\begin{aligned} &\left|\partial_Y^\beta \left[(\partial^\alpha E)(X-Y) - P_X^{m-1,\alpha}(Y)\right]\right| \\ &\quad = \left|(\partial^{\alpha+\beta} E)(X-Y) - P_X^{m-1-|\beta|,\alpha+\beta}(Y)\right| \\ &\quad \le C \left(\sup_{Z\in[0,Y]} \sum_{|\gamma|=m+|\alpha|} |(\partial^\gamma E)(X-Z)||Y|^{m-1-|\beta|} \right). \end{aligned} \tag{6.38}$$

In the context of the single multi-layer potential, this estimate is used with $Y \in \partial\Omega$ and X can be thought as being large in comparison to Y. Hence, (6.35) follows now from (6.38) and (4.29). □

We are ready now to present the injectivity result mentioned earlier.

Theorem 6.8. *Suppose that Ω is a bounded Lipschitz domain in $\mathbb{R}^n$ with connected complement, and assume that L is an S-elliptic homogeneous differential operator of order $2m$ with (complex) matrix-valued constant coefficients. Then the operator*

$$\tfrac{1}{2}I + \dot{K}^* : \left(\dot{B}^{2,2}_{m-1,1/2}(\partial\Omega)\right)^* \longrightarrow \left(\dot{B}^{2,2}_{m-1,1/2}(\partial\Omega)\right)^* \quad \textit{is one-to-one.} \tag{6.39}$$

Proof. Consider

$$\Lambda \in \left(\dot{B}^{2,2}_{m-1,1/2}(\partial\Omega)\right)^* \quad \text{such that} \quad \left(\tfrac{1}{2}I + \dot{K}^*\right)\Lambda = \dot{0} \tag{6.40}$$

and set $u := \dot{\mathscr{S}}\Lambda$ in $\Omega_\pm$ (cf. (2.11)). Assume that the differential operator L is expressed as in (5.92), for some choice of the coefficient tensor $A = \left(A_{\alpha\beta}\right)_{|\alpha|=|\beta|=m}$. Hence, by (5.5),

$$Lu = 0 \ \text{ in } \ \mathbb{R}^n \setminus \partial\Omega. \tag{6.41}$$

Let $\partial^A_{\nu_\pm}$ stand for the conormal operators acting from Ω_+ and Ω_-, respectively (cf. (2.11)). Then from the jump formula (5.152) and (6.40) (cf. also (5.147))

$$\begin{aligned} &\partial^A_{\nu_-}(\psi u, L(\psi u)) = \left(\tfrac{1}{2}I + \dot{K}^*\right)\Lambda = 0, \\ &\forall\, \psi \in C_c^\infty(\mathbb{R}^n) \ \text{ with } \ \psi \equiv 1 \ \text{ near } \ \partial\Omega. \end{aligned} \tag{6.42}$$

Moreover, for every $\psi \in C_c^\infty(\mathbb{R}^n)$,

$$\psi u \in B_m^{2,2}(\mathbb{R}^n) = F_m^{2,2}(\mathbb{R}^n) = W^{m,2}(\mathbb{R}^n), \tag{6.43}$$

by Proposition 5.11. Furthermore, according to (6.35) in Lemma 6.7 (whose validity is ensured by the last condition in (6.40) and Lemma 6.6), given $\alpha \in \mathbb{N}_0^n$ we have for large $|X|$:

$$\big|(\partial^\alpha u)(X)\big| \leq \begin{cases} C|X|^{m-n-|\alpha|}\log|X| \text{ if } & 0 \leq |\alpha| \leq m-n, \\ C|X|^{m-n-|\alpha|} & \text{if } |\alpha| > m-n, \text{ or } n \text{ odd or } n > 2m, \end{cases} \tag{6.44}$$

In particular,

$$\big|(\partial^\alpha u)(X)\big| \leq \frac{C}{|X|^n}, \quad \text{if } |X| \text{ is large, whenever } \alpha \in \mathbb{N}_0^n \text{ has } |\alpha| = m. \tag{6.45}$$

To proceed, fix a sufficiently large $R > 0$ such that $\overline{\Omega} \subset B_R(0)$. Also, select a function

$$\psi \in C_c^\infty(\mathbb{R}^n), \quad \psi \equiv 1 \text{ in a neighborhood of } \overline{B_R(0)}. \tag{6.46}$$

Let $\operatorname{tr}_{m-1}^\pm$ stand, respectively, for the multi-trace operators acting from $\Omega_\pm$. Also, denote by $\operatorname{tr}_{m-1}^R$ the multi-trace operator from $B_R^- := \mathbb{R}^n \setminus \overline{B_R(0)}$, and by $\partial_{\nu_R}^A$ the conormal associated with L in the domain B_R^-. In this notation, we claim that

$$\begin{aligned}(-1)^m & \sum_{|\alpha|=|\beta|=m} \int_{B_R(0)\setminus\overline{\Omega}} \Big\langle A_{\alpha\beta}\, \partial^\beta u, \overline{\partial^\alpha u}\Big\rangle\, dX \\ &= \Big\langle \partial_{\nu_-}^A(\psi u, L(\psi u)), \overline{\operatorname{tr}_{m-1}^- u}\Big\rangle - \int_{\partial B_R(0)} \Big\langle \partial_{\nu_R}^A u, \overline{\operatorname{tr}_{m-1}^R u}\Big\rangle\, d\sigma_R,\end{aligned} \tag{6.47}$$

where σ_R stands for the surface measure of the sphere $\partial B_R(0)$. To justify this claim, note first that by (6.41) and (6.46),

$$L(\psi u) \in C_c^\infty(B_R^-) \subseteq C_c^\infty(\Omega_-), \tag{6.48}$$

then use Proposition 5.26 to write

$$\begin{aligned}\Big\langle \partial_{\nu_-}^A(\psi u, L(\psi u)), \overline{\operatorname{tr}_{m-1}^- u}\Big\rangle &= (-1)^m \sum_{|\alpha|=|\beta|=m} \int_{\Omega_-} \Big\langle A_{\alpha\beta}\, \partial^\beta(\psi u), \overline{\partial^\alpha u}\Big\rangle\, dX \\ &\quad - \int_{\Omega_-} \langle L(\psi u), u\rangle\, dX.\end{aligned} \tag{6.49}$$

In a similar manner, working in the domain B_R^-, we may write

$$\int_{\partial B_R(0)} \left\langle \partial^A_{\nu_R} u, \overline{\mathrm{tr}^R_{m-1} u} \right\rangle d\sigma_R = (-1)^m \sum_{|\alpha|=|\beta|=m} \int_{B_R^-} \left\langle A_{\alpha\beta}\, \partial^\beta(\psi u), \overline{\partial^\alpha u} \right\rangle dX$$

$$- \int_{B_R^-} \langle L(\psi u), u \rangle \, dX. \tag{6.50}$$

Taking the difference between (6.49) and (6.50) then yields (6.47).

Next, we claim that

$$\lim_{R\to\infty} \int_{\partial B_R(0)} \left\langle \partial^A_{\nu_R} u, \overline{\mathrm{tr}^R_{m-1} u} \right\rangle d\sigma_R = 0. \tag{6.51}$$

In this regard, let us first note that for every multi-index $\gamma \in \mathbb{N}_0^n$ with $|\gamma| \le m-1$ we have (cf. (5.87))

$$\left(\partial^A_{\nu_R} u\right)_\gamma = \sum_{|\theta|=2m-|\gamma|-1} \sum_{j=1}^n \nu_j^R B_{j\theta} \left(\partial^\theta u\right)\Big|_{\partial B_R(0)}, \tag{6.52}$$

where ν_j^R is the j-th component of ν^R, the outward unit normal to the ball $B_R(0)$, and $B_{j\theta} \in \mathbb{C}$. Then, since

$$\left\langle \partial^A_{\nu_R} u, \overline{\mathrm{tr}^R_{m-1} u} \right\rangle = \sum_{|\gamma|\le m-1} \left(\partial^A_{\nu_R} u\right)_\gamma \overline{\partial^\gamma u}\Big|_{\partial B_R(0)}, \tag{6.53}$$

we conclude, based on (6.52)–(6.53) and (6.44), that

$$\left|\left\langle \partial^A_{\nu_R} u, \overline{\mathrm{tr}^R_{m-1} u} \right\rangle\right| \le C \sum_{|\gamma|\le m-1} R^{m-n-|\gamma|} (\log R)\, R^{m-n-2m+|\gamma|+1}$$

$$\le C R^{1-2n} \log R \quad \text{on} \quad \partial B_R(0). \tag{6.54}$$

Since $\sigma_R\big(\partial B_R(0)\big) \approx R^{n-1}$, and $R^{n-1} \cdot R^{1-2n} \log R = R^{-n} \log R \to 0$ as $R \to \infty$, the proof of (6.51) is completed.

Going further, (6.45) and (6.43) ensure that $\sum_{|\alpha|=|\beta|=m} \left\langle A_{\alpha\beta}\, \partial^\beta u, \overline{\partial^\alpha u} \right\rangle$ is absolutely integrable in Ω_-. Based on this, Lebesgue's Dominated Convergence Theorem, and (6.47), we may write

$$(-1)^m \sum_{|\alpha|=|\beta|=m} \int_{\Omega_-} \left\langle A_{\alpha\beta}\, \partial^\beta u, \overline{\partial^\alpha u} \right\rangle dX$$

$$= (-1)^m \lim_{R\to\infty} \sum_{|\alpha|=|\beta|=m} \int_{B_R(0)\setminus\overline{\Omega}} \left\langle A_{\alpha\beta}\, \partial^\beta u, \overline{\partial^\alpha u} \right\rangle dX$$

$$
\begin{aligned}
&= \langle (\tfrac{1}{2}I + \dot{K}^*)\,\Lambda\,,\, \overline{\mathrm{tr}_{m-1}^{-} u}\rangle - \lim_{R\to\infty} \int_{\partial B_R(0)} \Big\langle \partial^A_{\nu_R} u, \overline{\mathrm{tr}^R_{m-1} u}\Big\rangle \, d\sigma_R \\
&= 0, \qquad (6.55)
\end{aligned}
$$

where the last equality uses (6.42) and (6.51). Altogether, the above calculation shows that, on the one hand,

$$
\sum_{|\alpha|=|\beta|=m} \int_{\Omega_-} \Big\langle A_{\alpha\beta}\, \partial^\beta u, \overline{\partial^\alpha u}\Big\rangle \, dX = 0. \qquad (6.56)
$$

On the other hand, the S-ellipticity condition on L implies (cf. (4.16))

$$
\begin{aligned}
\mathrm{Re} \sum_{|\alpha|=|\beta|=m} \Big\langle A_{\alpha\beta}\, \partial^\beta u, \overline{\partial^\alpha u}\Big\rangle &= \mathrm{Re}\left(\sum_{|\alpha|=|\beta|=m} \sum_{j,k=1}^{M} a^{\alpha\beta}_{jk}\, \partial^\alpha u_j \overline{\partial^\beta u_k} \right) \\
&\geq C \sum_{|\alpha|=m} \sum_{j=1}^{M} \frac{\alpha!}{m!} |\partial^\alpha u_j|^2. \qquad (6.57)
\end{aligned}
$$

In concert, (6.56) and (6.57) give that $\partial^\alpha u \equiv 0$ in Ω_- for every multi-index $\alpha \in \mathbb{N}_0^n$ with $|\alpha| = m$. As such, an elementary reasoning based on Taylor's formula gives that u is a polynomial of degree at most $m-1$ in Ω_- (which, by assumption, is a connected set). Since from estimate (6.44) we know that $u(X) = \mathscr{O}\big(|X|^{m-n} \log |X|\big)$ as $|X| \to \infty$, we ultimately conclude that

$$
u \text{ is a polynomial of degree at most } m-n \text{ in } \Omega_-. \qquad (6.58)
$$

In particular, this and (5.44) force

$$
\mathrm{tr}_{m-1}(u|_{\Omega_+}) = \mathrm{tr}_{m-1}(u|_{\Omega_-}) \in \dot{\mathscr{P}}_{m-1}(\partial\Omega). \qquad (6.59)
$$

Appealing to (5.136) (for the interior domain $\Omega_+ = \Omega$) yields, in view of (6.41), that

$$
\begin{aligned}
(-1)^{m+1} \sum_{|\alpha|=|\beta|=m} \int_{\Omega} \Big\langle A_{\alpha\beta}\, \partial^\beta u, \overline{\partial^\alpha u}\Big\rangle \, dX &= \Big\langle \partial^A_{\nu_+} u, \overline{\mathrm{tr}_{m-1}(u|_{\Omega_+})}\Big\rangle \\
&= \Big\langle (-\tfrac{1}{2}I + \dot{K}^*)\,\Lambda\,,\, \overline{\mathrm{tr}_{m-1}(u|_{\Omega_+})}\Big\rangle \\
&= \Big\langle (-\tfrac{1}{2}I + \dot{K}^*)\,\Lambda\,,\, \overline{\mathrm{tr}_{m-1}(u|_{\Omega_-})}\Big\rangle \\
&= -\big\langle \Lambda\,,\, \overline{\mathrm{tr}_{m-1}(u|_{\Omega_-})}\big\rangle = 0. \qquad (6.60)
\end{aligned}
$$

In the above calculation, we have used the jump-formula (5.152) in the second equality, the fact that $\mathrm{tr}_{m-1}(u|_{\Omega_+}) = \mathrm{tr}_{m-1}(u|_{\Omega_-})$ in the third equality, the last formula in (6.40) in the fourth equality, and the membership in (6.59) together with (6.33) for the last equality. Hence,

$$\sum_{|\alpha|=|\beta|=m} \int_{\Omega} \left\langle A_{\alpha\beta}\, \partial^{\beta} u, \overline{\partial^{\alpha} u} \right\rangle dX = 0 \tag{6.61}$$

which together with (6.57) imply that

$$\partial^{\alpha} u \equiv 0 \ \text{ in } \Omega_+, \ \text{for every } \alpha \in \mathbb{N}_0^n \text{ with } |\alpha| \leq m-1. \tag{6.62}$$

Consequently, u is locally a polynomial of degree $\leq m-1$ in Ω_+, hence the function $u \in C^{\infty}(\overline{\Omega})$, and

$$\partial_{\nu_+}^{A} u = 0 \tag{6.63}$$

since the definition of the conormal from (5.87) involves taking at least m derivatives on u before restricting to the boundary. In turn, (6.63) and the jump-formula (5.152) imply

$$\left(-\tfrac{1}{2}I + \dot{K}^*\right)\Lambda = 0. \tag{6.64}$$

Collectively, the second formula in (6.40) and (6.64) readily imply that, necessarily, $\Lambda = 0$. This completes the proof of the injectivity of the operator $\frac{1}{2}I + \dot{K}^*$ in (6.39) and finishes the proof of the theorem. □

We are now in a position to identify a context in which the principal-value double multi-layer is actually an invertible operator.

Theorem 6.9. *Let Ω be a bounded Lipschitz domain in $\mathbb{R}^n$ with connected complement, and assume that L is a self-adjoint, S-elliptic homogeneous differential operator of order* $2m$ *with (complex) matrix-valued constant coefficients. Then there exists $\varepsilon > 0$ with the property that whenever*

$$|p-2| + |q-2| + |s-1/2| < \varepsilon \tag{6.65}$$

the operators

$$\tfrac{1}{2}I + \dot{K}^* : \left(\dot{B}^{p,q}_{m-1,s}(\partial\Omega)\right)^* \longrightarrow \left(\dot{B}^{p,q}_{m-1,s}(\partial\Omega)\right)^*, \tag{6.66}$$

$$\tfrac{1}{2}I + \dot{K} : \dot{B}^{p,q}_{m-1,s}(\partial\Omega) \longrightarrow \dot{B}^{p,q}_{m-1,s}(\partial\Omega), \tag{6.67}$$

are invertible.

Proof. The invertibility of the operator in (6.66) when $p = q = 2$ and $s = 1/2$ is a consequence of Theorems 6.4 and 6.8. With this in hand, the fact that this operator continues to be invertible for the larger range of indices as in (6.65) follows from

Theorem 3.38 and [63, Theorem 2.7]. Finally, the claim about the operator (6.67) follows by duality and (3.86). This finishes the proof of the theorem. □

The following companion result to Theorem 6.8 addresses the issue of the injectivity of the single multi-layer. We do so under the assumption that the dimension of the ambient Euclidean space is larger than the order of the differential operator involved. Such an assumption is natural in view of well-known phenomena related to capacity (for example, the harmonic single multi-layer operator in the plane may fail to be injective for certain special values of the logarithmic capacity associated with the domain in question).

Theorem 6.10. *Assume that Ω is a bounded Lipschitz domain in $\mathbb{R}^n$, and suppose that L is a homogeneous differential operator of order $2m$ with (complex) matrix-valued constant coefficients, which satisfies the Legendre–Hadamard ellipticity condition* (4.15).

Then, if $n > 2m$,

$$\text{the operator } \dot{S} : \left(\dot{B}^{2,2}_{m-1,1/2}(\partial\Omega)\right)^* \longrightarrow \dot{B}^{2,2}_{m-1,1/2}(\partial\Omega) \text{ is one-to-one.} \tag{6.68}$$

Proof. The proof parallels that of Theorem 6.8. To get started, consider

$$\Lambda \in \left(\dot{B}^{2,2}_{m-1,1/2}(\partial\Omega)\right)^* \quad \text{such that} \quad \dot{S}\Lambda = \dot{0} \tag{6.69}$$

and introduce $u := \dot{\mathscr{S}}\Lambda$ in $\Omega_\pm$ (cf. (2.11)). By Proposition 5.11, for every function $\psi \in C_c^\infty(\mathbb{R}^n)$ we have

$$\psi u \in B^{2,2}_m(\mathbb{R}^n) = F^{2,2}_m(\mathbb{R}^n) = W^{m,2}(\mathbb{R}^n). \tag{6.70}$$

Moreover, by (5.5), (5.44), and (6.69),

$$Lu = 0 \text{ in } \mathbb{R}^n \setminus \partial\Omega, \text{ and} \tag{6.71}$$

$$\operatorname{tr}^+_{m-1} u = \operatorname{tr}^-_{m-1} u = 0, \tag{6.72}$$

where $\operatorname{tr}^\pm_{m-1}$ denote the multi-trace operators from $\Omega_\pm$. Finally, from (4.29) and (5.1) we deduce that for each $\alpha \in \mathbb{N}_0^n$ there exists $C_\alpha > 0$ such that

$$|\partial^\alpha u(X)| \leq \begin{cases} \dfrac{C_\alpha}{|X|^{n-2m+|\alpha|}} & \text{if either } n \text{ is odd, or } n > 2m, \text{ or if } |\alpha| > 2m-n, \\[2ex] \dfrac{C_\alpha(1+|\log|X||)}{|X|^{n-2m+|\alpha|}} & \text{if } 0 \leq |\alpha| \leq 2m-n, \end{cases} \tag{6.73}$$

uniformly for $|X|$ large. In particular,

$$\big|(\partial^\alpha u)(X)\big| \leq \frac{C}{|X|^{n-m}}, \qquad \text{if } |X| \text{ is large, whenever } \alpha \in \mathbb{N}_0^n \text{ has } |\alpha| = m. \quad (6.74)$$

Moving on, assume that the differential operator L is expressed as in (5.92), and fix a sufficiently large $R > 0$. Then (6.47) holds for every function ψ as in (6.46), and we make the claim that

$$\lim_{R\to\infty} \int_{\partial B_R(0)} \Big\langle \partial^A_{\nu_R} u, \overline{\operatorname{tr}^R_{m-1} u} \Big\rangle \, d\sigma_R = 0. \quad (6.75)$$

To justify this claim, we first note that for every multi-index $\gamma \in \mathbb{N}_0^n$ with $|\gamma| \leq m-1$ we have (cf. (5.87))

$$\big(\partial^A_{\nu_R} u\big)_\gamma = \sum_{|\theta|=2m-|\gamma|-1} \sum_{j=1}^n \nu^R_j B_{j\theta} \left(\partial^\theta u\right)\Big|_{\partial B_R(0)}, \quad (6.76)$$

where ν^R_j is the j-th component of ν^R, the outward unit normal to the ball $B_R(0)$, and $B_{j\theta} \in \mathbb{C}$. Then, since

$$\big\langle \partial^A_{\nu_R} u, \overline{\operatorname{tr}^R_{m-1} u} \big\rangle = \sum_{|\gamma|\leq m-1} \big(\partial^A_{\nu_R} u\big)_\gamma \overline{\partial^\gamma u}\Big|_{\partial B_R(0)}, \quad (6.77)$$

we deduce from on (6.76)–(6.77) and (6.73), that

$$\begin{aligned} \Big|\big\langle \partial^A_{\nu_R} u, \overline{\operatorname{tr}^R_{m-1} u}\big\rangle\Big| &\leq C \sum_{|\gamma|\leq m-1} R^{2m-n-|\gamma|} (\log R)\, R^{2m-n-2m+|\gamma|+1} \\ &\leq C R^{2m-2n+1} \log R \quad \text{on} \quad \partial B_R(0). \end{aligned} \quad (6.78)$$

Now, $\sigma_R\big(\partial B_R(0)\big) \approx R^{n-1}$, and $R^{n-1} \cdot R^{2m-2n+1} \log R = R^{2m-n} \log R$. Since we are currently assuming that $n > 2m$, we have $R^{2m-n} \log R \to 0$ as $R \to \infty$, and the proof of (6.75) is completed.

To proceed, observe that

$$\sum_{|\alpha|=|\beta|=m} \big\langle A_{\alpha\beta}\, \partial^\beta u, \overline{\partial^\alpha u}\big\rangle \text{ is absolutely integrable in } \Omega_-, \quad (6.79)$$

thanks to (6.74), (6.70) and the fact that we are assuming $n > 2m$. Based on this, Lebesgue's Dominated Convergence Theorem, and (6.47), we may write

$$
\begin{aligned}
(-1)^m & \sum_{|\alpha|=|\beta|=m} \int_{\Omega_-} \Big\langle A_{\alpha\beta}\, \partial^\beta u, \overline{\partial^\alpha u}\Big\rangle\, dX \\
&= (-1)^m \lim_{R\to\infty} \sum_{|\alpha|=|\beta|=m} \int_{B_R(0)\setminus\overline{\Omega}} \Big\langle A_{\alpha\beta}\, \partial^\beta u, \overline{\partial^\alpha u}\Big\rangle\, dX \\
&= \big\langle \big(\tfrac{1}{2}I + \dot{K}^*\big)\Lambda\,,\, \overline{\mathrm{tr}_{m-1}^{-} u}\big\rangle - \lim_{R\to\infty} \int_{\partial B_R(0)} \Big\langle \partial^A_{\nu_R} u, \overline{\mathrm{tr}_{m-1}^{R} u}\Big\rangle\, d\sigma_R \\
&= 0, \qquad\qquad (6.80)
\end{aligned}
$$

where the last equality uses (6.72) and (6.75). The above calculation shows

$$
\sum_{|\alpha|=|\beta|=m} \int_{\Omega_-} \Big\langle A_{\alpha\beta}\, \partial^\beta u, \overline{\partial^\alpha u}\Big\rangle\, dX = 0. \qquad (6.81)
$$

At this stage, fix a function $\psi \in C^\infty\big(B_{2R}(0)\big)$ such that $\psi \equiv 1$ near $\overline{B_R(0)}$ and, for each $j \in \mathbb{N}$, define $\psi_j(X) := \psi(X/j)$ for each $X \in \mathbb{R}^n$. Then, for each $j \in \mathbb{N}$, Leibniz's formula yields

$$
\sum_{|\alpha|=|\beta|=m} \int_{\Omega_-} \Big\langle A_{\alpha\beta}\, \partial^\beta(\psi_j u), \overline{\partial^\alpha(\psi_j u)}\Big\rangle\, dX = I_j + II_j, \qquad (6.82)
$$

where

$$
I_j := \sum_{|\alpha|=|\beta|=m} \int_{\Omega_-} \Big\langle A_{\alpha\beta}\, \partial^\beta u, \overline{\partial^\alpha u}\Big\rangle\, \psi_j^2\, dX \qquad (6.83)
$$

and

$$
II_j := \sum_{\substack{|\alpha|=m\\|\beta|=m}} \sum_{\substack{\gamma_1+\delta_1=\beta\\|\gamma_1|\geq 1}} \sum_{\substack{\gamma_2+\delta_2=\alpha\\|\gamma_2|\geq 1}} \frac{\alpha!}{\gamma_1!\delta_1!}\frac{\beta!}{\gamma_2!\delta_2!} \int_{\Omega_-} \Big\langle A_{\alpha\beta}\, \partial^{\gamma_1}\psi_j\, \partial^{\delta_1} u, \overline{\partial^{\gamma_2}\psi_j\, \partial^{\delta_2} u}\Big\rangle\, dX. \qquad (6.84)
$$

Making use of (6.79), Lebesgue's Dominated Convergence Theorem, and (6.81) we may then write

$$
\lim_{j\to\infty} I_j = \sum_{|\alpha|=|\beta|=m} \int_{\Omega_-} \Big\langle A_{\alpha\beta}\, \partial^\beta u, \overline{\partial^\alpha u}\Big\rangle\, dX = 0. \qquad (6.85)
$$

Also, using that

$$\begin{aligned} &\left|\partial^\gamma \psi_j\right| \le C_{\gamma,\psi,R}\, j^{-|\gamma|} \quad \text{and} \quad \operatorname{supp}\left(\partial^\gamma \psi_j\right) \subseteq B_{2jR}(0) \setminus B_{jR}(0), \\ &\qquad \text{for every} \quad \gamma \in \mathbb{N}_0^n \ \text{ with } \ |\gamma| \ge 1 \ \text{ and each } \ j \in \mathbb{N}, \end{aligned} \tag{6.86}$$

for each $j \in \mathbb{N}$ we may estimate

$$\begin{aligned} \left|II_j\right| \le C_{A,m,\psi,R} \sum_{|\alpha|=|\beta|=m} \sum_{\substack{\gamma_1+\delta_1=\beta \\ |\gamma_1|\ge 1}} \sum_{\substack{\gamma_2+\delta_2=\alpha \\ |\gamma_2|\ge 1}} \frac{\alpha!}{\gamma_1!\delta_1!} \frac{\beta!}{\gamma_2!\delta_2!} j^{-|\gamma_1|} j^{-|\gamma_2|} \times \\ \times \int_{B_{2jR}(0)\setminus B_{jR}(0)} \left|\partial^{\delta_1} u(X)\right|\left|\partial^{\delta_1} u(X)\right| dX. \end{aligned} \tag{6.87}$$

Keeping in mind the decay of u described in (6.73) this further yields

$$\left|II_j\right| \le C_{A,m,\psi,R}\, j^{2m-n} |\log j|^2, \qquad \forall\, j \in \mathbb{N}. \tag{6.88}$$

Hence, since we are currently assuming that $n > 2m$,

$$\lim_{j\to\infty} II_j = 0. \tag{6.89}$$

Together, (6.85) and (6.89) imply that

$$\lim_{j\to\infty} \sum_{|\alpha|=|\beta|=m} \int_{\Omega_-} \left\langle A_{\alpha\beta}\, \partial^\beta(\psi_j u), \overline{\partial^\alpha(\psi_j u)}\right\rangle dX = 0. \tag{6.90}$$

Let us momentarily fix an arbitrary $j \in \mathbb{N}$. From (6.70) we know that the function $\psi_j u \in W^{m,p}(\Omega_-)$, while (6.72) guarantees that $\operatorname{tr}^-_{m-1}(\psi_j u) = 0$. As such, Theorem 3.17 ensures the existence of a sequence

$$\{v_{ji}\}_{i\in\mathbb{N}} \subset C_c^\infty(\Omega_-) \ \text{ such that } \ v_{ji} \to \psi_j u \ \text{ in } \ W^{m,2}(\Omega_-) \ \text{ as } \ i \to \infty. \tag{6.91}$$

Based on (6.91) and (6.11), we may therefore write

$$\begin{aligned} &\operatorname{Re} \sum_{|\alpha|=|\beta|=m} \int_{\Omega_-} \left\langle A_{\alpha\beta}\, \partial^\beta(\psi_j u), \overline{\partial^\alpha(\psi_j u)}\right\rangle dX \\ &\qquad = \lim_{i\to\infty} \operatorname{Re} \sum_{|\alpha|=|\beta|=m} \int_{\Omega_-} \left\langle A_{\alpha\beta}\, \partial^\beta v_{ji}, \overline{\partial^\alpha v_{ji}}\right\rangle dX \\ &\qquad = \lim_{i\to\infty} \operatorname{Re} \sum_{|\alpha|=|\beta|=m} \int_{\mathbb{R}^n} \left\langle A_{\alpha\beta}\, \partial^\beta v_{ji}, \overline{\partial^\alpha v_{ji}}\right\rangle dX \end{aligned}$$

$$\geq C \liminf_{i\to\infty} \int_{\mathbb{R}^n} \sum_{|\gamma|=m} |(\partial^\gamma v_{ji})(X)|^2\, dX$$

$$= C \liminf_{i\to\infty} \int_{\Omega_-} \sum_{|\gamma|=m} |(\partial^\gamma v_{ji})(X)|^2\, dX$$

$$= C \int_{\Omega_-} \sum_{|\gamma|=m} |\partial^\gamma(\psi_j u)(X)|^2\, dX. \tag{6.92}$$

In turn, from (6.92) and (6.90) we deduce that

$$\lim_{j\to\infty} \int_{\Omega_-} \sum_{|\gamma|=m} |\partial^\gamma(\psi_j u)(X)|^2\, dX = 0. \tag{6.93}$$

Writing $\partial^\gamma(\psi_j u)$ as $\psi_j \partial^\gamma u$ plus error terms when at least one partial derivative falls on ψ_j, a reasoning very similar to the one that has led to (6.89) shows that

$$\lim_{j\to\infty} \int_{\Omega_-} \sum_{|\gamma|=m} |\partial^\gamma(\psi_j u)(X)|^2\, dX = \int_{\Omega_-} \sum_{|\gamma|=m} |\partial^\gamma u(X)|^2\, dX. \tag{6.94}$$

Altogether, (6.93) and (6.94) prove that

$$\int_{\Omega_-} \sum_{|\gamma|=m} |\partial^\gamma u(X)|^2\, dX = 0. \tag{6.95}$$

This, of course, forces $\partial^\gamma u \equiv 0$ in Ω_- for every $\gamma \in \mathbb{N}_0^n$ with $|\gamma| = m$. Much as in similar situations in the past, this implies that u locally coincides with polynomials of degree at most $m-1$ in the set Ω_-. Granted this, and recalling (6.72), we may finally conclude that

$$u \equiv 0 \ \text{ in } \ \Omega_-. \tag{6.96}$$

At this stage, we appeal to (5.136), (5.152) and (6.71)–(6.72) in order to write that

$$(-1)^{m+1} \sum_{|\alpha|=|\beta|=m} \int_\Omega \Big\langle A_{\alpha\beta}\, \partial^\beta u, \overline{\partial^\alpha u}\Big\rangle\, dX$$

$$= \Big\langle \big(-\tfrac{1}{2}I + \dot{K}^*\big)\, \Lambda\, , \, \overline{\mathrm{tr}^+_{m-1} u}\Big\rangle = 0. \tag{6.97}$$

Hence,

$$\sum_{|\alpha|=|\beta|=m} \int_\Omega \Big\langle A_{\alpha\beta}\, \partial^\beta u, \overline{\partial^\alpha u}\Big\rangle\, dX = 0. \tag{6.98}$$

With this in hand, a similar argument to the one used to prove (6.95) now implies that

$$\partial^\gamma u \equiv 0 \ \text{ in } \Omega_+, \ \text{ for every } \gamma \in \mathbb{N}_0^n \text{ with } |\gamma| \le m-1. \tag{6.99}$$

Consequently, u is locally a polynomial of degree $\le m-1$ in Ω_+ and, thanks to (6.72), we may conclude that

$$u \equiv 0 \text{ in } \Omega_+. \tag{6.100}$$

Finally, from (5.152), (6.96), (6.100), and the definition of the conormal from (5.87), we obtain that

$$\Lambda = \partial^A_{\nu_-} u - \partial^A_{\nu_+} u = 0. \tag{6.101}$$

This proves that the operator $\dot{S}$ in (6.68) is injective. □

At this stage we are able to formulate and prove the following invertibility result for the single multi-layer operator.

Theorem 6.11. *Let Ω be a bounded Lipschitz domain in $\mathbb{R}^n$ and assume that the differential operator L, of order $2m$, is as in* (4.1)–(4.2), *and satisfies the Legendre–Hadamard ellipticity condition* (4.15). *In addition, suppose that L is self-adjoint and that $n > 2m$. Then there exists $\varepsilon > 0$ with the property that*

$$\begin{gathered} \dot{S} : \left(\dot{B}^{p,q}_{m-1,s}(\partial\Omega)\right)^* \longrightarrow \dot{B}^{p',q'}_{m-1,1-s}(\partial\Omega) \ \textit{ is invertible if} \\ |p-2| + |q-2| + \left|s - \tfrac{1}{2}\right| < \varepsilon \ \textit{ and } \ \tfrac{1}{p} + \tfrac{1}{p'} = \tfrac{1}{q} + \tfrac{1}{q'} = 1. \end{gathered} \tag{6.102}$$

Proof. The invertibility of the operator in (6.102) when $p = q = 2$ and $s = 1/2$ is a consequence of Theorems 6.3 and 6.10. Having established this, the fact that this operator continues to be invertible for the larger range of indices as in (6.102) follows from Theorem 3.38 and [63, Theorem 2.7]. □

Under additional geometric hypotheses on the Lipschitz domain, the range of indices for which the invertibility result for the single multi-layer operator from Theorem 6.11 holds increases, as indicated in our next result.

Theorem 6.12. *Let Ω be a bounded Lipschitz domain in $\mathbb{R}^n$ whose outward unit normal ν has the property that*

$$\nu \in \mathrm{vmo}\,(\partial\Omega). \tag{6.103}$$

Also, assume that the differential operator L, of order $2m$, is as in (4.1)–(4.2), *and satisfies the Legendre–Hadamard ellipticity condition* (4.15). *In addition, suppose that L is self-adjoint and that $n > 2m$. Then*

$$\begin{gathered} \dot{S} : \left(\dot{B}^{p,q}_{m-1,s}(\partial\Omega)\right)^* \longrightarrow \dot{B}^{p',q'}_{m-1,1-s}(\partial\Omega) \ \textit{ is invertible} \\ \textit{for every } \ s \in (0,1) \ \textit{ and } \ p,q,p',q' \in (1,\infty) \ \textit{ with } \ \tfrac{1}{p} + \tfrac{1}{p'} = \tfrac{1}{q} + \tfrac{1}{q'} = 1. \end{gathered} \tag{6.104}$$

Proof. The starting point is to recall that, under the condition formulated in (6.103), the inhomogeneous Dirichlet boundary value problem

$$\begin{cases} u \in W^{m,p}_{s-1/p}(\Omega), \\ Lu = v \in \left(\overset{\circ}{W}{}^{m,p'}_{s-1/p'}(\Omega)\right)^*, \\ \operatorname{tr}_{m-1} u = \dot{f} \in \dot{B}^{p,p}_{m-1,1-s}(\partial\Omega), \end{cases} \tag{6.105}$$

has been proved in [76] to be well-posed whenever

$$s \in (0,1) \text{ and } p, p' \in (1,\infty) \text{ satisfy } \tfrac{1}{p} + \tfrac{1}{p'} = 1. \tag{6.106}$$

Fix s, p, p' as in (6.106). As a consequence of the result just quoted, for every bounded Lipschitz domain $\Omega \subset \mathbb{R}^n$ as in the statement of the theorem, there exists a finite geometric constant $C > 0$ with the property that

$$\|u\|_{W^{m,p}_{s-1/p}(\Omega)} \le C\Big(\|Lu\|_{\left(\overset{\circ}{W}{}^{m,p'}_{s-1/p'}(\Omega)\right)^*} + \|\operatorname{tr}_{m-1} u\|_{\dot{B}^{p,p}_{m-1,1-s}(\partial\Omega)}\Big), \tag{6.107}$$

for every function $u \in W^{m,p}_{s-1/p}(\Omega)$. On the other hand, the fact that the conormal derivative operator is bounded in the context described in Proposition 5.26 yields the estimate

$$\left\|\partial^A_\nu(u,w)\right\|_{\left(\dot{B}^{p',p'}_{m-1,1-s}(\partial\Omega)\right)^*} \le C\Big(\|u\|_{W^{m,p}_{s-1/p}(\Omega)} + \|w\|_{\left(W^{m,p'}_{s-1/p'}(\Omega)\right)^*}\Big) \tag{6.108}$$

for every pair

$$\begin{gathered} (u,w) \in W^{m,p}_{s-1/p}(\Omega) \oplus \left(W^{m,p'}_{s-1/p'}(\Omega)\right)^* \\ \text{such that } Lu = w \text{ as distributions in } \Omega. \end{gathered} \tag{6.109}$$

To proceed, fix a function $\psi \in C^\infty_c(\mathbb{R}^n)$ with the property that $\psi \equiv 1$ in a neighborhood of $\overline{\Omega}$, and select $R > 0$ such that $\operatorname{supp}\psi \subset B_R(0)$. Also, define

$$\Omega^R_- := B_R(0) \setminus \overline{\Omega}. \tag{6.110}$$

Let us now fix an arbitrary Whitney array $\Lambda \in \left(\dot{B}^{p,p}_{m-1,s}(\partial\Omega)\right)^*$ and set (cf. (2.11))

$$u^+ := \dot{\mathscr{S}}^+\Lambda \text{ in } \Omega_+, \tag{6.111}$$

$$u^- := \psi\,\dot{\mathscr{S}}^-\Lambda \text{ in } \Omega^R_-. \tag{6.112}$$

From Corollary 5.10, it follows that

$$u^+ \in W^{m,p}_{s-1/p}(\Omega) \text{ and } \left\|u^+\right\|_{W^{m,p}_{s-1/p}(\Omega)} \le C\|\Lambda\|_{\left(\dot{B}^{p,p}_{m-1,s}(\partial\Omega)\right)^*}, \tag{6.113}$$

$$u^- \in W^{m,p}_{s-1/p}(\Omega^R_-) \text{ and } \left\|u^-\right\|_{W^{m,p}_{s-1/p}(\Omega^R_-)} \le C\|\Lambda\|_{\left(\dot{B}^{p,p}_{m-1,s}(\partial\Omega)\right)^*}, \tag{6.114}$$

for some finite $C > 0$ independent of Λ. Set

$$w^+ := Lu^+ = 0, \text{ and } w^- := Lu^- \in C^\infty_c(\Omega^R_-). \tag{6.115}$$

In particular, the pairs $(u^\pm, w^\pm)$ are as in (6.109) relative to the domains Ω_+ and Ω^R_-. Moreover, as is easily seen from the membership in (6.115), the assignment

$$\left(\dot{B}^{p,p}_{m-1,s}(\partial\Omega)\right)^* \ni \Lambda \mapsto w^- \in \left(W^{m,p'}_{s-1/p'}(\Omega^R_-)\right)^* \text{ is a compact operator,} \tag{6.116}$$

subsequently denoted simply by Comp. We may then estimate

$$\begin{aligned}
\|\Lambda\|_{\left(\dot{B}^{p,p}_{m-1,s}(\partial\Omega)\right)^*} &= \left\|\partial^A_{\nu_+}(u^+, w^+) - \partial^A_{\nu_-}(u^-, w^-)\right\|_{\left(\dot{B}^{p,p}_{m-1,s}(\partial\Omega)\right)^*} \\
&\le \left\|\partial^A_{\nu_+}(u^+, w^+)\right\|_{\left(\dot{B}^{p,p}_{m-1,s}(\partial\Omega)\right)^*} + \left\|\partial^A_{\nu_-}(u^-, w^-)\right\|_{\left(\dot{B}^{p,p}_{m-1,s}(\partial\Omega)\right)^*} \\
&\le C\|u^+\|_{W^{m,p}_{s-1/p}(\Omega_+)} + C\|u^-\|_{W^{m,p}_{s-1/p}(\Omega^R_-)} + C\|w^-\|_{\left(W^{m,p'}_{s-1/p'}(\Omega^R_-)\right)^*} \\
&\le C\left\|\mathrm{tr}^+_{m-1}u^+\right\|_{\dot{B}^{p,p}_{m-1,1-s}(\partial\Omega)} + C\left\|\mathrm{tr}^-_{m-1}u^-\right\|_{\dot{B}^{p,p}_{m-1,1-s}(\partial\Omega)} \\
&\quad + C\|w^-\|_{\left(W^{m,p'}_{s-1/p'}(\Omega^R_-)\right)^*} \\
&= C\left\|\dot{S}\Lambda\right\|_{\dot{B}^{p,p}_{m-1,1-s}(\partial\Omega)} + \|\mathrm{Comp}\,\Lambda\|.
\end{aligned} \tag{6.117}$$

Above, the first equality is the jump formula (5.153), the second inequality follows from the estimates for the conormal derivative operator corresponding to (6.108) written for Ω_+ and Ω^R_-, the third inequality is a consequence of the "well-posedness" estimate (6.107) written, again, for Ω_+ and Ω^R_- (it is important to note that Ω^R_- continues to be a bounded Lipschitz domain whose outward unit normal has vanishing mean oscillations), while the last equality uses (5.44) and (6.116).

In summary, the proof so far shows that there exist a constant $C \in (0, \infty)$ along with a Banach space-valued compact operator Comp defined on $\left(\dot{B}^{p,p}_{m-1,s}(\partial\Omega)\right)^*$, with the property that for every functional $\Lambda \in \left(\dot{B}^{p,p}_{m-1,s}(\partial\Omega)\right)^*$ there holds

$$\|\dot{S}\Lambda\|_{\dot{B}^{p,p}_{m-1,1-s}(\partial\Omega)} + \|\mathrm{Comp}\,\Lambda\| \geq C\|\Lambda\|_{\left(\dot{B}^{p,p}_{m-1,s}(\partial\Omega)\right)^*}. \tag{6.118}$$

Consequently, for each $s \in (0,1)$ and $p, p' \in (1,\infty)$ with $1/p + 1/p' = 1$, the operator

$$\dot{S} : \left(\dot{B}^{p,p}_{m-1,s}(\partial\Omega)\right)^* \longrightarrow \dot{B}^{p',p'}_{m-1,1-s}(\partial\Omega) \tag{6.119}$$

has closed range and finite dimensional kernel. Since Theorem 6.11 implies that this operator also has a dense range, we ultimately deduce that the single multi-layer in (6.119) is onto. At this stage, given that $\dot{S}$ in (6.119) is always surjective, acts between two complex interpolation scales (cf. Theorem 3.38), and is invertible when $p = 2$ and $s = 1/2$ (cf. Theorem 6.11), the global extrapolation result proved in [63, Theorem 2.10] applies and gives that the claim in (6.104) is valid when $p = q$. In its most general form, this result then follows from what we have just proved and the real interpolation formula from Theorem 3.38. □

Let us now record the following significant corollary of Theorem 6.12.

Corollary 6.13. *Let Ω be a bounded C^1 domain in $\mathbb{R}^n$, and assume that the differential operator L, of order $2m$, is as in* (4.1)–(4.2), *and satisfies the Legendre–Hadamard ellipticity condition* (4.15). *In addition, suppose that L is self-adjoint and that $n > 2m$. Then*

$$\dot{S} : \left(\dot{B}^{p,q}_{m-1,s}(\partial\Omega)\right)^* \longrightarrow \dot{B}^{p',q'}_{m-1,1-s}(\partial\Omega) \text{ is invertible} \tag{6.120}$$

for every $s \in (0,1)$ and $p, q, p', q' \in (1,\infty)$ with $\frac{1}{p} + \frac{1}{p'} = \frac{1}{q} + \frac{1}{q'} = 1$.

Proof. This follows directly from Theorem 6.12 after observing that, the outward unit normal ν to a bounded C^1 domain $\Omega \subset \mathbb{R}^n$ is a continuous function and, as such, condition (6.103) is satisfied. □

As indicated below, Theorem 6.12 may be further refined, by targeting a specific triplet of indices p, q, s for which the boundary single multi-layer is invertible in the context described in the first line of (6.104).

Theorem 6.14. *Let Ω be a bounded Lipschitz domain in $\mathbb{R}^n$. Assume that the differential operator L, of order $2m$, is as in* (4.1)–(4.2), *and satisfies the Legendre–Hadamard ellipticity condition* (4.15). *In addition, suppose that L is self-adjoint and that $n > 2m$. Finally, suppose that*

$$s \in (0,1) \text{ and } p, q, p', q' \in (1,\infty) \text{ satisfy } \frac{1}{p} + \frac{1}{p'} = \frac{1}{q} + \frac{1}{q'} = 1. \tag{6.121}$$

Then there exists $\varepsilon > 0$, depending only on the indices s, p, q, the Lipschitz character of Ω, and the differential operator L, with the property that if the outward unit normal ν to Ω satisfies

$$\mathrm{dist}_{\mathrm{bmo}(\partial\Omega)}\Big(\nu\,,\,\mathrm{vmo}\,(\partial\Omega)\Big) < \varepsilon, \tag{6.122}$$

where

$$\mathrm{dist}_{\mathrm{bmo}(\partial\Omega)}\Big(\nu\,,\,\mathrm{vmo}\,(\partial\Omega)\Big) := \inf\Big\{\|\nu-\phi\|_{\mathrm{bmo}(\partial\Omega)} : \phi\in\mathrm{vmo}\,(\partial\Omega)\Big\}, \tag{6.123}$$

then

$$\dot{S} : \Big(\dot{B}^{p,q}_{m-1,s}(\partial\Omega)\Big)^* \longrightarrow \dot{B}^{p',q'}_{m-1,1-s}(\partial\Omega) \ \ \textit{is invertible.} \tag{6.124}$$

As observed in [76], an alternative equivalent way of expressing the condition formulated in (6.122) is

$$\limsup_{r\to 0^+}\Big\{\sup_{X\in\partial\Omega} ⨍_{B(X,r)\cap\partial\Omega}⨍_{B(X,r)\cap\partial\Omega}\big|\nu(Y)-\nu(Z)\big|\,d\sigma(Y)d\sigma(Z)\Big\} < \varepsilon. \tag{6.125}$$

Proof of Theorem 6.14. The proof largely parallels that of Theorem 6.12 keeping in mind that, as has been shown in [76], the well-posedness of the inhomogeneous Dirichlet boundary value problem (6.105) for a particular choice of s, p, q still holds provided the weaker condition (6.122) is satisfied (rather than (6.103)), in the manner described in the statement of the theorem. □

The following companion result to Theorem 6.8 describes the null-space of the operator $-\frac{1}{2}I+\dot{K}$.

Theorem 6.15. *Let Ω be a bounded, connected, Lipschitz domain in $\mathbb{R}^n$, whose complement is connected, and assume that L is a homogeneous, S-elliptic, differential operator of order $2m$, $m\in\mathbb{N}$, with constant (complex) matrix-valued coefficients. Then*

$$\Big\{\dot{f}\in\dot{B}^{2,2}_{m-1,1/2}(\partial\Omega) : \big(-\tfrac{1}{2}I+\dot{K}\big)\dot{f}=0\Big\} = \dot{\mathscr{P}}_{m-1}(\partial\Omega). \tag{6.126}$$

Moreover, the operator

$$-\tfrac{1}{2}I+\dot{K} : \dot{B}^{2,2}_{m-1,1/2}(\partial\Omega)\Big/\dot{\mathscr{P}}_{m-1}(\partial\Omega) \longrightarrow \dot{B}^{2,2}_{m-1,1/2}(\partial\Omega)\Big/\dot{\mathscr{P}}_{m-1}(\partial\Omega) \tag{6.127}$$

is well-defined, linear, bounded and injective. As a corollary, the operator

$$\begin{aligned}-\tfrac{1}{2}I+\dot{K}^* : \Big(\dot{B}^{2,2}_{m-1,1/2}(\partial\Omega)\Big/\dot{\mathscr{P}}_{m-1}(\partial\Omega)\Big)^*&\\ \longrightarrow \Big(\dot{B}^{2,2}_{m-1,1/2}(\partial\Omega)\Big/\dot{\mathscr{P}}_{m-1}(\partial\Omega)\Big)^*&\end{aligned} \tag{6.128}$$

is well-defined, linear, bounded and has dense range.

Proof. The right-to-left inclusion in (6.127) has been already established in Proposition 4.17. To get started in the opposite direction, denote by $A = (A_{\alpha\beta})_{|\alpha|=|\beta|=m}$ a strictly positive definite coefficient tensor for the operator L. Also, assume that

$$\dot{f} \in \dot{B}^{2,2}_{m-1,1/2}(\partial\Omega) \quad \text{is such that} \quad \big(-\tfrac{1}{2}I + \dot{K}\big)\dot{f} = 0, \tag{6.129}$$

and introduce $u := \dot{\mathscr{D}}^{\pm}\dot{f}$ in $\Omega_{\pm}$ (cf. (2.11)), where $\dot{\mathscr{D}}^{\pm}$ are the versions of the double multi-layer associated with L in $\Omega_{\pm}$. Let $\mathrm{tr}^{\pm}_{m-1}$ be the multi-trace operators acting from $\Omega_{\pm}$, and denote by $\partial^{A}_{\nu_{\pm}}$ the conormals acting from $\Omega_{\pm}$. Then, by (4.58), (4.250), and (6.129),

$$Lu = 0 \ \text{ in } \mathbb{R}^n \setminus \partial\Omega, \tag{6.130}$$

$$\mathrm{tr}^{-}_{m-1}u = 0. \tag{6.131}$$

Furthermore, by (5.175),

$$\partial^{A}_{\nu_{+}}\dot{\mathscr{D}}^{+}\dot{f} = \partial^{A}_{\nu_{-}}\dot{\mathscr{D}}^{-}\dot{f} \ \text{ in } \Big(\dot{B}^{2,2}_{m-1,1/2}(\partial\Omega)\Big)^{*}. \tag{6.132}$$

In addition, Theorem 4.20 gives that for every $\psi \in C^{\infty}_{c}(\mathbb{R}^n)$ there holds

$$\psi u \in B^{2,2}_{m}(\mathbb{R}^n) = F^{2,2}_{m}(\mathbb{R}^n) = W^{m,2}(\mathbb{R}^n). \tag{6.133}$$

Finally, by (4.57) and (4.29), given $\alpha \in \mathbb{N}^n_0$ we have for large $|X|$:

$$\big|(\partial^{\alpha}u)(X)\big| \leq \begin{cases} C|X|^{m-n-|\alpha|}\log|X| & \text{if } \ 0 \leq |\alpha| \leq m-n, \\ C|X|^{m-n-|\alpha|} & \text{if } |\alpha| > m-n, \ \text{or } n \text{ odd, or } n > 2m, \end{cases} \tag{6.134}$$

Granted the decay of u^{-} from (6.134), the same type of argument that has led to (6.55) gives

$$\sum_{|\alpha|=|\beta|=m} \int_{\Omega_{-}} \Big\langle A_{\alpha\beta}\,\partial^{\beta}u, \overline{\partial^{\alpha}u}\Big\rangle\, dX = \Big\langle \partial^{A}_{\nu_{-}}\dot{\mathscr{D}}^{-}\dot{f}\,,\, \overline{\mathrm{tr}^{-}_{m-1}u}\Big\rangle = 0, \tag{6.135}$$

where the last equality uses (6.131). Hence,

$$\sum_{|\alpha|=|\beta|=m} \int_{\Omega_{-}} \Big\langle A_{\alpha\beta}\,\partial^{\beta}u, \overline{\partial^{\alpha}u}\Big\rangle\, dX = 0 \tag{6.136}$$

which, in light of (6.57) (itself a consequence of the S-ellipticity of L), implies that

$$u \text{ is a polynomial of degree at most } m-1 \text{ in } \Omega_{-}. \tag{6.137}$$

As such, $\partial^A_{\nu_-} u = 0$, and by (6.132) we also obtain

$$\partial^A_{\nu_+} u = 0. \tag{6.138}$$

Next, working in Ω_+, a similar argument based on (5.132) and (6.138) shows that

$$u \text{ is a polynomial of degree at most } m-1 \text{ in } \Omega_+. \tag{6.139}$$

At this stage, based on (6.137), (6.139), and (4.250), we may write

$$\dot{f} = \mathrm{tr}^+_{m-1} u - \mathrm{tr}^-_{m-1} u \in \dot{\mathscr{P}}_{m-1}(\partial\Omega) - \dot{\mathscr{P}}_{m-1}(\partial\Omega) \subseteq \dot{\mathscr{P}}_{m-1}(\partial\Omega), \tag{6.140}$$

as desired. This completes the proof of (6.126).

Having established (6.126), we may then conclude that the operator in (6.127) is well-defined, linear and bounded. There remains to show that this operator is also injective. To this end, it suffices to prove that

$$\left.\begin{array}{c} \dot{f} \in \dot{B}^{2,2}_{m-1,1/2}(\partial\Omega) \\ \text{and} \\ \left(-\tfrac{1}{2}I + \dot{K}\right)\dot{f} = 0 \end{array}\right\} \Longrightarrow \dot{f} \in \dot{\mathscr{P}}_{m-1}(\partial\Omega). \tag{6.141}$$

Let $\dot{f}$ be as in the left-hand side of (6.141) and consider $u := \dot{\mathscr{D}}^{\pm}\dot{f}$ in $\Omega_{\pm}$. Thus, there exists $P \in \mathscr{P}_{m-1}$ with the property that

$$\mathrm{tr}^-_{m-1} u = \mathrm{tr}^-_{m-1} P \in \dot{\mathscr{P}}_{m-1}(\partial\Omega). \tag{6.142}$$

Then (6.132)–(6.134) continue to hold in the present setting, and we may write

$$\begin{aligned} \sum_{|\alpha|=|\beta|=m} \int_{\Omega_-} \left\langle A_{\alpha\beta}\, \partial^\beta u, \overline{\partial^\alpha u}\right\rangle dX &= \left\langle \partial^A_{\nu_-} \dot{\mathscr{D}}^- \dot{f}\,,\, \overline{\mathrm{tr}^-_{m-1} u}\right\rangle \\ &= \left\langle \partial^A_{\nu_+} \dot{\mathscr{D}}^+ \dot{f}\,,\, \overline{\mathrm{tr}^-_{m-1} P}\right\rangle \\ &= \sum_{|\alpha|=|\beta|=m} \int_{\Omega_+} \left\langle A_{\alpha\beta}\, \partial^\beta u, \overline{\partial^\alpha P}\right\rangle dX \\ &= 0, \end{aligned} \tag{6.143}$$

where the first equality above follows from the first equality in (6.135), the second equality above is a consequence of (6.132) and (6.142), the third equality above uses Proposition 5.24 (cf. (5.136)), and the last equality above is implied by the fact that $P \in \mathscr{P}_{m-1}$. As before, (6.143) allows us to conclude that (6.137) holds and, further, that (6.138) holds. Having established this, then (6.139) follows as before which, in turn, yields (6.140), as desired. This finishes the proof of the injectivity of the operator (6.127). □

In the last part of this section we estimate the spectral radius of the operator $\dot{K}$ in an appropriate context.

Lemma 6.16. *Let Ω be a bounded Lipschitz domain in $\mathbb{R}^n$, and assume that L is a homogeneous, differential operator of order $2m$, $m \in \mathbb{N}$, associated as in* (4.1) *with a constant (complex) matrix-valued tensor coefficient which is self-adjoint and satisfies the Legendre–Hadamard ellipticity condition* (4.15)*, as well as the semi-positivity condition* (4.20).

Suppose that $\lambda \in \mathbb{C}$ is such that there exists

$$\Lambda \in \left(\dot{B}^{2,2}_{m-1,1/2}(\partial\Omega)\right)^*, \quad \Lambda \neq 0, \text{ with } \left(\lambda I - \dot{K}^*\right)\Lambda = 0. \tag{6.144}$$

Then necessarily $\lambda \in [-\frac{1}{2}, \frac{1}{2}]$.

Proof. If $\lambda = \frac{1}{2}$ then there is nothing to prove, so assume in what follows that $\lambda \in \mathbb{C} \setminus \{\frac{1}{2}\}$. Granted this, and assuming that Λ is as in (6.144), we deduce from Lemma 6.6 that

$$\langle \Lambda, \dot{f} \rangle = 0 \quad \text{for every} \quad \dot{f} \in \dot{\mathscr{P}}_{m-1}(\partial\Omega). \tag{6.145}$$

To proceed, set $u := \dot{\mathscr{S}}\Lambda$ in $\Omega_\pm$ (cf. (2.11)). By Proposition 5.11, for every function $\psi \in C^\infty_c(\mathbb{R}^n)$ we have

$$\psi u \in B^{2,2}_m(\mathbb{R}^n) = F^{2,2}_m(\mathbb{R}^n) = W^{m,2}(\mathbb{R}^n). \tag{6.146}$$

Moreover, Lemmas 6.6, 6.7 and the current assumption on λ ensure that u satisfies the decay condition formulated in (6.44). In turn, this decay condition allows us to write, much as in (6.80), that

$$(-1)^m \sum_{|\alpha|=|\beta|=m} \int_{\Omega_-} \left\langle A_{\alpha\beta}\, \partial^\beta u, \overline{\partial^\alpha u} \right\rangle dX = \left\langle \left(\tfrac{1}{2}I + \dot{K}^*\right)\Lambda\,, \overline{\mathrm{tr}^-_{m-1} u} \right\rangle. \tag{6.147}$$

Furthermore, as in (6.60), we have

$$(-1)^{m+1} \sum_{|\alpha|=|\beta|=m} \int_{\Omega_+} \left\langle A_{\alpha\beta}\, \partial^\beta u, \overline{\partial^\alpha u} \right\rangle dX = \left\langle \left(-\tfrac{1}{2}I + \dot{K}^*\right)\Lambda\,, \overline{\mathrm{tr}^+_{m-1} u} \right\rangle. \tag{6.148}$$

Using the last equality in (6.144) along with the fact that $\mathrm{tr}^+_{m-1} u = \mathrm{tr}^-_{m-1} u = \dot{S}\Lambda$, we may then write

$$
\begin{aligned}
0 &= \Big\langle (\lambda I - \dot{K}^*)\,\Lambda\,,\,\overline{\dot{S}\Lambda}\Big\rangle \\
&= -(\lambda + \tfrac{1}{2})\Big\langle (-\tfrac{1}{2}I + \dot{K}^*)\,\Lambda\,,\,\overline{\mathrm{tr}^{+}_{m-1}u}\Big\rangle + (\lambda - \tfrac{1}{2})\big\langle (\tfrac{1}{2}I + \dot{K}^*)\,\Lambda\,,\,\overline{\mathrm{tr}^{-}_{m-1}u}\big\rangle \\
&= (-1)^{m+1}(\lambda + \tfrac{1}{2}) \sum_{|\alpha|=|\beta|=m} \int_{\Omega_-} \Big\langle A_{\alpha\beta}\,\partial^\beta u, \overline{\partial^\alpha u}\Big\rangle\, dX \\
&\quad +(-1)^{m+1}(\lambda - \tfrac{1}{2}) \sum_{|\alpha|=|\beta|=m} \int_{\Omega_+} \Big\langle A_{\alpha\beta}\,\partial^\beta u, \overline{\partial^\alpha u}\Big\rangle\, dX.
\end{aligned}
\tag{6.149}
$$

The fact that the operator L is self-adjoint implies that the integrands are (pointwise) real. As such, taking the imaginary parts of the most extreme sides of (6.149) yields (after multiplication by $(-1)^{m+1}$)

$$
\mathrm{Im}\,\lambda \sum_{|\alpha|=|\beta|=m} \int_{\mathbb{R}^n} \Big\langle A_{\alpha\beta}\,\partial^\beta u, \overline{\partial^\alpha u}\Big\rangle\, dX = 0. \tag{6.150}
$$

To proceed observe that, thanks to (6.45),

$$
\sum_{|\alpha|=|\beta|=m} \big\langle A_{\alpha\beta}\,\partial^\beta u, \overline{\partial^\alpha u}\big\rangle \text{ is absolutely integrable in } \mathbb{R}^n. \tag{6.151}
$$

As in the past, fix a function $\psi \in C^\infty\big(B_{2R}(0)\big)$ such that $\psi \equiv 1$ near $\overline{B_R(0)}$ and, for each $j \in \mathbb{N}$, define $\psi_j(X) := \psi(X/j)$ for each $X \in \mathbb{R}^n$. Then, for each $j \in \mathbb{N}$, Leibniz's formula yields

$$
\mathrm{Im}\,\lambda \sum_{|\alpha|=|\beta|=m} \int_{\mathbb{R}^n} \Big\langle A_{\alpha\beta}\,\partial^\beta (\psi_j u), \overline{\partial^\alpha (\psi_j u)}\Big\rangle\, dX = I_j + II_j, \tag{6.152}
$$

where

$$
I_j := \mathrm{Im}\,\lambda \sum_{|\alpha|=|\beta|=m} \int_{\mathbb{R}^n} \Big\langle A_{\alpha\beta}\,\partial^\beta u, \overline{\partial^\alpha u}\Big\rangle\, \psi_j^2\, dX \tag{6.153}
$$

and

$$
\begin{aligned}
II_j &:= \mathrm{Im}\,\lambda \times \\
&\times \sum_{\substack{|\alpha|=m\\|\beta|=m}} \sum_{\substack{\gamma_1+\delta_1=\beta\\|\gamma_1|\geq 1}} \sum_{\substack{\gamma_2+\delta_2=\alpha\\|\gamma_2|\geq 1}} \frac{\alpha!}{\gamma_1!\delta_1!}\frac{\beta!}{\gamma_2!\delta_2!} \int_{\mathbb{R}^n} \Big\langle A_{\alpha\beta}\,\partial^{\gamma_1}\psi_j\,\partial^{\delta_1}u, \overline{\partial^{\gamma_2}\psi_j\,\partial^{\delta_2}u}\Big\rangle\, dX.
\end{aligned}
\tag{6.154}
$$

Making use of (6.151), Lebesgue's Dominated Convergence Theorem, and (6.150) we may then write

$$\lim_{j\to\infty} I_j = \operatorname{Im}\lambda \sum_{|\alpha|=|\beta|=m} \int_{\mathbb{R}^n} \Big\langle A_{\alpha\beta}\, \partial^\beta u, \overline{\partial^\alpha u}\Big\rangle\, dX = 0. \tag{6.155}$$

Also, using that

$$\begin{gathered} \big|\partial^\gamma \psi_j\big| \leq C_{\gamma,\psi,R}\, j^{-|\gamma|} \quad \text{and} \quad \operatorname{supp}\big(\partial^\gamma \psi_j\big) \subseteq B_{2jR}(0) \setminus B_{jR}(0), \\ \text{for every} \quad \gamma \in \mathbb{N}_0^n \quad \text{with} \quad |\gamma| \geq 1 \quad \text{and each} \quad j \in \mathbb{N}, \end{gathered} \tag{6.156}$$

for each $j \in \mathbb{N}$ we may estimate

$$\begin{aligned} \big|II_j\big| \leq C_{A,m,\psi,R}\, |\operatorname{Im}\lambda| &\sum_{|\alpha|=|\beta|=m} \sum_{\substack{\gamma_1+\delta_1=\beta \\ |\gamma_1|\geq 1}} \sum_{\substack{\gamma_2+\delta_2=\alpha \\ |\gamma_2|\geq 1}} \frac{\alpha!}{\gamma_1!\delta_1!} \frac{\beta!}{\gamma_2!\delta_2!} j^{-|\gamma_1|} j^{-|\gamma_2|} \times \\ &\times \int_{B_{2jR}(0)\setminus B_{jR}(0)} \big|\partial^{\delta_1} u(X)\big|\big|\partial^{\delta_1} u(X)\big|\, dX. \end{aligned} \tag{6.157}$$

Keeping in mind the decay of u described in (6.44) this further yields

$$\big|II_j\big| \leq C_{A,m,\psi,R}\, |\operatorname{Im}\lambda|\, j^{-n} |\log j|^2, \qquad \forall\, j \in \mathbb{N}, \tag{6.158}$$

hence

$$\lim_{j\to\infty} II_j = 0. \tag{6.159}$$

Together, (6.155) and (6.159) imply that

$$\lim_{j\to\infty} \left\{ |\operatorname{Im}\lambda| \sum_{|\alpha|=|\beta|=m} \int_{\mathbb{R}^n} \Big\langle A_{\alpha\beta}\, \partial^\beta(\psi_j u), \overline{\partial^\alpha(\psi_j u)}\Big\rangle\, dX \right\} = 0. \tag{6.160}$$

Since from (6.146) we know that $\psi_j u \in W^{m,p}(\mathbb{R}^n)$ for every $j \in \mathbb{N}$, it follows from (6.160) and (6.12) that

$$\lim_{j\to\infty} \left\{ |\operatorname{Im}\lambda| \int_{\mathbb{R}^n} \sum_{|\gamma|=m} |\partial^\gamma(\psi_j u)(X)|^2\, dX \right\} = 0. \tag{6.161}$$

Writing $\partial^\gamma(\psi_j u)$ as $\psi_j \partial^\gamma u$ plus error terms when at least one partial derivative falls on ψ_j, a reasoning very similar to the one that has led to (6.159) shows that

$$\lim_{j\to\infty}\int_{\mathbb{R}^n}\sum_{|\gamma|=m}|\partial^\gamma(\psi_j u)(X)|^2\,dX=\int_{\mathbb{R}^n}\sum_{|\gamma|=m}|\partial^\gamma u(X)|^2\,dX. \tag{6.162}$$

In concert, (6.161) and (6.162) prove that

$$|\operatorname{Im}\lambda|\int_{\mathbb{R}^n}\sum_{|\gamma|=m}|\partial^\gamma u(X)|^2\,dX=0. \tag{6.163}$$

In the case when $\operatorname{Im}\lambda\neq 0$, this forces $\partial^\gamma u\equiv 0$ in $\mathbb{R}^n$ for every $\gamma\in\mathbb{N}_0^n$ with $|\gamma|=m$. Much as in similar situations in the past, this implies that u is a polynomial of degree at most $m-1$ in $\mathbb{R}^n$. Granted this, we may finally conclude that

$$\Lambda=\partial^A_{\nu_-}u-\partial^A_{\nu_+}u=0\ \text{ in }\ \left(\dot{B}^{2,2}_{m-1,1/2}(\partial\Omega)\right)^*, \tag{6.164}$$

in contradiction with the second condition in (6.144). This argument shows that necessarily $\operatorname{Im}\lambda=0$, i.e., that

$$\lambda\in\mathbb{R}. \tag{6.165}$$

If we now define

$$\Phi_\pm:=\sum_{|\alpha|=|\beta|=m}\int_{\Omega_\pm}\left\langle A_{\alpha\beta}\,\partial^\beta u,\overline{\partial^\alpha u}\right\rangle dX, \tag{6.166}$$

it follows that

$$\Phi_+\text{ and }\Phi_-\text{ cannot be simultaneously zero} \tag{6.167}$$

since otherwise the same type of argument, starting with (6.150) and leading to (6.164), would yield a contradiction as before. Moreover, the self-adjointness of the operator L together with the semi-positivity condition (4.20) imply that, on the one hand,

$$\Phi_\pm\in[0,\infty). \tag{6.168}$$

On the other hand, (6.149) entails

$$(\lambda+\tfrac{1}{2})\Phi_-+(\lambda-\tfrac{1}{2})\Phi_+=0. \tag{6.169}$$

Note that if $\lambda\in\mathbb{R}\setminus[-\frac{1}{2},\frac{1}{2}]$ then $\lambda+\frac{1}{2}$ and $\lambda-\frac{1}{2}$ are nonzero real numbers of the same sign. However, in light of (6.169), (6.168) and (6.167), this is an impossibility. Together with (6.165), this contraction proves that necessarily $\lambda\in[-\frac{1}{2},\frac{1}{2}]$, as desired. □

Our main spectral result for the operator $\dot{K}$ is contained in the next theorem.

Theorem 6.17. *Let Ω be a bounded, connected, Lipschitz domain in $\mathbb{R}^n$, whose complement is connected. Also, assume that L is a homogeneous differential operator of order 2m, $m \in \mathbb{N}$, associated with a constant (complex) matrix-valued tensor coefficient as in (4.1) which is self-adjoint and satisfies the S-ellipticity condition (4.16). Then*

$$\begin{gathered}\textit{the spectrum of } \dot{K} \textit{ acting on } \dot{B}^{2,2}_{m-1,1/2}(\partial\Omega)\big/\dot{\mathscr{P}}_{m-1}(\partial\Omega) \\ \textit{is included in the interval } (-1/2, 1/2).\end{gathered} \tag{6.170}$$

Consequently,

$$\begin{gathered}\textit{the spectral radius of } \dot{K} \textit{ acting on } \dot{B}^{2,2}_{m-1,1/2}(\partial\Omega)\big/\dot{\mathscr{P}}_{m-1}(\partial\Omega) \\ \textit{is strictly less than } \tfrac{1}{2}.\end{gathered} \tag{6.171}$$

Moreover, there exists some small $\varepsilon > 0$ with the property that whenever

$$|p-2| + |q-2| + |s-1/2| < \varepsilon \tag{6.172}$$

one has

$$\Big\{\dot{f} \in \dot{B}^{p,q}_{m-1,s}(\partial\Omega) : \big(-\tfrac{1}{2}I + \dot{K}\big)\dot{f} = 0\Big\} = \dot{\mathscr{P}}_{m-1}(\partial\Omega), \tag{6.173}$$

and the operators

$$\pm\tfrac{1}{2}I + \dot{K} : \dot{B}^{p,q}_{m-1,s}(\partial\Omega)\big/\dot{\mathscr{P}}_{m-1}(\partial\Omega) \longrightarrow \dot{B}^{p,q}_{m-1,s}(\partial\Omega)\big/\dot{\mathscr{P}}_{m-1}(\partial\Omega), \tag{6.174}$$

and

$$\begin{aligned}\pm\tfrac{1}{2}I + \dot{K}^* : \Big(\dot{B}^{p,q}_{m-1,s}(\partial\Omega)\big/\dot{\mathscr{P}}_{m-1}(\partial\Omega)\Big)^* \\ \longrightarrow \Big(\dot{B}^{p,q}_{m-1,s}(\partial\Omega)\big/\dot{\mathscr{P}}_{m-1}(\partial\Omega)\Big)^*\end{aligned} \tag{6.175}$$

are invertible.

Proof. Assume that $\lambda \in \mathbb{C} \setminus (-\frac{1}{2}, \frac{1}{2})$ belongs to the spectrum of the operator

$$\dot{K}^* : \Big(\dot{B}^{2,2}_{m-1,1/2}(\partial\Omega)\big/\dot{\mathscr{P}}_{m-1}(\partial\Omega)\Big)^* \to \Big(\dot{B}^{2,2}_{m-1,1/2}(\partial\Omega)\big/\dot{\mathscr{P}}_{m-1}(\partial\Omega)\Big)^*. \tag{6.176}$$

The last part in Theorem 6.4 implies that

$$\lambda I - \dot{K}^* \text{ is Fredholm with index zero} \\ \text{on } \Big(\dot{B}^{2,2}_{m-1,1/2}(\partial\Omega)\Big/\dot{\mathscr{P}}_{m-1}(\partial\Omega)\Big)^* \tag{6.177}$$

which, keeping in mind the significance of λ, implies that

$$\lambda I - \dot{K}^* \text{ is not injective on } \Big(\dot{B}^{2,2}_{m-1,1/2}(\partial\Omega)\Big/\dot{\mathscr{P}}_{m-1}(\partial\Omega)\Big)^*. \tag{6.178}$$

Going further, let

$$\pi : \dot{B}^{2,2}_{m-1,1/2}(\partial\Omega) \longrightarrow \dot{B}^{2,2}_{m-1,1/2}(\partial\Omega)\Big/\dot{\mathscr{P}}_{m-1}(\partial\Omega) \tag{6.179}$$

denote the canonical projection, taking an arbitrary Whitney array $\dot{f} \in \dot{B}^{2,2}_{m-1,1/2}(\partial\Omega)$ into its equivalence class $[\dot{f}] \in \dot{B}^{2,2}_{m-1,1/2}(\partial\Omega)\Big/\dot{\mathscr{P}}_{m-1}(\partial\Omega)$. Since π is surjective, its adjoint

$$\pi^* : \Big(\dot{B}^{2,2}_{m-1,1/2}(\partial\Omega)\Big/\dot{\mathscr{P}}_{m-1}(\partial\Omega)\Big)^* \longrightarrow \Big(\dot{B}^{2,2}_{m-1,1-s}(\partial\Omega)\Big)^* \tag{6.180}$$

is injective. Moreover, we have the commutative diagram

$$\begin{array}{ccc}
\Big(\dot{B}^{2,2}_{m-1,1-s}(\partial\Omega)\Big)^* & \xrightarrow{\ \lambda I - \dot{K}^*\ } & \Big(\dot{B}^{2,2}_{m-1,1-s}(\partial\Omega)\Big)^* \\
\Big\uparrow \pi^* & & \Big\uparrow \pi^* \\
\mathscr{X} & \xrightarrow[\ \lambda I - \dot{K}^*\ ]{} & \mathscr{X}
\end{array}$$

where

$$\mathscr{X} := \Big(\dot{B}^{2,2}_{m-1,1/2}(\partial\Omega)\Big/\dot{\mathscr{P}}_{m-1}(\partial\Omega)\Big)^*. \tag{6.181}$$

In light of (6.178) and the injectivity of π^*, this implies that there exists

$$\Lambda \in \Big(\dot{B}^{2,2}_{m-1,1/2}(\partial\Omega)\Big)^*, \quad \Lambda \neq 0, \text{ satisfying } \Big(\lambda I - \dot{K}^*\Big)\Lambda = 0. \tag{6.182}$$

As such, Lemma 6.16 applies and gives that, necessarily, $\lambda \in [-\frac{1}{2}, \frac{1}{2}]$. Given that we are assuming $\lambda \in \mathbb{C} \setminus (-\frac{1}{2}, \frac{1}{2})$ to begin with, this forces $\lambda \in \{\pm\frac{1}{2}\}$. On the other hand, when used in combination with the claim made in Theorem 6.15 about the operator in (6.128), the Fredholmness result recorded in (6.177) precludes λ from being equal to $\frac{1}{2}$. Hence, there remains to study the case when $\lambda = -\frac{1}{2}$.

To this end, we claim that

$$\left.\begin{array}{c} \dot{f} \in \dot{B}^{2,2}_{m-1,1/2}(\partial\Omega) \\ \text{and} \\ \left(\frac{1}{2}I + \dot{K}\right)\dot{f} \in \dot{\mathscr{P}}_{m-1}(\partial\Omega) \end{array}\right\} \Longrightarrow \dot{f} \in \dot{\mathscr{P}}_{m-1}(\partial\Omega). \tag{6.183}$$

To justify this claim, note that if $\dot{f}$ is as in the left-hand side of (6.183) then there exists $\dot{P} \in \dot{\mathscr{P}}_{m-1}(\partial\Omega)$ with the property that $\left(\frac{1}{2}I + \dot{K}\right)\dot{f} = \dot{P}$. However, $\left(\frac{1}{2}I + \dot{K}\right)\dot{P} = \dot{P}$ by Proposition 4.17 and, in particular, $\left(\frac{1}{2}I + \dot{K}\right)(\dot{f} - \dot{P}) = 0$. With this in hand, the conclusion in (6.183) follows from Theorem 6.9. In turn, (6.183) implies that the operator

$$\begin{array}{c} \frac{1}{2}I + \dot{K} : \dot{B}^{p,q}_{m-1,s}(\partial\Omega) \big/ \dot{\mathscr{P}}_{m-1}(\partial\Omega) \longrightarrow \dot{B}^{p,q}_{m-1,s}(\partial\Omega) \big/ \dot{\mathscr{P}}_{m-1}(\partial\Omega) \\ \text{is injective} \end{array} \tag{6.184}$$

hence, by duality,

$$\tfrac{1}{2}I + \dot{K}^* \ \text{ has dense range on } \left(\dot{B}^{p,q}_{m-1,s}(\partial\Omega) \big/ \dot{\mathscr{P}}_{m-1}(\partial\Omega)\right)^*. \tag{6.185}$$

In conjunction with (6.177) (which is valid for any $\lambda \in \mathbb{C} \setminus (-\frac{1}{2}, \frac{1}{2})$), this ultimately gives that

$$\tfrac{1}{2}I + \dot{K}^* \ \text{ is invertible on } \left(\dot{B}^{p,q}_{m-1,s}(\partial\Omega) \big/ \dot{\mathscr{P}}_{m-1}(\partial\Omega)\right)^*, \tag{6.186}$$

which is in contradiction with the fact that λ was assumed to belong to the spectrum of the operator $\frac{1}{2}I + \dot{K}^*$ on $\left(\dot{B}^{p,q}_{m-1,s}(\partial\Omega) \big/ \dot{\mathscr{P}}_{m-1}(\partial\Omega)\right)^*$.

In summary, the proof so far shows that the spectrum of the operator $\dot{K}^*$ acting on the space $\left(\dot{B}^{2,2}_{m-1,1/2}(\partial\Omega) \big/ \dot{\mathscr{P}}_{m-1}(\partial\Omega)\right)^*$ is contained in $(-\frac{1}{2}, \frac{1}{2})$. Via duality, this yields the claim made in (6.170), and (6.171) follows from the latter.

Moving on, formula (6.173) is a consequence of Theorem 6.4, (2.169), and Lemma 6.18 stated below. Finally, the fact that the operators in (6.174)–(6.175) are invertible when $p = q = 2$ and $s = 1/2$ is clear from (6.171). With this in hand, the extension of this invertibility result to the larger range specified in (6.172) may be done using (6.173), (3.314), and the stability theory developed in [63]. □

Here is the lemma invoked in the above proof.

Lemma 6.18. *Let X_j, Y_j, $j = 1, 2$, be two pairs of quasi-Banach spaces such that*

$$X_1 \hookrightarrow X_2 \ \textit{continuously}, \tag{6.187}$$

$$Y_1 \hookrightarrow Y_2 \ \textit{continuously and with dense range}. \tag{6.188}$$

If $T \in \mathscr{L}(X_1 \to Y_1) \cap \mathscr{L}(X_2 \to Y_2)$ has the property that $T : X_j \to Y_j$ is Fredholm for $j = 1, 2$, with the same index, then

$$\{f \in X_1 : Tf = 0\} = \{f \in X_2 : Tf = 0\}. \tag{6.189}$$

See [94, Lemma 11.40] for a proof.

6.2 Compactness Criteria for the Double Multi-Layer on Whitney–Besov Spaces

In this section, the goal is to identify a context in which the boundary double multi-layer operators associated with higher-order elliptic operators are compact when acting on Whitney–Besov spaces.

To set the stage, we begin by recalling several useful abstract interpolation results. In [29], M. Cwikel has proved the following remarkable one-sided compactness property for the real method of interpolation for (compatible) Banach couples.

Theorem 6.19. *Let X_j, Y_j, $j = 0, 1$, be two compatible Banach couples and suppose that the linear operator $T : X_j \to Y_j$ is bounded for $j = 0$ and compact for $j = 1$. Then $T : (X_0, X_1)_{\theta,q} \to (Y_0, Y_1)_{\theta,q}$ is compact for all $\theta \in (0, 1)$ and $q \in [1, \infty]$.*

The corresponding result for the complex method of interpolation remains open. However, in [29] M. Cwikel has shown that the property of being compact can be extrapolated on complex interpolation scales of Banach spaces:

Theorem 6.20. *Let X_j, Y_j, $j = 0, 1$, be two compatible Banach couples and suppose that $T : X_j \to Y_j$, $j = 0, 1$, is a bounded, linear operator with the property that there exists $\theta^* \in (0, 1)$ such that $T : [X_0, X_1]_{\theta^*} \to [Y_0, Y_1]_{\theta^*}$ is compact. Then the operator $T : [X_0, X_1]_\theta \to [Y_0, Y_1]_\theta$ is compact for all values of θ in $(0, 1)$.*

It is unclear whether a similar result holds for arbitrary compatible quasi-Banach couples. We shall show nonetheless show that such an extrapolation result holds for the entire scale of Besov spaces. More specifically, we have:

Theorem 6.21. *Let R be an open, convex subset of $\mathbb{R}\times\mathbb{R}_+\times\mathbb{R}_+$ and assume that T is a linear operator such that*

$$T : B_s^{p,q}(\mathbb{R}^n) \longrightarrow B_s^{p,q}(\mathbb{R}^n), \tag{6.190}$$

is bounded whenever $(s, 1/p, 1/q) \in R$. If there exists $(s^*, 1/p^*, 1/q^*) \in R$ such that T maps $B_{s^*}^{p^*,q^*}(\mathbb{R}^n)$ compactly into itself then the operator (6.190) is in fact compact for all $(s, 1/p, 1/q) \in R$.

Proof. Instead of the "continuous" scale of Besov space in $\mathbb{R}^n$ we find it convenient to work with its discrete version. That this is permissible is guaranteed by Theorem 7.1 in [45], according to which these spaces are isomorphic via the wavelet transform. For the benefit of the reader, we briefly elaborate on this idea. Let Φ and Ψ be, respectively, the Lemarié–Meyer "father" and a "mother" wavelet in $\mathbb{R}$ and set

$$\varphi(X) := \prod_{i=1}^{n} \Phi(x_i), \qquad X = (x_1, \ldots, x_n) \in \mathbb{R}^n, \tag{6.191}$$

$$\psi_J(X) := \Big(\prod_{i\in J} \Psi(x_j)\Big)\Big(\prod_{i\in J^c} \Phi(x_i)\Big), \tag{6.192}$$

where $J \subset \{1, \ldots, n\}$ is nonempty and $J^c := \{1, \ldots, n\} \setminus J$. Typically, the collection $\{\psi_J\}_J$ is relabeled $\{\psi_\ell\}_\ell$ with ℓ running from 1 to $2^n - 1$ (which is the number of such different sets J). Denote by $b_s^{p,q}(\mathbb{R}^n)$ the space of numerical sequences $\{c_k\}_{k\in\mathbb{Z}^n} \times \{c_{k,j,\ell}\}_{k\in\mathbb{Z}^n,\, j\in\mathbb{N}_0,\, 1\leq\ell\leq 2^n-1}$ with the property that

$$\Big\|\{c_k\}_{k\in\mathbb{Z}^n} \times \{c_{k,j,\ell}\}_{k\in\mathbb{Z}^n,\, j\in\mathbb{N}_0,\, 1\leq\ell\leq 2^n-1}\Big\|_{b_s^{p,q}(\mathbb{R}^n)} := \Big(\sum_{k\in\mathbb{Z}^n} |c_k|^p\Big)^{1/p}$$

$$+ \sum_{\ell=1}^{2^n-1}\Big[\sum_{j=0}^{\infty} 2^{jsq}\Big(\sum_{k\in\mathbb{Z}^n} 2^{-jn}|c_{k,j,\ell}|^p\Big)^{q/p}\Big]^{1/q} < \infty. \tag{6.193}$$

Here $0 < p, q \leq +\infty$ and $s \in \mathbb{R}$. Then, for this range of indices, the mapping

$$\mathscr{T} : \{c_k\}_{k\in\mathbb{Z}^n} \times \{c_{k,j,\ell}\}_{k\in\mathbb{Z}^n,\, j\in\mathbb{N}_0,\, 1\leq\ell\leq 2^n-1} \tag{6.194}$$

$$\longmapsto \sum_{k\in\mathbb{Z}^n} c_k\, \varphi(\cdot - k) + \sum_{\ell=1}^{2^n-1}\sum_{j=0}^{\infty}\sum_{k\in\mathbb{Z}^n} c_{k,j,\ell}\, \psi_\ell(2^j \cdot - k)$$

establishes an isomorphism between $b_s^{p,q}(\mathbb{R}^n)$ and $B_s^{p,q}(\mathbb{R}^n)$, whose inverse is given by

$$\mathscr{R}f := \Big\{\langle f, \varphi(\cdot - k)\rangle\Big\}_{k\in\mathbb{Z}^n} \times \Big\{\langle f, 2^{jn}\psi_\ell(2^j \cdot - k)\rangle\Big\}_{k\in\mathbb{Z}^n,\, j\in\mathbb{N}_0,\, 1\leq\ell\leq 2^n-1}. \tag{6.195}$$

The proof of the theorem then proceeds as follows. For each $N \in \mathbb{N}$ we now introduce

$$P_N\Big(\{c_k\}_{k\in\mathbb{Z}^n} \times \{c_{k,j,\ell}\}_{k\in\mathbb{Z}^n,\, j\in\mathbb{N}_0,\, 1\leq\ell\leq 2^n-1}\Big) \tag{6.196}$$
$$:= \{\tilde{c}_k\}_{k\in\mathbb{Z}^n} \times \{\tilde{c}_{k,j,\ell}\}_{k\in\mathbb{Z}^n,\, j\in\mathbb{N}_0,\, 1\leq\ell\leq 2^n-1}$$

where we have set $\tilde{c}_k := c_k$, $\tilde{c}_{k,j,\ell} := c_{k,j,\ell}$ if $|k| + |j| \leq N$, and zero otherwise. Thus, $P_N : b^{p,q}_s(\mathbb{R}^n) \to b^{p,q}_s(\mathbb{R}^n)$, $N \in \mathbb{N}$, are linear operators, of finite rank which also satisfy

$$\sup_{N\in\mathbb{N}} \|P_N\|_{b^{p,q}_s(\mathbb{R}^n)\to b^{p,q}_s(\mathbb{R}^n)} < \infty, \quad \text{and} \tag{6.197}$$

$$P_N \longrightarrow I \text{ pointwise on } b^{p,q}_s(\mathbb{R}^n) \text{ as } N \to \infty. \tag{6.198}$$

Let $\mathscr{O}$ be an arbitrary relatively compact subset of $b^{p,q}_s(\mathbb{R}^n)$. From (6.197)–(6.198) plus a standard argument, based on covering $\mathscr{O}$ with finitely many balls of sufficiently small radii, it follows that

$$P_N \longrightarrow I \text{ as } N \to \infty, \text{ uniformly on } \mathscr{O}. \tag{6.199}$$

In particular, if

$$T : b^{p_0,q_0}_{s_0}(\mathbb{R}^n) \longrightarrow b^{p_0,q_0}_{s_0}(\mathbb{R}^n) \text{ is linear and bounded, and} \tag{6.200}$$

$$T : b^{p_1,q_1}_{s_1}(\mathbb{R}^n) \longrightarrow b^{p_1,q_1}_{s_1}(\mathbb{R}^n) \text{ is linear and compact,} \tag{6.201}$$

then there exists a finite constant $C = C(T) > 0$ and, for each $\varepsilon > 0$, an integer $N(\varepsilon) \in \mathbb{N}$ such that

$$\|T - P_N T\|_{b^{p_0,q_0}_{s_0}(\mathbb{R}^n)\to b^{p_0,q_0}_{s_0}(\mathbb{R}^n)} \leq C, \quad \forall\, N \in \mathbb{N}, \tag{6.202}$$

$$\|T - P_N T\|_{b^{p_1,q_1}_{s_1}(\mathbb{R}^n)\to b^{p_1,q_1}_{s_1}(\mathbb{R}^n)} \leq \varepsilon, \quad \forall\, N \geq N(\varepsilon). \tag{6.203}$$

Since for each $\theta \in (0,1)$ it has been established in [77] that

$$\Big[b^{p_0,q_0}_{s_0}(\mathbb{R}^n)\,,\, b^{p_1,q_1}_{s_1}(\mathbb{R}^n)\Big]_\theta = b^{p,q}_s(\mathbb{R}^n) \tag{6.204}$$

provided $1/p := (1-\theta)/p_0 + \theta/p_1$, $1/q := (1-\theta)/q_0 + \theta/q_1$, $s := (1-\theta)s_0 + \theta s_1$, we may conclude from this and (6.202)–(6.203) that

$$\|T - P_N T\|_{b_s^{p,q}(\mathbb{R}^n) \to b_s^{p,q}(\mathbb{R}^n)} \leq C^{1-\theta} \varepsilon^\theta \tag{6.205}$$

granted that $N \geq N(\varepsilon)$. Thus, T is compact as an operator on $b_s^{p,q}(\mathbb{R}^n)$ since it can be approximated in the strong operator norm by linear operators of finite rank. □

We shall next adapt the extrapolation result from Theorem 6.21 to Whitney–Besov spaces on Lipschitz surfaces. As a preamble, we need to establish a procedure that allows us to "lift" compactness from the named surface to the entire Euclidean space. More specifically, we have the following result.

Proposition 6.22. *Suppose that Ω is a bounded Lipschitz domain in $\mathbb{R}^n$ and assume that $\frac{n-1}{n} < p \leq \infty$, $0 < q \leq \infty$ and $(n-1)(1/p - 1)_+ < s < 1$. Consider next a linear and continuous operator*

$$T : \dot{B}^{p,q}_{m-1,s}(\partial\Omega) \longrightarrow \dot{B}^{p,q}_{m-1,s}(\partial\Omega). \tag{6.206}$$

If $\mathscr{E}$ stands for the extension operator from (3.103)*, and $\mathscr{R}_\Omega$ is the operator restricting distributions from $\mathbb{R}^n$ to Ω, set*

$$\widetilde{T} := \mathscr{E} \circ T \circ \mathrm{tr}_{m-1} \circ \mathscr{R}_\Omega : B^{p,q}_{m-1+s+1/p}(\mathbb{R}^n) \to B^{p,q}_{m-1+s+1/p}(\mathbb{R}^n), \tag{6.207}$$

where $\mathrm{tr}_{m-1} : B^{p,q}_{m-1+s+1/p}(\Omega) \to \dot{B}^{p,q}_{m-1,s}(\partial\Omega)$ is the multi-trace operator from § 3.3. Then

$$\widetilde{T} \text{ compact on } B^{p,q}_{m-1+s+1/p}(\mathbb{R}^n) \Longleftrightarrow T \text{ compact on } \dot{B}^{p,q}_{m-1,s}(\partial\Omega). \tag{6.208}$$

Proof. If we set $\widehat{T} := T \circ \mathrm{tr}_{m-1}$, then the following implications hold:

$$\widetilde{T} \text{ compact} \Longrightarrow \widehat{T} \text{ compact} \Longrightarrow T \text{ compact}. \tag{6.209}$$

In the light of the boundedness of the operators tr_{m-1} and $\mathscr{E}$ established in Theorem 3.9, the sequence of implications (6.208) immediately follows once we notice that $T = \widehat{T} \circ \mathscr{E}$. □

Here is the extrapolation result advertised earlier.

Theorem 6.23. *Assume that Ω is a bounded Lipschitz domain in $\mathbb{R}^n$ and consider the open, convex region in $\mathbb{R} \times \mathbb{R}_+ \times \mathbb{R}_+$ given by*

$$\mathscr{O} := \Big\{(s, 1/p, 1/q) : \tfrac{n-1}{n} < p < \infty,\ 0 < q < \infty \tag{6.210}$$

$$\text{and } (n-1)\big(\tfrac{1}{p} - 1\big)_+ < s < 1\Big\}.$$

Let L be a constant coefficient W-elliptic differential operator of order $2m$, $m \in \mathbb{N}$, and recall the principal-value double multi-layer $\dot{K}$, associated with Ω and L as in Definition 4.13.

If there exists $(p_o, q_o, s_o) \in \mathcal{O}$ with the property that

$$\dot{K} : \dot{B}^{p_o,q_o}_{m-1,s_o}(\partial\Omega) \longrightarrow \dot{B}^{p_o,q_o}_{m-1,s_o}(\partial\Omega) \text{ is compact,} \tag{6.211}$$

then

$$\begin{gathered} \dot{K} : \dot{B}^{p,q}_{m-1,s}(\partial\Omega) \longrightarrow \dot{B}^{p,q}_{m-1,s}(\partial\Omega) \\ \text{is compact whenever } (s, 1/p, 1/q) \in \mathcal{O}. \end{gathered} \tag{6.212}$$

Proof. This follows directly from Proposition 6.22, Theorem 6.21, and the boundedness of the operator in (4.249) established in Theorem 4.21. □

Our next theorem indicates how compactness for the principal-value double multi-layer acting on one Whitney–Lebesgue space implies compactness for this operator on the entire Whitney–Besov scale.

Theorem 6.24. *Assume that Ω is a bounded Lipschitz domain in $\mathbb{R}^n$ and consider a constant coefficient W-elliptic differential operator L of order $2m$, $m \in \mathbb{N}$. Finally, recall the principal-value double multi-layer $\dot{K}$, associated with Ω and L as in Definition 4.13.*

If there exists $p_o \in (1, \infty)$ with the property that

$$\dot{K} : \dot{L}^{p_o}_{m-1,0}(\partial\Omega) \longrightarrow \dot{L}^{p_o}_{m-1,0}(\partial\Omega) \text{ is compact,} \tag{6.213}$$

then

$$\begin{gathered} \dot{K} : \dot{B}^{p,q}_{m-1,s}(\partial\Omega) \longrightarrow \dot{B}^{p,q}_{m-1,s}(\partial\Omega) \text{ is compact whenever} \\ 0 < p, q < \infty, \quad (n-1)\big(\tfrac{1}{p} - 1\big)_+ < s < 1. \end{gathered} \tag{6.214}$$

Proof. Recall the set $\mathcal{O}$ from (6.210). In a first stage, Theorem 6.19, Corollary 3.40, and Theorem 4.14 give, under the current hypotheses, that there exist s_o, q_o such that $(s_o, 1/p_o, 1/q_o) \in \mathcal{O}$ and (6.211) holds. Having established this, Corollary 6.23 applies and yields the desired conclusion. □

In [25] Cohen and Gosselin have used a writing for the biharmonic operator Δ^2 as in (4.50) for a special choice of the coefficient tensor $A = (A_{\alpha\beta})_{|\alpha|=|\beta|=2}$, which goes back to more general work of Agmon in [3], in the two dimensional setting, which has the property that the associated principal-value biharmonic double multi-layer, call it $\dot{K}_{\Delta^2}$, satisfies

$$\begin{aligned}&\dot{K}_{\Delta^2} : \dot{L}^p_{1,0}(\partial\Omega) \longrightarrow \dot{L}^p_{1,0}(\partial\Omega) \text{ is compact, for each } p \in (1,\infty),\\&\qquad\text{provided } \Omega \text{ is a bounded } C^1 \text{ domain in } \mathbb{R}^2.\end{aligned} \tag{6.215}$$

By relying on the recent results established in [56] it is possible to further relax the assumptions on the domain Ω in (6.215) and obtain the following generalization of Cohen–Gosselin's compactness result:

$$\begin{aligned}&\dot{K}_{\Delta^2} : \dot{L}^p_{1,0}(\partial\Omega) \longrightarrow \dot{L}^p_{1,0}(\partial\Omega) \text{ is compact, for each } p \in (1,\infty),\\&\text{provided } \Omega \text{ is a bounded Lipschitz domain in } \mathbb{R}^2 \text{ with } \nu \in \mathrm{vmo}(\partial\Omega).\end{aligned} \tag{6.216}$$

Above, $\mathrm{vmo}\,(\partial\Omega)$ denotes Sarason's space of functions of vanishing mean oscillation on $\partial\Omega$ (cf. (2.447)). In turn, this permits us to state the following theorem.

Theorem 6.25. *Assume that Ω is a bounded Lipschitz domain in $\mathbb{R}^2$ whose unit normal belongs to $\mathrm{vmo}(\partial\Omega)$. In this context, let $\dot{K}_{\Delta^2}$ be the principal-value biharmonic double multi-layer introduced by Cohen and Gosselin in [25]. Then*

$$\begin{aligned}&\dot{K}_{\Delta^2} : \dot{B}^{p,q}_{1,s}(\partial\Omega) \longrightarrow \dot{B}^{p,q}_{1,s}(\partial\Omega) \text{ is compact whenever}\\&\qquad 0 < p,q < \infty, \quad \left(\tfrac{1}{p} - 1\right)_+ < s < 1.\end{aligned} \tag{6.217}$$

As a corollary,

$$\begin{aligned}&\pm\tfrac{1}{2}I + \dot{K}_{\Delta^2} : \dot{B}^{p,q}_{1,s}(\partial\Omega) \longrightarrow \dot{B}^{p,q}_{1,s}(\partial\Omega) \text{ is Fredholm with index zero}\\&\qquad \text{provided } 0 < p,q < \infty, \quad \left(\tfrac{1}{p} - 1\right)_+ < s < 1.\end{aligned} \tag{6.218}$$

Finally, if in addition to the conditions on Ω stipulated so far, it is also assumed that $\partial\Omega$ is connected, then in fact

$$\begin{aligned}&\tfrac{1}{2}I + \dot{K}_{\Delta^2} : \dot{B}^{p,q}_{1,s}(\partial\Omega) \longrightarrow \dot{B}^{p,q}_{1,s}(\partial\Omega) \text{ is invertible}\\&\qquad \text{whenever } 0 < p,q < \infty, \quad \left(\tfrac{1}{p} - 1\right)_+ < s < 1.\end{aligned} \tag{6.219}$$

Proof. The compactness of $\dot{K}_{\Delta^2}$ in (6.217) is seen by combining the compactness result recorded in (6.216) with the general extrapolation result from Theorem 6.24. Having established this, (6.218) follows from standard Fredholm theory. As far as (6.219) is concerned, in a first stage the work in [25] ensures (under the assumption that $\partial\Omega$ is connected) that

$$\tfrac{1}{2}I + \dot{K}_{\Delta^2} : \dot{L}^p_{1,0}(\partial\Omega) \longrightarrow \dot{L}^p_{1,0}(\partial\Omega) \text{ is invertible for each } p \in (1,\infty). \tag{6.220}$$

In turn, this implies that the operator in (6.219) has dense range hence, ultimately, is onto (thanks to (6.218)). With this in hand, Theorem 6.9 and the global extrapolation result proved in [63, Theorem 2.10] applies and gives that the claim in (6.219) is valid as stated. □

We conclude this section with yet another result pertaining to the invertibility of the boundary biharmonic double multi-layer potential on Whitney–Besov spaces in planar domains.

Theorem 6.26. *Let Ω be a bounded Lipschitz domain in $\mathbb{R}^2$ with connected boundary and whose unit normal belongs to* vmo$(\partial\Omega)$. *As before, let $\dot{K}_{\Delta^2}$ be the principal-value biharmonic double multi-layer introduced by Cohen and Gosselin in [25]. Then the operators*

$$\pm\tfrac{1}{2}I + \dot{K}_{\Delta^2} : \dot{B}^{p,q}_{1,s}(\partial\Omega)\Big/\dot{\mathscr{P}}_1(\partial\Omega) \longrightarrow \dot{B}^{p,q}_{1,s}(\partial\Omega)\Big/\dot{\mathscr{P}}_1(\partial\Omega), \tag{6.221}$$

$$\pm\tfrac{1}{2}I + \dot{K}^*_{\Delta^2} : \Big(\dot{B}^{p,q}_{1,s}(\partial\Omega)\Big/\dot{\mathscr{P}}_1(\partial\Omega)\Big)^* \longrightarrow \Big(\dot{B}^{p,q}_{1,s}(\partial\Omega)\Big/\dot{\mathscr{P}}_1(\partial\Omega)\Big)^* \tag{6.222}$$

are invertible whenever

$$0 < p,q < \infty \ \text{ and } \ \big(\tfrac{1}{p}-1\big)_+ < s < 1. \tag{6.223}$$

As a corollary, the above invertibility results hold for any bounded C^1 domain $\Omega \subset \mathbb{R}^2$ with connected boundary.

Proof. This is established by arguing much as in the proof of Theorem 6.17, using the compactness of $\dot{K}_{\Delta^2}$ formulated in (6.217). □

6.3 Uniqueness for the Dirichlet Problem with Data in Whitney–Lebesgue Spaces

In this section we shall consider the issue of uniqueness for the Dirichlet problem formulated for a higher-order differential operator and data in the appropriate Whitney–Lebesgue space. To get started, we describe a notion of Green function which is going to be relevant shortly.

Definition 6.27. Assume that Ω is a bounded Lipschitz domain in $\mathbb{R}^n$ and consider a constant coefficient W-elliptic differential operator L of order $2m$, $m \in \mathbb{N}$. Also, fix $p \in (1,\infty)$. In this context, we say that `property` G_p `holds` provided for each $X \in \Omega$ there exists a function $G(X,\cdot) \in C^\infty\big(\Omega \setminus \{X\}\big)$ satisfying (with δ_X denoting the Dirac distribution with mass at X)

$$\begin{cases} L_Y G(X,Y) = \delta_X(Y), \\ G(X,\cdot)\Big|_{\partial\Omega}^{m-1} = 0, \\ \mathcal{N}\Big(\nabla_Y^m G(X,\cdot)\Big) \in L^p(\partial\Omega), \end{cases} \tag{6.224}$$

where the non-tangential maximal operator $\mathcal{N}$ is considered with respect to a regular family of cones truncated at height much smaller than the distance from X to $\partial\Omega$.

For example, if

$$\tfrac{1}{2}I + \dot{K} : \dot{L}^p_{m-1,1}(\partial\Omega) \longrightarrow \dot{L}^p_{m-1,1}(\partial\Omega) \text{ is invertible,} \tag{6.225}$$

a Green function with the properties stipulated in (6.224) may be constructed by considering, for each $X, Y \in \Omega$ with $X \neq Y$,

$$G(X,Y) := E(X-Y) - \dot{\mathscr{D}}\Big[\Big(\tfrac{1}{2}I + \dot{K}\Big)^{-1}(\mathrm{tr}_{m-1}E(X-\cdot))\Big](Y), \tag{6.226}$$

where E is the fundamental solution of the operator L discussed in Theorem 4.2. To see that this is indeed the case, note that since for each point $X \in \Omega$ fixed we have $\mathrm{tr}_{m-1}E(X-\cdot) \in \dot{L}^p_{m-1,1}(\partial\Omega)$, and using (6.225) and (4.64) we obtain

$$\mathcal{N}\Big(\nabla^m \dot{\mathscr{D}}\Big[\Big(\tfrac{1}{2}I + \dot{K}\Big)^{-1}(\mathrm{tr}_{m-1}E(X-\cdot))\Big]\Big) \in L^p(\partial\Omega) \tag{6.227}$$

hence, ultimately,

$$\mathcal{N}\Big(\nabla_Y^m G(X,\cdot)\Big) \in L^p(\partial\Omega) \tag{6.228}$$

if G is as in (6.226). Furthermore, (6.226) and (4.164) ensure that the middle condition in (6.224) holds as well. Finally, the first condition in (6.224) is clear from the design of G.

A specific case when these considerations are relevant is presented next (see also Corollary 6.51 for another relevant scenario).

Proposition 6.28. *Consider the biharmonic operator, $L = \Delta^2$, in the plane and assume that Ω is a bounded Lipschitz domain in $\mathbb{R}^2$ with connected boundary, and whose unit normal belongs to* vmo$(\partial\Omega)$. *Then, in this context, property G_p holds for every $p \in (1,\infty)$.*

Proof. Fix an arbitrary $p \in (1,\infty)$. Also, as before, let $\dot{K}_{\Delta^2}$ be the principal-value biharmonic double multi-layer introduced by Cohen and Gosselin in [25]. A careful inspection of their work reveals that $\dot{K}_{\Delta^2}$ is compact on $\dot{L}^p_{1,1}(\partial\Omega)$ (see also [87] where the authors prove much more general results of this nature), hence

$$\tfrac{1}{2}I + \dot{K}_{\Delta^2} : \dot{L}^p_{1,1}(\partial\Omega) \longrightarrow \dot{L}^p_{1,1}(\partial\Omega) \quad \text{is invertible,} \tag{6.229}$$

thanks to (6.220). Thus, (6.225) holds in this case (recall that, currently, $m = 2$), and the desired conclusion follows from the earlier discussion. □

After this preamble we are ready to state the main uniqueness result of this section.

Theorem 6.29. *Let Ω be a bounded Lipschitz domain in $\mathbb{R}^n$ and consider a constant coefficient W-elliptic differential operator L of order $2m$, $m \in \mathbb{N}$. Assume that indices $p, p' \in (1,\infty)$ are such that $1/p + 1/p' = 1$, and that property $G_{p'}$ holds in this context.*

If u is a solution of the homogeneous Dirichlet boundary value problem

$$\begin{cases} Lu = 0 \quad \text{in} \quad \Omega, \\ \mathcal{N}(\nabla^{m-1}u) \in L^p(\partial\Omega), \\ u\Big|_{\partial\Omega}^{m-1} = 0, \end{cases} \tag{6.230}$$

then necessarily $u \equiv 0$ in Ω.

Proof. For each $\varepsilon > 0$ set

$$\Omega_\varepsilon := \{X \in \Omega : \operatorname{dist}(X, \partial\Omega) > \varepsilon\}. \tag{6.231}$$

We claim that there exist a family of functions $\Phi_\varepsilon \in C_c^\infty(\mathbb{R}^n)$, indexed by $\varepsilon \in (0,1)$, and two constants $0 < C_1 < C_2 < \infty$ such that

$$\Phi_\varepsilon \equiv 1 \quad \text{on} \quad \Omega_{C_2\varepsilon} \quad \text{and} \quad \Phi_\varepsilon \equiv 0 \quad \text{on} \quad \mathbb{R}^n \setminus \overline{\Omega_{C_1\varepsilon}} \tag{6.232}$$

and with the property that for each multi-index $\alpha \in \mathbb{N}_0^n$ there exists $C(\alpha) \in (0,\infty)$ such that

$$|\partial^\alpha \Phi_\varepsilon| \le \frac{C_\alpha}{\varepsilon^{|\alpha|}}, \qquad \forall\, \varepsilon \in (0,1). \tag{6.233}$$

To establish th existence of such a family, consider $\psi \in C^\infty(\mathbb{R})$ with the property that $\psi \equiv 0$ on $(-\infty, 1)$ and $\psi \equiv 1$ on $(2, +\infty)$. Then, if $\rho_{\rm reg}$ denotes the regularized distance to $\mathbb{R}^2 \setminus \Omega$ (in the sense of Theorem 2, p. 171 in [119]), we may take

$$\Phi_\varepsilon(X) := \psi(\varepsilon^{-1}\rho_{\rm reg}(X)), \qquad X \in \mathbb{R}^2. \tag{6.234}$$

Given that

$$C_1 \operatorname{dist}(X, \partial\Omega) \leq \rho_{\text{reg}}(X) \leq C_2 \operatorname{dist}(X, \partial\Omega), \tag{6.235}$$

$$|\partial^\alpha \rho_{\text{reg}}(X)| \leq C_\alpha \operatorname{dist}(X, \partial\Omega)^{1-|\alpha|}, \quad \forall\, \alpha \in \mathbb{N}_0^n, \tag{6.236}$$

the conditions listed in (6.232)–(6.233) follow.

Fix next a point $X \in \Omega$ and pick a number $\varepsilon > 0$ small enough to guarantee that $X \in \Omega_\varepsilon$. Since we are assuming that property $G_{p'}$ holds, there exists a Green function $G(X, \cdot) \in C^\infty(\Omega \setminus \{X\})$ satisfying

$$\begin{cases} L_Y G(X, Y) = \delta_X(Y), \\ G(X, \cdot)\Big|_{\partial\Omega}^{m-1} = 0. \\ \mathcal{N}\Big(\nabla_Y^m G(X, \cdot)\Big) \in L^{p'}(\partial\Omega), \end{cases} \tag{6.237}$$

To proceed, assume that the function u is a solution of the homogeneous Dirichlet problem (6.230). The starting point is the identity

$$u(X) = (u\Phi_\varepsilon)(X) = \int_\Omega L_Y G(X, Y)\Phi_\varepsilon(Y)u(Y)\, dY. \tag{6.238}$$

Integrating by parts and utilizing the support conditions on Φ_ε we further obtain

$$\begin{aligned} u(X) &= \int_\Omega G(X, Y) L_Y(\Phi_\varepsilon u)(Y)\, dY \\ &= \int_\Omega \sum_{|\alpha|=|\beta|=m} G(X, Y) A_{\alpha\beta}\, \partial^{\alpha+\beta}(\Phi_\varepsilon u)(Y)\, dY. \end{aligned} \tag{6.239}$$

Also, Leibniz's formula gives

$$\partial^{\alpha+\beta}(\Phi_\varepsilon u) = \sum_{\alpha+\beta=\gamma+\delta} C_{\gamma\delta}^{\alpha\beta}\, \partial^\gamma \Phi_\varepsilon \partial^\delta u. \tag{6.240}$$

In turn, using formula (6.240) and the fact that

$$\sum_{|\alpha|=|\beta|=m} A_{\alpha\beta} C_{0(\alpha+\beta)}^{\alpha\beta}\, \partial^{\alpha+\beta} u = Lu = 0, \tag{6.241}$$

since $C^{\alpha\beta}_{0(\alpha+\beta)} = 1$, we conclude that

$$\sum_{|\alpha|=|\beta|=m} A_{\alpha\beta}\, \partial^{\alpha+\beta}(\Phi_\varepsilon u) = \sum_{|\alpha|=|\beta|=m} A_{\alpha\beta} \sum_{\substack{\alpha+\beta=\gamma+\delta\\ \gamma\neq 0}} C^{\alpha\beta}_{\gamma\delta}\, \partial^\gamma \Phi_\varepsilon \partial^\delta u. \tag{6.242}$$

Next, we split the sum in the right-hand side of (6.242) over the set of multi-indices δ in $\mathbb{N}_0^n$ of length $\leq m-1$ and over the set of multi-indices δ in $\mathbb{N}_0^n$ of length $\geq m$. In the latter case we write $\delta = \mu + \theta$ with $\mu, \theta \in \mathbb{N}_0^n$ and $|\mu| = m-1$, then integrate by parts in order to move ∂_Y^θ from u to $G(X,\cdot)$. Since

$$G(X,\cdot)\Big|^{m-1}_{\partial\Omega} = 0, \tag{6.243}$$

this does not create any new boundary terms. Thus, starting with (6.239), (6.242), and then carrying out this program yields the representation

$$u(X) = I_\varepsilon(X) + II_\varepsilon(X), \tag{6.244}$$

where

$$I_\varepsilon(X) := \int_\Omega G(X,Y) \sum_{|\alpha|=|\beta|=m} \sum_{\substack{\alpha+\beta=\gamma+\delta\\ \gamma\neq 0,\, |\delta|\leq m-1}} A_{\alpha\beta} C^{\alpha\beta}_{\gamma\delta} (\partial^\gamma \Phi_\varepsilon)(Y)(\partial^\delta u)(Y)\, dY, \tag{6.245}$$

and

$$II_\varepsilon(X) := \int_\Omega \sum_{|\alpha|=|\beta|=m} \sum_{\substack{\alpha+\beta=\gamma+\delta\\ \gamma\neq 0,\, |\delta|\geq m}} \sum_{\substack{\delta=\mu+\theta\\ |\mu|=m-1}} (-1)^{|\theta|} A_{\alpha\beta} C^{\alpha\beta}_{\gamma\delta}\, \partial_Y^\theta G(X,Y) \times$$
$$\times (\partial^\gamma \Phi_\varepsilon)(Y)(\partial^\mu u)(Y)\, dY. \tag{6.246}$$

Consider the term $II_\varepsilon(X)$. Notice that $|\theta| = m - |\gamma| + 1 \leq m$, since the summation is performed over $\gamma \neq 0$. Using again that $\gamma \neq 0$, one may replace Ω by $\Omega \setminus \Omega_\varepsilon$ as the domain of integration in (6.245) and (6.246). Going further, we break up the integral over sufficiently small domains $(U_i)_{1\leq i\leq N}$, each contained in a local coordinate system where $U_i \cap \Omega$ can be regarded as the upper-graph of a Lipschitz function $\phi_i : Q_i \to \mathbb{R}$, where Q_i is an open subset of $\mathbb{R}^{n-1}$. Based on these observations and (6.233), we may then write

$$|II_\varepsilon(X)| \leq C \sum_{i=1}^N \sum_{|\alpha|=|\beta|=m} \sum_{\substack{\alpha+\beta=\gamma+\delta\\ \gamma\neq 0,\, |\delta|\geq m}} \sum_{\substack{\delta=\mu+\theta\\ |\mu|=m-1}} \int_{Q_i} \int_0^{C\varepsilon} \frac{1}{\varepsilon^{|\gamma|}} |(\partial^\mu u)(y', t+\phi_i(y'))| \times$$
$$\times |(\partial_Y^\theta G)(X, (y', t+\phi_i(y')))|\, dt\, dy', \tag{6.247}$$

for some finite constant $C = C(\Omega) > 0$. Let us now focus on the factor involving $\partial_Y^\theta G$, for some arbitrary but fixed multi-index θ of length $\leq m$, appearing in the right-hand side of (6.247). A simple application of the Fundamental Theorem of Calculus gives that, for each i,

$$(\partial_Y^\theta G)(X,(y',t+\phi_i(y'))) = -\int_0^t (\partial_Y^{\theta+e_n} G)(X,(y',t_1+\phi_i(y')))\,dt_1 \quad \text{whenever } y' \in Q_i \tag{6.248}$$

thanks to (6.243). Subsequent iterations of the Fundamental Theorem of Calculus allow us (making repeated use of (6.243)) to obtain m derivatives on the function G in the right-hand side of (6.248), i.e., write

$$\begin{aligned}(\partial_Y^\theta G)(X,(y',t+\phi_i(y'))) &= (-1)^{m-|\theta|}\times \\ \times \int_0^t\int_0^{t_1}\cdots\int_0^{t_{m-|\theta|-1}} &(\partial_Y^{\theta+(m-|\theta|)e_n} G)(X,(y',r+\phi_i(y')))\,dr\,dt_{m-|\theta|-1}\ldots dt_1,\end{aligned} \tag{6.249}$$

whenever $y' \in Q_i$. Using now that for each $j \in \{1,\ldots,m-|\theta|-1\}$ we have $|t_j| < C\varepsilon$, we may further conclude that for each $y' \in Q_i$, $1 \leq i \leq N$,

$$\begin{aligned}\big|(\partial_Y^\theta G)(X,(y',t+\phi_i(y')))\big| &\leq \varepsilon^{m-|\theta|} \sup_{0<r<C\varepsilon} |(\nabla_Y^m G)(X,(y',y'+r\phi_i(y')))| \\ &\leq \varepsilon^{m-|\theta|} \mathcal{N}(\nabla_Y^m G(X,\cdot))(y',\phi_i(y')).\end{aligned} \tag{6.250}$$

In connection with this, it is important to note that

$$\text{the map } Q_i \ni y' \longmapsto \mathcal{N}\big(\nabla_Y^m G(X,\cdot)\big)(y',\phi_i(y')) \in [0,+\infty] \quad \text{belongs to } L^{p'}(Q_i) \tag{6.251}$$

Continuing our analysis of $II_\varepsilon(X)$ we note that, since $m-|\theta|-|\gamma| = -1$, we have from (6.250)

$$\begin{aligned}|II_\varepsilon(X)| \leq C \sum_{|\mu|=m-1}\sum_{i=1}^N \int_{Q_i}\Big(\frac{1}{\varepsilon}\int_0^{C\varepsilon} &|(\partial^\mu u)(y',t+\phi_i(y'))|\times \\ &\times \mathcal{N}((\nabla_Y^m G)(X-\cdot))(y',\phi_i(y'))\,dt\Big)\,dy'.\end{aligned} \tag{6.252}$$

The idea now is to employ Lebesgue's Dominated Convergence Theorem in order to show that the last integral above converges to zero as $\varepsilon \to 0^+$. To this end, we first observe that for each $i \in \{1,\ldots,N\}$ and each multi-index μ with $|\mu| = m-1$, we have

$$\left|\frac{1}{\varepsilon}\int_0^{C\varepsilon} |(\partial^\mu u)(y', t+\phi_i(y'))|\,dt\right| \le C\mathcal{N}(\nabla^{m-1}u)(y',\phi_i(y')) \ \text{ for } \ y'\in Q_i. \tag{6.253}$$

Let us also recall that, by hypothesis,

$$\begin{array}{c}\text{the map } \ Q_i \ni y' \longmapsto \mathcal{N}(\nabla^{m-1}u)(y',\phi(y')) \in [0,+\infty] \\ \text{belongs to } L^{p'}(Q_i)\end{array} \tag{6.254}$$

Collectively, (6.253)–(6.254) and (6.251) ensure that the uniform pointwise domination part of Lebesgue's theorem is satisfied. As for the pointwise convergence to zero part of Lebesgue's theorem, we start by making the simple observation that if $f:(0,1)\to\mathbb{R}$ is a continuous function with the property that $\lim\limits_{t\to 0^+} f(t)=0$ then $\lim\limits_{\varepsilon\to 0^+}\frac{1}{\varepsilon}\int_0^\varepsilon f(t)\,dt = 0$ (as seen easily from an application of the Mean Value Theorem). Since, by hypothesis,

$$\lim_{t\to 0^+}(\partial^\mu u)(y',(t+\phi_i(y'))=0 \quad \text{for a.e. } \ y'\in Q_i, \quad 1\le i\le N, \tag{6.255}$$

the above observation applies and shows that, pointwise a.e., the integrand in (6.252) converges to zero. Thus, the Lebesgue Dominated Convergence Theorem gives

$$\lim_{\varepsilon\to 0} II_\varepsilon(X)=0. \tag{6.256}$$

Turning attention to $I_\varepsilon(X)$, repeated applications of the Fundamental Theorem of Calculus (as in the previous analysis) permit us to write, for each $i\in\{1,\dots,N\}$ and each $y'\in Q_i$,

$$\begin{aligned}(\partial^\delta u)(y',t+\phi_i(y')) &= -\int_0^t (\partial^{\delta+e_n}u)(y',t_1+\phi_i(y'))\,dt_1 \\ &= \dots = (-1)^{m-1-|\delta|}\times \\ &\quad\times\int_0^t\int_0^{t_1}\dots\int_0^{t_{m-2-|\delta|}}(\partial^{\delta+(m-1-|\delta|)e_n}u)\times \\ &\quad\times(y',r+\phi_i(y'))\,dr\,dt_{m-2-|\delta|}\dots dt_1.\end{aligned} \tag{6.257}$$

Note that for each $y'\in Q_i$, $1\le i\le N$,

$$\begin{aligned}\left|(\partial^\delta u)(y',t+\phi_i(y'))\right| &\le \varepsilon^{m-1-|\delta|}\sup_{0<r<C\varepsilon}\left|(\nabla^{m-1}u)(y',r+\phi_i(y'))\right| \\ &\le \varepsilon^{m-1-|\delta|}\,\mathcal{N}(\nabla^{m-1}u)(y',\phi_i(y')).\end{aligned} \tag{6.258}$$

Also, using (6.250), we have

$$|G(X,(y',t+\phi_i(y'))| \le \varepsilon^m \mathcal{N}(\nabla^m G(X,\cdot))(y',\phi_i(y')), \tag{6.259}$$

for each $y' \in Q_i$, $1 \le i \le N$. Therefore, for some $C = C(\Omega) > 0$,

$$|I_\varepsilon(X)| \le C \sum_{i=1}^{N} \int_{Q_i} \Big\{ \frac{1}{\varepsilon^{|\gamma|}} \int_0^{C\varepsilon} |G(X,(y',t+\phi_i(y'))| \times \tag{6.260}$$

$$\times \int_0^t \int_0^{t_1} \cdots \int_0^{t_{m-2-|\delta|}} \Big|(\partial^{\delta+(m-1-|\delta|)e_n} u)(y',r+\phi_i(y'))\Big|\, dr\, dt_{m-2-|\delta|} \dots dt_1\, dt \Big\}\, dy'.$$

Using (6.233) and (6.258)–(6.259), we see that the expression in the curly brackets in (6.260) is

$$\le C\Big[\varepsilon^{1-|\gamma|}\varepsilon^m \mathcal{N}(\nabla^m G(X,\cdot))(y',\phi_i(y'))\Big]\Big[\varepsilon^{m-|\delta|-1}\mathcal{N}(\nabla^{m-1}u)(y',\phi_i(y'))\Big]$$

$$= \mathcal{N}(\nabla^{m-1}u)(y',\phi_i(y'))\mathcal{N}(\nabla^m G(X,\cdot))(y',\phi_i(y')). \tag{6.261}$$

Given that from hypotheses, for each $i \in \{1,\dots,N\}$, we have

$$Q_i \ni y' \longmapsto \mathcal{N}(\nabla^{m-1}u)(y',\phi_i(y'))\mathcal{N}(\nabla^m G(X,\cdot))(y',\phi_i(y')) \quad \text{belongs to } L^1(Q_i), \tag{6.262}$$

the uniform domination condition in Lebesgue's theorem is satisfied. Also, as before, since for each $i \in \{1,\dots,N\}$, and a.e. $y' \in Q_i$,

$$\lim_{\varepsilon\to 0} \frac{1}{\varepsilon} \int_0^{C\varepsilon} (\partial^{\delta+(m-1-|\delta|)e_n} u)(y',r+\phi_i(y'))\, dr = 0, \tag{6.263}$$

Lebesgue's Dominated Convergence Theorem applies and gives that

$$\lim_{\varepsilon\to 0} I_\varepsilon(X) = 0. \tag{6.264}$$

Together, (6.264), (6.256) and (6.244) give that $u(X) = 0$, hence, $u \equiv 0$ in Ω. This finishes the proof of Theorem 6.29. □

We conclude this section by presenting a well-posedness and regularity result for the Dirichlet boundary value problem for the planar biharmonic operator. This extends the scope of the work in [25] by considering more general domains, by establishing uniqueness (thus answering the question asked by Cohen and Gosselin at the bottom of page 238 in [26]) and by proving a regularity result.

Theorem 6.30. *Assume that Ω is a bounded Lipschitz domain in $\mathbb{R}^2$ with connected boundary, and whose unit normal belongs to* vmo$(\partial\Omega)$. *Fix $p \in (1, \infty)$. Then the Dirichlet boundary value problem for the biharmonic operator*

$$\begin{cases} \Delta^2 u = 0 \quad \text{in} \quad \Omega, \\ \mathcal{N}(\nabla u) \in L^p(\partial\Omega), \\ \big(u\lfloor_{\partial\Omega}, \nabla u\lfloor_{\partial\Omega}\big) = \dot{f} \in \dot{L}^p_{1,0}(\partial\Omega), \end{cases} \tag{6.265}$$

has a unique solution. This solution has the integral representation formula

$$u(X) = \dot{\mathscr{D}}_{\Delta^2}\Big[\Big(\tfrac{1}{2}I + \dot{K}_{\Delta^2}\Big)^{-1}\dot{f}\Big](X), \qquad X \in \Omega, \tag{6.266}$$

where $\dot{\mathscr{D}}_{\Delta^2}$ and $\dot{K}_{\Delta^2}$ are the double multi-layer potential operators from [25], and satisfies

$$\|\mathcal{N}(\nabla u)\|_{L^p(\partial\Omega)} \leq C(\Omega, p)\|\dot{f}\|_{\dot{L}^p_{1,0}(\partial\Omega)}. \tag{6.267}$$

Moreover, the following regularity result holds (quantitatively):

$$\mathcal{N}(\nabla\nabla u) \in L^p(\partial\Omega) \iff \dot{f} \in \dot{L}^p_{1,1}(\partial\Omega). \tag{6.268}$$

Proof. In concert, Proposition 6.28 and Theorem 6.29 prove that the boundary value problem (6.265) has at most one solution. Next, the fact that the function u in (6.266) is a well-defined solution of the boundary value problem (6.265) which satisfies (6.267) is seen from (6.220) and the properties of generic double multi-layer operators established in § 4. Finally, the regularity result from (6.268) is a consequence of (6.266), (6.229), Proposition 2.15 and, once again, the results in § 4. □

6.4 Boundary Problems on Besov and Triebel–Lizorkin Spaces

In this section we shall treat a variety of boundary value problems for higher order operators in Lipschitz domains (occasionally satisfying additional conditions). As such, the discussion highlights the basic role played by the Calderón–Zygmund theory of multi-layer potentials developed in earlier chapters.

We begin with a general duality result.

Proposition 6.31. *Assume that Ω is a bounded Lipschitz domain in $\mathbb{R}^n$ and consider an arbitrary homogeneous, constant coefficient differential operator L of order $2m$, $m \in \mathbb{N}$. Fix $p, q \in (1, \infty)$ and $0 < s < 1$. Then the adjoint of the operator*

$$L : \mathring{B}^{p,q}_{m-1+s+1/p}(\Omega) \longrightarrow B^{p,q}_{-m-1+s+1/p}(\Omega) \tag{6.269}$$

is

$$L^* : \mathring{B}^{p',q'}_{m-1+s'+1/p'}(\Omega) \longrightarrow B^{p',q'}_{-m-1+s'+1/p'}(\Omega), \tag{6.270}$$

where $1/p + 1/p' = 1/q + 1/q' = 1$ *and* $s' := 1 - s$.

Moreover, a similar property is valid on the Triebel–Lizorkin scale.

Proof. Under the current assumptions on the indices, from Proposition 3.15 and (2.216)–(2.217) we have

$$\Big(\mathring{B}^{p,q}_{m-1+s+1/p}(\Omega)\Big)^* = \Big(B^{p,q}_{m-1+s+1/p,z}(\Omega)\Big)^* = B^{p',q'}_{-m-1+s'+1/p',z}(\Omega). \tag{6.271}$$

and

$$\Big(B^{p,q}_{-m-1+s+1/p}(\Omega)\Big)^* = B^{p',q'}_{m-1+s'+1/p',z}(\Omega) = \mathring{B}^{p',q'}_{m-1+s'+1/p',z}(\Omega). \tag{6.272}$$

Thus, the claim made in the statement of the proposition is proved as soon as we show that

$$\langle Lu, \overline{v}\rangle = \langle u, \overline{L^* v}\rangle, \quad \forall\, u \in \mathring{B}^{p,q}_{m-1+s+1/p}(\Omega), \ \forall\, v \in \mathring{B}^{p',q'}_{m-1+s'+1/p',z}(\Omega). \tag{6.273}$$

However, this is obvious when $u, v \in C^\infty_c(\Omega)$, so that (6.273) follows by density.

Finally, a similar reasoning shows that an analogous duality phenomenon holds on the Triebel–Lizorkin scale. □

Moving on, we next address the issue of the realization of a differential operator as a linear isomorphism between appropriate Besov and Triebel–Lizorkin spaces in Lipschitz domains.

Theorem 6.32. *Assume that the differential operator L of order $2m$, $m \in \mathbb{N}$, is as in* (4.1)–(4.2)*, and satisfies the Legendre–Hadamard ellipticity condition* (4.15)*. Also, suppose that Ω is a bounded Lipschitz domain in $\mathbb{R}^n$. Then there exists $\varepsilon > 0$ with the property that whenever*

$$|p-2| + |q-2| + |s-1/2| < \varepsilon \tag{6.274}$$

the operators

$$L : \mathring{F}^{p,q}_{m-1+s+1/p}(\Omega) \longrightarrow F^{p,q}_{-m-1+s+1/p}(\Omega), \tag{6.275}$$

$$L : \mathring{B}^{p,q}_{m-1+s+1/p}(\Omega) \longrightarrow B^{p,q}_{-m-1+s+1/p}(\Omega), \tag{6.276}$$

are isomorphisms.

Proof. Working in each component, there is no loss of generality in assuming that Ω is connected. Also, using Theorem 3.25 and invoking the fact that the property of being an isomorphism is stable on complex interpolation scales (cf. [63]), it is enough to treat just (6.275) in the case corresponding to $p = q = 2$ and $s = 1/2$.

Hence, consider $u \in \mathring{F}^{2,2}_m(\Omega) = \mathring{W}^{m,2}(\Omega)$ such that $Lu = 0$. From (2.219) we know that there exists a sequence of functions $u_j \in C^\infty_c(\Omega)$, $j \in \mathbb{N}$, with the property that $u_j \to u$ in $W^{m,2}(\Omega)$, as $j \to \infty$. Then, with $\langle \cdot, \cdot \rangle_{\mathscr{D}'(\Omega)-\mathscr{D}(\Omega)}$ denoting the standard distributional pairing in the open set Ω, we may write

$$\begin{aligned}
(-1)^m \int_\Omega \sum_{|\alpha|=|\beta|=m} &\langle A_{\alpha\beta}\, \partial^\alpha u(X)\,,\, \overline{\partial^\beta u(X)}\,\rangle\, dX \\
&= (-1)^m \lim_{j\to\infty} \int_\Omega \sum_{|\alpha|=|\beta|=m} \langle A_{\alpha\beta}\, \partial^\alpha u(X)\,,\, \overline{\partial^\beta u_j(X)}\,\rangle\, dX \\
&= (-1)^m \lim_{j\to\infty} \Big\langle \sum_{|\alpha|=|\beta|=m} A_{\alpha\beta}\, \partial^\alpha u\,,\, \overline{\partial^\beta u_j} \Big\rangle_{\mathscr{D}'(\Omega)-\mathscr{D}(\Omega)} \\
&= \lim_{j\to\infty} \Big\langle \sum_{|\alpha|=|\beta|=m} A_{\alpha\beta}\, \partial^\alpha \partial^\beta u\,,\, \overline{u_j} \Big\rangle_{\mathscr{D}'(\Omega)-\mathscr{D}(\Omega)} \\
&= \lim_{j\to\infty} \Big\langle Lu\,,\, \overline{u_j} \Big\rangle_{\mathscr{D}'(\Omega)-\mathscr{D}(\Omega)} = 0. \qquad (6.277)
\end{aligned}$$

On the other hand, based on formula (6.11) we may write

$$\begin{aligned}
\mathrm{Re} \int_\Omega \sum_{|\alpha|=|\beta|=m} &\langle A_{\alpha\beta}\, \partial^\alpha u(X)\,,\, \overline{\partial^\beta u(X)}\rangle\, dX \\
&= \lim_{j\to\infty} \int_\Omega \sum_{|\alpha|=|\beta|=m} \langle A_{\alpha\beta}\, \partial^\alpha u_j(X)\,,\, \overline{\partial^\beta u_j(X)}\rangle\, dX \\
&= \lim_{j\to\infty} \int_{\mathbb{R}^n} \sum_{|\alpha|=|\beta|=m} \langle A_{\alpha\beta}\, \partial^\alpha u_j(X)\,,\, \overline{\partial^\beta u_j(X)}\rangle\, dX \\
&\geq \lim_{j\to\infty} C \int_{\mathbb{R}^n} \sum_{|\gamma|=m} |\partial^\gamma u_j(X)|^2\, dX \\
&\geq \lim_{j\to\infty} C \int_\Omega \sum_{|\gamma|=m} |\partial^\gamma u_j(X)|^2\, dX \\
&= C \int_\Omega \sum_{|\gamma|=m} |\partial^\gamma u(X)|^2\, dX. \qquad (6.278)
\end{aligned}$$

From (6.277) and (6.278) we may therefore conclude that

$$\partial^\gamma u \equiv 0 \text{ in } \Omega, \quad \text{for each } \gamma \in \mathbb{N}_0^n \text{ with } |\gamma| = m. \tag{6.279}$$

In concert with Taylor's formula and the connectivity of Ω, (6.279) readily gives that u must be a polynomial of degree $\leq m-1$ in Ω. Given that we also know that $\text{tr}_{m-1}(u) = 0$ (cf. Corollary 3.16), this readily forces $u \equiv 0$ in Ω, as wanted. □

Next we indicate how the invertibility result from Theorem 6.32 may be combined with the multi-trace theory developed in (3.3) in order to prove well-posedness results for the inhomogeneous Dirichlet problem for higher-order operators.

Theorem 6.33. *Suppose that the differential operator L of order $2m$, $m \in \mathbb{N}$, is as in* (4.1)–(4.2), *and satisfies the Legendre–Hadamard ellipticity condition* (4.15). *Also, assume that Ω is a bounded Lipschitz domain in $\mathbb{R}^n$. Then there exists $\varepsilon > 0$ with the property that whenever*

$$|p-2| + |q-2| + |s-1/2| < \varepsilon \tag{6.280}$$

the inhomogeneous Dirichlet boundary value problems

$$\begin{cases} u \in B^{p,q}_{m-1+s+1/p}(\Omega), \\ Lu = v \in B^{p,q}_{-m-1+s+1/p}(\Omega), \\ \text{tr}_{m-1}u = \dot{f} \in \dot{B}^{p,q}_{m-1,s}(\partial\Omega), \end{cases} \tag{6.281}$$

and

$$\begin{cases} u \in F^{p,q}_{m-1+s+1/p}(\Omega), \\ Lu = v \in F^{p,q}_{-m-1+s+1/p}(\Omega), \\ \text{tr}_{m-1}u = \dot{f} \in \dot{B}^{p,p}_{m-1,s}(\partial\Omega), \end{cases} \tag{6.282}$$

are well-posed.

Proof. Assume that $\varepsilon > 0$ is as in Theorem 6.32. To prove existence for the boundary value problem (6.281) in this context, assume that $v \in B^{p,q}_{-m-1+s+1/p}(\Omega)$ and $\dot{f} \in \dot{B}^{p,q}_{m-1,s}(\partial\Omega)$ have been given. With the operator $\mathscr{E}$ as in Theorem 3.9, consider $\mathscr{E}\dot{f} \in B^{p,q}_{m-1+s+1/p}(\Omega)$ and note that $L(\mathscr{E}\dot{f}) \in B^{p,q}_{-m-1+s+1/p}(\Omega)$. By invoking Theorem 6.32 we may then find $w \in \mathring{B}^{p,q}_{m-1+s+1/p}(\Omega)$ with the property that $Lw = v - L(\mathscr{E}\dot{f})$. Hence, if we set $u := w + \mathscr{E}\dot{f}$ it follows that $u \in B^{p,q}_{m-1+s+1/p}(\Omega)$, $Lu = v$, and $\text{tr}_{m-1}u = \dot{f}$, by (3.171) and Theorem 3.9. Thus, u solves (6.281). The fact that this u also satisfies

$$\|u\|_{B^{p,q}_{m-1+s+1/p}(\Omega)} \le C\Big(\|v\|_{B^{p,q}_{-m-1+s+1/p}(\Omega)} + \|\dot{f}\|_{\dot{B}^{p,q}_{m-1,s}(\partial\Omega)}\Big), \tag{6.283}$$

for some finite constant $C = C(\Omega, L, p, q, s) > 0$ independent of $u, v, \dot{f}$, is implicit in the way u has been constructed. Finally, uniqueness for (6.281) is a direct consequence of (3.171) and Theorem 6.32. The same type of reasoning applies to the boundary value problem (6.282), and this completes the proof of the theorem. □

The following may be regarded as a companion result to Theorem 6.33. Compared to the latter, the main novelty is the inclusion of a larger range of indices for which the inhomogeneous Dirichlet problem is well-posed, under appropriate additional conditions on the underlying Lipschitz domain.

Theorem 6.34. *Let Ω be a bounded Lipschitz domain in $\mathbb{R}^n$. Assume that the differential operator L, of order $2m$, is as in (4.1)–(4.2), and satisfies the Legendre–Hadamard ellipticity condition (4.15). In addition, suppose that L is self-adjoint and that $n > 2m$. Finally, assume that*

$$s \in (0,1) \ \text{ and } \ p, q \in (1,\infty) \ \text{ are given.} \tag{6.284}$$

Then there exists $\varepsilon > 0$, depending only on the indices s, p, q, the Lipschitz character of Ω, and the differential operator L, with the property that if the outward unit normal ν to Ω satisfies

$$\limsup_{r\to 0^+} \Big\{ \sup_{X\in\partial\Omega} ⨍_{B(X,r)\cap\partial\Omega} ⨍_{B(X,r)\cap\partial\Omega} \big|\nu(Y) - \nu(Z)\big|\, d\sigma(Y) d\sigma(Z) \Big\} < \varepsilon, \tag{6.285}$$

then the inhomogeneous Dirichlet problem on Besov spaces

$$\begin{cases} u \in B^{p,q}_{m-1+s+1/p}(\Omega), \\ Lu = w \in B^{p,q}_{-m-1+s+1/p}(\Omega), \\ \operatorname{tr}_{m-1} u = \dot{f} \in \dot{B}^{p,q}_{m-1,s}(\partial\Omega), \end{cases} \tag{6.286}$$

is well-posed. That is, there exists a unique solution u of (6.286) and, for some finite constant $C = C(\Omega, p, q, s, L) > 0$, there holds

$$\|u\|_{B^{p,q}_{m-1+s+1/p}(\Omega)} \le C\Big(\|w\|_{B^{p,q}_{-m-1+s+1/p}(\Omega)} + \|\dot{f}\|_{\dot{B}^{p,q}_{m-1,s}(\partial\Omega)}\Big). \tag{6.287}$$

As a consequence, if the outward unit normal ν to Ω satisfies

$$\nu \in \mathrm{vmo}(\partial\Omega) \tag{6.288}$$

(hence, in particular, if Ω is actually a bounded C^1 domain), then the above well-posedness result is valid for every $s \in (0,1)$ and $p, q \in (1,\infty)$.

Analogous results to those stated above are valid for the inhomogeneous Dirichlet problem formulated on Triebel–Lizorkin spaces, i.e., for

$$\begin{cases} u \in F^{p,q}_{m-1+s+1/p}(\Omega), \\ Lu = w \in F^{p,q}_{-m-1+s+1/p}(\Omega), \\ \operatorname{tr}_{m-1} u = \dot{f} \in \dot{B}^{p,p}_{m-1,s}(\partial\Omega). \end{cases} \tag{6.289}$$

Finally, when $w = 0$*, the solution of* u *of either* (6.286) *or* (6.289) *admits the single multi-layer integral representation formula*

$$u = \dot{\mathscr{S}}\left(\dot{S}^{-1}\dot{f}\right) \quad in \ \ \Omega. \tag{6.290}$$

Proof. Retain the assumptions formulated in the first part of the statement. In particular, assume that $s \in (0,1)$ and $p, q \in (1,\infty)$ have been given, and consider

$$p', q' \in (1,\infty) \ \text{ such that } \ \tfrac{1}{p} + \tfrac{1}{p'} = \tfrac{1}{q} + \tfrac{1}{q'} = 1. \tag{6.291}$$

In this context, select $\varepsilon > 0$ as in Theorem 6.14, relative to the indices considered above. Such a choice guarantees (cf. (6.124)) that

$$\dot{S} : \left(\dot{B}^{p',q'}_{m-1,1-s}(\partial\Omega)\right)^* \longrightarrow \dot{B}^{p,q}_{m-1,s}(\partial\Omega) \ \text{ is invertible.} \tag{6.292}$$

Also, the general boundedness results established in Theorem 5.7 ensure that

$$\dot{\mathscr{S}} : \left(\dot{B}^{p',q'}_{m-1,1-s}(\partial\Omega)\right)^* \longrightarrow B^{p,q}_{m-1+s+1/p}(\Omega), \tag{6.293}$$

$$\dot{\mathscr{S}} : \left(\dot{B}^{p',p'}_{m-1,1-s}(\partial\Omega)\right)^* \longrightarrow F^{p,q}_{m-1+s+1/p}(\Omega), \tag{6.294}$$

continuously in each case.

Consider now arbitrary data $w \in B^{p,q}_{-m-1+s+1/p}(\Omega)$ and $\dot{f} \in \dot{B}^{p,q}_{m-1,s}(\partial\Omega)$. Based on (2.167) it is possible to find $v \in B^{p,q}_{-m-1+s+1/p}(\mathbb{R}^n)$ such that

$$\begin{gathered} v\Big|_{\Omega} = w, \quad \|v\|_{B^{p,q}_{-m-1+s+1/p}(\mathbb{R}^n)} \leq 2\|w\|_{B^{p,q}_{-m-1+s+1/p}(\Omega)}, \\ \text{and } \ \operatorname{supp} w \ \text{ is a compact subset of } \ \mathbb{R}^n. \end{gathered} \tag{6.295}$$

Then, with E denoting the fundamental solution for L constructed in Theorem 4.2, if we set

$$\omega := \left[E * v\right]\Big|_{\Omega}, \tag{6.296}$$

classical results imply that for some $C = C(L, n) \in (0, \infty)$,

$$\begin{aligned} &\omega \in B^{p,q}_{m-1+s+1/p}(\Omega), \quad L\omega = w \text{ in } \Omega, \\ &\|\omega\|_{B^{p,q}_{m-1+s+1/p}(\Omega)} \leq C\|w\|_{B^{p,q}_{-m-1+s+1/p}(\Omega)}. \end{aligned} \tag{6.297}$$

In particular,

$$\operatorname{tr}_{m-1}\omega \in \dot{B}^{p,q}_{m-1,s}(\partial\Omega) \text{ and } \left\|\operatorname{tr}_{m-1}\omega\right\|_{\dot{B}^{p,q}_{m-1,s}(\partial\Omega)} \leq C\|w\|_{B^{p,q}_{-m-1+s+1/p}(\Omega)}, \tag{6.298}$$

by (6.297) and Theorem 3.9. The idea is now to define

$$u := \omega + \dot{\mathscr{S}}\Big(\dot{S}^{-1}\Big[\dot{f} - \operatorname{tr}_{m-1}\omega\Big]\Big) \text{ in } \Omega, \tag{6.299}$$

and note that, by (6.292)–(6.293), this function is well-defined, belongs to the space $B^{p,q}_{m-1+s+1/p}(\Omega)$, and also satisfies (6.287), thanks to (6.298). Furthermore, we have $Lu = L\omega = w$ by (5.5) and (6.297), whereas by (5.44),

$$\begin{aligned} \operatorname{tr}_{m-1}u &= \operatorname{tr}_{m-1}\omega + \operatorname{tr}_{m-1}\dot{\mathscr{S}}\Big(\dot{S}^{-1}\Big[\dot{f} - \operatorname{tr}_{m-1}\omega\Big]\Big) \\ &= \operatorname{tr}_{m-1}\omega + \big(\dot{f} - \operatorname{tr}_{m-1}\omega\big) = \dot{f}, \end{aligned} \tag{6.300}$$

This proves existence and estimates for the problem (6.286). Also, the integral representation formula (6.299) reduces to (6.290) when $w = 0$ since, in this case, we may take $\omega = 0$ to begin with.

At this stage, as far as the well-posedness of the inhomogeneous Dirichlet problem (6.286) is concerned, there remains to establish uniqueness. To this end, observe that, generally speaking, existence for (6.286) implies (by taking $\dot{f} = 0$ and invoking Corollary 3.16) that

$$\text{the operator } L : \mathring{B}^{p,q}_{m-1+s+1/p}(\Omega) \longrightarrow B^{p,q}_{-m-1+s+1/p}(\Omega) \text{ is onto.} \tag{6.301}$$

Let us now assume that $\varepsilon > 0$ is small enough so that we may establish existence for (6.286), via the same type of argument as in (6.292)–(6.300), in the case when this boundary value problem is formulated using the Hölder conjugate exponents p', q' in place of the given p, q, and using $1 - s$ in lieu of the original s. Then, much as in the case of (6.301), we may now conclude that

$$\text{the operator } L : \mathring{B}^{p',q'}_{m-s+1/p'}(\Omega) \longrightarrow B^{p',q'}_{-m-s+1/p'}(\Omega) \text{ is onto.} \tag{6.302}$$

Based on this, Proposition 6.31, the assumption that L is self-adjoint, and standard functional analysis, we then deduce that

$$\text{the operator } L : \mathring{B}^{p,q}_{m-1+s+1/p}(\Omega) \longrightarrow B^{p,q}_{-m-1+s+1/p}(\Omega) \text{ is injective.} \tag{6.303}$$

In turn, this readily yields uniqueness for the problem (6.286) (again, by relying on Corollary 3.16). The treatment of the inhomogeneous Dirichlet problem (6.286) is therefore complete. That the boundary value problem (6.286) is well-posedness for every $s \in (0,1)$ and $p, q \in (1, \infty)$ under the stronger assumption (6.288) is seen from what we have proved so far and Theorem 6.12. Finally, the inhomogeneous Dirichlet problem (6.289) is treated similarly and this finishes the proof of the theorem. □

Our next theorem contains a more refined analysis in the case of the inhomogeneous Dirichlet problem for the biharmonic operator in planar domains. To state it, recall the principal-value biharmonic double multi-layer operator $\dot{K}_{\Delta^2}$ introduced in the discussion preceding Theorem 6.25, and denote by $\dot{\mathscr{D}}_{\Delta^2}$ the biharmonic double multi-layer defined as in (4.57) for the choice of the tensor coefficient $A = (A_{\alpha\beta})_{|\alpha|=|\beta|=2}$ used in the writing of $L = \Delta^2$ that has produced $\dot{K}_{\Delta^2}$.

Theorem 6.35. *Assume that Ω is a bounded Lipschitz domain in $\mathbb{R}^2$ with connected boundary, and whose unit normal belongs to* $\mathrm{vmo}(\partial\Omega)$. *Then the inhomogeneous Dirichlet problems*

$$\begin{cases} u \in B^{p,q}_{1+s+1/p}(\Omega), \\ \Delta^2 u = w \in B^{p,q}_{-3+s+1/p}(\Omega), \\ \big(\mathrm{Tr}\, u\,,\, \mathrm{Tr}(\nabla u)\big) = \dot{f} \in \dot{B}^{p,q}_{1,s}(\partial\Omega), \end{cases} \tag{6.304}$$

and

$$\begin{cases} u \in F^{p,q}_{1+s+1/p}(\Omega), \\ \Delta^2 u = w \in F^{p,q}_{-3+s+1/p}(\Omega), \\ \big(\mathrm{Tr}\, u\,,\, \mathrm{Tr}(\nabla u)\big) = \dot{f} \in \dot{B}^{p,p}_{1,s}(\partial\Omega), \end{cases} \tag{6.305}$$

are well-posed whenever

$$0 < p, q < \infty \quad \text{and} \quad \left(\tfrac{1}{p} - 1\right)_+ < s < 1. \tag{6.306}$$

Moreover, in both cases, the solution corresponding to the case when $w = 0$ may be expressed (with the conventions made earlier) as

$$u = \dot{\mathscr{D}}_{\Delta^2}\Big[\big(\tfrac{1}{2}I + \dot{K}_{\Delta^2}\big)^{-1}\dot{f}\Big] \quad \text{in } \Omega. \tag{6.307}$$

Proof. Fix indices p, q, s satisfying the conditions in (6.306) and assume that arbitrary data $w \in B^{p,q}_{-3+s+1/p}(\Omega)$ and $\dot{f} \in \dot{B}^{p,q}_{1,s}(\partial\Omega)$ have been given. Using (2.167) it is possible to select $v \in B^{p,q}_{-3+s+1/p}(\mathbb{R}^n)$ such that

$$v\Big|_{\Omega} = w, \quad \|v\|_{B^{p,q}_{-3+s+1/p}(\mathbb{R}^2)} \leq 2\|w\|_{B^{p,q}_{-3+s+1/p}(\Omega)}, \qquad (6.308)$$
$$\text{and } \operatorname{supp} w \text{ is a compact subset of } \mathbb{R}^2.$$

Recall that a fundamental solution for Δ^2 in the plane is given by

$$E(X) := \frac{1}{8\pi}|X|^2 \log|X|, \qquad X \in \mathbb{R}^2 \setminus \{0\}. \qquad (6.309)$$

With E as above, set $\omega := \big[E * v\big]\Big|_{\Omega}$. Then

$$\omega \in B^{p,q}_{1+s+1/p}(\Omega) \text{ and } \Delta^2\omega = w \text{ in } \Omega. \qquad (6.310)$$

Moreover, there exists $C \in (0, \infty)$, with the property that

$$\|\omega\|_{B^{p,q}_{1+s+1/p}(\Omega)} \leq C\|w\|_{B^{p,q}_{-3+s+1/p}(\Omega)}. \qquad (6.311)$$

In particular,

$$\operatorname{tr}_1\omega \in \dot{B}^{p,q}_{1,s}(\partial\Omega) \text{ and } \big\|\operatorname{tr}_1\omega\big\|_{\dot{B}^{p,q}_{1,s}(\partial\Omega)} \leq C\|w\|_{B^{p,q}_{-3+s+1/p}(\Omega)}, \qquad (6.312)$$

by (6.297) and Theorem 3.9. Define next

$$u = \omega + \dot{\mathscr{D}}_{\Delta^2}\Big[\big(\tfrac{1}{2}I + \dot{K}_{\Delta^2}\big)^{-1}(\dot{f} - \operatorname{tr}_1\omega)\Big] \text{ in } \Omega. \qquad (6.313)$$

It is then straightforward to check based on its definition, (4.58), (4.250), and (4.224) that u is a solution of (6.304) (satisfying a naturally accompanying estimate).

Turning our attention to proving uniqueness for the boundary problem (6.304) we start with the observation that, much as in the case of the uniqueness for the boundary problem (6.286) proved in Theorem 6.34, the fact that the problem (6.304) has a solution for every $p, q \in (1, \infty)$ and $s \in (0, 1)$ (which the proof so far guarantees) implies that uniqueness for the boundary problem (6.286) also holds for this range of indices.

As for the larger range specified in (6.306), observe first that if p, q, s are as in (6.306) then an elementary argument based on Proposition 3.8 shows that there exist $p^*, q^* \in (1, \infty)$ and $s^* \in (0, 1)$ such that

$$B^{p,q}_{1+s+1/p}(\Omega) \hookrightarrow B^{p^*,q^*}_{1+s^*+1/p^*}(\Omega). \qquad (6.314)$$

Then the uniqueness for (6.304) formulated using the indices p, q, s follows from the uniqueness for (6.304) formulated using the indices p^*, q^*, s^*, already treated in the previous paragraph.

This finishes the proof of the well-posedness for (6.304). Clearly, the integral representation formula (6.313) reduced to just (6.307) in the case when $w = 0$. Finally, the treatment of the problem (6.305) is similar to that for (6.304), and this completes the proof of Theorem 6.35. □

In the last part of this section we deal with the inhomogeneous Neumann boundary value problem. To state our first result in this regard, recall the manner in which the conormal operator has been introduced in Definition 5.20.

Theorem 6.36. *Assume that Ω is a bounded, connected, Lipschitz domain in $\mathbb{R}^n$, whose complement is connected. Also, assume that L is a homogeneous differential operator of order $2m$, $m \in \mathbb{N}$, associated with a constant (complex) matrix-valued tensor coefficient $A = (A_{\alpha\beta})_{|\alpha|=|\beta|=m}$ as in (4.1) which is self-adjoint and satisfies the S-ellipticity condition (4.16). Then there exists $\varepsilon > 0$ with the property that, whenever*

$$|p-2| + |q-2| + |s-1/2| < \varepsilon, \tag{6.315}$$

the inhomogeneous Neumann problem

$$\begin{cases} u \in B^{p,q}_{m-1+s+1/p}(\Omega), \\ Lu = w\big\lfloor_{\Omega}, \quad w \in \left(B^{p',q'}_{m-s+1/p'}(\Omega)\right)^*, \\ \partial^A_\nu(u,w) = \Lambda \in \left(\dot{B}^{p',q'}_{m-1,1-s}(\partial\Omega)\right)^*, \end{cases} \tag{6.316}$$

where $1/p + 1/p' = 1/q + 1/q' = 1$, and the boundary data satisfies the necessary compatibility condition

$$\langle \Lambda, \dot{P} \rangle = \langle w, P\big|_{\Omega} \rangle_{\Omega} \quad \text{for each} \quad P \in \mathscr{P}_{m-1}, \tag{6.317}$$

is well-posed (with uniqueness understood modulo polynomials of degree $\leq m-1$).

Moreover, a similar well-posedness result is valid for the inhomogeneous Neumann problem involving Triebel–Lizorkin spaces, i.e., for

$$\begin{cases} u \in F^{p,q}_{m-1+s+1/p}(\Omega), \\ Lu = w\big\lfloor_{\Omega}, \quad w \in \left(F^{p',q'}_{m-s+1/p'}(\Omega)\right)^*, \\ \partial^A_\nu u = \Lambda \in \left(\dot{B}^{p',p'}_{m-1,1-s}(\partial\Omega)\right)^*. \end{cases} \tag{6.318}$$

Proof. That the compatibility condition (6.317) is necessary is clear from formula (5.129) and simple degree considerations. We proceed by discussing a reduction step, in which it may be assumed that $p, q \in (1, \infty)$ and $s \in (0, 1)$ are arbitrary. As always, let p', q'denote the Hölder conjugate indices of p and q. Recall the restriction operator $\mathscr{R}_\Omega$ from (2.196). Given $w \in \left(B^{p',q'}_{m-s+1/p'}(\Omega)\right)^*$, observe that

$$
\begin{aligned}
w \circ \mathscr{R}_\Omega \in \left(B^{p',q'}_{m-s+1/p'}(\mathbb{R}^n)\right)^* &= B^{p,q}_{-m+s-1+1/p}(\mathbb{R}^n) \\
\text{and } \operatorname{supp}\left(w \circ \mathscr{R}_\Omega\right) &\subseteq \overline{\Omega},
\end{aligned}
\tag{6.319}
$$

hence

$$
w \circ \mathscr{R}_\Omega \in B^{p,q}_{-m+s-1+1/p,z}(\Omega). \tag{6.320}
$$

Let E be the fundamental solution associated with the operator L as in Theorem 4.2. Then

$$
E * \left(w \circ \mathscr{R}_\Omega\right) \in B^{p,q}_{m+s-1+1/p}(\mathbb{R}^n) \tag{6.321}
$$

and if

$$
\omega := \left[E * \left(w \circ \mathscr{R}_\Omega\right)\right]\Big|_\Omega \in \mathscr{D}'(\Omega), \tag{6.322}
$$

then there exists $C = C(L, n) \in (0, \infty)$ so that

$$
\begin{aligned}
&\omega \in B^{p,q}_{m-1+s+1/p}(\Omega), \quad L\omega = w\big\lfloor_\Omega, \\
&\|\omega\|_{B^{p,q}_{m-1+s+1/p}(\Omega)} \leq C\,\|w\|_{\left(B^{p',q'}_{m-s+1/p'}(\Omega)\right)^*}.
\end{aligned}
\tag{6.323}
$$

In particular,

$$
\begin{aligned}
&\partial^A_\nu(\omega, w) \in \left(\dot{B}^{p',q'}_{m-1,1-s}(\partial\Omega)\right)^* \quad \text{and} \\
&\left\|\partial^A_\nu(\omega, w)\right\|_{\left(\dot{B}^{p',q'}_{m-1,1-s}(\partial\Omega)\right)^*} \leq C\,\|w\|_{\left(B^{p',q'}_{m-s+1/p'}(\Omega)\right)^*},
\end{aligned}
\tag{6.324}
$$

by Definition 5.20. Furthermore, if $\Lambda \in \left(\dot{B}^{p',q'}_{m-1,1-s}(\partial\Omega)\right)^*$ satisfies the compatibility condition (6.317), then

$$
\begin{aligned}
&\Lambda - \partial^A_\nu(\omega, w) \in \left(\dot{B}^{p',q'}_{m-1,1-s}(\partial\Omega)\right)^* \quad \text{satisfies} \\
&\langle \Lambda - \partial^A_\nu(\omega, w), \dot{P}\rangle = 0 \quad \text{for each} \quad \dot{P} \in \dot{\mathscr{P}}_{m-1}(\partial\Omega),
\end{aligned}
\tag{6.325}
$$

thanks to (5.129). In summary, working with $u - \omega$ in place of u (and keeping in mind that $\partial_\nu^A(u-\omega,0) = \partial_\nu^A(u,w) - \partial_\nu^A(\omega,w)$), as far as the well-posedness of the inhomogeneous Neumann problem (6.316) is concerned, matters have been reduced to proving the well-posedness of (6.316) in the particular case when $w = 0$ and the compatibility condition (6.317) takes the form

$$\langle \Lambda, \dot{P} \rangle = 0 \quad \text{for each} \quad \dot{P} \in \dot{\mathscr{P}}_{m-1}(\partial\Omega). \tag{6.326}$$

Assume that this is the case and select $\varepsilon > 0$ as in Theorem 6.17. In addition, suppose now that p, q, s are as in (6.315). Given $\Lambda \in \left(\dot{B}^{p',q'}_{m-1,1-s}(\partial\Omega)\right)^*$ satisfying the compatibility condition (6.326), define

$$\widetilde{\Lambda} \in \left(\dot{B}^{p',q'}_{m-1,1-s}(\partial\Omega) \big/ \dot{\mathscr{P}}_{m-1}(\partial\Omega)\right)^* \tag{6.327}$$

by setting

$$\langle \widetilde{\Lambda}, [\dot{f}] \rangle := \langle \Lambda, \dot{f} \rangle, \qquad \forall\, \dot{f} \in \dot{B}^{p',q'}_{m-1,1-s}(\partial\Omega), \tag{6.328}$$

where $[\dot{f}]$ denotes the equivalence class of the Whitney array $\dot{f} \in \dot{B}^{p',q'}_{m-1,1-s}(\partial\Omega)$ in the quotient space $\dot{B}^{p',q'}_{m-1,1-s}(\partial\Omega) \big/ \dot{\mathscr{P}}_{m-1}(\partial\Omega)$. Thanks to (6.326), this definition is unambiguous.

Going further, let

$$\pi : \dot{B}^{p',q'}_{m-1,1-s}(\partial\Omega) \longrightarrow \dot{B}^{p',q'}_{m-1,1-s}(\partial\Omega) \big/ \dot{\mathscr{P}}_{m-1}(\partial\Omega) \tag{6.329}$$

denote the canonical projection operator, taking an arbitrary Whitney-Besov array $\dot{f} \in \dot{B}^{p',q'}_{m-1,1-s}(\partial\Omega)$ into $[\dot{f}] \in \dot{B}^{p',q'}_{m-1,1-s}(\partial\Omega) \big/ \dot{\mathscr{P}}_{m-1}(\partial\Omega)$. Its adjoint then becomes

$$\pi^* : \left(\dot{B}^{p',q'}_{m-1,1-s}(\partial\Omega) \big/ \dot{\mathscr{P}}_{m-1}(\partial\Omega)\right)^* \longrightarrow \left(\dot{B}^{p',q'}_{m-1,1-s}(\partial\Omega)\right)^*, \tag{6.330}$$

and we proceed to define

$$u = \dot{\mathscr{S}}\left(\pi^*\big(-\tfrac{1}{2}I + \dot{K}^*\big)^{-1}\widetilde{\Lambda}\right) \text{ in } \Omega, \tag{6.331}$$

where the inverse operator is understood in the sense of (6.175). Then we have that $u \in B^{p,q}_{m-1+s+1/p}(\Omega)$ and $Lu = 0$ in Ω by (5.43) and (5.5). Moreover, by the continuity of the operators involved,

$$\|u\|_{B^{p,q}_{m-1+s+1/p}(\Omega)} \leq C\|\Lambda\|_{\left(\dot{B}^{p',q'}_{m-1,1-s}(\partial\Omega)\right)^*} \tag{6.332}$$

for some finite constant $C = C(\Omega, L) > 0$ independent of Λ. Next, based on (6.331) and (5.152) we obtain

$$\partial_\nu^A u = (-\tfrac{1}{2}I + \dot{K}^*)\Big(\pi^*(-\tfrac{1}{2}I + \dot{K}^*)^{-1}\widetilde{\Lambda}\Big) \text{ in } \Big(\dot{B}^{p',q'}_{m-1,1-s}(\partial\Omega)\Big)^*. \quad (6.333)$$

To further understand the nature of the functional in the right-hand side of (6.333), pick an arbitrary $\dot{f} \in \dot{B}^{p',q'}_{m-1,1-s}(\partial\Omega)$ and compute

$$\begin{aligned}
&\Big\langle(-\tfrac{1}{2}I + \dot{K}^*)\Big(\pi^*(-\tfrac{1}{2}I + \dot{K}^*)^{-1}\widetilde{\Lambda}\Big), \dot{f}\Big\rangle \\
&\quad = \Big\langle \pi^*(-\tfrac{1}{2}I + \dot{K}^*)^{-1}\widetilde{\Lambda}, (-\tfrac{1}{2}I + \dot{K})\dot{f}\Big\rangle \\
&\quad = \Big\langle (-\tfrac{1}{2}I + \dot{K}^*)^{-1}\widetilde{\Lambda}, \pi(-\tfrac{1}{2}I + \dot{K})\dot{f}\Big\rangle \\
&\quad = \Big\langle (-\tfrac{1}{2}I + \dot{K}^*)^{-1}\widetilde{\Lambda}, [(-\tfrac{1}{2}I + \dot{K})\dot{f}]\Big\rangle \\
&\quad = \Big\langle (-\tfrac{1}{2}I + \dot{K}^*)^{-1}\widetilde{\Lambda}, (-\tfrac{1}{2}I + \dot{K})[\dot{f}]\Big\rangle \\
&\quad = \Big\langle \widetilde{\Lambda}, (-\tfrac{1}{2}I + \dot{K})^{-1}(-\tfrac{1}{2}I + \dot{K})[\dot{f}]\Big\rangle \\
&\quad = \langle \widetilde{\Lambda}, [\dot{f}]\rangle = \langle \Lambda, \dot{f}\rangle.
\end{aligned} \quad (6.334)$$

Hence, ultimately,

$$\partial_\nu^A u = \Lambda \text{ in } \Big(\dot{B}^{p',q'}_{m-1,1-s}(\partial\Omega)\Big)^*, \quad (6.335)$$

which finishes the proof of the fact that u in (6.331) is a solution for the Neumann problem (6.316) when $w = 0$.

Regarding the well-posedness of the problem (6.316), there remains to establish uniqueness (in the sense specified in the statement). To this end, if u is a solution of (6.316) with $w = 0$ and $\Lambda = 0$, then Green's formula (5.141) gives

$$u = \dot{\mathscr{D}}(\text{tr}_{m-1}u) - \dot{\mathscr{S}}(\partial_\nu^A u) = \dot{\mathscr{D}}(\text{tr}_{m-1}u) \text{ in } \Omega, \quad (6.336)$$

where for the second equality we have used that $\partial_\nu^A u = 0$. Applying tr_{m-1} to both sides of (6.336) and using (4.250) then yields

$$\text{tr}_{m-1}u = (\tfrac{1}{2}I + \dot{K})(\text{tr}_{m-1}u), \quad (6.337)$$

hence, $(-\frac{1}{2}I + \dot{K})(\text{tr}_{m-1}u) = 0$. Using (6.126) in Theorem 6.15 this further implies that $\text{tr}_{m-1}u \in \dot{\mathscr{P}}_{m-1}$. In particular, there exists a polynomial $P \in \mathscr{P}_{m-1}$ such that $\mathrm{tr}_{m-1}u = \mathrm{tr}_{m-1}P$. Combining this with (6.336) and (4.62) finally gives that $u = \dot{\mathscr{D}}(\mathrm{tr}_{m-1}u) = \dot{\mathscr{D}}(\mathrm{tr}_{m-1}P) = P$, as desired.

Finally, the well-posedness of the problem (6.318) is handled similarly, and this finishes the proof of the theorem.

A more nuanced well-posedness result for the inhomogeneous Neumann problem may be proved for the biharmonic operator in planar domains.

Theorem 6.37. *Assume that Ω is a bounded Lipschitz domain in $\mathbb{R}^2$ with connected boundary, and whose unit normal belongs to* $\mathrm{vmo}(\partial\Omega)$. *Let $A = (A_{\alpha\beta})_{|\alpha|=|\beta|=2}$ be the coefficient tensor in the writing of $L = \Delta^2$ which produces $\dot{K}_{\Delta^2}$ (from the discussion preceding Theorem 6.25). Then the Neumann problem for the biharmonic operator*

$$\begin{cases} u \in B^{p,q}_{1+s+1/p}(\Omega), \\ \Delta^2 u = w\big\lfloor_{\Omega}, \quad w \in \left(B^{p',q'}_{2-s+1/p'}(\Omega)\right)^*, \\ \partial^A_\nu u = \Lambda \in \left(\dot{B}^{p',q'}_{1,1-s}(\partial\Omega)\right)^*, \end{cases} \tag{6.338}$$

where the boundary datum satisfies the necessary compatibility condition

$$\langle \Lambda, \dot{P}\rangle = \left\langle w, P\big|_{\Omega}\right\rangle_{\Omega} \quad \textit{for each} \quad P \in \mathscr{P}_1, \tag{6.339}$$

is well-posed, with uniqueness understood modulo polynomials of degree ≤ 1, whenever

$$s \in (0,1) \ \text{ and } \ 1 < p,q,p',q' < \infty \ \text{ are such that } \ \tfrac{1}{p} + \tfrac{1}{p'} = \tfrac{1}{q} + \tfrac{1}{q'} = 1. \tag{6.340}$$

Moreover, a similar well-posedness result holds for the Neumann problem

$$\begin{cases} u \in F^{p,q}_{1+s+1/p}(\Omega), \\ \Delta^2 u = w\big\lfloor_{\Omega}, \quad w \in \left(B^{p',q'}_{2-s+1/p'}(\Omega)\right)^*, \\ \partial^A_\nu u = \Lambda \in \left(\dot{B}^{p',p'}_{1,1-s}(\partial\Omega)\right)^*. \end{cases} \tag{6.341}$$

In particular, the above well-posedness results hold for any bounded C^1 domain $\Omega \subset \mathbb{R}^2$ with connected boundary.

Proof. All claims may be established along the lines of the proof of Theorem 6.36, with the help of the invertibility results established in Theorem 6.26. □

We conclude this section by recording a well-posedness result for the inhomogeneous Dirichlet problem for the bi-Laplacian in Lipschitz domains in $\mathbb{R}^3$ (the higher-dimensional setting is considered separately, in the next section).

Theorem 6.38. *Assume that $\Omega \subset \mathbb{R}^3$ is a bounded Lipschitz domain (of arbitrary topology). Then there exists $\varepsilon = \varepsilon(\Omega) \in (0,1]$ with the following property. Suppose that $0 < q \leq \infty$ and that s, p are such that either of the following two conditions holds:*

$$\begin{aligned} &(I): \quad 0 \leq \tfrac{1}{p} < \tfrac{s}{2} + \tfrac{1+\varepsilon}{2} \quad \text{and} \quad 0 < s < \varepsilon, \\ &(II): \quad -\tfrac{\varepsilon}{2} < \tfrac{1}{p} - \tfrac{s}{2} < \tfrac{1+\varepsilon}{2} \quad \text{and} \quad \varepsilon \leq s < 1. \end{aligned} \tag{6.342}$$

Then the inhomogeneous Dirichlet problem

$$\begin{cases} u \in B^{p,q}_{s+\frac{1}{p}+1}(\Omega), \\ \Delta^2 u = w \in B^{p,q}_{s+\frac{1}{p}-3}(\Omega), \\ \big(\mathrm{Tr}\, u\,,\, \mathrm{Tr}(\nabla u)\big) = \dot{f} \in \dot{B}^{p,q}_{1,s}(\partial\Omega), \end{cases} \tag{6.343}$$

is well-posed. In particular, there exists a finite constant $C = C(\Omega, s, p) > 0$ with the property that the solution u of (6.343) satisfies

$$\|u\|_{B^{p,q}_{s+\frac{1}{p}+1}(\Omega)} \leq C\Big(\|w\|_{B^{p,q}_{s+\frac{1}{p}-3}(\Omega)} + \|\dot{f}\|_{\dot{B}^{p,p}_{1,s}(\partial\Omega)}\Big). \tag{6.344}$$

Furthermore, a similar well-posedness result is valid for the version of the above boundary problem formulated on Triebel–Lizorkin spaces, that is, for

$$\begin{cases} u \in F^{p,q}_{s+\frac{1}{p}+1}(\Omega), \\ \Delta^2 u = f \in F^{p,q}_{s+\frac{1}{p}-3}(\Omega), \\ \big(\mathrm{Tr}\, u\,,\, \mathrm{Tr}(\nabla u)\big) = \dot{f} \in \dot{B}^{p,p}_{1,s}(\partial\Omega), \end{cases} \tag{6.345}$$

granted that, this time, $\max\{p,q\} < \infty$.

This is the main result in [88], to which the reader is referred to for a proof.

6.5 Boundary Problems for the Bi-Laplacian in Higher Dimensions

The discussion in this section is largely motivated by the classical free-plate problem arising in the Kirchhoff–Love theory of thin plates. In the case of a domain Ω in the two dimensional setting, this problem reads as follows:

$$\Delta^2 u = 0 \ \text{ in } \Omega, \ \text{ with } Mu \text{ and } Nu \text{ prescribed on } \partial\Omega, \tag{6.346}$$

where the boundary operators M, N are defined by

$$\begin{aligned} Mu &:= \eta \Delta u + (1-\eta)\frac{\partial^2 u}{\partial \nu^2}, \\ Nu &:= \frac{\partial \Delta u}{\partial \nu} + (1-\eta)\frac{\partial^3 u}{\partial \nu \partial \tau^2}, \end{aligned} \tag{6.347}$$

where η is the Poisson coefficient of the plate, and ν, τ denote, respectively, the outward unit normal and unit tangent to $\partial\Omega$. See, e.g., [4], [10, (3.29)–(3.31), p. 679], [9, (10)–(11), p. 1237], [50, p. 6], [51], [97, Proposition 3.1], [99, (2.2)–(2.3), p. 24], [109, (2.12), p. 136], [32, pp. 420–423], [98] as well as the informative discussion in [96] where it is indicated that the above problem has been first solved by Gustav Kirchhoff in a variational sense. Indeed, it is now folklore that, for boundary data in appropriate function spaces (and by imposing suitable bounds on the Poisson coefficient), the problem (6.346) has a unique variational solution $u \in W^{2,2}(\Omega)/\mathscr{P}_1$.

One of our main goals is to study further regularity properties of such a solution, measured on Besov and Triebel–Lizorkin scales. We shall do so working in the higher dimensional setting and the starting point is to establish well-posedness results when the size of the solution is measured using the non-tangential maximal operator. The final results are then obtained via interpolation.

To set the stage, fix $n \in \mathbb{N}$ with $n \geq 2$ and, as in the past, denote by $\{e_j\}_{1\leq j\leq n}$ the standard orthonormal basis in $\mathbb{R}^n$. As before, we continue to canonically identify these vectors with multi-indices from $\mathbb{N}_0^n$. Given an arbitrary number $\theta \in \mathbb{R}$, consider the coefficient tensor

$$A_\theta := \big(A_{\alpha\beta}(\theta)\big)_{|\alpha|=|\beta|=2}, \tag{6.348}$$

with scalar entries, defined for every pair of multi-indices $\alpha, \beta \in \mathbb{N}_0^n$ with the property that $|\alpha| = |\beta| = 2$ by the formula

$$\begin{aligned} A_{\alpha\beta}(\theta) := \frac{1}{1+2\theta+n\theta^2} \sum_{i,j=1}^{n} \Big(\delta_{\beta(e_i+e_j)} + \theta\,\delta_{ij} \sum_{k=1}^{n} \delta_{\beta(2e_k)}\Big) \times \\ \times \Big(\delta_{\alpha(e_i+e_j)} + \theta\,\delta_{ij} \sum_{k=1}^{n} \delta_{\alpha(2e_k)}\Big) \end{aligned} \tag{6.349}$$

where, generally speaking,

$$\delta_{ab} := \begin{cases} 1 \text{ if } a = b, \\ 0 \text{ if } a \neq b, \end{cases} \tag{6.350}$$

stands for the usual Kronecker symbol. Next, consider a bounded Lipschitz domain $\Omega \subset \mathbb{R}^n$ and, in relation to the coefficient tensor (6.348)–(6.349), for each $\theta \in \mathbb{R}$ introduce the bilinear form (with $\Delta := \partial_1^2 + \cdots + \partial_n^2$ denoting, as usual, the Laplacian in $\mathbb{R}^n$)

$$\mathscr{B}_\theta(u,v) := \sum_{|\alpha|=|\beta|=2} \int_\Omega A_{\alpha\beta}(\theta)(\partial^\beta u)(X)(\partial^\alpha v)(X)\,dX \tag{6.351}$$

$$= \frac{1}{1+2\theta+n\theta^2} \sum_{i,j=1}^n \int_\Omega \big[(\partial_i\partial_j + \theta\,\delta_{ij}\,\Delta)u\big](X)\big[(\partial_i\partial_j + \theta\,\delta_{ij}\,\Delta)v\big](X)\,dX,$$

where u, v are any two reasonably behaved (real-valued) functions in Ω. See, e.g., [10, Lemma 3.4, p. 680], [50, p. 5], [99, (2.13), p. 25], [129, (10.2)]. Then it can be readily verified that for each $\theta \in \mathbb{R}$ the bi-Laplacian may be written as

$$\Delta^2 = \sum_{|\alpha|=|\beta|=2} \partial^\alpha A_{\alpha\beta}(\theta)\,\partial^\beta. \tag{6.352}$$

In particular, for each $\theta \in \mathbb{R}$ the bilinear form $\mathscr{B}_\theta(\cdot,\cdot)$ introduced in (6.351) satisfies

$$\mathscr{B}(u,v) = \int_\Omega (\Delta^2 u)(X)v(X)\,dX, \qquad \forall\, u, v \in C_c^\infty(\Omega). \tag{6.353}$$

Indeed, it is easy to check that

$$\frac{1}{1+2\theta+n\theta^2} \sum_{i,j=1}^n (\partial_i\partial_j + \theta\delta_{ij}\Delta)(\partial_i\partial_j + \theta\delta_{ij}\Delta) = \Delta^2. \tag{6.354}$$

Let us also note that

$$\Delta^2 \text{ is S-elliptic,} \tag{6.355}$$

since, as a direct calculation based on (6.349) shows,

$$\sum_{|\alpha|=|\beta|=2} A_{\alpha\beta}(\theta)\,\xi^{\alpha+\beta} = |\xi|^4, \quad \text{for each } \xi \in \mathbb{R}^n. \tag{6.356}$$

Going further, given a bounded Lipschitz domain $\Omega \subset \mathbb{R}^n$ with outward unit normal $\nu = (\nu_j)_{1\le j\le n}$ and a function $u \in C^1(\Omega)$, define the normal derivative, $\partial_\nu u$, of u on the boundary of Ω by the formula

$$\partial_\nu u := \sum_{i=1}^n \nu_i(\partial_i u)\big|_{\partial\Omega}, \tag{6.357}$$

whenever the boundary traces in the right-hand side are meaningful. Hence,

$$\partial_\nu u = \nu \cdot \big((\nabla u)\big|_{\partial\Omega}\big). \tag{6.358}$$

In a similar manner, let us agree that for a function $u \in C^2(\Omega)$,

$$\partial_\nu^2 u := \sum_{i,j=1}^{n} \nu_i \nu_j (\partial_i \partial_j u)\big|_{\partial\Omega}. \tag{6.359}$$

Finally, if $u \in C^3(\Omega)$, for each $\theta \in \mathbb{R}$ set (in analogy with (6.347), following [129])

$$N_\theta(u) := \partial_\nu(\Delta u) + \frac{1}{2(1+2\theta+n\theta^2)} \sum_{i,j=1}^{n} \partial_{\tau_{ij}} \Big(\sum_{k=1}^{n} \nu_k \partial_{\tau_{ij}} \partial_k u\Big),$$

$$M_\theta(u) := \frac{2\theta+n\theta^2}{1+2\theta+n\theta^2} \Delta u + \frac{1}{1+2\theta+n\theta^2} \sum_{j,k=1}^{n} \nu_j \nu_k \partial_j \partial_k u, \tag{6.360}$$

where all spacial partial derivatives of u in the right-hand sides are understood as being restricted (either in a non-tangential pointwise sense, or as tangential derivatives of such traces) to $\partial\Omega$. Simple algebraic manipulations show that the above operators may be alternatively expressed as

$$\begin{aligned} N_\theta(u) &= \partial_\nu(\Delta u) + \frac{1}{1+2\theta+n\theta^2} \sum_{i,j=1}^{n} \partial_{\tau_{ij}} \Big(\sum_{k=1}^{n} \nu_k \nu_i \partial_j \partial_k u\Big) \\ &= \partial_\nu(\Delta u) + \frac{1}{1+2\theta+n\theta^2} \sum_{i,j=1}^{n} \partial_{\tau_{ij}} \Big(\sum_{k=1}^{n} \nu_i \partial_{\tau_{kj}} \partial_k u\Big), \end{aligned} \tag{6.361}$$

and

$$M_\theta(u) = \Delta u + \frac{1}{1+2\theta+n\theta^2} \sum_{j,k=1}^{n} \nu_j \partial_{\tau_{kj}} \partial_k u. \tag{6.362}$$

The relationship between the operators N_θ, M_θ and the bilinear form $\mathscr{B}_\theta(\cdot,\cdot)$ is brought to prominence in the following result, describing a Green-type formula for the bi-Laplacian (cf. [10, Lemma 3.4, p. 680] and [99, (2.20), p. 26] for a proof in domains in $\mathbb{R}^2$, and [129, (10.2)] for a statement in the setting of biharmonic functions in domains in $\mathbb{R}^n$, $n \geq 2$).

Proposition 6.39. *Assume that $\Omega \subset \mathbb{R}^n$ is a bounded Lipschitz domain with outward unit normal $\nu = (\nu_j)_{1\leq j\leq n}$ and surface measure σ. Let $\theta \in \mathbb{R}$ and recall the operators N_θ and M_θ introduced in (6.360), relative to this setting. Then for any $u, v \in C^\infty(\overline{\Omega})$ there holds*

$$\mathscr{B}_\theta(u,v) = \int_{\partial\Omega} \Big\langle \big(M_\theta(u), N_\theta(u)\big), (\partial_\nu v, -v) \Big\rangle \, d\sigma + \int_\Omega (\Delta^2 u)(X) v(X)\, dX, \quad (6.363)$$

where $\langle\cdot,\cdot\rangle$ denotes the canonical pointwise scalar product between vector-valued functions.

In particular, if $v \in C^\infty(\overline{\Omega})$ and u is a reasonably behaved null-solution of the bi-Laplacian Δ^2 in Ω, then

$$\mathscr{B}_\theta(u,v) = \int_{\partial\Omega} \Big\langle \big(M_\theta(u), N_\theta(u)\big), (\partial_\nu v, -v) \Big\rangle \, d\sigma. \quad (6.364)$$

Proof. Integrating by parts and using Einstein's convention of summation over repeated indices, we may write

$$\begin{aligned}
&\int_\Omega (\partial_i\partial_j + \theta\delta_{ij}\Delta)u \cdot (\partial_i\partial_j + \theta\delta_{ij}\Delta)v\, dX \qquad (6.365)\\
&= -\int_\Omega \partial_i(\partial_i\partial_j + \theta\delta_{ij}\Delta)u \cdot \partial_j v\, dX + \int_{\partial\Omega} \nu_i \cdot (\partial_i\partial_j + \theta\delta_{ij}\Delta)u \cdot \partial_j v\, d\sigma\\
&\quad -\theta\delta_{ij} \cdot \Big\{\int_\Omega \partial_k(\partial_i\partial_j + \theta\delta_{ij}\Delta)u \cdot \partial_k v\, dX + \int_{\partial\Omega} (\partial_i\partial_j + \theta\delta_{ij}\Delta)u \cdot \partial_\nu v\, d\sigma\Big\}.
\end{aligned}$$

Integrating by parts one more time and using (6.354), identity (6.365) further implies

$$\begin{aligned}
&\int_\Omega (\partial_i\partial_j + \theta\delta_{ij}\Delta)u \cdot (\partial_i\partial_j + \theta\delta_{ij}\Delta)v\, dX - (1 + 2\theta + n\theta^2)\int_\Omega (\Delta^2 u)v\, dX\\
&= -\int_{\partial\Omega} \nu_j \cdot \partial_i(\partial_i\partial_j + \theta\delta_{ij}\Delta)u \cdot v\, d\sigma + \int_{\partial\Omega} \nu_i \cdot (\partial_i\partial_j + \theta\delta_{ij}\Delta)u \cdot \partial_j v\, d\sigma \qquad (6.366)\\
&\quad - \theta\delta_{ij} \cdot \Big\{\int_{\partial\Omega} \nu_k \cdot \partial_k(\partial_i\partial_j + \theta\delta_{ij}\Delta)u \cdot v\, d\sigma + \int_{\partial\Omega} (\partial_i\partial_j + \theta\delta_{ij}\Delta)u \cdot \partial_\nu v\, d\sigma\Big\}.
\end{aligned}$$

Using that $\partial_j = \nu_r\nu_r\partial_j = \nu_r\partial_{\tau_{rj}} + \nu_j\partial_\nu$ in the second term in the right-hand side of (6.366) allows us to express this as

$$\int_{\partial\Omega} \nu_i \cdot (\partial_i\partial_j + \theta\delta_{ij}\Delta)u \cdot \partial_j v\, d\sigma = \int_{\partial\Omega} (\partial_i\partial_j + \theta\delta_{ij}\Delta)u\, \nu_i\nu_j\partial_\nu v\, d\sigma \qquad (6.367)$$

$$+ \int_{\partial\Omega} \partial_{\tau_{jr}} \big[\nu_i\nu_r(\partial_i\partial_j + \theta\delta_{ij}\Delta)u\big]\, v\, d\sigma.$$

In turn, this and (6.366) give

$$\int_\Omega (\partial_i\partial_j + \theta\delta_{ij}\Delta)u \cdot (\partial_i\partial_j + \theta\delta_{ij}\Delta)v\, dX \qquad (6.368)$$

$$= (1 + 2\theta + n\theta^2) \int_\Omega (\Delta^2 u)(X) v(X)\, dX$$

$$+ \int_{\partial\Omega} \mathrm{I}(u) \cdot v\, d\sigma + \int_{\partial\Omega} \mathrm{II}(u) \cdot \partial_\nu v\, d\sigma,$$

where we have set

$$\mathrm{I}(u) := \partial_{\tau_{jr}}[\nu_i\nu_r(\partial_i\partial_j + \theta\delta_{ij}\Delta)u] - \nu_j\partial_i(\partial_i\partial_j + \theta\delta_{ij}\Delta)u$$

$$-\theta\delta_{ij}\nu_k\partial_k(\partial_i\partial_j + \theta\delta_{ij}\Delta)u, \qquad (6.369)$$

and

$$\mathrm{II}(u) := \theta\delta_{ij}(\partial_i\partial_j + \theta\delta_{ij}\Delta)u + \nu_i\nu_j(\partial_i\partial_j + \theta\delta_{ij}\Delta)u. \qquad (6.370)$$

Next, observe that

$$\partial_{\tau_{jr}}\big[\nu_i\nu_r\theta\delta_{ij}\Delta u\big] = \theta\partial_{\tau_{jr}}\big[\nu_j\nu_r\Delta u\big] = 0 \qquad (6.371)$$

by symmetry considerations, and that

$$\partial_{\tau_{jr}}\big[\nu_i\nu_r\partial_i\partial_j u\big] = \tfrac{1}{2}\Big\{\partial_{\tau_{jr}}\big[\nu_i\nu_r\partial_i\partial_j u\big] + \partial_{\tau_{rj}}\big[\nu_i\nu_j\partial_i\partial_r u\big]\Big\} \qquad (6.372)$$

$$= -\tfrac{1}{2}\partial_{\tau_{jr}}\big[\nu_i\partial_{\tau_{jr}}\partial_i u\big],$$

where the first identity in formula (6.372) follows from rewriting the expression $\partial_{\tau_{jr}}\big[\nu_i\nu_r\partial_i\partial_j u\big]$ as $\partial_{\tau_{rj}}\big[\nu_i\nu_j\partial_i\partial_r u\big]$ and the second one uses the definition of $\partial_{\tau_{jr}}$. Based on (6.369) and (6.371)–(6.372), straightforward algebraic manipulations yield

$$\mathrm{I}(u) = -(1 + 2\theta + n\theta^2)\partial_\nu\Delta u - \tfrac{1}{2}\partial_{\tau_{ij}}\big[\nu_k\partial_{\tau_{ij}}\partial_k u\big] \qquad (6.373)$$

$$= -(1 + 2\theta + n\theta^2)N_\theta(u).$$

Also, a simple inspection of (6.370) reveals that

$$\text{II}(u) = (2\theta + n\theta^2)\Delta u + \partial_\nu^2 u = (1 + 2\theta + n\theta^2)M_\theta(u). \tag{6.374}$$

At this stage, (6.364) follows form (6.368) and (6.373)–(6.374). □

In our next proposition we identify the formula for the conormal derivative associated with the writing of the bi-Laplacian as in (6.352) for the tensor coefficient given in (6.348)–(6.349). As a preamble, the reader is reminded that the trace operator of order one on the boundary of a Lipschitz domain is defined by (cf. (3.100))

$$\operatorname{tr}_1 u := \{\operatorname{Tr}[\partial^\alpha u]\}_{|\alpha|\le 1}, \tag{6.375}$$

whenever meaningful.

Proposition 6.40. *Let $\Omega \subset \mathbb{R}^n$ be a bounded Lipschitz domain with outward unit normal $\nu = (\nu_j)_{1\le j\le n}$ and surface measure σ. Pick $\theta \in \mathbb{R}$ and recall the operators N_θ and M_θ from* (6.360)*, corresponding to this setting. Then for any reasonably well-behaved biharmonic function u in Ω there holds:*

$$\begin{aligned}&\textit{coefficient tensor } A_\theta \textit{ as in } (6.348)\text{–}(6.349) \Rightarrow \partial_\nu^{A_\theta} u := \Big\{\Big(\partial_\nu^{A_\theta} u\Big)_r\Big\}_{0\le r\le n}\\ &\textit{where } \Big(\partial_\nu^{A_\theta} u\Big)_0 = -N_\theta(u) \textit{ and } \Big(\partial_\nu^{A_\theta} u\Big)_r = \nu_r M_\theta(u) \textit{ for } 1\le r\le n.\end{aligned} \tag{6.376}$$

Proof. Let u be as in the statement and pick and arbitrary function $v \in C^\infty(\overline{\Omega})$. Based on formula (5.91) (used here with $L = \Delta^2$) and identity (1.40) we may conclude that

$$\begin{aligned}\int_{\partial\Omega} \langle \partial_\nu^{A_\theta} u, \operatorname{tr}_1 v\rangle\, d\sigma &= \mathscr{B}_\theta(u,v) = \int_{\partial\Omega} \Big[M_\theta(u)\partial_\nu v - N_\theta(u) v\Big]\, d\sigma \\ &= \int_{\partial\Omega} \Big\langle \big(-N_\theta(u), \nu_1 M_\theta(u), \dots, \nu_n M_\theta(u)\big), \operatorname{tr}_1 v\Big\rangle\, d\sigma.\end{aligned} \tag{6.377}$$

Therefore, (6.376) follows. □

It is useful to record the explicit expressions of the components of the conormal. Indeed, making use of the first formula in (6.361) and the second formula in (6.360), it follows that the components of $\partial_\nu^{A_\theta} u$ described in (6.376) are (using the usual summation convention over repeated indices):

$$\begin{aligned}\Big(\partial_\nu^{A_\theta} u\Big)_0 &= -\partial_\nu(\Delta u) - \frac{1}{1+2\theta+n\theta^2}\partial_{\tau_{ij}}\Big(\nu_\ell \nu_i \partial_j \partial_\ell u\Big) \quad \text{and}\\ \Big(\partial_\nu^{A_\theta} u\Big)_r &= \frac{2\theta+n\theta^2}{1+2\theta+n\theta^2}\nu_r \Delta u + \frac{1}{1+2\theta+n\theta^2}\nu_r\nu_j\nu_\ell\partial_j\partial_\ell u \quad \text{for } 1\le r\le n,\end{aligned} \tag{6.378}$$

again, with the understanding that all derivatives in the right-hand sides are restricted to the boundary.

We continue by recording a result whose relevance is going to be apparent shortly.

Proposition 6.41. *Assume that Ω is a bounded Lipschitz domain in $\mathbb{R}^n$ and denote by $\nu = (\nu_j)_{1\leq j\leq n}$ its outward unit normal. Then for every $p \in (1,\infty)$, the mapping*

$$\Psi : \dot{L}^p_{1,0}(\partial\Omega) \longrightarrow L^p_1(\partial\Omega) \times L^p(\partial\Omega), \qquad \Psi(\dot{f}) := \Big(f_0\,, -\textstyle\sum_{j=1}^n \nu_j f_j\Big), \tag{6.379}$$
$$\text{for every Whitney array } \dot{f} = (f_0, f_1, \dots, f_n) \in \dot{L}^p_{1,0}(\partial\Omega),$$

is an isomorphism, whose inverse may be described as

$$\Psi^{-1} : L^p_1(\partial\Omega) \times L^p(\partial\Omega) \longrightarrow \dot{L}^p_{1,0}(\partial\Omega), \quad \Psi^{-1}(F,g) = \dot{f} := (f_0, f_1, \dots, f_n)$$
$$\text{where } f_0 := F \text{ and } f_j := -\nu_j g + \sum_{k=1}^n \nu_k \partial_{\tau_{kj}} F \text{ for } 1 \leq j \leq n,$$
$$\text{for every } (F,g) \in L^p_1(\partial\Omega) \times L^p(\partial\Omega). \tag{6.380}$$

Furthermore, if $v : \Omega \to \mathbb{R}$ is a function with the property that

$$\mathcal{N}(v), \mathcal{N}(\nabla v) \in L^p(\partial\Omega), \quad \exists\, v\big\lfloor_{\partial\Omega} \text{ and } \exists\, (\nabla v)\big\lfloor_{\partial\Omega}, \tag{6.381}$$

then

$$\Psi^{-1}\Big(v\big\lfloor_{\partial\Omega}, -\partial_\nu v\Big) = \big(v\big\lfloor_{\partial\Omega}, (\nabla v)\big\lfloor_{\partial\Omega}\big) = \mathrm{tr}_1 v. \tag{6.382}$$

Proof. This is a version of Proposition (3.5) corresponding to the case $m = 2$. □

We shall also need to use the adjoint of the operator Ψ. Its main properties are summarized below.

Proposition 6.42. *Retain the same background hypotheses as in Proposition 6.41 and denote by Ψ^* the adjoint of the operator Ψ defined in (6.379). Then, for each pair of indices $p, p' \in (1,\infty)$ with $1/p + 1/p' = 1$,*

$$\Psi^* : L^{p'}_{-1}(\partial\Omega) \times L^{p'}(\partial\Omega) \longrightarrow \big(\dot{L}^p_{1,0}(\partial\Omega)\big)^*, \tag{6.383}$$

is an isomorphism. Moreover, for each $(G, f) \in L^{p'}_{-1}(\partial\Omega) \times L^{p'}(\partial\Omega)$ one has

$$\Psi^*(G, f) = \big(G, -(\nu_j f)_{1\leq j\leq n}\big), \tag{6.384}$$

in the sense that

$$\langle \Psi^*(G,f),\dot{g}\rangle = \langle G, g_0\rangle - \sum_{j=1}^{n}\int_{\partial\Omega} \nu_j f g_j\,d\sigma,$$
$$\text{for all } \dot{g} = \big(g_0,(g_j)_{1\le j\le n}\big) \in \dot{L}^p_{1,0}(\partial\Omega). \tag{6.385}$$

Furthermore, the inverse of Ψ^* *in* (6.383) *may be described as*

$$\left(\Psi^*\right)^{-1}(\Lambda) = \Big(\eta_0 - \sum_{j,k=1}^{n} \partial_{\tau_{kj}}(\nu_k\eta_j),\, -\sum_{j=1}^{n}\nu_j\eta_j\Big) \tag{6.386}$$

if the functional $\Lambda \in \big(\dot{L}^p_{1,0}(\partial\Omega)\big)^*$ *is given by paring against the* $(n+1)$*-tuple* $\dot\eta$ *where* $\dot\eta = \big(\eta_0,(\eta_j)_{1\le j\le n}\big) \in L^{p'}_{-1}(\partial\Omega)\times\big[L^{p'}(\partial\Omega)\big]^n$.

Proof. This follows by unraveling definitions, in a straightforward manner. □

Moving on, let E be the canonical fundamental solution for Δ^2 in $\mathbb{R}^n$ given at each $X \in \mathbb{R}^n \setminus \{0\}$ by

$$E(X) := \begin{cases} \dfrac{1}{2(n-4)(n-2)\omega_{n-1}}|X|^{4-n} & \text{if } n=3, \text{ or if } n>4,\\[2mm] -\dfrac{1}{4\omega_3}\log|X| & \text{if } n=4,\\[2mm] -\dfrac{1}{8\pi}|X|^2(1-\log|X|) & \text{if } n=2, \end{cases} \tag{6.387}$$

where ω_{n-1} denotes the surface area of the unit sphere S^{n-1} in $\mathbb{R}^n$. In particular,

$$(\Delta E)(X) := \begin{cases} \dfrac{1}{(2-n)\omega_{n-1}}|X|^{2-n} & \text{if } n\ge 3,\\[2mm] \dfrac{1}{2\pi}\log|X| & \text{if } n=2. \end{cases} \tag{6.388}$$

In relation to the latter, given a bounded Lipschitz domain $\Omega \subset \mathbb{R}^n$ with outward unit normal ν and surface measure σ, recall that the harmonic double and single layer operators are, respectively, given by

$$\mathscr{D}_\Delta f(X) := \int_{\partial\Omega} \partial_{\nu(Y)}[(\Delta E)(X-Y)]f(Y)\,d\sigma(Y), \quad X \in \mathbb{R}^n\setminus\partial\Omega, \tag{6.389}$$

$$\mathscr{S}_\Delta f(X) := \int_{\partial\Omega} (\Delta E)(X-Y)f(Y)\,d\sigma(Y), \quad X \in \mathbb{R}^n\setminus\partial\Omega. \tag{6.390}$$

Based on definitions, if $f \in L^p(\partial\Omega)$ for some $p \in (1,\infty)$ and $\ell \in \{1,\dots,n\}$, then the following identities may be readily verified at each point $X \in \mathbb{R}^n \setminus \partial\Omega$:

$$\partial_\ell(\mathscr{D}_\Delta f)(X) = -\sum_{i=1}^{n} \partial_i \mathscr{S}_\Delta(\partial_{\tau_{i\ell}} f)(X), \tag{6.391}$$

$$\partial_\ell(\mathscr{S}_\Delta f)(X) = -(\mathscr{S}_\Delta(\partial_{\tau_{\ell i}}(\nu_i f))(X) - (\mathscr{D}_\Delta(\nu_\ell f))(X). \tag{6.392}$$

We are now prepared to make the following basic definition.

Definition 6.43. Let Ω be a bounded Lipschitz domain in $\mathbb{R}^n$ with outward unit normal $\nu = (\nu_j)_{1\le j\le n}$ and surface measure σ. Also, fix $\theta \in \mathbb{R}$. In this context, the biharmonic double multi-layer $\dot{\mathscr{D}}_\theta$ is defined according to the general recipe from Definition 4.4 implemented for the writing of Δ^2 as in (6.352) corresponding to the tensor coefficient $A_\theta = \big(A_{\alpha\beta}(\theta)\big)_{|\alpha|=|\beta|=2}$ from (6.349) (and with E as in (6.387)).

Specifically, for each Whitney array $\dot{f} = (f_0, f_1, \dots, f_n)$, define the biharmonic double multi-layer

$$\dot{\mathscr{D}}_\theta \dot{f}(X) := -\tfrac{1}{2} \sum_{\substack{|\alpha|=2\\|\beta|=2}} \alpha!\, A_{\beta\alpha}(\theta) \sum_{\substack{i,j \text{ so that}\\ e_i+e_j=\alpha}} \int_{\partial\Omega} \nu_j(Y)\Big\{(\partial^\beta E)(X-Y) f_i(Y)$$

$$+(\partial^{\beta+e_i} E)(X-Y) f_0(Y)\Big\}\, d\sigma(Y) \tag{6.393}$$

for $X \in \mathbb{R}^n \setminus \partial\Omega$. Also, as in the past, denote by $\dot{\mathscr{D}}_\theta^{\pm}$ the restrictions of the biharmonic double multi-layer (6.393) to $\Omega_\pm$.

Finally, denote by $\dot{K}_\theta$ the boundary biharmonic double multi-layer, defined as in Definition 4.13 for the writing of Δ^2 from (6.352) using the tensor coefficient $A_\theta = \big(A_{\alpha\beta}(\theta)\big)_{|\alpha|=|\beta|=2}$ from (6.349), and with E as in (6.387).

Concretely, the action of the boundary double multi-layer operator $\dot{K}_\theta$ on an arbitrary Whitney array $\dot{f} = (f_0, f_1, \dots, f_n) \in \dot{L}^p_{1,0}(\partial\Omega)$ is

$$\dot{K}_\theta \dot{f} := \Big((\dot{K}_\theta \dot{f})_0, (\dot{K}_\theta \dot{f})_1, \dots, (\dot{K}_\theta \dot{f})_n\Big) \tag{6.394}$$

where, for σ a.e. $X \in \partial\Omega$,

$$(\dot{K}_\theta \dot{f})_0(X) := \lim_{\varepsilon\to 0+} \int_{\substack{Y\in\partial\Omega\\ |X-Y|>\varepsilon}} \partial_{\nu(Y)}[(\Delta E)(X-Y)] f_0(Y)\, d\sigma(Y) \tag{6.395}$$

$$-\int_{\partial\Omega}(\Delta E)(X-Y)\sum_{k=1}^{n}\nu_k(Y)f_k(Y)\,d\sigma(Y)$$

$$+\frac{1}{1+2\theta+n\theta^2}\int_{\partial\Omega}\sum_{j,k=1}^{n}\partial_{\tau_{kj}(Y)}[(\partial_k E)(X-Y)]f_j(Y)\,d\sigma(Y),$$

while for each $\ell\in\{1,\dots,n\}$,

$$(\dot{K}_\theta\dot{f})_\ell(X):=\lim_{\varepsilon\to 0+}\int_{\substack{Y\in\partial\Omega\\|X-Y|>\varepsilon}}\Big\{\partial_{\nu(Y)}[(\Delta E)(X-Y)]f_\ell(Y)$$

$$+\sum_{i=1}^{n}\partial_{\tau_{i\ell}(Y)}[(\Delta E)(X-Y)]f_i(Y) \tag{6.396}$$

$$+\frac{1}{1+2\theta+n\theta^2}\sum_{j,k=1}^{n}\partial_{\tau_{kj}(Y)}[(\partial_\ell\partial_k E)(X-Y)]f_j(Y)\Big\}\,d\sigma(Y).$$

To explain the rationale behind the definition (6.394)–(6.396) of the boundary biharmonic double multi-layer $\dot{K}_\theta$ recall formula (4.164) from Theorem 4.14 which, in the context of the above definition, reads (with $1<p<\infty$)

$$\Big(\dot{\mathscr{D}}_\theta^{\pm}\dot{f}\big|_{\partial\Omega},\,(\nabla\dot{\mathscr{D}}_\theta^{\pm}\dot{f})\big|_{\partial\Omega}\Big)=(\pm\tfrac{1}{2}I+\dot{K}_\theta)\dot{f},\quad\forall\,\dot{f}\in\dot{L}^p_{1,0}(\partial\Omega). \tag{6.397}$$

This may be used to identify a concrete formula for $\dot{K}_\theta$, and the fact that formulas (6.394)–(6.396) are natural may be seen by combining (6.397) with (6.404) and (6.420) (proved later).

In addition to the operator $\dot{\mathscr{D}}_\theta$ defined above, with N_θ, M_θ as in (6.360), consider the integral operator acting on each pair $(F,g)\in L^p_1(\partial\Omega)\oplus L^p(\partial\Omega)$, where the index $p\in(1,\infty)$, according to the formula

$$\widetilde{\mathscr{D}}_\theta(F,g)(X):=\int_{\partial\Omega}\Big\{M_\theta[E(X-\cdot)](Y)g(Y) \tag{6.398}$$

$$+N_\theta[E(X-\cdot)](Y)F(Y)\Big\}\,d\sigma(Y),$$

at each $X\in\mathbb{R}^n\setminus\partial\Omega$. The goal now is to elaborate on the relationship between the operators $\dot{\mathscr{D}}_\theta$ and $\widetilde{\mathscr{D}}_\theta$ just introduced. In this vein, it is worth recalling the isomorphism Ψ described in Proposition 6.41.

Proposition 6.44. *Assume that Ω is a bounded Lipschitz domain in $\mathbb{R}^n$ and fix $\theta \in \mathbb{R}$. Then*

$$\dot{\mathscr{D}}_\theta = \widetilde{\mathscr{D}}_\theta \circ \Psi \quad in \ \mathbb{R}^n \setminus \partial\Omega, \tag{6.399}$$

when both operators are acting on arbitrary Whitney arrays from $\dot{L}^p_{1,0}(\partial\Omega)$ with $p \in (1, \infty)$.

Proof. Thanks to Definition 6.43 and the density result established in Proposition 3.3, it suffices to show that the two operators from (6.399) act identically on Whitney arrays of the form

$$\dot{f} = \mathrm{tr}_1 v, \qquad v \in C^\infty(\overline{\Omega}). \tag{6.400}$$

Assume that this is the case, i.e., $\dot{f} = (f_0, f_1, \ldots, f_n) = \Big(v|_{\partial\Omega}, (\partial_1 v)|_{\partial\Omega}, \ldots, (\partial_n v)|_{\partial\Omega}\Big)$, and introduce

$$F := f_0 \in L^p_1(\partial\Omega) \quad \text{and} \quad g := -\sum_{j=1}^n \nu_j f_j \in L^p(\partial\Omega), \tag{6.401}$$

where $\nu = (\nu_j)_{1 \le j \le n}$ denotes the outward unit normal to $\partial\Omega$. Hence,

$$\Psi(\dot{f}) = (F, g) = \big(v|_{\partial\Omega}, -\partial_\nu v\big). \tag{6.402}$$

Then, based on definition (6.398), Propositions 6.39, 4.5, and (6.400), for every point $X \in \mathbb{R}^n \setminus \partial\Omega$ we may write

$$\begin{aligned} v(X) - \widetilde{\mathscr{D}}_\theta(F,g)(X) &= v(X) - \int_{\partial\Omega} \Big\{ -M_\theta[E(X-\cdot)](Y)\partial_\nu v(Y) \\ &\qquad + N_\theta[E(X-\cdot)](Y) v(Y) \Big\}\, d\sigma(Y) \\ &= \mathscr{B}_\theta\big(E(X-\cdot), v\big) = v(X) - \dot{\mathscr{D}}_\theta(\mathrm{tr}_1 v)(X) \\ &= v(X) - (\dot{\mathscr{D}}_\theta \dot{f})(X). \end{aligned} \tag{6.403}$$

As such, in light of (6.402) we conclude that (6.399) holds. □

Next we take a closer look at the action of the biharmonic double multi-layer, originally introduced in Definition 6.43, on Whitney arrays.

Proposition 6.45. *Assume that Ω is a bounded Lipschitz domain in $\mathbb{R}^n$ with outward unit normal $\nu = (\nu_j)_{1 \le j \le n}$ and surface measure σ. Also, fix a number*

$\theta \in \mathbb{R}$. Then the action of the double multi-layer $\dot{\mathscr{D}}_\theta$ introduced in Definition 6.43 on a Whitney array $\dot{f} = (f_0, f_1, \dots, f_n)$ from $\dot{L}^p_{1,0}(\partial\Omega)$, with $1 < p < \infty$, may be described as

$$(\dot{\mathscr{D}}_\theta \dot{f})(X) = \int_{\partial\Omega} \partial_{\nu(Y)}[(\Delta E)(X-Y)] f_0(Y)\, d\sigma(Y) \tag{6.404}$$
$$- \int_{\partial\Omega} (\Delta E)(X-Y) \sum_{k=1}^{n} \nu_k(Y) f_k(Y)\, d\sigma(Y)$$
$$+ \frac{1}{1+2\theta+n\theta^2} \int_{\partial\Omega} \sum_{j,k=1}^{n} \partial_{\tau_{kj}(Y)}[(\partial_k E)(X-Y)] f_j(Y)\, d\sigma(Y),$$

for each $X \in \mathbb{R}^n \setminus \partial\Omega$.

In particular, using the notation introduced in (6.389)–(6.390),

$$(\dot{\mathscr{D}}_\theta \dot{f})(X) = (\mathscr{D}_\Delta f_0)(X) - \mathscr{S}_\Delta\Big(\sum_{k=1}^{n} \nu_k f_k\Big)(X) \tag{6.405}$$
$$+ \frac{1}{1+2\theta+n\theta^2} \int_{\partial\Omega} \sum_{j,k=1}^{n} \partial_{\tau_{kj}(Y)}[(\partial_k E)(X-Y)] f_j(Y)\, d\sigma(Y),$$

for each $X \in \mathbb{R}^n \setminus \partial\Omega$.

Proof. For every $\dot{f} = (f_0, f_1, \dots, f_n) \in \dot{L}^p_{1,0}(\partial\Omega)$, based on (6.399), (6.379), and (6.447), at every $X \in \mathbb{R}^n \setminus \partial\Omega$ we may write (using the summation convention over repeated indices)

$$(\dot{\mathscr{D}}_\theta \dot{f})(X) = \widetilde{\mathscr{D}}_\theta\big(f_0, -\nu_i f_i\big)(X) = \int_{\partial\Omega} \Big\{ -(\Delta E)(X-Y)$$
$$+ \frac{1}{1+2\theta+n\theta^2} \nu_j(Y) \partial_{\tau_{kj}(Y)}[(\partial_k E)(X-Y)] \Big\} \nu_i(Y) f_i(Y)\, d\sigma(Y)$$
$$+ \int_{\partial\Omega} \Big\{ \partial_{\nu(Y)}[(\Delta E)(X-Y)] \tag{6.406}$$
$$- \frac{1}{1+2\theta+n\theta^2} \partial_{\tau_{ij}(Y)}\Big(\nu_i(Y) \partial_{\tau_{kj}(Y)}[(\partial_k E)(X-Y)]\Big) \Big\} f_0(Y)\, d\sigma(Y)$$
$$= \int_{\partial\Omega} \Big\{ \partial_{\nu(Y)}[(\Delta E)(X-Y)] f_0(Y) - (\Delta E)(X-Y) \nu_i(Y) f_i(Y)$$
$$+ \frac{1}{1+2\theta+n\theta^2} \partial_{\tau_{kj}(Y)}[(\partial_k E)(X-Y)] \Big(\nu_i \partial_{\tau_{ij}} f_0 + \nu_j \nu_i f_i \Big)(Y) \Big\}\, d\sigma(Y),$$

thanks to (6.361)–(6.362) and an integration by parts on the boundary. Now, the claim made in (6.404) follows from (6.406) after observing that

$$\nu_i \partial_{\tau_{ij}} f_0 + \nu_j \nu_i f_i = \nu_i(\nu_i f_j - \nu_j f_i) + \nu_j \nu_i f_i = f_j, \tag{6.407}$$

by the compatibility conditions satisfied by the components of the Whitney array $\dot{f}$. □

It is also instructive to derive formula (6.404) directly from (6.351), based on the recipe from (4.59). Concretely, fix an arbitrary point $X \in \Omega$ and, with E as in (6.387), set $E^X := E(X - \cdot)$ in Ω. Then, given any $v \in C^\infty(\overline{\Omega})$, successive integrations by parts give (using the summation convention over repeated indices)

$$\begin{aligned}
&\int_\Omega (\partial_i \partial_j + \theta\, \delta_{ij} \Delta) E^X \cdot (\partial_i \partial_j + \theta\, \delta_{ij} \Delta) v \, dX \\
&= - \int_\Omega (\partial_i \partial_i \partial_j + \theta\, \delta_{ij} \partial_i \Delta) E^X \cdot \partial_j v \, dX \\
&\quad - \int_\Omega (\partial_k \partial_i \partial_j + \theta\, \delta_{ij} \partial_k \Delta) E^X \cdot \theta \delta_{ij} \partial_k v \, dX \\
&\quad + \int_{\partial\Omega} \nu_i (\partial_i \partial_j + \theta\, \delta_{ij} \Delta) E^X \cdot \partial_j v \, d\sigma + \int_{\partial\Omega} \nu_k (\partial_i \partial_j + \theta\, \delta_{ij} \Delta) E^X \cdot \theta \delta_{ij} \partial_k v \, d\sigma \\
&= -(1 + 2\theta + n\theta^2) \int_\Omega \partial_j \Delta E^X \cdot \partial_j v \, dX \\
&\quad + \int_{\partial\Omega} \partial_\nu \partial_j E^X \cdot \partial_j v \, d\sigma + (2\theta + n\theta^2) \int_{\partial\Omega} \Delta E^X \cdot \nu_j \partial_j v \, d\sigma \\
&= (1 + 2\theta + n\theta^2) v(X) - (1 + 2\theta + n\theta^2) \int_{\partial\Omega} \partial_\nu \Delta E^X \cdot v \, d\sigma \\
&\quad + \int_{\partial\Omega} \partial_\nu \partial_j E^X \cdot \partial_j v \, d\sigma + (2\theta + n\theta^2) \int_{\partial\Omega} \Delta E^X \cdot \nu_j \partial_j v \, d\sigma \\
&= (1 + 2\theta + n\theta^2) v(X) - (1 + 2\theta + n\theta^2) \int_{\partial\Omega} \partial_\nu \Delta E^X \cdot v \, d\sigma \\
&\quad + \int_{\partial\Omega} \partial_{\tau_{kj}} \partial_k E^X \cdot \partial_j v \, d\sigma + (1 + 2\theta + n\theta^2) \int_{\partial\Omega} \Delta E^X \cdot \nu_j \partial_j v \, d\sigma.
\end{aligned} \tag{6.408}$$

This shows that

$$\begin{aligned}\mathscr{B}_\theta\big(E^X,v\big) &= v(X) - \int_{\partial\Omega} \partial_\nu \Delta E^X \cdot v\, d\sigma \\ &\quad + \frac{1}{1+2\theta+n\theta^2}\int_{\partial\Omega} \partial_{\tau_{kj}}\partial_k E^X \cdot \partial_j v\, d\sigma + \int_{\partial\Omega} \Delta E^X \cdot \nu_j \partial_j v\, d\sigma. \\ &= v(X) - \dot{\mathscr{D}}_\theta \dot{f}(X), \end{aligned} \tag{6.409}$$

if $\dot{\mathscr{D}}_\theta$ is as in (6.404) and $\dot{f} = \big(v|_{\partial\Omega}, \partial_1 v|_{\partial\Omega}, \dots, \partial_n v|_{\partial\Omega}\big)$.

The next order of business is to study the mapping properties for the conormal derivative of the biharmonic double multi-layer. Our first result in this regard is the following theorem (we shall return to this topic later, after completing a necessary detour).

Theorem 6.46. *Let Ω be a bounded Lipschitz domain in $\mathbb{R}^n$ and fix $\theta \in \mathbb{R}$. Also, assume that $p, p' \in (1,\infty)$ are such that $1/p + 1/p' = 1$. Recall $\dot{\mathscr{D}}_\theta$ introduced in Definition 6.43 and the conormal $\partial_\nu^{A_\theta}$ from Proposition 6.40. Then the operator*

$$\partial_\nu^{A_\theta}\dot{\mathscr{D}}_\theta : \dot{L}^p_{1,1}(\partial\Omega) \longrightarrow \Big(\dot{L}^{p'}_{1,0}(\partial\Omega)\Big)^* \tag{6.410}$$

is well-defined, linear and bounded. Moreover, this operator further extends as a linear and bounded mapping in the context

$$\partial_\nu^{A_\theta}\dot{\mathscr{D}}_\theta : \dot{L}^p_{1,0}(\partial\Omega) \longrightarrow \Big(\dot{L}^{p'}_{1,1}(\partial\Omega)\Big)^*. \tag{6.411}$$

Proof. For each $\dot{f} \in \dot{L}^p_{1,1}(\partial\Omega)$ we know from Theorem 4.7 that

$$\begin{aligned}\|\mathscr{N}(\dot{\mathscr{D}}_\theta \dot{f})\|_{L^p(\partial\Omega)} + \|\mathscr{N}(\nabla\dot{\mathscr{D}}_\theta \dot{f})\|_{L^p(\partial\Omega)} \\ +\|\mathscr{N}(\nabla^2\dot{\mathscr{D}}_\theta \dot{f})\|_{L^p(\partial\Omega)} \leq C\|\dot{f}\|_{\dot{L}^p_{1,1}(\partial\Omega)}\end{aligned} \tag{6.412}$$

for some finite constant $C > 0$ independent of $\dot{f}$. Moreover, (4.66) ensures that

$$\begin{aligned}\partial^\gamma \dot{\mathscr{D}}_\theta \dot{f}\Big|_{\partial\Omega} &\text{ exists for every } \dot{f} \in \dot{L}^p_{1,1}(\partial\Omega) \\ &\text{whenever } \gamma \in \mathbb{N}_0^n \text{ satisfies } |\gamma| \leq 2.\end{aligned} \tag{6.413}$$

However, the conormal entails up to three derivatives on $\dot{\mathscr{D}}_\theta$. Indeed, as seen from (6.378), the components of $\partial_\nu^{A_\theta}\dot{\mathscr{D}}_\theta \dot{f}$ are given by (here and elsewhere the usual summation convention over repeated indices is used)

$$\left(\partial_\nu^{A_\theta} \dot{\mathscr{D}}_\theta \dot{f}\right)_0 = -\partial_\nu(\Delta \dot{\mathscr{D}}_\theta \dot{f}) - \frac{1}{1+2\theta+n\theta^2}\, \partial_{\tau_{ij}}\left(\nu_\ell \nu_i \partial_j \partial_\ell \dot{\mathscr{D}}_\theta \dot{f}\right), \tag{6.414}$$

and, for $1 \le r \le n$,

$$\left(\partial_\nu^{A_\theta} \dot{\mathscr{D}}_\theta \dot{f}\right)_r = \frac{2\theta+n\theta^2}{1+2\theta+n\theta^2}\, \nu_r \Delta \dot{\mathscr{D}}_\theta \dot{f} + \frac{1}{1+2\theta+n\theta^2} \nu_r \nu_j \nu_\ell \partial_j \partial_\ell \dot{\mathscr{D}}_\theta \dot{f} \tag{6.415}$$

with the understanding that all derivatives in the right-hand sides are restricted to the boundary. Note that, thanks to (6.412) and (6.413), the map

$$\dot{L}^p_{1,1}(\partial\Omega) \ni \dot{f} \longmapsto \partial_j \partial_\ell \dot{\mathscr{D}}_\theta \dot{f}\Big|_{\partial\Omega} \in L^p(\partial\Omega) \tag{6.416}$$

is well-defined, linear and bounded, for every $\ell, j \in \{1,\dots,n\}$. As such, the mapping

$$\dot{L}^p_{1,1}(\partial\Omega) \ni \dot{f} \longmapsto \partial_{\tau_{ij}}\left(\nu_\ell \nu_i \partial_j \partial_\ell \dot{\mathscr{D}}_\theta \dot{f}\right) \in L^p_{-1}(\partial\Omega) \tag{6.417}$$

is also well-defined, linear and bounded, for every $i, j, \ell \in \{1,\dots,n\}$.

We propose to take a closer look at the structure of the derivatives of the biharmonic double multi-layer operator. In a first stage, fix an arbitrary Whitney array $\dot{f} = (f_0, f_1, \dots, f_n) \in \dot{L}^p_{1,0}(\partial\Omega)$ then for every $\ell \in \{1,\dots,n\}$ and every $X \in \mathbb{R}^n \setminus \partial\Omega$ compute

$$\begin{aligned} \partial_\ell\big(\dot{\mathscr{D}}_\theta \dot{f}\big)(X) &= \partial_\ell\big(\mathscr{D}_\Delta f_0\big)(X) - \partial_\ell \mathscr{S}_\Delta\big(\nu_i f_i\big)(X) \\ &\quad + \frac{1}{1+2\theta+n\theta^2} \int_{\partial\Omega} \partial_{\tau_{kj}(Y)}[(\partial_\ell \partial_k E)(X-Y)] f_j(Y)\, d\sigma(Y). \end{aligned} \tag{6.418}$$

Upon observing that, for every $X \in \mathbb{R}^n \setminus \partial\Omega$, identity (6.391) and the compatibility conditions satisfied by the components of the Whitney array $\dot{f}$ allow us to write

$$\begin{aligned} &\partial_\ell\big(\mathscr{D}_\Delta f_0\big)(X) - \partial_\ell \mathscr{S}_\Delta\big(\nu_i f_i\big)(X) \\ &\quad = -\partial_i \mathscr{S}_\Delta\big(\partial_{\tau_{i\ell}} f_0\big)(X) - \partial_\ell \mathscr{S}_\Delta\big(\nu_i f_i\big)(X) \\ &\quad = -\partial_i \mathscr{S}_\Delta\big(\nu_i f_\ell\big)(X) + \partial_i \mathscr{S}_\Delta\big(\nu_\ell f_i\big)(X) - \partial_\ell \mathscr{S}_\Delta\big(\nu_i f_i\big)(X) \\ &\quad = \big(\mathscr{D}_\Delta f_\ell\big)(X) + \int_{\partial\Omega} \partial_{\tau_{i\ell}(Y)}[(\Delta E)(X-Y)] f_i(Y)\, d\sigma(Y) \\ &\quad = \big(\mathscr{D}_\Delta f_\ell\big)(X) + \mathscr{S}_\Delta\big(\partial_{\tau_{\ell i}} f_i\big)(X), \end{aligned} \tag{6.419}$$

we deduce from (6.418) that, at each point $X \in \mathbb{R}^n \setminus \partial\Omega$,

$$\partial_\ell(\dot{\mathscr{D}}_\theta \dot{f})(X) = (\mathscr{D}_\Delta f_\ell)(X) + \mathscr{S}_\Delta(\partial_{\tau_{\ell i}} f_i)(X) \tag{6.420}$$
$$+\frac{1}{1+2\theta+n\theta^2}\int_{\partial\Omega} \partial_{\tau_{kj}(Y)}[(\partial_\ell\partial_k E)(X-Y)] f_j(Y)\,d\sigma(Y).$$
$$= \int_{\partial\Omega}\Big\{\partial_{\nu(Y)}[(\Delta E)(X-Y)] f_\ell(Y) + \partial_{\tau_{i\ell}(Y)}[(\Delta E)(X-Y)] f_i(Y)$$
$$+\frac{1}{1+2\theta+n\theta^2}\partial_{\tau_{kj}(Y)}[(\partial_\ell\partial_k E)(X-Y)] f_j(Y)\Big\}\,d\sigma(Y).$$

In the case when the array $\dot{f} = (f_0, f_1, \dots, f_n)$ actually belongs to the Whitney–Sobolev space $\dot{L}^p_{1,1}(\partial\Omega)$, we may integrate by parts on the boundary in (6.418) in order to write, for every $\ell \in \{1,\dots,n\}$,

$$\partial_\ell(\dot{\mathscr{D}}_\theta \dot{f})(X) = (\mathscr{D}_\Delta f_\ell)(X) + \int_{\partial\Omega}\Big\{(\Delta E)(X-Y)(\partial_{\tau_{\ell i}} f_i)(Y) \tag{6.421}$$
$$+\frac{1}{1+2\theta+n\theta^2}(\partial_\ell\partial_k E)(X-Y)(\partial_{\tau_{ik}} f_i)(Y)\Big\}\,d\sigma(Y),$$

for each $X \in \mathbb{R}^n \setminus \partial\Omega$. In this scenario, we may take one extra derivative while still retaining control of the finiteness of the L^p-norm of the non-tangential maximal function. Concretely, for each $j, \ell \in \{1,\dots,n\}$ we obtain (with the help of (6.391))

$$\partial_j\partial_\ell(\dot{\mathscr{D}}_\theta \dot{f})(X) = -\partial_i\mathscr{S}_\Delta(\partial_{\tau_{ij}} f_\ell)(X) + \int_{\partial\Omega}\Big\{(\partial_j\Delta E)(X-Y)(\partial_{\tau_{\ell i}} f_i)(Y) \tag{6.422}$$
$$+\frac{1}{1+2\theta+n\theta^2}(\partial_j\partial_\ell\partial_k E)(X-Y)(\partial_{\tau_{ik}} f_i)(Y)\Big\}\,d\sigma(Y),$$

at each $X \in \mathbb{R}^n \setminus \partial\Omega$, whenever $\dot{f} = (f_0, f_1, \dots, f_n)$ belongs to the Whitney–Sobolev space $\dot{L}^p_{1,1}(\partial\Omega)$. Concisely, for every $\dot{f} = (f_0, f_1, \dots, f_n) \in \dot{L}^p_{1,1}(\partial\Omega)$ we have

$$\partial_j\partial_\ell\dot{\mathscr{D}}_\theta \dot{f} = -\partial_i\mathscr{S}_\Delta(\partial_{\tau_{ij}} f_\ell) + \partial_j\mathscr{S}_\Delta(\partial_{\tau_{\ell i}} f_i) \tag{6.423}$$
$$+\frac{1}{1+2\theta+n\theta^2}\int_{\partial\Omega}(\partial_j\partial_\ell\partial_k E)(\cdot - Y)(\partial_{\tau_{ik}} f_i)(Y)\,d\sigma(Y) \quad \text{in } \mathbb{R}^n \setminus \partial\Omega.$$

In particular, summing up over $j = \ell$ yields

$$\Delta\dot{\mathscr{D}}_\theta \dot{f} = \frac{1}{1+2\theta+n\theta^2}\,\partial_k \mathscr{S}_\Delta(\partial_{\tau_{ik}} f_i) \quad \text{in } \mathbb{R}^n \setminus \partial\Omega, \qquad \forall\, \dot{f} = (f_0, f_1, \dots, f_n) \in \dot{L}^p_{1,1}(\partial\Omega), \tag{6.424}$$

and, further,

$$\Big(\partial_\nu \Delta\dot{\mathscr{D}}_\theta \dot{f}\Big)(X) = \frac{1}{1+2\theta+n\theta^2} \times \partial_{\tau_{jk}(X)}\Big(\lim_{\varepsilon\to 0+} \int_{\substack{Y\in\partial\Omega\\ |X-Y|>\varepsilon}} (\partial_j \Delta E)(X-Y)(\partial_{\tau_{ik}} f_i)(Y)\, d\sigma(Y)\Big) \tag{6.425}$$

in $\mathbb{R}^n \setminus \partial\Omega$, for every $\dot{f} = (f_0, f_1, \dots, f_n) \in \dot{L}^p_{1,1}(\partial\Omega)$. Consequently,

$$\dot{L}^p_{1,1}(\partial\Omega) \ni \dot{f} \longmapsto \partial_\nu \Delta\dot{\mathscr{D}}_\theta \dot{f} \in L^p_{-1}(\partial\Omega) \tag{6.426}$$

is a well-defined, linear and bounded mapping.

In summary, from (6.414)–(6.417), (6.426) we deduce that the mapping

$$\dot{L}^p_{1,1}(\partial\Omega) \ni \dot{f} \longmapsto \Big(\big(\partial_\nu^{A_\theta}\dot{\mathscr{D}}_\theta \dot{f}\big)_0, \big(\partial_\nu^{A_\theta}\dot{\mathscr{D}}_\theta \dot{f}\big)_{1\le r\le n}\Big) \in L^p_{-1}(\partial\Omega) \oplus \big[L^p(\partial\Omega)\big]^n \tag{6.427}$$

is well-defined, linear and bounded. Furthermore, it is clear from (6.415) that

$$\big(\partial_\nu^{A_\theta}\dot{\mathscr{D}}_\theta \dot{f}\big)_{1\le r\le n} = \big(-\nu_1 f, \dots, -\nu_n f\big) \quad \text{for each } \dot{f} \in \dot{L}^p_{1,1}(\partial\Omega), \quad \text{where}$$
$$f := -\frac{2\theta+n\theta^2}{1+2\theta+n\theta^2}\,\Delta\dot{\mathscr{D}}_\theta \dot{f} - \frac{1}{1+2\theta+n\theta^2}\,\nu_j \nu_\ell \partial_j \partial_\ell \dot{\mathscr{D}}_\theta \dot{f} \in L^p(\partial\Omega). \tag{6.428}$$

At this stage, the fact that the operator $\partial_\nu^{A_\theta}\dot{\mathscr{D}}_\theta$ is well-defined, linear and bounded in the context of (6.410) follows from (6.427)–(6.428) and Proposition 6.42. Lastly, that this operator further extends as a linear and bounded mapping in the context of (6.411), follows from what we have proved so far, the second claim in the statement of Proposition 5.32, and duality. □

Let Ω be a bounded Lipschitz domain in $\mathbb{R}^n$ and fix $p \in (1,\infty)$. Recall from (5.1) that the bi-Laplacian single multi-layer operator $\dot{\mathscr{S}}$ acts on an arbitrary functional $\Lambda \in \big(\dot{L}^p_{1,0}(\partial\Omega)\big)^*$ according to the formula (with E as in (6.387))

$$(\dot{\mathscr{S}}\Lambda)(X) := \Big\langle \Big(E(X-\cdot)\big|_{\partial\Omega}, -(\nabla E)(X-\cdot)\big|_{\partial\Omega}\Big), \Lambda \Big\rangle, \quad X \in \mathbb{R}^n \setminus \partial\Omega, \tag{6.429}$$

where the expression in round parentheses is regarded as a Whitney array in $\dot{L}^p_{1,0}(\partial\Omega)$.

Proposition 6.47. *Let Ω be a bounded Lipschitz domain in $\mathbb{R}^n$ and assume that $p \in (1,\infty)$. Fix $\theta \in \mathbb{R}$ and recall the conormal $\partial_\nu^{A_\theta}$ from Proposition 6.40. Also, let $\dot{\mathscr{S}}$ stand for the bi-Laplacian single multi-layer operator from (6.429). Then*

$$\partial_\nu^{A_\theta}\dot{\mathscr{S}} : \Big(\dot{L}^p_{1,0}(\partial\Omega)\Big)^* \longrightarrow \Big(\dot{L}^p_{1,0}(\partial\Omega)\Big)^* \tag{6.430}$$

is a well-defined, linear and bounded operator.

As a corollary, for each $p \in (1,\infty)$,

$$\partial_\nu^{A_\theta}\dot{\mathscr{S}} = -\tfrac{1}{2}I + \dot{K}^*_\theta \ \text{ as operators on } \Big(\dot{L}^p_{1,0}(\partial\Omega)\Big)^*. \tag{6.431}$$

Proof. As usual, let p' denote the Hölder conjugate exponent of p. As far as the claim pertaining to the operator in (6.430) is concerned, the crux of the matter is establishing the estimate

$$\big\|\partial_\nu \Delta \dot{\mathscr{S}}\Lambda\big\|_{L^{p'}_{-1}(\partial\Omega)} \le C\|\Lambda\|_{\big(\dot{L}^p_{1,0}(\partial\Omega)\big)^*} \tag{6.432}$$

for some finite constant $C > 0$ independent of $\Lambda \in \Big(\dot{L}^p_{1,0}(\partial\Omega)\Big)^*$. Once this has been done, the same type of reasoning as in Theorem 6.46 (which also uses the non-tangential maximal function estimates for generic single multi-layers proved in Proposition 5.2) may then be used to complete the proof of the boundedness of the operator in (6.430).

Given an arbitrary $\Lambda \in \Big(\dot{L}^p_{1,0}(\partial\Omega)\Big)^*$, the reasoning from (5.8)–(5.10) leads to the conclusion that there exist $(g_0, g_1, \ldots, g_n) \in \big[L^{p'}(\partial\Omega)\big]^{n+1}$ with the property that

$$\sum_{j=0}^{n} \|g_j\|_{L^{p'}(\partial\Omega)} \le \|\Lambda\|_{(\dot{L}^p_{1,0}(\partial\Omega))^*}, \tag{6.433}$$

and such that, for each $X \in \mathbb{R}^n \setminus \partial\Omega$,

$$\begin{aligned} \dot{\mathscr{S}}\Lambda(X) &= \int_{\partial\Omega} E(X-Y) g_0(Y)\, d\sigma(Y) \\ &\quad - \sum_{j=1}^{n} \int_{\partial\Omega} (\partial_j E)(X-Y) g_j(Y)\, d\sigma(Y). \end{aligned} \tag{6.434}$$

As such, we may write

$$\Delta\dot{\mathscr{S}}\Lambda = \mathscr{S}_\Delta g_0 - \sum_{j=1}^{n} \partial_j \mathscr{S}_\Delta g_j \quad \text{in } \Omega, \tag{6.435}$$

where $\mathscr{S}_\Delta$ is the harmonic single multi-layer in Ω (cf. (6.390)). To proceed, recall that if S_Δ denotes the boundary harmonic single layer, then the identity

$$\partial_\nu \mathscr{D}_\Delta g = \tfrac{1}{2} \sum_{i,k=1}^{n} \partial_{\tau_{ki}} S_\Delta\big(\partial_{\tau_{ki}} g\big), \tag{6.436}$$

is valid for any $g \in L^p(\partial\Omega)$. Consequently, based on (6.392), (6.435), and (6.436), we may compute

$$\begin{aligned}
\partial_\nu \Delta\dot{\mathscr{S}}\Lambda &= \partial_\nu \mathscr{S}_\Delta g_0 + \sum_{i,j=1}^{n} \partial_\nu \mathscr{S}_\Delta(\partial_{\tau_{ji}}(\nu_i g_j)) + \sum_{j=1}^{n} \partial_\nu \mathscr{D}_\Delta(\nu_j g_j) \\
&= \big(-\tfrac{1}{2}I + K_\Delta^*\big)g_0 + \sum_{i,j=1}^{n} \big(-\tfrac{1}{2}I + K_\Delta^*\big)(\partial_{\tau_{ji}}(\nu_i g_j)) \\
&\quad + \tfrac{1}{2} \sum_{i,j,k=1}^{n} \partial_{\tau_{ki}} S_\Delta\big(\partial_{\tau_{ki}}(\nu_i g_j)\big),
\end{aligned} \tag{6.437}$$

where K_Δ^* is the adjoint of the boundary harmonic double layer from (1.3). Since the operators

$$\begin{aligned}
K_\Delta^* &: L^{p'}_{-1}(\partial\Omega) \to L^{p'}_{-1}(\partial\Omega), \\
K_\Delta^* &: L^{p'}(\partial\Omega) \to L^{p'}(\partial\Omega), \\
S_\Delta &: L^{p'}_{-1}(\partial\Omega) \to L^{p'}(\partial\Omega),
\end{aligned} \tag{6.438}$$

are bounded, estimate (6.432) now follows from (6.437) and (6.433). This completes the proof of the boundedness of the operator in (6.430).

With this in hand, identity (6.431) follows from (5.152) and a density argument. □

Moving on, assume that $\Omega \subset \mathbb{R}^n$ is a bounded Lipschitz domain, with outward unit normal $\nu = (\nu_j)_{1\le j\le n}$ and surface measure σ. Also, fix $\theta \in \mathbb{R}$ and $1 < p < \infty$. In this setting, consider the 2×2 matrix-valued singular integral operator

$$\widetilde{K}_\theta : L^p_1(\partial\Omega) \oplus L^p(\partial\Omega) \longrightarrow L^p_1(\partial\Omega) \oplus L^p(\partial\Omega), \tag{6.439}$$

$$\widetilde{K}_\theta := \begin{pmatrix} \mathscr{R}_\theta^{11} & \mathscr{R}_\theta^{12} \\ \mathscr{R}_\theta^{21} & \mathscr{R}_\theta^{22} \end{pmatrix}, \tag{6.440}$$

where the entries in the above matrix,

$$\begin{aligned} \mathscr{R}_\theta^{11} &: L_1^p(\partial\Omega) \longrightarrow L_1^p(\partial\Omega), \\ \mathscr{R}_\theta^{12} &: L^p(\partial\Omega) \longrightarrow L_1^p(\partial\Omega), \\ \mathscr{R}_\theta^{21} &: L_1^p(\partial\Omega) \longrightarrow L^p(\partial\Omega), \\ \mathscr{R}_\theta^{22} &: L^p(\partial\Omega) \longrightarrow L^p(\partial\Omega), \end{aligned} \tag{6.441}$$

are the principal-value singular integral operators given at σ-a.e. $X \in \partial\Omega$ by

$$\begin{aligned} \big(\mathscr{R}_\theta^{11} F\big)(X) := \lim_{\varepsilon\to 0+} \int\limits_{\substack{Y\in\partial\Omega \\ |X-Y|>\varepsilon}} & \Big\{\partial_{\nu(Y)}[(\Delta E)(X-Y)]F(Y) \\ & + \frac{1}{1+2\theta+n\theta^2}\partial_{\tau_{kj}(Y)}[(\partial_k E)(X-Y)]\nu_i(Y)(\partial_{\tau_{ij}} F)(Y)\Big\}\, d\sigma(Y), \end{aligned} \tag{6.442}$$

$$\begin{aligned} \big(\mathscr{R}_\theta^{12} g\big)(X) := \lim_{\varepsilon\to 0+} \int\limits_{\substack{Y\in\partial\Omega \\ |X-Y|>\varepsilon}} & \Big\{(\Delta E)(X-Y)g(Y) \\ & - \frac{1}{1+2\theta+n\theta^2}\partial_{\tau_{kj}(Y)}[(\partial_k E)(X-Y)]\nu_j(Y)g(Y)\Big\}\, d\sigma(Y), \end{aligned} \tag{6.443}$$

$$\begin{aligned} \big(\mathscr{R}_\theta^{21} F\big)(X) := \lim_{\varepsilon\to 0+} \int\limits_{\substack{Y\in\partial\Omega \\ |X-Y|>\varepsilon}} & \nu_\ell(X)\Big\{-(\partial_i \Delta E)(X-Y)(\partial_{\tau_{\ell i}} F)(Y) \\ & - \frac{1}{1+2\theta+n\theta^2}\partial_{\tau_{kj}(Y)}[(\partial_\ell\partial_k E)(X-Y)]\nu_i(Y)(\partial_{\tau_{ij}} F)(Y)\Big\}\, d\sigma(Y), \end{aligned} \tag{6.444}$$

and

$$\begin{aligned} \big(\mathscr{R}_\theta^{22} g\big)(X) := \lim_{\varepsilon\to 0+} \int\limits_{\substack{Y\in\partial\Omega \\ |X-Y|>\varepsilon}} & \nu_\ell(X)\Big\{-(\partial_\ell \Delta E)(X-Y)g(Y) \\ & + \frac{1}{1+2\theta+n\theta^2}\partial_{\tau_{kj}(Y)}[(\partial_\ell\partial_k E)(X-Y)]\nu_j(Y)g(Y)\Big\}\, d\sigma(Y), \end{aligned} \tag{6.445}$$

for each $F \in L_1^p(\partial\Omega)$ and each $g \in L^p(\partial\Omega)$. Here the summation convention over repeated indices has been used.

Proposition 6.48. *Retain the same setting as above, and recall the definition of the boundary biharmonic double multi-layer operator $\dot{K}_\theta$ on $\partial\Omega$ from Definition 6.43. Also, recall the mapping Ψ from Proposition 6.41. Then, for each $p \in (1, \infty)$, the following diagram is commutative:*

$$
\begin{array}{ccc}
L_1^p(\partial\Omega) \oplus L^p(\partial\Omega) & \xrightarrow{\widetilde{K}_\theta} & L_1^p(\partial\Omega) \oplus L^p(\partial\Omega) \\
\uparrow \Psi & & \uparrow \Psi \\
\dot{L}_{1,0}^p(\partial\Omega) & \xrightarrow[\dot{K}_\theta]{} & \dot{L}_{1,0}^p(\partial\Omega)
\end{array}
$$

Proof. Fix an arbitrary pair of functions, $(F, g) \in L_1^p(\partial\Omega) \oplus L^p(\partial\Omega)$, along with an arbitrary point $X \in \mathbb{R}^n \setminus \partial\Omega$. Based on (6.398) and (6.361)–(6.362) we may write, in a manner analogous to (6.406) (using the summation convention over repeated indices),

$$
\begin{aligned}
\widetilde{\mathscr{D}}_\theta(F, g)(X) = \int_{\partial\Omega} \Big\{ & (\Delta E)(X - Y) \\
& - \frac{1}{1 + 2\theta + n\theta^2} \nu_j(Y) \partial_{\tau_{kj}(Y)} [(\partial_k E)(X - Y)] \Big\} g(Y)\, d\sigma(Y) \\
+ \int_{\partial\Omega} \Big\{ & \partial_{\nu(Y)} [(\Delta E)(X - Y)] \\
& - \frac{1}{1 + 2\theta + n\theta^2} \partial_{\tau_{ij}(Y)} \Big(\nu_i(Y) \partial_{\tau_{kj}(Y)} [(\partial_k E)(X - Y)] \Big) \Big\} F(Y)\, d\sigma(Y),
\end{aligned}
\tag{6.446}
$$

where we have also integrated by parts on the boundary. Hence, for every pair of functions $(F, g) \in L_1^p(\partial\Omega) \oplus L^p(\partial\Omega)$ we have

$$
\begin{aligned}
\widetilde{\mathscr{D}}_\theta(F, g)(X) = {} & (\mathscr{D}_\Delta F)(X) + (\mathscr{S}_\Delta g)(X) \\
& + \frac{1}{1 + 2\theta + n\theta^2} \int_{\partial\Omega} \Big\{ \partial_{\tau_{kj}(Y)} [(\partial_k E)(X - Y)] \times \\
& \times \Big(\nu_i \partial_{\tau_{ij}} F - \nu_j g \Big)(Y) \Big\}\, d\sigma(Y),
\end{aligned}
\tag{6.447}
$$

at every $X \in \mathbb{R}^n \setminus \partial\Omega$. Consequently, for every number $\ell \in \{1, \dots, n\}$, at each point $X \in \mathbb{R}^n \setminus \partial\Omega$ we may write

$$\partial_\ell\Big(\widetilde{\mathscr{D}}_\theta(F,g)\Big)(X) \tag{6.448}$$

$$= \int_{\partial\Omega}\Big\{(\partial_i \Delta E)(X-Y)(\partial_{\tau_{\ell i}} F)(Y) + (\partial_\ell \Delta E)(X-Y)g(Y)$$

$$+\frac{1}{1+2\theta+n\theta^2}\,\partial_{\tau_{kj}(Y)}[(\partial_\ell\partial_k E)(X-Y)]\Big(\nu_i\partial_{\tau_{ij}} F - \nu_j g\Big)(Y)\Big\}\,d\sigma(Y),$$

based on (6.447) and (6.391).

From (6.447)–(6.448), on the one hand, and (6.439)–(6.445), on the other hand, we deduce, by also making use of the general jump-formula (2.530), that (cf. [129, (14.2) on p. 253])

$$\Big(\widetilde{\mathscr{D}}_\theta(F,g)\big\lfloor_{\partial\Omega}, -\partial_\nu\widetilde{\mathscr{D}}_\theta(F,g)\Big) = \big(\tfrac12 I + \widetilde{K}_\theta\big)(F,g),$$
$$\text{for each } F \in L^p_1(\partial\Omega) \text{ and each } g \in L^p(\partial\Omega). \tag{6.449}$$

As such, for every $\dot f \in \dot L^p_{1,0}(\partial\Omega)$ we may compute

$$\begin{aligned}
\big(\tfrac12 I + \widetilde{K}_\theta\big)\Psi(\dot f) &= \Big([\widetilde{\mathscr{D}}_\theta \circ \Psi(\dot f)]\big\lfloor_{\partial\Omega}, -\partial_\nu[\widetilde{\mathscr{D}}_\theta \circ \Psi(\dot f)]\Big)\\
&= \Big(\dot{\mathscr{D}}_\theta \dot f\big\lfloor_{\partial\Omega}, -\partial_\nu \dot{\mathscr{D}}_\theta \dot f\Big)\\
&= \Psi\Big(\mathrm{tr}_1 \dot{\mathscr{D}}_\theta \dot f\Big)\\
&= \Psi\Big(\big(\tfrac12 I + \dot K_\theta\big)\dot f\Big),
\end{aligned} \tag{6.450}$$

where the first equality is (6.449) written for $(F,g) := \Psi(\dot f)$, the second equality has been established in Proposition 6.44, the third equality makes use of (6.382), and the fourth equality is a consequence of Theorem 4.14. Now the claim about the commutativity of the diagram in the statement of the proposition readily follows from (6.450). □

We are now prepared to state and prove a basic invertibility result, extending work in [118, 129].

Theorem 6.49. *Assume that $\Omega \subset \mathbb{R}^n$, with $n \geq 2$, is a bounded Lipschitz domain with connected boundary, and fix $\theta \in \mathbb{R}$ with $\theta > -\frac{1}{n}$. Also, recall the boundary biharmonic double multi-layer operator $\dot K_\theta$ on $\partial\Omega$ from Definition 6.43. Then there exists $\varepsilon > 0$ with the property that*

$$\begin{aligned}&\tfrac{1}{2}I + \dot{K}_\theta : \dot{L}^p_{1,0}(\partial\Omega) \longrightarrow \dot{L}^p_{1,0}(\partial\Omega) \text{ is an isomorphism}\\&\quad \text{whenever } p \in \big(2-\varepsilon, \tfrac{2(n-1)}{n-3}+\varepsilon\big) \text{ if } n \geq 4,\\&\quad \text{and whenever } p \in \big(2-\varepsilon, \infty\big) \text{ if } n \in \{2,3\},\end{aligned} \tag{6.451}$$

and

$$\begin{aligned}&\tfrac{1}{2}I + \dot{K}_\theta : \dot{L}^p_{1,1}(\partial\Omega) \longrightarrow \dot{L}^p_{1,1}(\partial\Omega) \text{ is an isomorphism}\\&\quad \text{whenever } p \in \big(\tfrac{2(n-1)}{n+1}-\varepsilon, 2+\varepsilon\big) \text{ if } n \geq 4,\\&\quad \text{and whenever } p \in \big(1, 2+\varepsilon\big) \text{ if } n \in \{2,3\}.\end{aligned} \tag{6.452}$$

In addition, the inverses of the isomorphisms in (6.451) *and* (6.452) *act in a compatible manner on the intersection of their domains.*

Furthermore,

$$\begin{aligned}&-\tfrac{1}{2}I + \dot{K}_\theta : \dot{L}^p_{1,0}(\partial\Omega)\big/\dot{\mathscr{P}}_1(\partial\Omega) \longrightarrow \dot{L}^p_{1,0}(\partial\Omega)\big/\dot{\mathscr{P}}_1(\partial\Omega)\\&\qquad\text{is an isomorphism}\\&\text{if } p \in \big(2-\varepsilon, \tfrac{2(n-1)}{n-3}+\varepsilon\big) \text{ for } n \geq 4, \text{ and } p \in \big(2-\varepsilon, \infty\big) \text{ for } n \in \{2,3\},\end{aligned} \tag{6.453}$$

and

$$\begin{aligned}&-\tfrac{1}{2}I + \dot{K}_\theta : \dot{L}^p_{1,1}(\partial\Omega)\big/\dot{\mathscr{P}}_1(\partial\Omega) \longrightarrow \dot{L}^p_{1,1}(\partial\Omega)\big/\dot{\mathscr{P}}_1(\partial\Omega)\\&\qquad\text{is an isomorphism}\\&\text{if } p \in \big(\tfrac{2(n-1)}{n+1}-\varepsilon, 2+\varepsilon\big) \text{ for } n \geq 4, \text{ and } p \in \big(1, 2+\varepsilon\big) \text{ for } n \in \{2,3\},\end{aligned} \tag{6.454}$$

and once again the inverses of the isomorphisms in (6.453) *and* (6.454) *act in a compatible manner on the intersection of their domains.*

Proof. Given $\varepsilon \in (0,1)$, consider the open intervals

$$I_\varepsilon := \begin{cases} \Big(\frac{2(n-1)}{n+1+\varepsilon}, \frac{2(n-1)}{n-1-\varepsilon}\Big) & \text{if } n \geq 4,\\[6pt] \Big(1, \frac{2(n-1)}{n-1-\varepsilon}\Big) & \text{if } n \in \{2,3\},\end{cases} \tag{6.455}$$

and

$$I'_\varepsilon := \begin{cases} \left(\frac{2(n-1)}{n-1+\varepsilon}, \frac{2(n-1)}{n-3-\varepsilon}\right) & \text{if } n \geq 4, \\ \left(\frac{2(n-1)}{n-1+\varepsilon}, \infty\right) & \text{if } n \in \{2,3\}. \end{cases} \tag{6.456}$$

Hence, for any $p, p' \in (1,\infty)$ with $\frac{1}{p} + \frac{1}{p'} = 1$ we have

$$p \in I'_\varepsilon \Longleftrightarrow p' \in I_\varepsilon. \tag{6.457}$$

The starting point is the result asserting that there exists $\varepsilon \in (0,1)$ such that, with $\widetilde{K}_\theta$ as in (6.439)–(6.445), the operators

$$\begin{gathered} \pm\tfrac{1}{2}I + \widetilde{K}_\theta : L^p_1(\partial\Omega) \oplus L^p(\partial\Omega) \longrightarrow L^p_1(\partial\Omega) \oplus L^p(\partial\Omega) \\ \text{are Fredholm with index zero whenever } \ p \in I'_\varepsilon. \end{gathered} \tag{6.458}$$

This result (which uses $\theta > -\frac{1}{n}$) has been established first when $p \in (2-\varepsilon, 2+\varepsilon)$ by G. Verchota in [129], and the extension to the larger range $p \in I'_\varepsilon$ is due to Z. Shen in [118]. Moreover, it has been established in [118] that, for some $\varepsilon \in (0,1)$,

$$\begin{gathered} \tfrac{1}{2}I + \widetilde{K}_\theta : L^p_1(\partial\Omega) \oplus L^p(\partial\Omega) \longrightarrow L^p_1(\partial\Omega) \oplus L^p(\partial\Omega) \\ \text{is an isomorphism whenever } \ p \in I'_\varepsilon. \end{gathered} \tag{6.459}$$

Granted this, the invertibility of the operator in (6.451) then follows from (6.459), Propositions 6.48, and 6.41. Concerning the operator in (6.453), in a first stage the same circle of ideas give, based on (6.458), that

$$\begin{gathered} \text{the operator } \ -\tfrac{1}{2}I + \dot{K}_\theta : \dot{L}^p_{1,0}(\partial\Omega) \longrightarrow \dot{L}^p_{1,0}(\partial\Omega) \\ \text{is Fredholm with index zero for each } \ p \in I'_\varepsilon. \end{gathered} \tag{6.460}$$

In turn, thanks to (6.460) and Proposition 4.17, we also have that

$$\begin{gathered} -\tfrac{1}{2}I + \dot{K}_\theta : \dot{L}^p_{1,0}(\partial\Omega)\Big/\mathscr{P}_1(\partial\Omega) \longrightarrow \dot{L}^p_{1,0}(\partial\Omega)\Big/\mathscr{P}_1(\partial\Omega) \\ \text{is Fredholm with index zero whenever } \ p \in I'_\varepsilon. \end{gathered} \tag{6.461}$$

Given that the embedding $\dot{B}^{2,2}_{1,1/2}(\partial\Omega) \hookrightarrow \dot{L}^p_{1,0}(\partial\Omega)$ is well-defined, continuous and with dense range, for each $p \in I_\varepsilon$ provided that $\varepsilon > 0$ is small enough, we deduce from the invertibility of (6.174), Lemma 6.18, (6.461), and Corollary 6.5 (keeping in mind (6.355)) that

$$-\tfrac{1}{2}I + \dot{K}_\theta : \dot{L}^p_{1,0}(\partial\Omega)\Big/\dot{\mathscr{P}}_1(\partial\Omega) \longrightarrow \dot{L}^p_{1,0}(\partial\Omega)\Big/\dot{\mathscr{P}}_1(\partial\Omega)$$
$$\text{is an injective operator for each } \ p \in I'_\varepsilon. \tag{6.462}$$

In passing, let us also note that the same type of reasoning, based on (6.126), Lemma 6.18 and (6.460), gives

$$\Big\{\dot{f} \in \dot{L}^p_{1,0}(\partial\Omega) : \big(-\tfrac{1}{2}I + \dot{K}_\theta\big)\dot{f} = 0\Big\} = \dot{\mathscr{P}}_1(\partial\Omega), \quad \forall\, p \in I'_\varepsilon. \tag{6.463}$$

This is going to be useful later on.

Let us now consider the claims made in (6.452) and (6.454). To this end, we first note that, thanks to (6.458), Propositions 6.48, and 6.41, we have that

$$\pm\tfrac{1}{2}I + \dot{K}_\theta : \dot{L}^p_{1,0}(\partial\Omega) \longrightarrow \dot{L}^p_{1,0}(\partial\Omega)$$
$$\text{are Fredholm with index zero for each } \ p \in I'_\varepsilon. \tag{6.464}$$

Thus, by duality,

$$\pm\tfrac{1}{2}I + \dot{K}^*_\theta : \big(\dot{L}^p_{1,0}(\partial\Omega)\big)^* \longrightarrow \big(\dot{L}^p_{1,0}(\partial\Omega)\big)^*$$
$$\text{are Fredholm with index zero for each } \ p \in I'_\varepsilon. \tag{6.465}$$

Let us also remark that, as seen from (6.464), the composition

$$\big(\tfrac{1}{2}I + \dot{K}_\theta\big) \circ \big(-\tfrac{1}{2}I + \dot{K}_\theta\big) : \dot{L}^p_{1,0}(\partial\Omega) \longrightarrow \dot{L}^p_{1,0}(\partial\Omega)$$
$$\text{is Fredholm with index zero for each } \ p \in I'_\varepsilon, \tag{6.466}$$

hence by duality,

$$\big(\tfrac{1}{2}I + \dot{K}^*_\theta\big) \circ \big(-\tfrac{1}{2}I + \dot{K}^*_\theta\big) : \big(\dot{L}^p_{1,0}(\partial\Omega)\big)^* \longrightarrow \big(\dot{L}^p_{1,0}(\partial\Omega)\big)^*$$
$$\text{is Fredholm with index zero for each } \ p \in I'_\varepsilon. \tag{6.467}$$

To proceed, denote by $\dot{S}$ the boundary version of the biharmonic single multi-layer (associated with $L = \Delta^2$ as in (5.19)). Hence, if $p, p' \in (1, \infty)$ are such that $1/p + 1/p' = 1$, then Theorem 5.4 ensures that

$$\dot{S} : \big(\dot{L}^{p'}_{1,1}(\partial\Omega)\big)^* \longrightarrow \dot{L}^p_{1,0}(\partial\Omega) \ \text{ boundedly, and} \tag{6.468}$$

$$\dot{S} : \big(\dot{L}^p_{1,0}(\partial\Omega)\big)^* \longrightarrow \dot{L}^{p'}_{1,1}(\partial\Omega) \ \text{ boundedly.} \tag{6.469}$$

Based on these and Theorem 6.46 we may therefore conclude that, for each index $p \in (1, \infty)$, the operators

$$\partial_\nu^{A_\theta} \dot{\mathscr{D}}_\theta \circ \dot{S} : \left(\dot{L}^p_{1,0}(\partial\Omega)\right)^* \longrightarrow \left(\dot{L}^p_{1,0}(\partial\Omega)\right)^*, \tag{6.470}$$

$$\partial_\nu^{A_\theta} \dot{\mathscr{D}}_\theta \circ \dot{S} : \left(\dot{L}^p_{1,1}(\partial\Omega)\right)^* \longrightarrow \left(\dot{L}^p_{1,1}(\partial\Omega)\right)^*, \tag{6.471}$$

$$\dot{S} \circ \partial_\nu^{A_\theta} \dot{\mathscr{D}}_\theta : \dot{L}^p_{1,0}(\partial\Omega) \longrightarrow \dot{L}^p_{1,0}(\partial\Omega), \tag{6.472}$$

$$\dot{S} \circ \partial_\nu^{A_\theta} \dot{\mathscr{D}}_\theta : \dot{L}^p_{1,1}(\partial\Omega) \longrightarrow \dot{L}^p_{1,1}(\partial\Omega), \tag{6.473}$$

are well-defined, linear and bounded (where, as usual, $1/p + 1/p' = 1$). Having established these boundedness results, we may then conclude from (6.470), (6.467), formula (5.176), and the density results proved in Propositions 3.3 and 3.7, that

$$\begin{gathered}\partial_\nu^{A_\theta} \dot{\mathscr{D}}_\theta \circ \dot{S} : \left(\dot{L}^p_{1,0}(\partial\Omega)\right)^* \longrightarrow \left(\dot{L}^p_{1,0}(\partial\Omega)\right)^* \\ \text{is Fredholm with index zero for each } p \in I'_\varepsilon.\end{gathered} \tag{6.474}$$

In turn, (6.474), (6.469), and (6.410) readily imply that

$$\begin{gathered}\text{the operator } \partial_\nu^{A_\theta} \dot{\mathscr{D}}_\theta : \dot{L}^{p'}_{1,1}(\partial\Omega) \longrightarrow \left(\dot{L}^p_{1,0}(\partial\Omega)\right)^* \text{ has} \\ \text{closed range, of finite codimension, if } p \in I'_\varepsilon \text{ and } \tfrac{1}{p} + \tfrac{1}{p'} = 1.\end{gathered} \tag{6.475}$$

With this in hand and availing ourselves of the fact that the operators in (6.410) and (6.411) are adjoint to one another, it follows from (6.475) and duality that

$$\begin{gathered}\text{the operator } \partial_\nu^{A_\theta} \dot{\mathscr{D}}_\theta : \dot{L}^p_{1,0}(\partial\Omega) \longrightarrow \left(\dot{L}^{p'}_{1,1}(\partial\Omega)\right)^* \text{ has} \\ \text{finite dimensional kernel, if } p \in I'_\varepsilon \text{ and } \tfrac{1}{p} + \tfrac{1}{p'} = 1.\end{gathered} \tag{6.476}$$

Next we claim that

$$\forall\, p' \in I_\varepsilon \quad \exists\, q \in I'_\varepsilon \text{ such that } \dot{L}^{p'}_{1,1}(\partial\Omega) \hookrightarrow \dot{L}^q_{1,0}(\partial\Omega). \tag{6.477}$$

To justify this claim, assume first that $n \geq 4$ and note that this forces $p' \in (1, n-1)$ for any $p' \in I_\varepsilon$. As such, the embedding $\dot{L}^{p'}_{1,1}(\partial\Omega) \hookrightarrow \dot{L}^q_{1,0}(\partial\Omega)$ holds whenever $q := \left(\frac{1}{p'} - \frac{1}{n-1}\right)^{-1}$. On the other hand, it may be verified without difficulty that

$$\left\{\left(\tfrac{1}{p'} - \tfrac{1}{n-1}\right)^{-1} : p' \in I_\varepsilon\right\} = I'_\varepsilon. \tag{6.478}$$

This, of course, proves the claim in (6.477) when $n \geq 4$. When $n = 2$, the embedding in (6.477) holds for any $p', q \in (1, \infty)$, while when $n = 3$ is obviously true whenever indices $p' \in [2, \infty)$ and $q \in (1, \infty)$. Finally, in the remaining case, i.e., for $n = 3$ and $p' \in (1, 2)$, we may take $q := \left(\frac{1}{p'} - \frac{1}{2}\right)^{-1} \in (2, \infty) \subseteq I'_\varepsilon$. This finishes the proof of (6.477).

Moving on, we may then deduce from (6.476), (6.477), and (6.475), that

$$\begin{gathered}\text{the operator } \partial_\nu^{A_\theta} \dot{\mathscr{D}}_\theta : \dot{L}^{p'}_{1,1}(\partial\Omega) \longrightarrow \left(\dot{L}^p_{1,0}(\partial\Omega)\right)^* \text{ has both} \\ \text{closed range, of finite codimension, and finite dimensional kernel} \\ \text{whenever } p \in I'_\varepsilon \text{ and } \tfrac{1}{p} + \tfrac{1}{p'} = 1.\end{gathered} \tag{6.479}$$

In other words,

$$\begin{gathered}\partial_\nu^{A_\theta} \dot{\mathscr{D}}_\theta : \dot{L}^{p'}_{1,1}(\partial\Omega) \longrightarrow \left(\dot{L}^p_{1,0}(\partial\Omega)\right)^* \text{ is Fredholm} \\ \text{provided } p \in I'_\varepsilon \text{ and } \tfrac{1}{p} + \tfrac{1}{p'} = 1.\end{gathered} \tag{6.480}$$

In particular, the operator $\partial_\nu^{A_\theta} \dot{\mathscr{D}}_\theta$ has, in the above context, a quasi-inverse. In concrete terms, this means that whenever $p \in I'_\varepsilon$ and $\frac{1}{p} + \frac{1}{p'} = 1$, there exist a Fredholm operator $R : \left(\dot{L}^p_{1,0}(\partial\Omega)\right)^* \to \dot{L}^{p'}_{1,1}(\partial\Omega)$, and a linear compact operator Comp mapping $\dot{L}^{p'}_{1,1}(\partial\Omega)$ into some Banach space $\mathscr{X}$, with the property that

$$R \circ \partial_\nu^{A_\theta} \dot{\mathscr{D}}_\theta = I + \text{Comp} \text{ on } \dot{L}^{p'}_{1,1}(\partial\Omega). \tag{6.481}$$

Composing the Fredholm operator in (6.474) to the left with the Fredholm operator R just considered, and keeping in mind that the class of Fredholm operators is closed under composition as well as additive compact perturbations, we arrive at the conclusion that

$$\begin{gathered}\dot{S} : \left(\dot{L}^p_{1,0}(\partial\Omega)\right)^* \longrightarrow \dot{L}^{p'}_{1,1}(\partial\Omega) \text{ is a Fredholm operator} \\ \text{whenever } p \in I'_\varepsilon \text{ and } \tfrac{1}{p} + \tfrac{1}{p'} = 1.\end{gathered} \tag{6.482}$$

In light of the self-adjointness of the single multi-layer (cf. (5.22)), we may take the dual of (6.482) in order to also obtain that

$$\begin{gathered}\dot{S} : \left(\dot{L}^{p'}_{1,1}(\partial\Omega)\right)^* \longrightarrow \dot{L}^p_{1,0}(\partial\Omega) \text{ is a Fredholm operator} \\ \text{whenever } p \in I'_\varepsilon \text{ and } \tfrac{1}{p} + \tfrac{1}{p'} = 1.\end{gathered} \tag{6.483}$$

At this stage, taking the composition of the Fredholm operators in (6.480) and (6.482) leads to the conclusion that

$$\begin{gathered}\dot{S}\circ\partial_\nu^{A_\theta}\dot{\mathscr{D}}_\theta : \dot{L}^{p'}_{1,1}(\partial\Omega)\longrightarrow\dot{L}^{p'}_{1,1}(\partial\Omega)\ \text{ is a Fredholm operator}\\ \text{whenever }\ p\in I'_\varepsilon\ \text{ and }\ \tfrac{1}{p}+\tfrac{1}{p'}=1.\end{gathered}\tag{6.484}$$

Granted this, from (5.177) and the density results proved in Propositions 3.3 and 3.7, we deduce that

$$\begin{gathered}\big(\tfrac{1}{2}I+\dot{K}_\theta\big)\circ\big(-\tfrac{1}{2}I+\dot{K}_\theta\big) : \dot{L}^{p'}_{1,1}(\partial\Omega)\longrightarrow\dot{L}^{p'}_{1,1}(\partial\Omega)\ \text{ is Fredholm}\\ \text{whenever }\ p\in I'_\varepsilon\ \text{ and }\ \tfrac{1}{p}+\tfrac{1}{p'}=1.\end{gathered}\tag{6.485}$$

In turn, this readily implies that the operators $\pm\frac{1}{2}I+\dot{K}_\theta$ (which commute with one another) have both closed ranges of finite codimension and finite dimensional kernels, thus, ultimately,

$$\begin{gathered}\pm\tfrac{1}{2}I+\dot{K}_\theta : \dot{L}^{p'}_{1,1}(\partial\Omega)\longrightarrow\dot{L}^{p'}_{1,1}(\partial\Omega)\ \text{ are Fredholm operators}\\ \text{whenever }\ p\in I'_\varepsilon\ \text{ and }\ \tfrac{1}{p}+\tfrac{1}{p'}=1.\end{gathered}\tag{6.486}$$

Going further, we make use of Proposition 5.30 (and the same type of boundedness and density results as before) in order to obtain the following intertwining identity

$$\dot{S}\circ\big(\pm\tfrac{1}{2}I+\dot{K}^*_\theta\big)=\big(\pm\tfrac{1}{2}I+\dot{K}_\theta\big)\circ\dot{S}\quad\text{on }\ \Big(\dot{L}^{p}_{1,0}(\partial\Omega)\Big)^*,\tag{6.487}$$

in which $\dot{S}$ is as in (6.482), $\dot{K}_\theta$ acts on $\dot{L}^{p'}_{1,1}(\partial\Omega)$, and $\dot{K}^*_\theta$ acts on $\Big(\dot{L}^{p}_{1,0}(\partial\Omega)\Big)^*$. From (6.487), (6.486), (6.482), (6.465), and the additivity law for the Fredholm index, we eventually obtain (cf. also (6.457)) that

$$\begin{gathered}\text{the operators }\ \pm\tfrac{1}{2}I+\dot{K}_\theta : \dot{L}^{p'}_{1,1}(\partial\Omega)\longrightarrow\dot{L}^{p'}_{1,1}(\partial\Omega)\\ \text{are Fredholm with index zero if }\ p'\in I_\varepsilon.\end{gathered}\tag{6.488}$$

In particular,

$$\begin{gathered}\pm\tfrac{1}{2}I+\dot{K}_\theta\ \text{ are Fredholm with index zero on }\dot{L}^{p'}_{1,1}(\partial\Omega)\Big/\mathscr{P}_1(\partial\Omega)\\ \text{whenever }\ p'\in I_\varepsilon.\end{gathered}\tag{6.489}$$

Using this, the embedding in (6.477), and the injectivity of the operator in (6.451) it follows that the operator in (6.452) is also injective, thus ultimately invertible by (6.488). This takes care of the claim made in (6.452). Finally, the same type of reasoning, based on the embedding (6.477) and the injectivity of the operator in (6.453), shows that the operator in (6.454) is also injective, thus ultimately invertible by (6.489).

Let us now prove that the inverses of the isomorphisms in (6.451) and (6.452) act in a compatible manner on the intersection of their domains. With this goal in mind, assume that

$$\begin{aligned} &\dot{f}_0 \in \dot{L}^{p_0}_{1,0}(\partial\Omega) \text{ with } p_0 \in I'_\varepsilon \text{ and } \dot{f}_1 \in \dot{L}^{p_1}_{1,1}(\partial\Omega) \text{ with } p_1 \in I_\varepsilon \\ &\qquad \text{are such that } \big(\tfrac{1}{2}I + \dot{K}_\theta\big)\dot{f}_0 = \big(\tfrac{1}{2}I + \dot{K}_\theta\big)\dot{f}_1. \end{aligned} \tag{6.490}$$

By (6.477), there exists $q \in I'_\varepsilon$ with the property that $\dot{L}^{p_1}_{1,1}(\partial\Omega) \hookrightarrow \dot{L}^{q}_{1,0}(\partial\Omega)$, hence if we now set $p := \min\{p_0, q\}$ then

$$p \in I'_\varepsilon \text{ and } \dot{L}^{p_0}_{1,0}(\partial\Omega) \cap \dot{L}^{p_1}_{1,1}(\partial\Omega) \hookrightarrow \dot{L}^{p}_{1,0}(\partial\Omega). \tag{6.491}$$

From (6.490)–(6.491) and the fact that $\frac{1}{2}I + \dot{K}_\theta$ is invertible on $\dot{L}^{p}_{1,0}(\partial\Omega)$, it follows that $\dot{f}_0 = \dot{f}_1$, as wanted.

Finally, the compatibility of the inverses of the isomorphisms in (6.453) and (6.454) on the intersection of their domains is established analogously, completing the proof of the theorem. □

We now proceed to record several significant consequences of Theorem 6.49 (and its proof).

Corollary 6.50. *Let $\Omega \subset \mathbb{R}^n$, with $n \geq 2$, be a bounded Lipschitz domain with connected boundary, and fix $\theta \in \mathbb{R}$ with $\theta > -\frac{1}{n}$. Also, recall the boundary biharmonic double multi-layer operator $\dot{K}_\theta$ on $\partial\Omega$ from Definition 6.43. Then there exists $\varepsilon > 0$ with the property that*

$$\Big\{\dot{f} \in \dot{L}^{p}_{1,1}(\partial\Omega) : \big(-\tfrac{1}{2}I + \dot{K}_\theta\big)\dot{f} = 0\Big\} = \dot{\mathscr{P}}_1(\partial\Omega) \tag{6.492}$$

whenever

$$\begin{aligned} &p \in \big(\tfrac{2(n-1)}{n+1} - \varepsilon, 2 + \varepsilon\big) \ \textit{if } n \geq 4, \\ &\textit{and } \ p \in \big(1, 2 + \varepsilon\big) \ \textit{if } n \in \{2, 3\}. \end{aligned} \tag{6.493}$$

Proof. This is a consequence of the formula in (6.463) and the embedding result recorded in (6.477). □

Our second result pertains to the existence, uniqueness, integral representation in terms of the multi-layers introduced in this monograph, and regularity (measured on the Besov scale), of the solution of the Dirichlet problem for the bi-Laplacian with boundary data from Whitney–Lebesgue spaces in Lipschitz domains. As such, this completes and refines work in [118, 129]. Before reading the statement, the reader is advised to recall Definition 6.27.

Corollary 6.51. *Consider the biharmonic operator, $L = \Delta^2$, in $\mathbb{R}^n$ with $n \geq 2$, and assume that Ω is a bounded Lipschitz domain in $\mathbb{R}^n$ with connected boundary. Then, in this context, there exists $\varepsilon > 0$ such that property G_p holds*

$$\begin{aligned} &\textit{whenever } \ p \in \left(\tfrac{2(n-1)}{n+1} - \varepsilon, 2 + \varepsilon\right) \ \textit{if } \ n \geq 4, \\ &\textit{and whenever } \ p \in \left(1, 2 + \varepsilon\right) \ \textit{if } \ n \in \{2, 3\}. \end{aligned} \tag{6.494}$$

As a consequence, there exists $\varepsilon > 0$ such that if

$$\begin{aligned} &p \in \left(2 - \varepsilon, \tfrac{2(n-1)}{n-3} + \varepsilon\right) \ \textit{if } \ n \geq 4, \\ &\textit{and } \ p \in \left(2 - \varepsilon, \infty\right) \ \textit{if } \ n \in \{2, 3\}, \end{aligned} \tag{6.495}$$

then Dirichlet boundary value problem for the bi-Laplacian with data from Whitney–Lebesgue spaces,

$$\begin{cases} \Delta^2 u = 0 \quad \textit{in} \quad \Omega, \\ \mathcal{N}(\nabla u) \in L^p(\partial\Omega), \\ \left(u\big\lfloor_{\partial\Omega}, (\nabla u)\big\lfloor_{\partial\Omega}\right) = \dot{f} \in \dot{L}^p_{1,0}(\partial\Omega), \end{cases} \tag{6.496}$$

has a unique solution which, for every $\theta \in \mathbb{R}$ with $\theta > -\frac{1}{n}$, may be represented as

$$u(X) = \dot{\mathscr{D}}_\theta\Big[\left(\tfrac{1}{2}I + \dot{K}_\theta\right)^{-1}\dot{f}\Big](X), \qquad \forall\, X \in \Omega. \tag{6.497}$$

In particular, the solution of (6.496) *satisfies*

$$\|u\|_{B^{p,p\vee 2}_{1+1/p}(\Omega)} \leq C\,\|\dot{f}\|_{\dot{L}^p_{1,0}(\partial\Omega)} \tag{6.498}$$

for some finite constant $C = C(\Omega, p, \theta, n) > 0$.

Proof. The fact that property G_p holds for p as in (6.494) is seen from Theorem 6.49 and the discussion in (6.225)–(6.228). As such, the uniqueness part in the well-posedness of the boundary problem (6.496) follows from what we have just proved and Theorem 6.29. Next, u in (6.497) is well-defined in light of (6.451), and solves (6.496) thanks to (4.58), Theorem 4.7, and (4.164). Finally, that u satisfies (6.498) follows from the integral representation formula (6.497) and (4.85). □

Our third result deals with the role of multi-layer potentials in the solvability of the so-called regularity problem for the bi-Laplacian in Lipschitz domains. As a preamble, we first recall the following estimate of Hardy-type.

Lemma 6.52. *Let L be a homogeneous, constant coefficient, W-elliptic operator, and assume that $\Omega \subset \mathbb{R}^n$ is a bounded Lipschitz domain. In this context, suppose that u is a null-solution of L in Ω which satisfies $\mathcal{N}(\nabla u) \in L^p(\partial\Omega)$ for some $p \in (0, n-1)$. Then*

$$\mathcal{N} u \in L^{p^*}(\partial\Omega) \ \ \textit{where} \ \ p^* := \left(\tfrac{1}{p} - \tfrac{1}{n-1}\right)^{-1}. \tag{6.499}$$

See [94, Lemma 11.9] for a proof. Here is the well-posedness result advertised earlier, which refines earlier work in [69, 118, 129].

Theorem 6.53. *Assume that $\Omega \subset \mathbb{R}^n$, with $n \geq 2$, is a bounded Lipschitz domain with connected boundary, and fix $\theta \in \mathbb{R}$ with $\theta > -\frac{1}{n}$. As before, let $\dot{\mathcal{D}}_\theta$ and $\dot{K}_\theta$ denote the biharmonic double multi-layer operators (relative to Ω) introduced in Definition 6.43.*

Then there exists $\varepsilon > 0$ with the property that whenever $p \in (1, \infty)$ satisfies

$$\begin{aligned} & p \in \left(\tfrac{2(n-1)}{n+1} - \varepsilon, 2 + \varepsilon\right) \ \ \textit{if} \ \ n \geq 4, \\ & \textit{and} \ \ p \in \big(1, 2 + \varepsilon\big) \ \ \textit{if} \ \ n \in \{2, 3\}, \end{aligned} \tag{6.500}$$

the Dirichlet boundary value problem for the bi-Laplacian with data from Whitney–Sobolev spaces,

$$\begin{cases} \Delta^2 u = 0 \quad \textit{in} \quad \Omega, \\ \mathcal{N}(\nabla^2 u) \in L^p(\partial\Omega), \\ \big(u\big|_{\partial\Omega}, (\nabla u)\big|_{\partial\Omega}\big) = \dot{f} \in \dot{L}^p_{1,1}(\partial\Omega), \end{cases} \tag{6.501}$$

has a unique solution, which actually admits the integral representation formula

$$u(X) = \dot{\mathcal{D}}_\theta\Big[\big(\tfrac{1}{2}I + \dot{K}_\theta\big)^{-1} \dot{f}\Big](X), \qquad \forall\, X \in \Omega. \tag{6.502}$$

In particular, the solution of (6.501) *satisfies*

$$\|u\|_{B^{p,p\vee 2}_{2+1/p}(\Omega)} \leq C \|\dot{f}\|_{\dot{L}^p_{1,1}(\partial\Omega)} \tag{6.503}$$

for some finite constant $C = C(\Omega, p, \theta, n) > 0$.

Proof. Let $\varepsilon > 0$ be as in Theorem 6.49. That the function u given by (6.502) is well-defined whenever p is as in (6.500) follows from the invertibility result recorded in (6.452). Also, the fact that this u actually solves (6.501) is clear from (4.58), Theorem 4.7, and (4.164). As regards uniqueness, suppose that u solves the homogeneous version of the boundary problem (6.501) for some p as in (6.500). Given the nature of the conclusion we seek, there is no loss of generality in assuming that the exponent p also satisfies $p < n-1$. Granted this, if p^* is defined as in (6.499) then (much as it was the case in the proof of Theorem 6.49) p^* satisfies the conditions listed in (6.495). Furthermore, Lemma 6.52 applied to ∇u ensures that $\mathcal{N}(\nabla u) \in L^{p^*}(\partial\Omega)$, since we are assuming that $\mathcal{N}(\nabla^2 u) \in L^p(\partial\Omega)$ to begin with. As such, the uniqueness result established in the second part of Corollary 6.51 applies and yields that $u \equiv 0$ in Ω, as wanted. Finally, that u satisfies (6.503) follows from the integral representation formula (6.502) and (4.86). □

Further invertibility results for multi-layers, complementing those established in Theorem 6.49, are discussed below.

Corollary 6.54. *Suppose that $\Omega \subset \mathbb{R}^n$, with $n \geq 2$, is a bounded Lipschitz domain with connected boundary, and fix $\theta \in \mathbb{R}$ with $\theta > -\frac{1}{n}$. As before, let $\dot{K}_\theta$ denote the boundary biharmonic double multi-layer operator on $\partial\Omega$ considered in Definition 6.43. Also, let $\dot{S}$ denote the boundary version of the biharmonic single multi-layer associated with $L = \Delta^2$ as in (5.19).*

Then there exists $\varepsilon > 0$ with the property that

$$\tfrac{1}{2}I + \dot{K}^*_\theta : \big(L^p_{1,0}(\partial\Omega)\big)^* \longrightarrow \big(L^p_{1,0}(\partial\Omega)\big)^* \text{ is an isomorphism}$$
$$\text{whenever } p \in \big(2-\varepsilon, \tfrac{2(n-1)}{n-3}+\varepsilon\big) \text{ if } n \geq 4, \tag{6.504}$$
$$\text{and whenever } p \in (2-\varepsilon, \infty) \text{ if } n \in \{2,3\},$$

and

$$\tfrac{1}{2}I + \dot{K}^*_\theta : \big(L^p_{1,1}(\partial\Omega)\big)^* \longrightarrow \big(L^p_{1,1}(\partial\Omega)\big)^* \text{ is an isomorphism}$$
$$\text{whenever } p \in \big(\tfrac{2(n-1)}{n+1}-\varepsilon, 2+\varepsilon\big) \text{ if } n \geq 4, \tag{6.505}$$
$$\text{and whenever } p \in (1, 2+\varepsilon) \text{ if } n \in \{2,3\}.$$

In addition, the inverses of the isomorphisms in (6.504) and (6.505) act in a compatible manner on the intersection of their domains.

Moreover,

$$-\tfrac{1}{2}I + \dot{K}^*_\theta \text{ is an isomorphism on } \Big(L^p_{1,0}(\partial\Omega)\Big/\dot{\mathcal{P}}_1(\partial\Omega)\Big)^* \tag{6.506}$$
$$\text{if } p \in \big(2-\varepsilon, \tfrac{2(n-1)}{n-3}+\varepsilon\big) \text{ for } n \geq 4, \text{ and } p \in (2-\varepsilon, \infty) \text{ for } n \in \{2,3\},$$

and

$$-\tfrac{1}{2}I + \dot{K}^*_\theta \text{ is an isomorphism on } \Big(L^p_{1,1}(\partial\Omega)\Big/\dot{\mathcal{P}}_1(\partial\Omega)\Big)^* \tag{6.507}$$
$$\text{if } p \in \big(\tfrac{2(n-1)}{n+1}-\varepsilon, 2+\varepsilon\big) \text{ for } n \geq 4, \text{ and } p \in (1, 2+\varepsilon) \text{ for } n \in \{2,3\}.$$

In addition, the inverses of the isomorphisms in (6.506) *and* (6.507) *act in a compatible fashion on the intersection of their domains.*

Finally, if $n \geq 3$ *and* $n \neq 4$, *then also*

$$\begin{aligned} &\dot{S} : \left(\dot{L}^p_{1,0}(\partial\Omega)\right)^* \longrightarrow \dot{L}^{p'}_{1,1}(\partial\Omega) \text{ is an isomorphism provided} \\ &p \in \left(2-\varepsilon, \tfrac{2(n-1)}{n-3}+\varepsilon\right) \text{ for } n \geq 5, \text{ and } p \in \left(2-\varepsilon, \infty\right) \text{ for } n=3, \end{aligned} \tag{6.508}$$

and

$$\begin{aligned} &\dot{S} : \left(\dot{L}^{p'}_{1,1}(\partial\Omega)\right)^* \longrightarrow \dot{L}^{p}_{1,0}(\partial\Omega) \text{ is an isomorphism provided} \\ &p \in \left(2-\varepsilon, \tfrac{2(n-1)}{n-3}+\varepsilon\right) \text{ for } n \geq 5, \text{ and } p \in \left(2-\varepsilon, \infty\right) \text{ for } n=3, \end{aligned} \tag{6.509}$$

and the inverses of the isomorphisms in (6.508) *and* (6.509) *are compatible on the intersection of their domains.*

Proof. The invertibility (and compatibility) claims concerning the boundary biharmonic double multi-layer operator are direct consequence of Theorem 6.49 and duality. As regards (6.508)–(6.509), these follow from (6.482)–(6.483), Theorem 6.11 when $n > 4$, and [129, Theorem 17.5] when $n = 3$, by reasoning as before. □

It is instructive to formulate and solve the Neumann problem for the bi-Laplacian with boundary data from the dual of Whitney–Lebesgue spaces. This parallels work in [118, 129] where a different formulation is emphasized.

Theorem 6.55. *Suppose that* $\Omega \subset \mathbb{R}^n$, *with* $n \geq 2$, *is a bounded Lipschitz domain with connected boundary, and fix* $\theta \in \mathbb{R}$ *with* $\theta > -\frac{1}{n}$. *As before, let* $\dot{K}_\theta$ *denote the biharmonic double multi-layer operators (relative to* Ω*) introduced in Definition 6.43. Finally, recall the biharmonic single multi-layer* $\dot{\mathscr{S}}$ *from* (6.429) *and the conormal derivative* $\partial_\nu^{A_\theta}$ *from Proposition 6.40.*

Then there exists $\varepsilon > 0$ *with the property that whenever* $p \in (1,\infty)$ *satisfies*

$$\begin{aligned} &p \in \left(2-\varepsilon, \tfrac{2(n-1)}{n-3}+\varepsilon\right) \text{ if } n \geq 4, \\ &\text{and } p \in \left(2-\varepsilon, \infty\right) \text{ if } n \in \{2,3\}, \end{aligned} \tag{6.510}$$

the Neumann boundary value problem for the bi-Laplacian with data from duals of Whitney–Lebesgue spaces,

$$\begin{cases} \Delta^2 u = 0 \quad \text{in} \quad \Omega, \\ \mathcal{N}(\nabla^2 u) \in L^{p'}(\partial\Omega), \\ \partial_\nu^{A_\theta} u = \Lambda \in \left(\dot{L}^p_{1,0}(\partial\Omega)\right)^* \end{cases} \tag{6.511}$$

where $1/p + 1/p' = 1$ *and the boundary data satisfies the necessary compatibility condition*

$$\langle \Lambda, \dot{P} \rangle = 0 \quad \textit{for each} \quad \dot{P} \in \dot{\mathscr{P}}_1(\partial\Omega), \tag{6.512}$$

is well-posed (with uniqueness understood modulo polynomials of degree ≤ 1*). Moreover, a solution may be given by the integral formula*

$$u(X) = \dot{\mathscr{S}}\Big[\pi^*\big(-\tfrac{1}{2}I + \dot{K}_\theta^*\big)^{-1}\widetilde{\Lambda}\Big](X), \qquad \forall\, X \in \Omega, \tag{6.513}$$

where

$$\widetilde{\Lambda} \in \Big(\dot{L}^p_{1,0}(\partial\Omega)\Big/\dot{\mathscr{P}}_1(\partial\Omega)\Big)^* \tag{6.514}$$

is defined by setting

$$\big\langle \widetilde{\Lambda}, [\dot{f}] \big\rangle := \langle \Lambda, \dot{f} \rangle, \qquad \forall\, \dot{f} \in \dot{L}^p_{1,0}(\partial\Omega), \tag{6.515}$$

with $[\dot{f}]$ *denoting the equivalence class of the Whitney array* $\dot{f} \in \dot{L}^p_{1,0}(\partial\Omega)$ *in the quotient space* $\dot{L}^p_{1,0}(\partial\Omega)\big/\dot{\mathscr{P}}_1(\partial\Omega)$*, and*

$$\pi^* : \Big(\dot{L}^p_{1,0}(\partial\Omega)\Big/\dot{\mathscr{P}}_1(\partial\Omega)\Big)^* \longrightarrow \Big(\dot{L}^p_{1,0}(\partial\Omega)\Big)^* \tag{6.516}$$

is the adjoint of the canonical projection

$$\pi : \dot{L}^p_{1,0}(\partial\Omega) \longrightarrow \dot{L}^p_{1,0}(\partial\Omega)\Big/\dot{\mathscr{P}}_1(\partial\Omega), \tag{6.517}$$

taking an arbitrary Whitney array $\dot{f} \in \dot{L}^p_{1,0}(\partial\Omega)$ *into* $[\dot{f}] \in \dot{L}^p_{1,0}(\partial\Omega)\big/\dot{\mathscr{P}}_1(\partial\Omega)$.

Proof. That the compatibility condition (6.512) is necessary is clear from formula (5.129) and degree considerations. As regards existence, let $\varepsilon > 0$ be as in Corollary 6.54. Then the function u given by (6.513) is well-defined whenever p is as in (6.510) follows from the invertibility result recorded in (6.506). By (5.5), Proposition 5.2, and (6.431) it may be checked that u solves (6.511) (by reasoning as in (6.334)). Thus, as far as the well-posedness of the problem (6.511) is concerned, there remains to establish uniqueness (in the sense specified in the statement of the theorem). To this end, assume that u is a solution of (6.511) with $\Lambda = 0$, and set

$$\dot{f} := \Big(u\big\lfloor_{\partial\Omega}, (\nabla u)\big\lfloor_{\partial\Omega}\Big) \in \dot{L}^{p'}_{1,1}(\partial\Omega). \tag{6.518}$$

Then Green's formula (5.141), suitably interpreted in the present setting, gives

$$u = \dot{\mathscr{D}}_\theta \dot{f} - \dot{\mathscr{S}}\big(\partial_\nu^{A_\theta} u\big) = \dot{\mathscr{D}}_\theta \dot{f} \quad \text{in } \Omega, \tag{6.519}$$

using $\partial_\nu^{A_\theta} u = 0$. Taking the first-order non-tangential boundary trace of both sides of (6.519) and using (6.397) then yields

$$\dot{f} = \big(\tfrac{1}{2}I + \dot{K}_\theta\big)\dot{f}, \tag{6.520}$$

which ultimately shows that $\big(-\frac{1}{2}I + \dot{K}_\theta\big)\dot{f} = 0$. From this and Corollary 6.50 we deduce that there exists $P \in \mathscr{P}_1$ such that $\dot{f} = \dot{P}$. Returning with this back in (6.519) and making use of (4.62) finally gives that $u = \dot{\mathscr{D}}_\theta \dot{P} = P$ in Ω, as desired. □

Our next goal is to explain how the invertibility results for the biharmonic layer potentials, as well as the well-posedness results for the various boundary problems for the bi-Laplacian, improve (in the sense that the range of exponents involved becomes larger) under additional regularity assumptions on the Lipschitz domain in question. This requires some preparations and we start by recalling that, given two quasi-Banach spaces $\mathscr{X}$, $\mathscr{Y}$, the space of all bounded linear operators mapping $\mathscr{X}$ into $\mathscr{Y}$ is denoted by $\mathscr{L}(\mathscr{X} \to \mathscr{Y})$. This becomes a quasi-Banach itself when equipped with the canonical operator norm

$$\|T\|_{\mathscr{L}(\mathscr{X}\to\mathscr{Y})} := \sup\{\|Tx\|_{\mathscr{Y}} : x \in \mathscr{X},\ \|x\|_{\mathscr{X}} \leq 1\}, \quad \forall\, T \in \mathscr{L}(\mathscr{X} \to \mathscr{Y}). \tag{6.521}$$

Moreover, let us also define

$$\mathrm{Comp}\,(\mathscr{X} \to \mathscr{Y}) := \text{ the space of all linear compact operators from } \mathscr{X} \text{ into } \mathscr{Y}, \tag{6.522}$$

and note that $\mathrm{Comp}\,(\mathscr{X}, \mathscr{Y})$ is a closed subspace of $\mathscr{L}(\mathscr{X}, \mathscr{Y})$. Finally, abbreviate

$$\mathscr{L}(\mathscr{X}) := \mathscr{L}(\mathscr{X} \to \mathscr{X}), \qquad \mathrm{Comp}(\mathscr{X}) := \mathrm{Comp}(\mathscr{X} \to \mathscr{X}). \tag{6.523}$$

The following is a particular case of a much more general result proved in [56, Theorem 4.36].

Theorem 6.56. *Let $\Omega \subset \mathbb{R}^n$ be a bounded Lipschitz domain. Denote by σ and ν, respectively, the surface measure and outward unit normal on $\partial\Omega$. Also, fix an arbitrary $p \in (1, \infty)$. Then for every $\varepsilon > 0$ the following holds. Given a function k satisfying*

$$k : \mathbb{R}^n \setminus \{0\} \to \mathbb{R} \text{ is smooth, even, and homogeneous of degree } -n \tag{6.524}$$

to which one associates the principal-value singular integral operator

$$Tf(X) := \lim_{\eta\to 0+} \int_{Y\in\partial\Omega,\,|X-Y|>\eta} \langle X-Y, \nu(Y)\rangle\, k(X-Y) f(Y)\, d\sigma(Y), \quad X \in \partial\Omega, \tag{6.525}$$

there exists $\delta > 0$*, depending only on* ε*, the geometric characteristics of* Ω*,* n*,* p *and* $\|k|_{S^{n-1}}\|_{C^N}$ *(where the integer* $N = N(n)$ *is sufficiently large) with the property that*

$$\operatorname{dist}(\nu\,, \operatorname{vmo}(\partial\Omega)) < \delta \Rightarrow \begin{cases} T \text{ is well-defined, belongs to } \mathscr{L}\big(L^p(\partial\Omega)\big) \\ \text{and } \operatorname{dist}\Big(T\,, \operatorname{Comp}\big(L^p(\partial\Omega)\big)\Big) < \varepsilon, \end{cases} \tag{6.526}$$

where the distance in the left-hand side is measured in $\operatorname{bmo}(\partial\Omega)$*, and the distance in the right-hand side is measured in* $\mathscr{L}\big(L^p(\partial\Omega)\big)$*.*

In particular, under the same background hypotheses, for every $p \in (1,\infty)$ *one has*

$$\nu \in \operatorname{vmo}(\partial\Omega) \Longrightarrow T : L^p(\partial\Omega) \longrightarrow L^p(\partial\Omega) \text{ is compact.} \tag{6.527}$$

Finally, the same claims remain valid when made for the operator

$$T^{\#} f(X) := \lim_{\eta\to 0+} \int_{\substack{Y\in\partial\Omega \\ |X-Y|>\eta}} \langle X-Y, \nu(X)\rangle\, k(X-Y) f(Y)\, d\sigma(Y), \quad X \in \partial\Omega, \tag{6.528}$$

with k *as in* (6.524)*, as well as for the operator*

$$\widetilde{T} f(X) := \lim_{\eta\to 0+} \int_{\substack{Y\in\partial\Omega \\ |X-Y|>\eta}} (\nu(X)-\nu(Y))\widetilde{k}(X-Y) f(Y)\, d\sigma(Y), \quad X \in \partial\Omega, \tag{6.529}$$

this time provided that

$$\widetilde{k} : \mathbb{R}^n \setminus \{0\} \to \mathbb{R} \text{ is smooth, odd, and homogeneous of degree } 1-n. \tag{6.530}$$

The following theorem augments earlier work in this section (compare with Theorem 6.49, Corollaries 6.54, 6.51, Theorems 6.53, and 6.55).

Theorem 6.57. *Assume that* $\Omega \subset \mathbb{R}^n$*, with* $n \geq 2$*, is a bounded Lipschitz domain with connected boundary, and fix* $\theta \in \mathbb{R}$ *with* $\theta > -\frac{1}{n}$*. Then given any* $p \in (1,\infty)$

there exists $\varepsilon > 0$, depending only on p, the Lipschitz character of Ω, n, and θ, with the property that if the outward unit normal ν to Ω satisfies

$$\limsup_{r\to 0+}\left\{\sup_{X\in\partial\Omega} \fint_{B(X,r)\cap\partial\Omega}\fint_{B(X,r)\cap\partial\Omega} |\nu(Y)-\nu(Z)|\,d\sigma(Y)d\sigma(Z)\right\} < \varepsilon, \quad (6.531)$$

the following claims are true:

(i) All invertibility results from Theorem 6.49 and Corollary 6.54 hold for the given p;

(ii) The well-posedness results from Corollary 6.51, Theorem 6.53, and Theorem 6.55 hold for the given p.

As a consequence, all results mentioned above actually hold for any $p \in (1,\infty)$ if

$$\nu \in \mathrm{vmo}(\partial\Omega) \quad (6.532)$$

hence, in particular, if Ω is a C^1 domain.

Proof. The crux of the matter is establishing that

$$\pm\tfrac{1}{2}I + \dot{K}_\theta = R_0 + R_1 \quad \text{as operators on } \dot{L}^p_{1,0}(\partial\Omega), \quad (6.533)$$

where $R_0, R_1 \in \mathscr{L}\Big(\dot{L}^p_{1,0}(\partial\Omega)\Big)$ satisfy

$$R_0 \text{ is an invertible operator on } \dot{L}^p_{1,0}(\partial\Omega), \quad (6.534)$$

and

$$\mathrm{dist}\Big(R_1\,,\,\mathrm{Comp}\big(\dot{L}^p_{1,0}(\partial\Omega)\big)\Big) < \|R_0\|_{\mathscr{L}\left(\dot{L}^p_{1,0}(\partial\Omega)\right)}, \quad (6.535)$$

where the distance in the left-hand side is taken in $\mathscr{L}\Big(\dot{L}^p_{1,0}(\partial\Omega)\Big)$. The significance of the decomposition in (6.533) is that, granted (6.534)–(6.535), this readily implies that

$$\pm\tfrac{1}{2}I + \dot{K}_\theta \ \text{ is a Fredholm operator with index zero on } \dot{L}^p_{1,0}(\partial\Omega). \quad (6.536)$$

With this in hand, earlier arguments then lead to the same type of invertibility results as in (6.451), (6.453) for the given p. In turn, the same type of analysis as in the proof of Theorem 6.49 then permits us to also establish analogous invertibility results to those stated in (6.452) and (6.454). Once these results are available, it is straightforward to complete the proof of the claim made in part (i) of the statement of the theorem. Then the claim made in part *(ii)* of the statement of the theorem becomes a consequence of the invertibility results from part (i), by reasoning as before.

Turning to the justification of the claims made in (6.533)–(6.535), there are two basic aspects we wish to emphasize. First, with equivalence constants depending only on the Lipschitz character of Ω,

$$\begin{aligned}&\operatorname{dist}(\nu\,,\operatorname{vmo}(\partial\Omega)) &(6.537)\\&\approx \limsup_{r\to 0+}\left\{\sup_{X\in\partial\Omega}\fint_{B(X,r)\cap\partial\Omega}\fint_{B(X,r)\cap\partial\Omega}\big|\nu(Y)-\nu(Z)\big|\,d\sigma(Y)d\sigma(Z)\right\},\end{aligned}$$

where the distance in the left-hand side is measured in bmo$(\partial\Omega)$. A proof of this claim may be found in [56, 85] . Hence, the smallness of the infinitesimal mean oscillation of the unit normal (defined as the limit in the left-hand side of (6.531)) forces the distance from the unit normal $\nu\in L^\infty(\partial\Omega)$ to the closed subspace vmo$(\partial\Omega)$, measured in bmo$(\partial\Omega)$, to be appropriately small. In turn, this opens the door for the close-to-compact criteria described in Theorem 6.56 to apply.

In the implementation of the aforementioned close-to-compact criteria, we find it useful to revert from the operator $\dot K_\theta$, considered on $\dot L^p_{1,0}(\partial\Omega)$, to the operator $\widetilde K_\theta$ introduced in (6.439)–(6.445), considered on $L^p_1(\partial\Omega)\oplus L^p(\partial\Omega)$. That this is permissible is ensured by the intertwining result proved in Proposition 6.48, keeping in mind the invertibility of the mapping Ψ established in Proposition 6.41. Thus, the goal becomes identifying various expressions from the makeup of the integral kernel of $\widetilde K_\theta$ which have the desired algebraic structure (indicated in Theorem 6.56).

According to the arguments in [129, § 11], there are four types of integral operators on $L^p(\partial\Omega)$ whose kernels must be shown to have the algebraic structure described in Theorem 6.56, namely:

$$\partial_{\nu(Y)}[(\Delta E)(X-Y)], \tag{6.538}$$

$$\nu_i(Y)\nu_j(Y)\nu_k(Y)(\partial_i\partial_j\partial_k E)(X-Y), \tag{6.539}$$

$$\begin{aligned}\partial_{\tau_{ij}(Y)}\partial_{\tau_{k\ell}(Y)}\partial_{\nu(Y)}[E(X-Y)] := {}&\nu_i(Y)\nu_k(Y)\nu_r(Y)(\partial_j\partial_\ell\partial_r E)(X-Y) &(6.540)\\&-\nu_i(Y)\nu_\ell(Y)\nu_r(Y)(\partial_j\partial_k\partial_r E)(X-Y)\\&-\nu_j(Y)\nu_k(Y)\nu_r(Y)(\partial_i\partial_\ell\partial_r E)(X-Y)\\&+\nu_j(Y)\nu_\ell(Y)\nu_r(Y)(\partial_i\partial_k\partial_r E)(X-Y),\end{aligned}$$

and

$$\nu_i(Y)\nu_j(Y)\partial_{\tau_{k\ell}(Y)}[(\partial_i\partial_j E)(X-Y)]-\tfrac{1}{2}\partial_{\tau_{k\ell}(Y)}[(\Delta E)(X-Y)]. \tag{6.541}$$

Concerning the kernel in (6.538), observe that (with c_n denoting a dimensional constant)

$$\partial_{\nu(Y)}[(\Delta E)(X-Y)] = c_n \frac{\langle \nu(Y), Y-X\rangle}{|X-Y|^n} \tag{6.542}$$

and this kernel gives rise to a principal-value singular integral operator T of the type described in (6.525) with $k(X) := c_n|X|^{-n}$. Such a function is as in (6.524), so this integral operator satisfies (6.526).

Regarding the kernel in (6.539) we first note that, for each triplet of numbers $a, b, c \in \{1, \ldots, n\}$ and each point $X = (x_1, \ldots, x_n) \in \mathbb{R}^n \setminus \{0\}$,

$$(\partial_a \partial_b \partial_c E)(X) = \frac{c_n}{|X|^n}\Big[\delta_{bc}x_a + \delta_{ac}x_b + \delta_{ab}x_c - n\frac{x_a x_b x_c}{|X|^2}\Big].$$

Based on this, we may then compute

$$\begin{aligned} &\nu_i(Y)\nu_j(Y)\nu_k(Y)(\partial_i\partial_j\partial_k E)(X-Y) \\ &\quad = c_n \frac{\langle \nu(Y), X-Y\rangle}{|X-Y|^n}\Big[3 - n\frac{\nu_i(Y)\nu_j(Y)(x_i-y_i)(x_j-y_j)}{|X-Y|^2}\Big]. \end{aligned} \tag{6.543}$$

As such, this kernel gives rise to a principal-value singular integral operator T of the form

$$T = T_0 + \sum_{i,j=1}^{n} T_{ij} \circ M_{\nu_i\nu_j}, \tag{6.544}$$

where, generally speaking, M_η denotes the multiplication by the function η, and T_0, $T_{ij}, i, j \in \{1, \ldots, n\}$ are principal-value singular integral operators with kernels

$$3c_n \frac{\langle \nu(Y), X-Y\rangle}{|X-Y|^n} \tag{6.545}$$

and

$$-nc_n \frac{\langle \nu(Y), X-Y\rangle}{|X-Y|^n}\frac{(x_i-y_i)(x_j-y_j)}{|X-Y|^2}, \quad i, j \in \{1, \ldots, n\}, \tag{6.546}$$

respectively. Since $M_{\nu_i\nu_j}$ is a bounded operator on $L^p(\partial\Omega)$, and since the functions

$$k_0(X) := 3c_n|X|^{-n} \quad \text{and} \quad k_{ij}(X) := -nc_n x_i x_j |X|^{-n-2}, \quad i, j \in \{1, \ldots, n\}, \tag{6.547}$$

are as in (6.524), the principal-value singular integral operator associated with the kernel (6.539) also satisfies (6.526).

Finally, a similar (tedious, but straightforward) analysis shows that the principal-value singular integral operators associated with the kernels from (6.540) and (6.541) fit in the class of operators treated in Theorem 6.56 as well, and this finishes the proof of the theorem. □

In the theorem below, the multi-layers $\dot{K}_\theta$ and $\dot{S}$ are associated with the bi-Laplacian, Δ^2, as before (cf. the statement of Corollary 6.54).

Theorem 6.58. *Assume that $\Omega \subset \mathbb{R}^n$, with $n \geq 2$, is a bounded Lipschitz domain with connected boundary, and fix $\theta \in \mathbb{R}$ with $\theta > -\frac{1}{n}$. Then there exists $\varepsilon > 0$ with the property that the operators*

$$\tfrac{1}{2}I + \dot{K}_\theta : \dot{B}^{p,q}_{1,s}(\partial\Omega) \longrightarrow \dot{B}^{p,q}_{1,s}(\partial\Omega), \tag{6.548}$$

$$-\tfrac{1}{2}I + \dot{K}_\theta : \dot{B}^{p,q}_{1,s}(\partial\Omega)\Big/\dot{\mathscr{P}}_1(\partial\Omega) \longrightarrow \dot{B}^{p,q}_{1,s}(\partial\Omega)\Big/\dot{\mathscr{P}}_1(\partial\Omega), \tag{6.549}$$

are isomorphisms whenever $0 < q \leq \infty$ and the indices $p \in (1,\infty)$ and $s \in (0,1)$ satisfy

$$\begin{aligned} &\tfrac{n-3-\varepsilon}{2} < \tfrac{n-1}{p} - s < \tfrac{n-1+\varepsilon}{2} \quad \text{when } n \geq 4, \\ &0 < \tfrac{1}{p} - \left(\tfrac{1-\varepsilon}{2}\right)s < \tfrac{1+\varepsilon}{2} \quad \text{when } n \in \{2,3\}. \end{aligned} \tag{6.550}$$

Moreover, if indices p, p', q, q', s satisfy $1 \leq q, q' \leq \infty$, $p, p' \in (1,\infty)$, $s \in (0,1)$, as well as $1/p + 1/p' = 1/q + 1/q' = 1$ and (6.550) holds, then the operators

$$\tfrac{1}{2}I + \dot{K}^*_\theta : \left(\dot{B}^{p,q}_{1,s}(\partial\Omega)\right)^* \longrightarrow \left(\dot{B}^{p,q}_{1,s}(\partial\Omega)\right)^*, \tag{6.551}$$

$$-\tfrac{1}{2}I + \dot{K}^*_\theta : \left(\dot{B}^{p,q}_{1,s}(\partial\Omega)\Big/\dot{\mathscr{P}}_1(\partial\Omega)\right)^* \longrightarrow \left(\dot{B}^{p,q}_{1,s}(\partial\Omega)\Big/\dot{\mathscr{P}}_1(\partial\Omega)\right)^*, \tag{6.552}$$

$$\dot{S} : \left(\dot{B}^{p',q'}_{1,1-s}(\partial\Omega)\right)^* \longrightarrow \dot{B}^{p,q}_{1,s}(\partial\Omega) \quad \text{if } n \geq 3 \text{ and } n \neq 4, \tag{6.553}$$

are also isomorphisms.

Finally, given any $p \in (1,\infty)$, $q \in (0,\infty]$, $s \in (0,1)$ there exists $\varepsilon > 0$, depending only on p, the Lipschitz character of Ω, n, and θ, with the property that if the outward unit normal ν to Ω satisfies

$$\limsup_{r\to 0+} \left\{ \sup_{X\in\partial\Omega} ⨍_{B(X,r)\cap\partial\Omega} ⨍_{B(X,r)\cap\partial\Omega} \big|\nu(Y) - \nu(Z)\big|\, d\sigma(Y) d\sigma(Z) \right\} < \varepsilon, \tag{6.554}$$

all operators in (6.548)–(6.553) *are invertible (assuming* $q \geq 1$ *in* (6.551)–(6.552) *and* $1/p + 1/p' = 1/q + 1/q' = 1$ *in* (6.553)*). As a consequence, all operators in* (6.548)–(6.553) *are invertible for any* $p \in (1, \infty)$, $q \in (0, \infty]$, *and* $s \in (0, 1)$ *(with the same conventions as above on* q, p', q'*) if*

$$\nu \in \mathrm{vmo}(\partial\Omega) \tag{6.555}$$

hence, in particular, if Ω *is a* C^1 *domain.*

Proof. Fix $\varepsilon > 0$ as in the proof of Theorem 6.49 and let I_ε and I'_ε be as in (6.455) and (6.456), respectively. From (6.451)–(6.452) and the compatibility of inverses stated just below (6.452) we obtain that

$$\begin{aligned}
&\big(\tfrac{1}{2}I + \dot{K}_\theta\big)^{-1} : \dot{L}^{p_0}_{1,0}(\partial\Omega) \longrightarrow \dot{L}^{p_0}_{1,0}(\partial\Omega) \ \text{ boundedly, for all } \ p_0 \in I'_\varepsilon,\\
&\big(\tfrac{1}{2}I + \dot{K}_\theta\big)^{-1} : \dot{L}^{p_1}_{1,1}(\partial\Omega) \longrightarrow \dot{L}^{p_1}_{1,1}(\partial\Omega) \ \text{ boundedly, for all } \ p_1 \in I_\varepsilon.
\end{aligned} \tag{6.556}$$

Based on this and Theorem 3.48 we eventually deduce that

$$\begin{aligned}
&\big(\tfrac{1}{2}I + \dot{K}_\theta\big)^{-1} : \dot{B}^{p,q}_{1,s}(\partial\Omega) \longrightarrow \dot{B}^{p,q}_{1,s}(\partial\Omega) \ \text{ is bounded}\\
&\qquad \text{for every } \ q \in (0, \infty] \ \text{ and } \ p, s \ \text{ as in } \ (6.550).
\end{aligned} \tag{6.557}$$

Since $\frac{1}{2}I + \dot{K}_\theta : \dot{B}^{p,q}_{1,s}(\partial\Omega) \to \dot{B}^{p,q}_{1,s}(\partial\Omega)$ is also bounded, thanks to Theorem 4.21, we finally arrive at the conclusion that the operator in (6.548) is an isomorphism whenever $q \in (0, \infty]$ and p, s are as in (6.550).

In fact, all other claims pertaining to (6.549)–(6.553) may be handled analogously. Finally, under the additional assumption that (6.554) holds, we reason similarly, starting with the invertibility results proved in Theorem 6.57. □

The invertibility results established in Theorem 6.58 are the key ingredients in the proofs of the well-posedness theorems discussed in the remaining portion of this section. We begin by treating the inhomogeneous Dirichlet problem for the bi-Laplacian with boundary data from Whitney–Besov spaces.

Theorem 6.59. *Assume that* $\Omega \subset \mathbb{R}^n$, *with* $n \geq 2$, *is a bounded Lipschitz domain with connected boundary, and fix* $\theta \in \mathbb{R}$ *with* $\theta > -\frac{1}{n}$. *Then there exists* $\varepsilon > 0$ *such that the inhomogeneous Dirichlet problem*

$$\begin{cases}
u \in B^{p,q}_{s+\frac{1}{p}+1}(\Omega),\\
\Delta^2 u = w \in B^{p,q}_{s+\frac{1}{p}-3}(\Omega),\\
\big(\mathrm{Tr}\, u\,,\, \mathrm{Tr}(\nabla u)\big) = \dot{f} \in \dot{B}^{p,q}_{1,s}(\partial\Omega),
\end{cases} \tag{6.558}$$

is well-posed whenever $0 < q \le \infty$ *while* $p \in (1, \infty)$ *and* $s \in (0,1)$ *satisfy*

$$\begin{aligned} &\tfrac{n-3-\varepsilon}{2} < \tfrac{n-1}{p} - s < \tfrac{n-1+\varepsilon}{2} \quad \textit{when } n \ge 4, \\ &0 < \tfrac{1}{p} - \left(\tfrac{1-\varepsilon}{2}\right)s < \tfrac{1+\varepsilon}{2} \quad \textit{when } n \in \{2,3\}. \end{aligned} \tag{6.559}$$

Moreover, if $w = 0$ *then the unique solution* u *of* (6.558) *admits the following integral representation*

$$u(X) = \dot{\mathscr{D}}_\theta\Big[\big(\tfrac12 I + \dot{K}_\theta\big)^{-1}\dot{f}\Big](X), \qquad \forall\, X \in \Omega. \tag{6.560}$$

Furthermore, given any $p \in (1,\infty)$, $q \in (0,\infty]$, $s \in (0,1)$ *there exists* $\varepsilon > 0$, *depending only on* p, q, s, *the Lipschitz character of* Ω, n, *and* θ, *with the property that if the outward unit normal* ν *to* Ω *satisfies*

$$\limsup_{r\to 0+}\left\{\sup_{X\in\partial\Omega} \fint_{B(X,r)\cap\partial\Omega}\fint_{B(X,r)\cap\partial\Omega} \big|\nu(Y)-\nu(Z)\big|\,d\sigma(Y)d\sigma(Z)\right\} < \varepsilon \tag{6.561}$$

then the problem (6.558) *is well-posed. As a consequence, the problem* (6.558) *is well-posed for any* $p \in (1,\infty)$, $q \in (0,\infty]$, *and* $s \in (0,1)$ *if*

$$\nu \in \mathrm{vmo}(\partial\Omega), \tag{6.562}$$

hence, in particular, if Ω *is a* C^1 *domain.*

Finally, similar results are valid for the inhomogeneous Dirichlet problem on Triebel–Lizorkin spaces, i.e., for

$$\begin{cases} u \in F^{p,q}_{s+\frac1p+1}(\Omega), \\ \Delta^2 u = w \in F^{p,q}_{s+\frac1p-3}(\Omega), \\ \big(\mathrm{Tr}\,u,\ \mathrm{Tr}(\nabla u)\big) = \dot{f} \in \dot{B}^{p,p}_{1,s}(\partial\Omega). \end{cases} \tag{6.563}$$

Proof. All well-posedness claims may be proved by reasoning as in the case of Theorem 6.35, this time, by relying on the invertibility results from Theorem 6.58. □

There are three corollaries to the above theorem which we wish to single out. To state the first, recall the weighted Sobolev spaces $W^{k,p}_a(\Omega)$ from (2.248).

Corollary 6.60. *Suppose that* $\Omega \subset \mathbb{R}^n$, *with* $n \ge 2$, *is a bounded Lipschitz domain with connected boundary. Then there exists* $\varepsilon > 0$ *with the property that whenever* $0 < q \le \infty$ *and* $p \in (1,\infty)$, $s \in (0,1)$ *satisfy* (6.559), *one has*

$$\|u\|_{B^{p,q}_{s+\frac1p+1}(\Omega)} \approx \|\mathrm{Tr}\,u\|_{\dot{B}^{p,q}_{1,s}(\partial\Omega)} + \|\mathrm{Tr}(\nabla u)\|_{\dot{B}^{p,q}_{1,s}(\partial\Omega)}, \tag{6.564}$$

uniformly for biharmonic functions u belonging to $B^{p,q}_{s+\frac{1}{p}+1}(\Omega)$, *and*

$$\|u\|_{W^{2,p}_{1-s-\frac{1}{p}}(\Omega)} \approx \|u\|_{F^{p,q}_{s+\frac{1}{p}+1}(\Omega)} \approx \|\mathrm{Tr}\, u\|_{\dot{B}^{p,p}_{1,s}(\partial\Omega)} + \|\mathrm{Tr}(\nabla u)\|_{\dot{B}^{p,p}_{1,s}(\partial\Omega)}, \tag{6.565}$$

uniformly for biharmonic functions u belonging to $F^{p,q}_{s+\frac{1}{p}+1}(\Omega)$.

Proof. The stated equivalences are consequences of Theorems 6.59, 3.9, and Corollary 2.43.

Here is the second corollary alluded to above.

Corollary 6.61. *Assume that* $\Omega \subset \mathbb{R}^n$, *with* $n \geq 2$, *is a bounded Lipschitz domain with connected boundary. Then there exists* $\varepsilon > 0$ *with the property that if the number* $p \in (1, \infty)$ *satisfies*

$$\begin{aligned} \tfrac{2n}{n+1+\varepsilon} < p < \tfrac{2n}{n-1-\varepsilon} \quad &\text{if } n \geq 3, \\ \tfrac{3}{2+\varepsilon} < p < \tfrac{3}{1-\varepsilon} \quad &\text{if } n = 2, \end{aligned} \tag{6.566}$$

one can find a finite constant $C = C(\Omega, p) > 0$ *such that (with* $p' \in (1, \infty)$*, such that* $1/p + 1/p' = 1$*),*

$$\|v\|_{W^{2,p}(\Omega)} \leq C \sup \Big\{ \int_\Omega \Delta v\, \Delta u\, dX : u \in C^\infty_c(\Omega) \ \text{with} \ \|u\|_{W^{2,p'}(\Omega)} \leq 1 \Big\} \tag{6.567}$$

for every function $v \in \mathring{W}^{2,p}(\Omega)$.

Moreover, if the outward unit normal ν *to* Ω *belongs to* $\mathrm{vmo}(\partial\Omega)$ *(hence, in particular, if* $\partial\Omega \in C^1$*), it follows that* (6.567) *holds for any* $p \in (1, \infty)$.

Proof. Let $\varepsilon > 0$ be as in Theorem 6.59 and assume that the exponent p is as in (6.566) and that $1/p + 1/p' = 1$. Finally, pick an arbitrary function $v \in \mathring{W}^{2,p}(\Omega)$. Note that

$$\Big(\mathring{W}^{2,p}(\Omega)\Big)^* = W^{-2,p'}(\Omega) = F^{p',2}_{-2}(\Omega), \tag{6.568}$$

where the last equality follows from (2.202). Hence, with $\langle \cdot, \cdot \rangle$ standing for a natural duality pairing, there exists a finite constant $C = C(\Omega, p) > 0$ with the property that

$$\|v\|_{W^{2,p}(\Omega)} \leq C \sup \Big\{ \langle v, h \rangle : h \in F^{p',2}_{-2}(\Omega) \ \text{with} \ \|h\|_{F^{p',2}_{-2}(\Omega)} \leq 1 \Big\}. \tag{6.569}$$

Fix an some $h \in F^{p',2}_{-2}(\Omega)$ with $\|h\|_{F^{p',2}_{-2}(\Omega)} \leq 1$. The incisive observation is that, together, $p' \in (1, \infty)$ and $s := 1/p \in (0, 1)$ satisfy the conditions in (6.559) and, as such, Theorem 6.59 guarantees the existence of $u \in \mathring{F}_2^{p',2}(\Omega) = \mathring{W}^{2,p'}(\Omega)$ with the property that

$$\Delta^2 u = h \quad \text{and} \quad \|u\|_{W^{2,p'}(\Omega)} \leq C(\Omega, p). \tag{6.570}$$

Consequently,

$$\langle v, h\rangle = \langle v, \Delta^2 u\rangle = \langle \Delta v, \Delta u\rangle = \int_\Omega \Delta v \Delta u \, dX. \tag{6.571}$$

At this stage, (6.567) follows from (6.569)–(6.571), and the density of $C_c^\infty(\Omega)$ in $\mathring{W}^{2,p'}(\Omega)$. □

It is instructive to formulate the well-posedness results from Theorem 6.59 in a fashion which emphasizes the smoothing properties of the Green operator for the inhomogeneous Dirichlet problem for the bi-Laplacian. Recall that this Green operator, call it **G**, is formally defined as

$$\mathbf{G}w := u, \tag{6.572}$$

where u solves

$$\Delta^2 u = w \text{ in } \Omega, \quad u = \partial_\nu u = 0 \text{ on } \partial\Omega. \tag{6.573}$$

Variational considerations based on the Lax–Milgram lemma and trace results ultimately yield that

$$\mathbf{G} : W^{-2,2}(\Omega) \longrightarrow \mathring{W}^{2,2}(\Omega) \text{ isomorphically}, \tag{6.574}$$

and we wish to explore the extent to which the Green operator continues to be smoothing of order 4 when considered on more general scales of Besov and Triebel–Lizorkin spaces. In this regard, we have the following result.

Corollary 6.62. *Assume that $\Omega \subset \mathbb{R}^n$, with $n \geq 2$, is a bounded Lipschitz domain with connected boundary. Then there exists some $\varepsilon = \varepsilon(\Omega) > 0$ such that the Green operators*

$$\mathbf{G} : B^{p,q}_{s+\frac{1}{p}-3}(\Omega) \longrightarrow \mathring{B}^{p,q}_{s+\frac{1}{p}+1}(\Omega), \tag{6.575}$$

$$\mathbf{G} : F^{p,q}_{s+\frac{1}{p}-3}(\Omega) \longrightarrow \mathring{F}^{p,q}_{s+\frac{1}{p}+1}(\Omega), \tag{6.576}$$

are isomorphisms whenever $p \in (1,\infty)$ *and* $s \in (0,1)$ *satisfy*

$$\begin{aligned} &\tfrac{n-3-\varepsilon}{2} < \tfrac{n-1}{p} - s < \tfrac{n-1+\varepsilon}{2} \quad \textit{if } n \geq 4, \\ &0 < \tfrac{1}{p} - \left(\tfrac{1-\varepsilon}{2}\right)s < \tfrac{1+\varepsilon}{2} \quad \textit{if } n \in \{2,3\}, \end{aligned} \tag{6.577}$$

and $0 < q \leq \infty$ *for the Besov scale, and* $\min\{p,1\} \leq q < \infty$ *for the Triebel–Lizorkin scale.*

In particular,

$$\mathbf{G} : W^{-2,p}(\Omega) \longrightarrow \mathring{W}^{2,p}(\Omega) \ \textit{isomorphically}, \tag{6.578}$$

provided

$$\begin{aligned} &\tfrac{2n}{n+1} - \varepsilon < p < \tfrac{2n}{n-1} + \varepsilon \quad \textit{if } n \geq 3, \\ &\tfrac{3}{2} - \varepsilon < p < 3 + \varepsilon \quad \textit{if } n = 2. \end{aligned} \tag{6.579}$$

Furthermore, given any $p \in (1,\infty)$, $q \in (0,\infty)$, *and* $s \in (0,1)$ *there exists* $\varepsilon > 0$, *depending only on* p, q, s, *and the Lipschitz character of* Ω *with the property that if the outward unit normal* ν *to* Ω *satisfies*

$$\limsup_{r\to 0+} \left\{ \sup_{X\in\partial\Omega} ⨍_{B(X,r)\cap\partial\Omega} ⨍_{B(X,r)\cap\partial\Omega} \big|\nu(Y) - \nu(Z)\big| \, d\sigma(Y) d\sigma(Z) \right\} < \varepsilon \tag{6.580}$$

then the operators (6.575)–(6.576) *are isomorphisms (also assuming the inequality* $\min\{p,1\} \leq q < \infty$ *in the case of* (6.576)*). As a consequence, the operators* (6.575)–(6.576) *are isomorphisms for any* $p \in (1,\infty)$, $q \in (0,\infty)$, *and* $s \in (0,1)$ *(also assuming that* $1 = \min\{p,1\} \leq q < \infty$ *in the case of* (6.576)*) if*

$$\nu \in \mathrm{vmo}(\partial\Omega), \tag{6.581}$$

hence, in particular, if Ω *is a* C^1 *domain.*

Proof. The fact that (6.575) is an isomorphism follows from the well-posedness of (6.558), the definition of the Green operator in (6.572)–(6.573), formula (3.205), and Corollary 3.16. The argument for (6.576) is similar, relying on the well-posedness of (6.563) and formula (3.207). Having proved this, (6.578) follows by specializing (6.576) to the case when $s + 1/p = 1$ and $q = 2$ (keeping in the identification result from Proposition 2.30). The remaining claims in the statement of the corollary are established similarly, making use of appropriate well-posedness results from Theorem 6.64. □

Regarding the optimality of Theorem 6.59, we have the following.

Proposition 6.63. *In the class of Lipschitz domains in* $\mathbb{R}^n$, *the range of indices* p, s *in* (6.559) *for which the inhomogeneous Dirichlet problems* (6.558), (6.563) *are well-posed is sharp when* $n \in \{4, 5\}$.

Proof. We begin by recording the following consequence of [102, Theorem 2.6, p. 623]: If $n \in \{2, 3, 4, 5\}$ then for each $\theta \in (0, \pi)$ there exist a bounded Lipschitz domain Ω_θ in $\mathbb{R}^n$, with connected boundary, such that $0 \in \partial\Omega_\theta$ and

$$\Omega_\theta \cap B(0,1) = \Big\{X = (x_1, \dots, x_n) \in B(0,1) : \qquad (6.582)$$

$$x_n < (\cot\theta)\sqrt{x_1^2 + \cdots + x_{n-1}^2}\Big\},$$

along with a non-zero function $u : \Omega_\theta \to \mathbb{R}$ satisfying

$$u \in C^\infty \text{ in } \overline{\Omega_\theta} \text{ away from the origin,} \qquad (6.583)$$

$$u(X) \equiv |X|^{\lambda(\theta)}\varphi(X/|X|) \text{ for } X \text{ near } 0, \qquad (6.584)$$

$$\varphi \in C^\infty(S^{n-1}) \text{ and } \lambda(\theta) \searrow \tfrac{5-n}{2} \text{ as } \theta \searrow 0, \qquad (6.585)$$

$$\Delta^2 u \in C^\infty(\overline{\Omega_\theta}), \quad u = \partial_\nu u = 0 \text{ on } \partial\Omega_\theta. \qquad (6.586)$$

Note that, in concert with Lemma 2.18, conditions (6.583)–(6.586) ensure that the function $u \in W^{2,2}(\Omega)$. Hence, if we set $f := \Delta^2 u \in C^\infty(\overline{\Omega_\theta})$, then $\mathbf{G}f = u$. On the other hand, (6.583)–(6.586) and Lemma 2.18 give that for any $p, q \in (0, \infty)$ and $s > n(1/p - 1)_+$

$$u \in F^{p,q}_{1+s+1/p}(\Omega) \Longleftrightarrow 1 + s + \tfrac{1}{p} < \tfrac{n}{p} + \lambda(\theta) \Longleftrightarrow 1 - \lambda(\theta) < \tfrac{n-1}{p} - s, \qquad (6.587)$$

and note that, by (6.585),

$$1 - \lambda(\theta) \nearrow \tfrac{n-3}{2} \text{ as } \theta \searrow 0. \qquad (6.588)$$

This proves that, when $n \in \{4, 5\}$, the lower bound for $\frac{n-1}{p} - s$ in (6.559) is sharp as far as the well-posedness of (6.563) is concerned. In fact, the same argument also shows that the aforementioned lower bound is optimal in relation to $\mathbf{G}$ being boundedly invertible in the context of (6.576) if $n \in \{4, 5\}$. In the later setting, by relying on the self-adjointness of the operator Green operator $\mathbf{G}$, it follows by duality that the upper bound for $\frac{n-1}{p} - s$ in (6.577) is also sharp when $n \in \{4, 5\}$. Ultimately, this result implies that the range in (6.559) is sharp as far as the well-posedness of (6.563) is concerned. Finally, the argument on the scale of Besov spaces is similar.

We conclude with a well-posedness result for the inhomogeneous Neumann problem for the bi-Laplacian with boundary data from duals of Whitney–Besov spaces.

Theorem 6.64. *Assume that $\Omega \subset \mathbb{R}^n$, with $n \geq 2$, is a bounded Lipschitz domain with connected boundary, and fix $\theta \in \mathbb{R}$ with $\theta > -\frac{1}{n}$. Then there exists $\varepsilon > 0$ such that the inhomogeneous Neumann problem for the biharmonic operator*

$$\begin{cases} u \in B^{p,q}_{1+s+1/p}(\Omega), \\ \Delta^2 u = w\big\lfloor_{\Omega}, \quad w \in \left(B^{p',q'}_{2-s+1/p'}(\Omega)\right)^*, \\ \partial_\nu^{A_\theta}(u,w) = \Lambda \in \left(\dot{B}^{p',q'}_{1,1-s}(\partial\Omega)\right)^*, \end{cases} \tag{6.589}$$

where the boundary datum satisfies the necessary compatibility condition

$$\langle \Lambda, \dot{P} \rangle = \langle w, P\big|_{\Omega} \rangle_{\Omega} \quad \text{for each} \quad P \in \mathscr{P}_1, \tag{6.590}$$

is well-posed, with uniqueness understood modulo polynomials of degree ≤ 1, whenever $s \in (0,1)$ and $p, p', q, q' \in (1,\infty)$, satisfy $1/p + 1/p' = 1/q + 1/q' = 1$, and

$$\begin{aligned} &\tfrac{n-3-\varepsilon}{2} < \tfrac{n-1}{p} - s < \tfrac{n-1+\varepsilon}{2} \quad \text{if } n \geq 4, \\ &0 < \tfrac{1}{p} - \left(\tfrac{1-\varepsilon}{2}\right)s < \tfrac{1+\varepsilon}{2} \quad \text{if } n \in \{2,3\}, \end{aligned} \tag{6.591}$$

Moreover, if $w = 0$ then a solution u of (6.589) is given by the following integral formula

$$u(X) = \dot{\mathscr{S}}\left[\pi^*\left(-\tfrac{1}{2}I + \dot{K}_\theta^*\right)^{-1}\widetilde{\Lambda}\right](X), \qquad \forall\, X \in \Omega, \tag{6.592}$$

where π^* and $\widetilde{\Lambda}$ are as in (6.329)–(6.330) and (6.327)–(6.328), respectively.

Furthermore, given any $p \in (1,\infty)$, $q \in (1,\infty)$, and $s \in (0,1)$ there exists $\varepsilon > 0$, depending only on p, q, s, the Lipschitz character of Ω, n, and θ, with the property that if the outward unit normal ν to Ω satisfies

$$\limsup_{r\to 0^+}\left\{\sup_{X\in\partial\Omega} ⨍_{B(X,r)\cap\partial\Omega} ⨍_{B(X,r)\cap\partial\Omega} \big|\nu(Y) - \nu(Z)\big|\, d\sigma(Y) d\sigma(Z)\right\} < \varepsilon \tag{6.593}$$

then the problem (6.589) is well-posed. As a consequence the problem (6.589) is well-posed for any $p \in (1,\infty)$, $q \in (1,\infty)$, and $s \in (0,1)$ if

$$\nu \in \mathrm{vmo}(\partial\Omega), \tag{6.594}$$

hence, in particular, if Ω is a C^1 domain.

Finally, similar results hold for the inhomogeneous Neumann problem formulated in Triebel–Lizorkin spaces, i.e., for

$$\begin{cases} u \in F^{p,q}_{1+s+1/p}(\Omega), \\ \Delta^2 u = w\big\lfloor_{\Omega}, \quad w \in \left(F^{p',q'}_{2-s+1/p'}(\Omega)\right)^*, \\ \partial_\nu^{A_\theta}(u,w) = \Lambda \in \left(\dot{B}^{p',p'}_{1,1-s}(\partial\Omega)\right)^*, \end{cases} \tag{6.595}$$

where the boundary datum satisfies the necessary compatibility condition (6.590).

Proof. The well-posedness claims formulated in the statement of the theorem may be justified via the same approach as in the proof of Theorem 6.36, making use of the invertibility results from Theorem 6.58. □

References

1. R.A. Adams, J.J.F. Fournier, in *Sobolev Spaces*. Pure and Applied Mathematics, vol. 140, 2nd edn. (Academic, New York, 2003)
2. V. Adolfsson, J. Pipher, The inhomogeneous Dirichlet problem for Δ^2 in Lipschitz domains. J. Funct. Anal. **159**(1), 137–190 (1998)
3. S. Agmon, Multiple layer potentials and the Dirichlet problem for higher order elliptic equations in the plane, I. Comm. Pure Appl. Math. **10**, 179–239 (1957)
4. S. Agmon, Remarks on self-adjoint and semi-bounded elliptic boundary value problems, in *Proceedings of the International Symposium on Linear Spaces*, Jerusalem, 1960 (Jerusalem Academic, Jerusalem, 1960), pp. 1–13
5. S. Agmon, A. Douglis, L. Nirenberg, Estimates near the boundary for solutions of elliptic partial differential equations satisfying general boundary conditions, II. Comm. Pure Appl. Math. **17**, 35–92 (1964)
6. R. Alvarado, I. Mitrea, M. Mitrea, Whitney-type extensions in geometrically doubling quasi-metric spaces, in *Communications in Pure and Applied Analysis* **12**(1) (2013)
7. I. Asekritova, N. Krugljak, Real interpolation of vector-valued spaces in non-diagonal case. Proc. Am. Math. Soc. **133**(6), 1665–1675 (2005) (electronic)
8. J. Bergh, J. Löfström, *Interpolation Spaces. An introduction* (Springer, Berlin, 1976)
9. C. Bourgeois, S. Nicaise, Prewavelet approximations for a system of boundary integral equations for plates with free edges on polygons. Math. Meth. Appl. Sci. **21**, 1233–1267 (1998)
10. M. Bourlard, S. Nicaise, Abstract Green formula and applications to boundary integral equations. Numer. Funct. Anal. Optim. **18**, 667–689 (1997)
11. M. Bownik, Boundedness of operators on Hardy spaces via atomic decompositions. Proc. Am. Math. Soc. **133**(12), 3535–3542 (2005)
12. F. Brackx, R. Delanghe, F. Sommen, in *Clifford Analysis*. Research Notes in Mathematics, vol. 76 (Pitman, Boston, 1982)
13. K. Brewster, I. Mitrea, M. Mitrea, Stein's extension operator on weighted Sobolev spaces on Lipschitz domains and applications to interpolation, in *Contemporary Mathematics* **581**, 13–38 (2012)
14. R.M. Brown, W. Hu, Boundary value problems for higher order parabolic equations. Trans. Am. Math. Soc. **353**(2), 809–838 (2001) (electronic)
15. V.I. Burenkov, Extension theory for Sobolev spaces on open sets with Lipschitz boundaries, in *Nonlinear Analysis, Function Spaces and Applications*, vol. 6 (Acad. Sci. Czech Repub., Prague, 1999), pp. 1–49
16. A. Calderón, Lebesgue spaces of differentiable functions and distributions, in *Proceedings of the Symposium on Pure Mathematics*, vol. IV (1961), pp. 33–49

I. Mitrea and M. Mitrea, *Multi-Layer Potentials and Boundary Problems*, Lecture Notes in Mathematics 2063, DOI 10.1007/978-3-642-32666-0,

17. A.P. Calderón, Intermediate spaces and interpolation, the complex method. Studia Math. **24**, 113–190 (1964)
18. A.P. Calderón, Cauchy integrals on Lipschitz curves and related operators. Proc. Natl. Acad. Sci. USA **74**(4), 1324–1327 (1977)
19. A.P. Calderón, Commutators, singular integrals on Lipschitz curves and applications, in *Proceedings of the International Congress of Mathematicians*, Helsinki, 1978 (Acad. Sci. Fennica, Helsinki, 1980), pp. 85–96
20. A.P. Calderón, C.P. Calderón, E. Fabes, M. Jodeit, N.M. Rivière, Applications of the Cauchy integral on Lipschitz curves. Bull. Am. Math. Soc. **84**(2), 287–290 (1978)
21. M. Christ, in *Lectures on Singular Integral Operators*. CBMS Regional Conference Series in Mathematics, vol. 77 (American Mathematical Society, Providence, 1990)
22. D.-C. Chang, S.G. Krantz, E.M. Stein, H^p theory on a smooth domain in $\mathbb{R}^N$ and elliptic boundary value problems. J. Funct. Anal. **114**(2), 286–347 (1993)
23. D.-C. Chang, S.G. Krantz, E.M. Stein, Hardy spaces and elliptic boundary value problems, in *The Madison Symposium on Complex Analysis*, Madison, WI, 1991. Contemporary Mathematics, vol. 137 (American Mathematical Society, Providence, 1992), pp. 119–131
24. J. Cohen, BMO estimates for biharmonic multiple layer potentials. Studia Math. **91**(2), 109–123 (1988)
25. J. Cohen, J. Gosselin, The Dirichlet problem for the biharmonic equation in a C^1 domain in the plane. Indiana Univ. Math. J. **32**(5), 635–685 (1983)
26. J. Cohen, J. Gosselin, Adjoint boundary value problems for the biharmonic equation on C^1 domains in the plane. Arkiv för Math. **23**(2), 217–240 (1985)
27. R.R. Coifman, A. McIntosh, Y. Meyer, Lintégrale de Cauchy définit un opérateur borné sur L^2 pour les courbes lipschitziennes. Ann. Math. **116**, 361–387 (1982)
28. R. Coifman, G. Weiss, Extensions of Hardy spaces and their use in analysis. Bull. Am. Math. Soc. **83**(4), 569–645 (1977)
29. M. Cwikel, Real and complex interpolation and extrapolation of compact operators. Duke Math. J. **65**(2), 333–343 (1992)
30. B.E. Dahlberg, C.E. Kenig, G.C. Verchota, The Dirichlet problem for the biharmonic equation in a Lipschitz domain. Ann. Inst. Fourier (Grenoble) **36**(3), 109–135 (1986)
31. B.E. Dahlberg, C.E. Kenig, J. Pipher, G.C. Verchota, Area integral estimates for higher order elliptic equations and systems, Ann. Inst. Fourier (Grenoble) **47**(5), 1425–1461 (1997)
32. R. Dautray, J.-L. Lions, in *Mathematical Analysis and Numerical Methods for Science and Technology, Vol. 2: Functional and Variational Methods* (Springer, Berlin, 2000)
33. G. David, S. Semmes, Singular Integrals and rectifiable sets in $\mathbb{R}^n$: Beyond Lipschitz graphs. Astérisque **193**, 152 (1991)
34. G. David, S. Semmes, in *Analysis of and on Uniformly Rectifiable Sets*. Mathematical Surveys and Monographs (AMS Series, Providence, 1993)
35. R.A. DeVore, R.C. Sharpley, Maximal Functions Measuring Smoothness. Mem. Am. Math. Soc. **47**, 293 (1984)
36. M. Dindoš, M.Mitrea, Semilinear Poisson problems in Sobolev-Besov spaces on Lipschitz domains. Publ. Math. **46**(2), 353–403 (2002)
37. L. Diomeda, B. Lisena, Boundary problems for the biharmonic operator in a square with L^p-data. SIAM J. Math. Anal. **15**(6), 1153–1168 (1984)
38. L. Diomeda, B. Lisena, Dirichlet and oblique derivative problems for the Δ^2-operator in a plane sector. (Italian) Rend. Math. (7) **2**(1), 189–217 (1982)
39. S. Dispa, Intrinsic characterizations of Besov spaces on Lipschitz domains. Math. Nachr. **260**, 21–33 (2003)
40. E.B. Fabes, M. Jodeit Jr., N.M. Rivière, Potential techniques for boundary value problems on C^1-domains. Acta Math. **141**(3–4), 165–186 (1978)
41. E. Fabes, O. Mendez, M. Mitrea, Boundary layers on Sobolev-Besov spaces and Poisson's equation for the Laplacian in Lipschitz domains. J. Funct. Anal. **159**(2), 323–368 (1998)
42. E.B. Fabes, C.E. Kenig, On the Hardy space H^1 of a C^1 domain. Ark. Math. **19**(1), 1–22 (1981)

43. C. Fefferman, E.M. Stein, H^p spaces of several variables. Acta Math. **129**(3–4), 137–193 (1972)
44. J. Franke, T. Runst, Regular elliptic boundary value problems in Besov-Triebel-Lizorkin spaces. Math. Nachr. **174**, 113–149 (1995)
45. M. Frazier, B. Jawerth, Decomposition of Besov spaces. Indiana Univ. Math. J. **34**(4), 777–799 (1985)
46. M. Frazier, B. Jawerth, A discrete transform and decompositions of distribution spaces. J. Funct. Anal. **93** (1), 34–170 (1990)
47. M. Frazier, B. Jawerth, G. Weiss, in *Littlewood-Paley Theory and the Study of Function Spaces*. CBMS Regional Conference Series in Mathematics, vol. 79 (American Mathematical Society, Providence, 1991)
48. I. Fredholm, Sur lintégrale fondamentale dune équation différentielle elliptique à coefficients constants. Rend. Circ. Math. Palermo **25**, 346–351 (1908)
49. S.J. Fromm, Potential space estimates for Green potentials in convex domains. Proc. Am. Math. Soc. **119**(1), 225–233 (1993)
50. F. Gazzola, H-.C. Grunau, G. Sweers, in *Polyharmonic Boundary Value Problems*. Lecture Notes in Mathematics, vol. 1991 (Springer, Berlin, 2010)
51. J. Giroire, J.-C. Nédélec, A new system of boundary integral equations for plates with free edges. Math. Meth. Appl. Sci. **18**(10), 755–772 (1995)
52. D. Goldberg, A local version of real Hardy spaces. Duke Math. J. **46**, 27–42 (1979)
53. P. Grisvard, *Elliptic Problems in Nonsmooth Domains, Monographs and Studies in Mathematics*, vol. 24 (Pitman, Boston, 1985)
54. L.I. Hedberg, Spectral synthesis in Sobolev spaces, and uniqueness of solutions of the Dirichlet problem. Acta Math. **147**, 237–264 (1981)
55. S. Hofmann, D. Mitrea, M. Mitrea, A.J. Morris, *L^p-Square Function Estimates on Spaces of Homogeneous Type and on Uniformly Rectifiable Sets*, book manuscript (2012)
56. S. Hofmann, M. Mitrea, M. Taylor, Singular integrals and elliptic boundary problems on regular Semmes-Kenig-Toro domains. Int. Math. Res. Not. **2010**, 2567–2865 (2010)
57. L. Hörmander, in *The Analysis of Linear Partial Differential Operators. I. Distribution Theory and Fourier Analysis*. Classics in Mathematics, Reprint of the second (1990) edition (Springer, Berlin, 2003)
58. D. Jerison, C. Kenig, The inhomogeneous Dirichlet problem in Lipschitz domains. J. Funct. Anal. **130**(1), 161–219 (1995)
59. D. Jerison, C.E. Kenig, Boundary behavior of harmonic functions in nontangentially accessible domains. Adv. Math. **46**(1), 80–147 (1982)
60. F. John, *Plane Waves and Spherical Means Applied to Partial Differential Equations* (Interscience Publishers, New York, 1955)
61. P.W. Jones, Quasiconformal mappings and extendability of functions in Sobolev spaces. Acta Math. **147**(1–2), 71–88 (1981)
62. A. Jonsson, H. Wallin, *Function Spaces on Subsets of* $\mathbb{R}^n$ (Harwood Academic, New York, 1984)
63. N. Kalton, M. Mitrea, Stability results on interpolation scales of quasi-Banach spaces and applications. Trans. Am. Math. Soc. **350**(10), 3903–3922 (1998)
64. N. Kalton, S. Mayboroda, M. Mitrea, Interpolation of Hardy-Sobolev-Besov-Triebel-Lizorkin spaces and applications to problems in partial differential equations, in *Interpolation Theory and Applications*. Contemporary Mathematics, vol. 445 (American Mathematical Society, Providence, 2007), pp. 121–177
65. N. Kalton, N.T. Peck, J. Roberts, in *An F-Space Sampler*. London Mathematical Society Lecture Notes, vol. 89 (Cambridge University Press, Cambridge, 1984)
66. G.A. Kalyabin, Theorems on extension, multipliers and diffeomorphisms for generalized Sobolev-Liouville classes in domains with a Lipschitz boundary. Proc. Steklov Inst. Math. **3**, 191–205 (1985)
67. T. Kato, *Perturbation Theory for Linear Operators*, Reprint of the 1980 edotion (Springer, Berlin, 1995)

68. C.E. Kenig, in *Harmonic Analysis Techniques for Second Order Elliptic Boundary Value Problems*. CBMS Regional Conference Series in Mathematics, vol. 83 (American Mathematical Society, Providence, 1994)
69. J. Kilty, Z. Shen, A bilinear estimate for biharmonic functions in Lipschitz domains. Math. Ann. **349**, 367-394 (2011)
70. V.A. Kozlov, V.G. Maz'ya, J. Rossmann, in *Spectral Problems Associated with Corner Singularities of Solutions to Elliptic Equations*. Mathematical Surveys and Monographs, vol. 85 (American Mathematical Society, Providence, 2001)
71. P. Koskela, E. Saksman, Pointwise characterizations of Hardy-Sobolev functions, Math. Res. Lett. **15**(4), 727–744 (2008)
72. J.-L. Lions, E. Magenes, in *Problèmes aux limites non homogènes et applications*, vol. 1. Travaux et Recherches Mathématiques, vol. 17 (Dunod, Paris, 1968)
73. S. Mayboroda, M. Mitrea, Sharp estimates for Green potentials on non-smooth domains. Math. Res. Lett. **11**(4), 481-492 (2004)
74. V. Maz'ya, *Sobolev Spaces* (Springer, Berlin, 1985)
75. S. Mayboroda, V. Maz'ya, Boundedness of the gradient of a solution and Wiener test of order one for the biharmonic equation. Inventiones. Math. **175**(2), 287–334 (2009)
76. V. Maz'ya, M. Mitrea, T. Shaposhnikova, The Dirichlet problem in Lipschitz domains with boundary data in Besov spaces for higher order elliptic systems with rough coefficients. J. d'Analyse Math. **110**(1), 167–239 (2010)
77. O. Mendez, M. Mitrea, The Banach envelopes of Besov and Triebel-Lizorkin spaces and applications to partial differential equations. J. Fourier Anal. Appl. **6**, 503–531 (2000)
78. Y. Meyer, in *Ondelettes et opérateurs, II, Opérateurs de Calderón-Zygmund*. Actualités Mathématiques (Hermann, Paris, 1990)
79. A. Miyachi, H^p spaces over open subsets of $\mathbb{R}^n$. Studia Math. **95**(3), 205–228 (1990)
80. A. Miyachi, Extension theorems for real variable Hardy and Hardy-Sobolev spaces, in *Harmonic Analysis* (Sendai, 1990). ICM-90 Satellite Conference Proceedings (Springer, Tokyo, 1991), pp. 170–182
81. N. Miller, Weighted Sobolev spaces and pseudodifferential operators with smooth symbols. Trans. Am. Math. Soc. **269**(1), 91–109 (1982)
82. D. Mitrea, I. Mitrea, On the Besov regularity of conformal maps and layer potentials on nonsmooth domains. J. Funct. Anal. **201**(2), 380–429 (2003)
83. D. Mitrea, M. Mitrea, S. Monniaux, The Poisson problem for the exterior derivative operator with Dirichlet boundary condition on nonsmooth domains. Comm. Pure Appl. Anal. **7**(6), 1295–1333 (2008)
84. D. Mitrea, I. Mitrea, M. Mitrea, The Divergence Theorem with non-tangential poitwise traces on the boundary, preprint (2012)
85. D. Mitrea, I. Mitrea, M. Mitrea, *A Treatise on the Theory of Elliptic Boundary Value Problems, Singular Integral Operators, and Smoothness Spaces in Rough Domains*, book manuscript (2012)
86. D. Mitrea, I. Mitrea, M. Mitrea, L. Yan, *On the Geometry of Domains satisfying Uniform Ball Conditions*, book manuscript (2012)
87. I. Mitrea, M. Mitrea, Higher order elliptic boundary value problems in two-dimensional chord-arc domains, preprint (2012)
88. I. Mitrea, M. Mitrea, M. Wright, Optimal estimates for the inhomogeneous problem for the bi-Laplacian in three-dimensional Lipschitz domains. J. Math. Sci. **172**(1), 24–134 (2011)
89. I. Mitrea, G. Verchota, *The Spectrum of Biharmonic Layer Potentials on Curvilinear Polygons* (in preparation)
90. M. Mitrea, in *Clifford Wavelets, Singular Integrals, and Hardy Spaces*. Lecture Notes in Mathematics, vol. 1575 (Springer, Berlin, 1994)
91. M. Mitrea, M. Taylor, Potential theory on Lipschitz domains in Riemannian manifolds: Sobolev-Besov space results and the Poisson problem. J. Funct. Anal. **176**(1), 1–79 (2000)
92. M. Mitrea, M. Taylor, Sobolev and Besov space estimates for solutions to second-order PDE on Lipschitz domains in manifolds with Dini or Hölder continuous metric tensors. Comm. Partial Differ. Equat. **30**, 1–37 (2005)

93. M. Mitrea, M. Taylor, The Poisson problem in weighted Sobolev spaces on Lipschitz domains. Indiana Univ. Math. J. **55**(3), 1063–1089 (2006)
94. M. Mitrea, M. Wright, Boundary value problems for the Stokes system in arbitrary Lipschitz domains, in *Astérisque*, vol. 344 (Societé Mathématique de France, 2012)
95. C.B. Morrey, Second order elliptic systems of differential equations. Contributions to the theory of partial differential equations. Ann. Math. Stud. **33**, 101–159 (1954)
96. A. Nadai, *Theory of Flow and Fracture of Solids*, vol. II (McGraw-Hill, New York, 1963)
97. C. Nazaret, A system of boundary integral equations for polygonal plates with free edges. Math. Meth. Appl. Sci. **21** (2), 165-185 (1998)
98. J. Nečas, in *Les Méthodes Directes en Théorie des Équations Elliptiques*, Masson et Cie, Éditeurs, Paris (Academia, Éditeurs, Prague, 1967)
99. S. Nicaise, Polygonal interface problems for the biharmonic operator. Math. Meth. Appl. Sci. **17**, 21–39 (1994)
100. N. Ortner, P. Wagner, A short proof of the Malgrange-Ehrenpreis theorem, in *Functional Analysis*, ed. by S. Dierolf, S. Dineen, P. Dománski. Proceedings of the 1st International Workshop, September 1994, University of Trier (de Gruyter, Berlin, 1996), pp. 343-352
101. A. Pelczyński, M. Wojciechowski, Sobolev spaces, in *Handbook of the Geometry of Banach Spaces*, ed. by W.B. Johnson, J. Lindenstrauss, vol. 2 (Elsevier, Amsterdam, 2003), pp. 1362–1423
102. J. Pipher, G.C. Verchota, Maximum principles for the polyharmonic equation on Lipschitz domains. Potential Anal. **4** (6), 615–636 (1995)
103. J. Pipher, G. Verchota, A maximum principle for biharmonic functions in Lipschitz and C^1 domains. Comment. Math. Helv. **68**(3), 385–414 (1993)
104. J. Pipher, G.C. Verchota, Dilation invariant estimates and the boundary Gårding inequality for higher order elliptic operators. Ann. Math. (2) **142**(1), 1–38 (1995)
105. R. Pisani, M. Tucci, Dirichlet problem for the operator Δ^N in a plane sector. (Italian) Rend. Acad. Sci. Fis. Mat. Napoli (4) **48**, 317–327 (1982) (1980/1981)
106. N. Rivière, Some open questions. Proc. Symp. Pure Appl. Math. **XXXV**(Part 1), xvii (1978)
107. T. Runst, W. Sickel, *Sobolev Spaces of Fractional Order, Nemytskij Operators, and Nonlinear Partial Differential Operators* (de Gruyter, Berlin, 1996)
108. V. Rychkov, On restrictions and extensions of the Besov and Triebel-Lizorkin spaces with respect to Lipschitz domains. J. London Math. Soc. (2) **60**(1), 237–257 (1999)
109. G. Schmidt, Boundary integral operators for plate bending in domains with corners, Zeitschrift für Analysis und ihre Anwendungen. J. Anal. Appl. **20** (1), 131–154 (2001)
110. A. Seeger, A note on Triebel-Lizorkin spaces, in *Approximation and Function Spaces*, Banach Center Publ., No. 22 (PWN, Warsaw, 1989), pp. 391–400
111. R. Selvaggi, I. Sisto, An existence theorem for the Dirichlet problem with respect to the operator Δ^2 in certain domains of class C^1. (Italian) Boll. Un. Mat. Ital. B (5) **18** (2), 473–483 (1981)
112. S.W. Semmes, A criterion for the boundedness of singular integrals on hypersurfaces. Trans. Am. Math. Soc. **311**(2), 501–513 (1989)
113. J.K. Seo, Regularity for solutions of biharmonic equation on Lipschitz domain. Bull. Korean Math. Soc. **33**(1), 17–28 (1996)
114. Z. Shapiro, On elliptical systems of partial differential equations. C. R. (Doklady) Acad. Sci. URSS (N. S.) **46**, 133–135 (1945)
115. Z. Shen, The L^p Dirichlet problem for elliptic systems on Lipschitz domains. Math. Res. Lett. **13**, 143–159 (2006)
116. Z. Shen, Necessary and sufficient conditions for the solvability of the L^p Dirichlet problem on Lipschitz domains. Math. Ann. **336**, 697–725 (2006)
117. Z. Shen, The L^p Dirichlet problem for elliptic systems on Lipschitz domains. Math. Res. Lett. **13**(1), 143-159 (2006)
118. Z. Shen, The L^p Boundary value problems on Lipschitz domains. Adv. Math. **216**, 212–254 (2007)

119. E.M. Stein, in *Singular Integrals and Differentiability Properties of Functions*. Princeton Mathematical Series, vol. 30 (Princeton University Press, Princeton, 1970)
120. E. Straube, Interpolation between Sobolev and between Lipschitz spaces of analytic functions on star-shaped domains. Trans. Am. Math. Soc. **316**(2), 653–671 (1989)
121. M.H. Taibleson, G. Weiss, The molecular characterization of certain Hardy spaces. Representation theorems for Hardy spaces, in *Astérisque*, vol. 77 (Soc. Math. France, Paris, 1980), pp. 67–149
122. H. Tanabe, in *Functional Analytic Methods for Partial Differential Equations*. Monographs and Textbooks in Pure and Applied Mathematics, vol. 204 (Marcel Dekker, New York, 1997)
123. H. Triebel, Function spaces on Lipschitz domains and on Lipschitz manifolds. Characteristic functions as pointwise multipliers. Rev. Mat. Complut. **15**, 475–524 (2002)
124. H. Triebel, in *Theory of Function Spaces, II*. Monographs in Mathematics, vol. 84 (Birkhäuser, Basel, 1992)
125. G. Verchota, Layer potentials and regularity for the Dirichlet problem for Laplace's equation in Lipschitz domains. J. Funct. Anal. **59**(3), 572–611 (1984)
126. G. Verchota, The Dirichlet problem for the biharmonic equation in C^1 domains. Indiana Univ. Math. J. **36**(4), 867–895 (1987)
127. G. Verchota, The Dirichlet problem for the polyharmonic equation in Lipschitz domains. Indiana Univ. Math. J. **39**(3), 671–702 (1990)
128. G.C. Verchota, Potentials for the Dirichlet problem in Lipschitz domains, in `Potential Theory - ICPT 94` (de Gruyter, Berlin, 1996), pp. 167–187
129. G. Verchota, The biharmonic Neumann problem in Lipschitz domains. Acta Math. **194**, 217–279 (2005)
130. P. Wagner, On the fundamental solutions of a class of elliptic quartic operators in dimension 3. J. Math. Pures Appl. **81**, 1191–1206 (2002)
131. H. Whitney, Analytic extensions of differentiable functions defined in closed sets. Trans. Am. Math. Soc. **36**(1), 63–89 (1934)

Symbol Index

I. Mitrea and M. Mitrea, *Multi-Layer Potentials and Boundary Problems*, Lecture Notes in Mathematics 2063, DOI 10.1007/978-3-642-32666-0,

Subject Index

I. Mitrea and M. Mitrea, *Multi-Layer Potentials and Boundary Problems*, Lecture Notes in Mathematics 2063, DOI 10.1007/978-3-642-32666-0,

Theorem Index

I. Mitrea and M. Mitrea, *Multi-Layer Potentials and Boundary Problems*, Lecture Notes in Mathematics 2063, DOI 10.1007/978-3-642-32666-0,

Author Index

I. Mitrea and M. Mitrea, *Multi-Layer Potentials and Boundary Problems*, Lecture Notes in Mathematics 2063, DOI 10.1007/978-3-642-32666-0,

LECTURE NOTES IN MATHEMATICS Springer

Edited by J.-M. Morel, B. Teissier; P.K. Maini

Editorial Policy (for the publication of monographs)

1. Lecture Notes aim to report new developments in all areas of mathematics and their applications - quickly, informally and at a high level. Mathematical texts analysing new developments in modelling and numerical simulation are welcome.
   Monograph manuscripts should be reasonably self-contained and rounded off. Thus they may, and often will, present not only results of the author but also related work by other people. They may be based on specialised lecture courses. Furthermore, the manuscripts should provide sufficient motivation, examples and applications. This clearly distinguishes Lecture Notes from journal articles or technical reports which normally are very concise. Articles intended for a journal but too long to be accepted by most journals, usually do not have this "lecture notes" character. For similar reasons it is unusual for doctoral theses to be accepted for the Lecture Notes series, though habilitation theses may be appropriate.

2. Manuscripts should be submitted either online at www.editorialmanager.com/lnm to Springer's mathematics editorial in Heidelberg, or to one of the series editors. In general, manuscripts will be sent out to 2 external referees for evaluation. If a decision cannot yet be reached on the basis of the first 2 reports, further referees may be contacted: The author will be informed of this. A final decision to publish can be made only on the basis of the complete manuscript, however a refereeing process leading to a preliminary decision can be based on a pre-final or incomplete manuscript. The strict minimum amount of material that will be considered should include a detailed outline describing the planned contents of each chapter, a bibliography and several sample chapters.
   Authors should be aware that incomplete or insufficiently close to final manuscripts almost always result in longer refereeing times and nevertheless unclear referees' recommendations, making further refereeing of a final draft necessary.
   Authors should also be aware that parallel submission of their manuscript to another publisher while under consideration for LNM will in general lead to immediate rejection.

3. Manuscripts should in general be submitted in English. Final manuscripts should contain at least 100 pages of mathematical text and should always include
   - a table of contents;
   - an informative introduction, with adequate motivation and perhaps some historical remarks: it should be accessible to a reader not intimately familiar with the topic treated;
   - a subject index: as a rule this is genuinely helpful for the reader.

   For evaluation purposes, manuscripts may be submitted in print or electronic form (print form is still preferred by most referees), in the latter case preferably as pdf- or zipped psfiles. Lecture Notes volumes are, as a rule, printed digitally from the authors' files. To ensure best results, authors are asked to use the LaTeX2e style files available from Springer's web-server at:

   ftp://ftp.springer.de/pub/tex/latex/svmonot1/ (for monographs) and
   ftp://ftp.springer.de/pub/tex/latex/svmultt1/ (for summer schools/tutorials).
   Additional technical instructions, if necessary, are available on request from lnm@springer.com.

4. Careful preparation of the manuscripts will help keep production time short besides ensuring satisfactory appearance of the finished book in print and online. After acceptance of the manuscript authors will be asked to prepare the final LaTeX source files and also the corresponding dvi-, pdf- or zipped ps-file. The LaTeX source files are essential for producing the full-text online version of the book (see http://www.springerlink.com/openurl.asp?genre=journal&issn=0075-8434 for the existing online volumes of LNM). The actual production of a Lecture Notes volume takes approximately 12 weeks.

5. Authors receive a total of 50 free copies of their volume, but no royalties. They are entitled to a discount of 33.3 % on the price of Springer books purchased for their personal use, if ordering directly from Springer.

6. Commitment to publish is made by letter of intent rather than by signing a formal contract. Springer-Verlag secures the copyright for each volume. Authors are free to reuse material contained in their LNM volumes in later publications: a brief written (or e-mail) request for formal permission is sufficient.

Addresses:

Professor J.-M. Morel, CMLA,
École Normale Supérieure de Cachan,
61 Avenue du Président Wilson, 94235 Cachan Cedex, France
E-mail: morel@cmla.ens-cachan.fr

Professor B. Teissier, Institut Mathématique de Jussieu,
UMR 7586 du CNRS, Équipe "Géométrie et Dynamique",
175 rue du Chevaleret
75013 Paris, France
E-mail: teissier@math.jussieu.fr

For the "Mathematical Biosciences Subseries" of LNM:

Professor P. K. Maini, Center for Mathematical Biology,
Mathematical Institute, 24-29 St Giles,
Oxford OX1 3LP, UK
E-mail: maini@maths.ox.ac.uk

Springer, Mathematics Editorial, Tiergartenstr. 17,
69121 Heidelberg, Germany,
Tel.: +49 (6221) 4876-8259

Fax: +49 (6221) 4876-8259
E-mail: lnm@springer.com